Progress in Colloid and Polymer Science · Volume 118 · 2001

Springer-Verlag Berlin Heidelberg GmbH

Progress in Colloid and Polymer Science

Editors: F. Kremer, Leipzig and G. Lagaly, Kiel

Volume 118 · 2001

Trends in Colloid and Interface Science XV

Volume Editor:
P.G. Koutsoukos

Springer

ISBN 978-3-662-14667-5   ISBN 978-3-540-45725-1 (eBook)
DOI 10.1007/978-3-540-45725-1

The series Progress in Colloid and Polymer Science is also available electronically (ISSN 1437-8027)

– Access to tables of contents and abstracts is *free* for everybody.
– Scientists affiliated with departments/institutes subscribing to Progress in Colloid and Polymer Science
  as a whole also have full access to all papers in PDF form. Point your librarian to the LINK access registration form
  at http://link.springer.de/series/pcps/reg-form.htm

ISSN 0340-255X

Typesetting: SPS, Madras, India
Cover Design: Estudio Calamar,
F. Steinen-Broo, Pau/Girona, Spain
Cover Production: design & production, D-69121 Heidelberg
SPIN: 10753249

Printed on acid-free paper

Progr Colloid Polym Sci (2001) 118: V
© Springer-Verlag 2001

PREFACE

The present volume includes selected contributions presented either orally or as posters at the 14th Conference of the European Society of the Colloid and Interface Society, which took place in September 2000 in Patras, Greece. The various contributions cover a very broad spectrum of interest both to researchers in the field of Colloid and Interface Science and to the industry which draws largely from the advances of scientific knowledge in colloid and interface science (inks, adhesives, paints, foodstuff, pharmaceuticals etc). New advances in understanding at the molecular level the interactions between polymers and surfactants using both modern and traditional techniques techniques are presented. Methods and techniques of characterization of colloidal suspensions provide a better understanding and extend the various and numerous practical applications of the colloidal systems. Applications of colloids in pharmaceutical products and in biological systems are presented, while better understanding of the behavior and of the measurements done in concentrated suspensions may be obtained through the models and methods presented.

More specifically the contribution in this volume were divided in nine thematic areas: Molecular interactions in thin films, Polymer-Surfactant Interactions, Structure and dynamics at interfaces, Biocolloids, Colloids in Pharmaceutical and Biological Applications, New trends in colloid and interface science techniques, Rheology, Self assembly of amphiphiles and Measurements in concentrated suspensions. Although the contributions are not evenly distributed in these areas a balanced presentation of the state of the art is presented.

Professor Petros G. Koutsoukos
University of Patras
Department of Chemical Engineering
GR26500, Patras GREECE

Progr Colloid Polym Sci (2001) 118: VI–IX
© Springer-Verlag 2001

CONTENTS

Biocolloids

Colloids in Pharmaceutical and Biological Applications

New Trends in Colloids and Interface Science Techniques

Rheology

Progr Colloid Polym Sci (2001) 118: 1–4
© Springer-Verlag 2001

Arnaud Saint-Jalmes
Thomas Zemb
Dominique Langevin

Water/oil/water thin films: construction and applications

A. Saint-Jalmes (✉) · D. Langevin
Laboratoire de Physique des Solides
Université Paris-Sud, 91405 Orsay Cédex
France
e-mail: saint-jalmes@lps.u-psud.fr

A. Saint-Jalmes · T. Zemb · D. Langevin
Service de Chimie Moléculaire
CEA Saclay, 91191 Gif sur Yvette Cédex
France

Abstract With the classical "thin-film balance" apparatus, one can study the properties of thin soap films (air/water/air films). Here, we present a new version of that apparatus allowing us to build a single thin oil film horizontally held on a frame and completely immersed in water. The frame used here is a glass frit: to make it suitable for holding oil films in water and to overcome wetting problems, we have developed a special surface treatment by silanization of the frit. With that device, we can directly and simultaneously control, change and measure both the film thickness and the disjoining pressure in these water/oil/water films. Related to structural, dynamical and stability issues, the range of studies and applications is wide with the new experimental configuration. We present experimental tests on the validity of the setup, also showing typical thickness instabilities which appears to be important in these kind of films. We also discuss information provided on the stability of thin liquid films in liquid–liquid extraction problems.

Key words Thin liquid films · Inverse emulsion · Liquid–liquid extraction · Thickness instability · Disjoining pressure

Introduction

Liquid–liquid extraction techniques are used for specific treatments of nuclear wastes. Appropriate molecules can selectively extract ions from complex solutions. In the industrial DIAMEX process, an oil solution of an extractant molecule (dimethyldibutyltetradecylmalonamide) is mixed with an aqueous solution containing the ions to extract [1]. The mixing process, creating an inverse emulsion (water in oil), results in the transfer of the ions from the aqueous phase to the oil one. The properties of the organic phase have recently been studied [1]. It turns out that it can be considered as a micellar solution of the extractant molecules, and that many features of the process can be explained through that picture. However, only little is known on the emulsion itself and on the mechanisms and dynamics of the transfer. Studying the stability and the properties of the water/oil/water (w/o/w) thin films separating two water droplets (where ion transfers mainly occur) is thus very interesting since it can provide new information on the emulsion microscopic structure and on the possible role and organization of the micelles in these films. Foam surfactant films have been studied extensively, especially with the thin-film balance (TFB) [2]. There are also numerous studies on pseudoemulsion films (oil/water/air films) and on emulsion films (oil/water/oil, o/w/o films) [3–8]; however, surfactant films of oil in water have almost never been studied in terms of disjoining pressures, π (force per area unit between the interfaces), versus film thickness, h. Only recently, a study of decane and polymeric surfactant films in water was reported [9]. In order to investigate w/o/w films, and especially those involved in the liquid–liquid extraction DIAMEX process, we have developed a new version of the TFB, which allows us to control and change the film thickness and to directly measure disjoining pressures.

Adapting the classical TFB method

The classical TFB allows the study of a single thin foam film, held horizontally on a support [2]. The support is a glass frit, and the film is made inside a small hole (diameter of 1–2 mm) in the frit. The frit is placed in a closed box where the internal pressure can be increased in order to apply normal forces on the film and to change its thickness (measured by interferometry). The frit, acting like a reservoir for the film, is connected to a reference pressure by a capillary tube. It is thus possible to get $\pi(h)$ curves as well as direct film observations by videomicroscopy. The main issue for adapting the classical TFB to make w/o/w films concerns the film holder. Glass frits are naturally hydrophilic; thus, if one tries to put such a frit full of oil in water, oil comes out of the frit and is replaced by water, and the experiment cannot be performed correctly. This technical problem is maybe one of the reasons why w/o/w films have not been studied so much. Our major breakthrough has been to develop a chemical treatment (silanization) of the frit to make it hydrophobic. Alkyl trichlorosilane molecules can covalently bind to the frit surface, making it hydrophobic via the grafted alkyl chains. In order to obtain the treatment inside all the frit pores, the silanization has to be done in the gaseous phase of the silane solution (using short alkyl chains, like C6 or C8) inside a sealed box under vacuum. This provides highly hydrophobic frits which, once full of oil, can be dropped in water without any problems. Other important improvements have been necessary in the setup: use of an immersion objective, installation of a suction system for changing the film thickness. Details can be seen in Fig. 1. Also, procedures have been developed, involving microsyringes, for the initial positioning of an oil droplet in the frit hole with water below and above it.

Results

We first tested our setup with the same system as in Ref. [9]: decane (or dodecane) and ABIL EM-90 films in water. ABIL EM-90 is a large comb copolymer known for strongly stabilizing inverse emulsions. It has a rigid poly(dimethylsiloxane) backbone with hydrophobic alkyl and hydrophilic ethylene/propylene oxide grafts. The molecular weight is 62,000 for the purified product. Note that neither the molecular weights of the various groups nor their distribution on the backbone is known.

Highly stable films have been observed and a $\pi(h)$ curve measured (Fig. 2). This curve is in good agreement with that in Ref. [9]. Note that in Ref. [9], disjoining pressures were not obtained directly, but via conductivity measurements. Here, the pressures are simply extracted from a manometer in the suction system. As already said in Ref. [9], understanding the microscopic structure within the film (with regard to the measured thickness) is not easy since the molecule structure is not well known; however, such measured thickness are consistent with the alkyl chains elongated orthogonally to the surface. With these films, we detected a spontaneous cyclic instability of the thickness (Fig. 3). The mean thickness rapidly decreases to around 10 nm in a few seconds, while many small droplets get trapped in the films (dimples). These dimples grow and coalesce; finally the initial thickness is recovered. The mechanism of this instability is not yet understood, but it can probably be connected to the observations of "cyclic dimpling" seen in direct emulsion films [7]; a phenomenon explained both by surfactant redistribution between the phases and by Marangoni effects.

We also tested a smaller and simpler molecule: SPAN80 (monooleate sorbitan) in dodecane. This is also

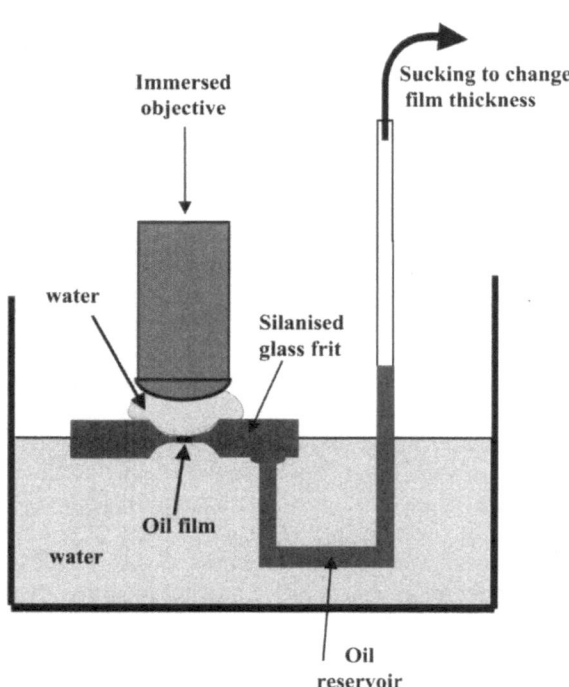

Fig. 1 Setup for water/oil/water thin films. The glass frit became hydrophobic by silanization

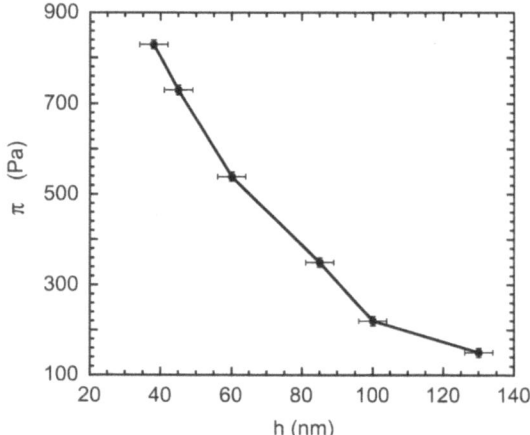

Fig. 2 Disjoining pressure versus film thickness for a dodecane and ABIL EM-90 film in water

Fig. 3 Spontaneous and cyclic thickness instability in a dodecane and ABIL EM-90 film in water. After step *4*, the homogeneous film, as in step *1*, is recovered

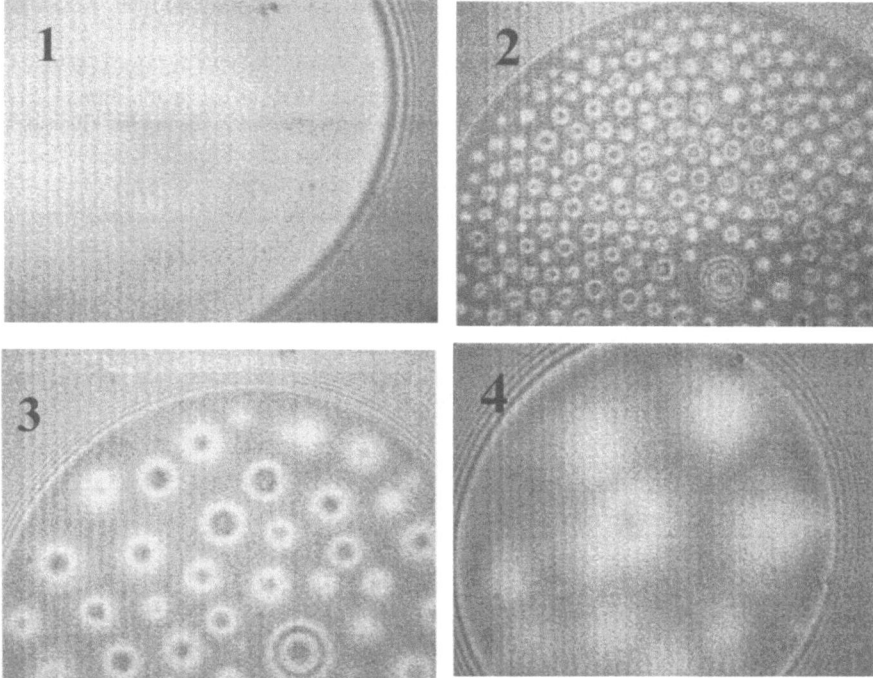

a well-known system for making stable inverse emulsion. Once again, very stables films were observed, showing a dynamical thinning by layers of micelles at a high concentration of SPAN80 and a single transition to a very thin black film at low concentration. Note also that all these w/o/w films show strong dimpling instability in their formation: a large amount of liquid can be trapped in the center of the film, inducing a nonflat film shape. This hydrodynamic effect means that the velocity of the film opening (describing how fast a droplet spread on another one) is much higher than the draining one within that film. Initial dimples also occur in foam films, but they usually disappear in a few seconds, while here they can last for minutes (and can eventually lead to the rupture of the films).

Finally, we studied the dodecane and extractant films in water. Thin films can be formed, but whatever the concentration of extractant used (0.05–1 M) and the temperature, no stable films can be found. The films also show strong initial dimpling. In the case where dimples escape out of the film, the films reach a flat shape and gently thin to a typical rupture thickness of 50 nm (Fig. 4). Again, this rupture thickness does not seem to depend on the extractant concentration. However, the thinning dynamics depends on it and on the temperature (Fig. 4). Actually, the more viscous the solution, the slower it drains. Just before rupture and for the thinnest films, some darker domains in the films are often seen (for high extractant concentration), indicating a thickness jump owing to structuralization of micelles near the surface (the measured thickness of the jump roughly

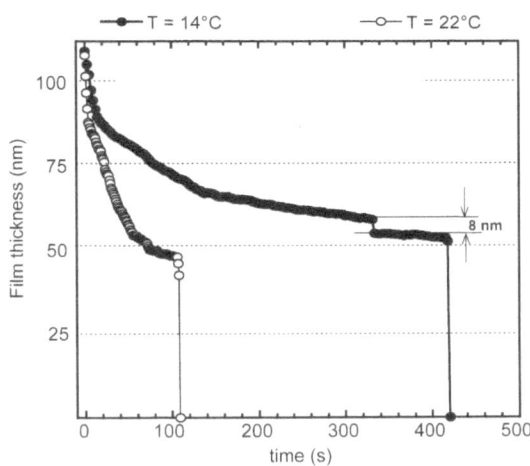

Fig. 4 Thinning dynamics for a dodecane and extractant film in water. An increase in extractant concentration produces the same features as a temperature decrease

corresponds to the calculated distance between the micelles, Fig. 4).

Conclusions

We have developed a new version of the TFB for studying thin oil films in water. The major improvement is related to the inversion of the wetting properties of the frit. Tests have shown that for systems, known for making stable inverse emulsions, stable thin films are observed. This new setup is the first, to our knowledge, that allows the complete study of this type of film. Note that with this setup, we are able to visualize thin oil films of only a few nanometer thickness in water, while the refractive index of water and oil are almost identical and we are able to measure forces between liquid interfaces separated by oil. The results also show that for these types of films, involving liquid–liquid interfaces, the film shapes can be complicated and can be unstable. This is different from foam films, where higher surface tensions make more flat surfaces. Coupling studies of w/o/w and o/w/o films could probably give new insight into the hydrophile–lipophile balance scale at this thin film level. Regarding the liquid–liquid extraction process, we have obtained new information on the emulsion structure. It turns out that the w/o/w thin films are always unstable. In this way, the ion transfer appears to be facilitated by continuous creation and destruction of interfaces and films, in order to have a high actual exchange surface. Thus, the important parameters appear to be the life time of the film and its typical range of thickness. Efficient extractant molecules should not make very stable emulsions, but should not make fast droplets coalescence either. Also, the typical thickness is that where extractant micelles get confined near the interfaces, showing that the transfer can be enhanced by this confinement.

References

1. Erlinger C, Belloni L, Zemb T, Madic C (1999) Langmuir 15:2290
2. (a) Mysels KJ, Jones MN (1966) Discuss Faraday Soc 42:42; (b) Scheludko A (1967) Adv Colloid Interface Sci 1:391; (c) Exerowa D, Scheludko A (1971) Chim Phys 24:47; (d) Bergeron V (1999) J Phys Condens Matter 11:R215
3. (a) Sonntag H, Buske N, Fruhner H (1972) Kolloid Z Z Polym 250:330; (b) Sonntag H, Netzel J, Klare H (1966) Kolloid Z Z Polym 211:121
4. Fisher LR, Mitchell EE, Parker NS (1989) J Colloid Interface Sci 128:35
5. Bergeron V, Fagan ME, Radke CJ (1993) Langmuir 9:1704
6. (a) Aveyard R, Binks BP, Cho WG, Fisher LR, Fletcher PDI, Klinkhammer F (1996) Langmuir 12:6561; (b) Binks BP, Cho WG, Fletcher PDI (1997) Langmuir 13:7180
7. (a) Velev OD, Gurkov TD, Borwankar RP (1993) J Colloid Interface Sci 159:497; (b) Velev OD, Gurkov TD, Ivanov IB, Borwankar RP (1995) Phys Rev Lett 75:264
8. (a) Lobo L, Wasan DT (1993) Langmuir 9:1668; (b) Koczo K, Nikolov AD, Wasan DT, Borwankar RP, Gonsalves A (1996) J Colloid Interface Sci 178:694; (c) Kim YH, Koczo K, Wasan DT (1997) J Colloid Interface Sci 187:29
9. Anklam MR, Saville DA, Prud'homme RK (1999) Langmuir 15:7299

Progr Colloid Polym Sci (2001) 118: 5–10
© Springer-Verlag 2001

Langmuir monolayers from substituted aromatic carboxylic acids

Patrycja Dynarowicz-Łaka
Katarzyna Kita
Piotr Milart
Anantharaman Dhanabalan
Ailton Cavalli
Demétrio A. da Silva Filho
Osvaldo N. Oliveira Jr

P. Dynarowicz-Łaka (✉) · K. Kita
P. Milart
Jagiellonian University
Faculty of Chemistry
Ingardena 3, 30-060 Kraków,
Poland

A. Dhanabalan
Eindhoven University of Technology
Laboratory for Macromolecular and
Organic Chemistry
P.O. Box 513, 5600 MB
Eindhoven, The Netherlands

A. Cavalli
Instituto de Biociências
Letras e Ciência Exatas
Universidade do Estado de São Paulo
São José do Rio Preto, SP, Brazil

D. A. da Silva Filho
Universidade Estadual de Campinas –
Instituto de Física Gleb Wataghin
Campinas, SP, Brazil

O. N. Oliveira Jr
Instituto de Física de São Carlos
Universidade de São Paulo, CP 369, CEP
13560-970, São Carlos, SP, Brazil

Abstract A series of 5′-phenyl-*m*-terphenyl carboxylic acid derivatives with methyl, phenyl, chloro, *p*-chlorophenyl and fluoro substituents have been characterised as Langmuir monolayers at the air/water interface by measuring the surface pressure and the electric surface potential upon monolayer compression. The three-layer capacitor model proposed by Demchak and Fort [(1974) *J Colloid Interface Sci* 46:191] is employed to relate the experimental surface potentials of the monolayers investigated to the molecular dipole moment calculated using semiempirical quantum methods. The local dielectric permittivity in the vicinity of hydrophobic groups was calculated to be 4.8. By adopting a dielectric constant of 7.6 in the vicinity of the hydrophilic groups, the contribution from the water reorientation was found to be 0.15 D, very close to that estimated for small aromatic molecules forming Gibbs monolayers at the air/water interface.

Key words Langmuir monolayers · Aromatic carboxylic acids · Surface pressure · Electric surface potential · Effective dipole moments

Introduction

Studies of monomolecular films at the air/water interface (so-called Langmuir monolayers) have mostly been carried out with amphiphiles that contain an apolar part usually comprising an alkyl (polymethylene) chain. Surprisingly, only little research [1, 2] has been done on monolayers formed by purely aromatic analogues. However, the compounds investigated, namely *p*-terphenyl derivatives, owing to their solubility in water, were found to be capable of monolayer formation only under drastic experimental conditions (on an aqueous subphase containing 4 M NaCl and at 4 °C). Our previous investigations [3, 4] have shown that the presence of at least four phenyl groups in the nonpolar part of the molecule is required for the proper hydrophilic–hydrophobic balance that enables stable monolayer formation at the air/water interface. Since aromatic amphiphiles represent an unexplored group of film-forming compounds, we have initiated a project in which a number of carboxylic acids with a polyphenyl core were synthesised and investigated as Langmuir monolayers [4, 5]. Such compounds can be treated as aromatic analogues of alkanoic amphiphiles. Their characteristics at the air/water interface, however, were found to differ significantly from those observed for model aliphatic compounds. The pressure/area (π/A) isotherms of aromatic carboxylic acids exhibit a broad plateau which spans the region corresponding to a decrease in area by a factor of about 2 [3, 4, 5, 6] and has been attributed to a gradual inclination of film molecules upon compression, followed by the formation of multilayer structures [6]. In our previous articles systematic studies of the basic compound, namely 5'-phenyl-1,1':3',1''-terphenyl-4-carboxylic acid (PTCA, Scheme 1, compound **a**,) [3, 7] and its 4' derivatives [5, 6] have shown that the nature (hydrophilic or hydrophobic) and bulkiness of the substituent influence the monolayer properties significantly. In this work we have furthered our research into PTCA derivatives containing methyl groups (compound **b**) and halogen atoms (F and Cl) at both side phenyl rings (compounds **c** and **d**). The influence of the introduction of a Cl atom in the para position of the 4'-phenyl PTCA (4',5'-diphenyl-1,1':3',1''-terphenyl-4-carboxylic acid, DTCA) (compounds **e** and **f**) is also studied. The experiments are based on surface pressure and electric surface potential measurements upon monolayer compression. Assuming the parallel plate condenser model of the interface, the electrical properties of the monolayers investigated (effective dipole moments of film-forming molecules, group dipole moments, as well as local dielectric permittivities in the vicinity of fuctional groups) have been calculated.

Experimental

Synthesis

The synthesis procedure of PTCA (**a**) and its derivatives (**b**–**e**) has already been described [3, 4, 5, 8].

4'-(4-chlorophenyl)-5'-phenyl-1,1',3',1''-terphenyl-4carboxylic acid (Schemes 1, 2, compound **f**) was prepared by refluxing the

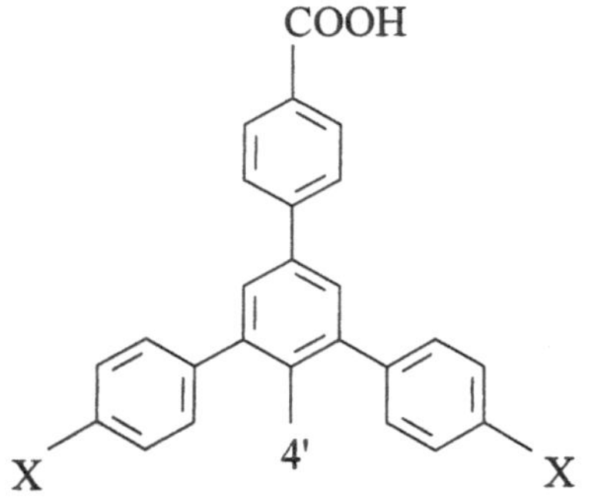

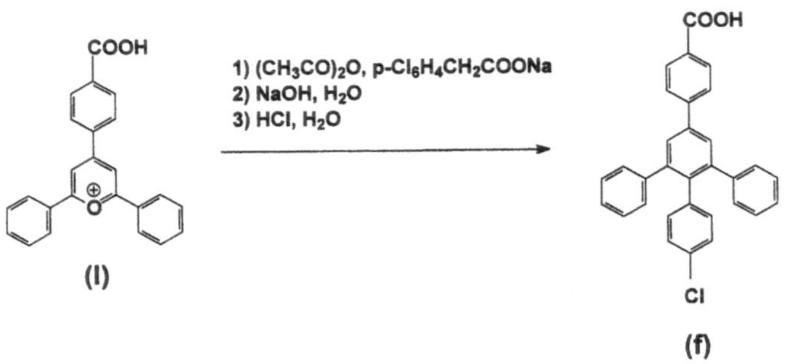

Scheme 2 Synthetic path to *p*-chloro-4',5'-diphenyl-1,1':3',1''-terphenyl-4-carboxylic acid

pyrylium salt (Scheme 2, compound **I**) (2.26 g, 5 mmol) and anhydrous sodium 4-chlorophenylacetate (3.85 g, 20 mmol) in acetic anhydride (15 ml) for 2 h with constant stirring. The mixture was cooled to room temperature. The colourless precipitate of sodium perchlorate and unreacted sodium 4-chlorophenylacetate was filtered off, washed with ethyl acetate and discarded. The mixture of solvents was removed under reduced pressure and the residue was boiled for 1 h with 40 ml 25% aqueous NaOH solution. The sodium salt which precipitated upon cooling for a few hours was separated by suction. The almost colourless salt was converted into an acid by treating its hot suspension in 50 ml water with 1:1 hydrochloric acid. The colourless product was separated, washed with water and dried in air. The crude acid was purified by boiling in acetic acid (about 30 ml) for 20 min, then recrystallized twice from nitromethane and dried in a vacuum (120°, 1 mm Hg). Yield: 0.58 g (25.2%). mp 277-278 °C. Anal. calcd for $C_{31}H_{21}ClO_2$: C, 80.76; H, 4.60. Found: C, 80.94; H, 4.53. IR (KBr) (1/cm) 3,437, 3,029, 2,866, 2,665, 2,540 (OH + CH); 1,694 (C = O); 1,609 (Ar rings); 1,496, 1,425, 1,245. ^1H NMR (CDCl$_3$) δ (ppm) 8.21 and 7.81 (2d, 4H, AA'BB', J_{ortho} = 8.2 Hz, protons of the ring with COOH group); 7.71 (s, 2H, protons 2' and 6'); 7.24–7.19 (m, 6H, protons meta and para of rings at 3' and 5'); 7.14–7.11 (m, 4H, protons ortho of rings at 3' and 5'); 6.98 and 6.79 (2d, 4H, AA'BB', J_{ortho} = 8.4 Hz, proton of the chlorosubstituted ring).

Surface pressure and surface potential isotherms

Spreading solutions for all the compounds investigated except the fluoro derivative (compound **c**) were prepared by dissolving the compounds in spectroscopic grade chloroform. Since the fluorosubstituted PTCA was found to be insoluble in pure chloroform, it was dissolved in a mixture of chloroform/absolute ethanol (95:5 v/ v %). Typical concentrations of the spreading solutions were about 0.4–0.5 g/l. The monolayers were spread on the subphase of ultrapure water, produced by a Nanopure (Infinity) water purification system coupled to a Milli-Q water purification system (resistivity of 18.2 MΩcm). The subphase temperature was controlled thermostatically to within 0.10 °C by a circulating water system from Neslab. Monolayer studies were carried out with a KSV-5000 LB trough (total area of 730.5 cm^2) placed on an antivibration table in a class 10,000 clean room. After spreading, the monolayers were left for 5 min for the solvent to evaporate prior to compression. The surface pressure of the floating monolayer was measured to within 0.1 mNm^{-1} using a Wilhelmy plate (made of chromatography paper, ashless Whatman Chr 1) connected to an electrobalance. Simultaneously, the surface potential was recorded using a vibrating plate located about 2 mm above the water surface. The reference electrode, made from platinum foil, was placed in the water subphase. The surface potential measurements were reproducible to ± 10 mV. The

Compound	4'	X
a (PTCA)	-H	-H
b (methyl-PTCA)	-H	-CH₃
c (fluoro-PTCA)	-H	-F
d (chloro-PTCA)	-H	-Cl
e (DTCA)	-C₆H₅	-H
f (4'-chloro DTCA)	-*p*-Cl-C₆H₄	-H

Scheme 1 Chemical structures of the compounds investigated

monolayers were usually compressed with a barrier speed of 25 mm/min (equivalent to a compression rate of 7.5×10^{17} Å^2/min) unless otherwise specified.

Apparatus

Melting points were determined with a Mel-Temp II melting point apparatus in open capillaries and were not corrected. IR spectra were recorded using a Bruker IFS48 spectrometer as KBr pellets. ^{1}H NMR spectra were taken at 500.13 MHz with a Bruker AMX500 spectrometer using CDCl$_3$ as a solvent and tetramethylsilane as an internal standard. Elemental analyses were performed by the Regional Laboratory of Physico-Chemical Analyses, Kraków.

Results and discussion

The π/A isotherms of the aromatic carboxylic acids investigated (**a–f**) spread on 10^{-3} M HCl aqueous solution at 20 °C are shown in Figs. 1 and 2. Also shown are the surface potential (ΔV)–A isotherms, which will be discussed later. All the films investigated exhibit a characteristic transition region in the course of the π/A

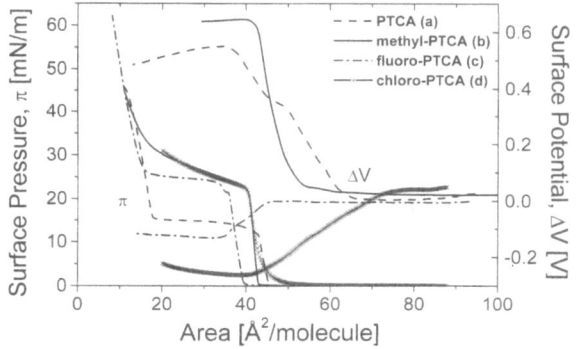

Fig. 1 Surface pressure (π)– and surface potential (ΔV)–area (A) isotherms of carboxylic acids 5′-phenyl-1,1′:3′,1″-terphenyl-4-carboxylic acid (*PTCA*), the methyl derivative, the fluoro derivative and the chloro derivative on a 10^{-3} M HCl aqueous subphase at 20 °C; compression speed 25 mm/min

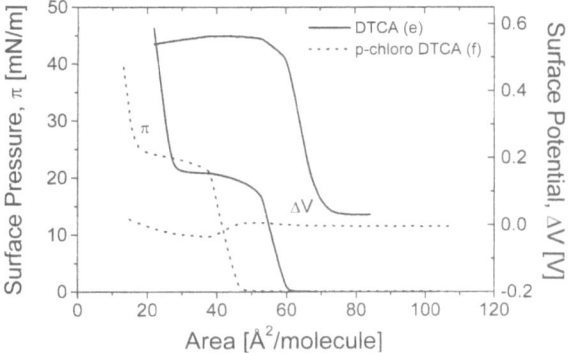

Fig. 2 π–A isotherms of 4′,5′-diphenyl-1,1′:3′,1″-terphenyl-4-carboxylic acid (*DTCA*) and p-chloro DTCA a on 10^{-3} M HCl aqueous subphase at 20 °C; compression speed 25 mm/min

isotherms. While the π/A isotherm of both 4′ derivatives (**e** and **f**) as well as the fluoro-substituted compound (**c**) resemble that of PTCA, with two liquid-condensed regions separated by a distinct broad plateau with an almost twofold decrease in the area per molecule, in the case of the methyl and chloro derivatives (**b**, **d**) the surface pressure increases gradually upon compression, without clear monolayer collapse. When comparing the π/A isotherms on acidic (10^{-3} M aqueous HCl) solution (Figs. 1, 2) to those recorded on a water subphase (results not shown here), the pressure lift-off is shifted slightly towards larger molecular areas on an acidic subphase only for the basic PTCA (which has been discussed previously [7]), while for all the other derivatives investigated the isotherms recorded on both subphases coincide within experimental error. In a set of systematic experiments with the substituted PTCA derivatives, it was observed that a change in the experimental conditions, such as speed of compression (5–100 mm/min), spreading volume (80–200 μl), time interval between monolayer deposition and the beginning of compression (5–30 min), does not influence significantly the characteristics of the monolayers. The change in the subphase temperature, however, has a profound effect on the isotherm pattern, with a decrease in both the onset area and the plateau surface pressure with the increase in the subphase temperature from 20 to 30 °C, a trend similar to that observed for both PTCA [7] and its 4′ derivatives [6].

For PTCA derivatives with substituents at both side phenyl rings (Fig. 1), the transition is attained at nearly the same surface pressure of about 25 mN/m, while for the unsubstituted compound (PTCA) the plateau occurs at a significantly lower π value. The limiting mean molecular areas (A_o) (obtained by extrapolating the surface pressure at the first liquid condensed region to zero value) are slightly higher for both the methyl and the chloro derivatives, compared to PTCA, while the value of A_0 for the fluoro-substituted compound appears to be smaller. In fact, larger A_o values for the methyl and chloro derivatives can be expected owing to their larger projected cross-sectional area. However, the contracted area for the fluoro derivative is quite surprising since fluorocarbon amphiphiles, in general, are known to occupy a larger area on the water surface than their hydrocarbon analogues [9–11], possibly owing to electrostatic repulsion forces between film-forming molecules caused by the presence of strong electronegative fluorine substituents. Interestingly, the mean molecular area for p-chloro DTCA is also shifted towards smaller values in comparison to DTCA (Fig. 2). It is supposed that in these two cases the introduction of a halogen atom into the polyphenyl core causes the molecules to exhibit strong cohesion, which results in the formation of a more closely packed monolayer than is observed for their unsubstituted analogues.

The electric surface potentials (ΔV) show an abrupt change at molecular areas larger than surface pressure lift-off areas (Figs. 1, 2). The surface potentials for all the halogen-substituted aromatic acids investigated here (**c**, **d**, **f**) show negative values upon monolayer compression, owing to the negative contribution from the C–Cl and C–F dipoles, in contrast to the methyl derivative (**b**) and DTCA (**e**), which exhibit positive surface potentials, similarly to PTCA. This leads to a most probable molecular orientation at the air/water interface in which strongly polar –COOH head groups are anchored at the water surface and the hydrophobic core with terminally attached methyl groups (**b**) or a phenyl ring (**e**) (contributing positively to the surface potential) or halogen substituents (introducing large negative contribution) are exposed to the air. It is interesting to point out that the beginning of the plateau on the π/A isotherms coincides with the maximum $|\Delta V|$ values. In contrast to the π/A isotherms, the electric surface potentials show different values on acidic subphases and pure water (Table 1) owing to the double-layer potential, Ψ_0, which appears for ionized monolayers [12] (the carboxylic acids investigated are partially ionized on a pure water subphase of pH~6).

For the comparative quantitative analysis it is useful to plot the effective dipole moment (μ_n)–area isotherm together with surface pressure and electric surface potential data. To eliminate the double-layer contribution, the effective dipole moments were calculated for monolayers spread on acidic subphases. Since the overall behaviour is essentially the same for all the compounds investigated, the effective dipole moment plotted together with the surface pressure and the electric surface potential–area isotherms are represented by the results of the methyl derivative (**b**) (Fig. 3). The effective dipole moment, which is the normal (with respect to the water surface) component of the dipole moment of the film molecule, was calculated from the surface potential data by applying the Helmholtz (parallel-plate condenser) model [12]. According to this approach, the effective dipole moment of the film-forming molecule is expressed by the following formula:

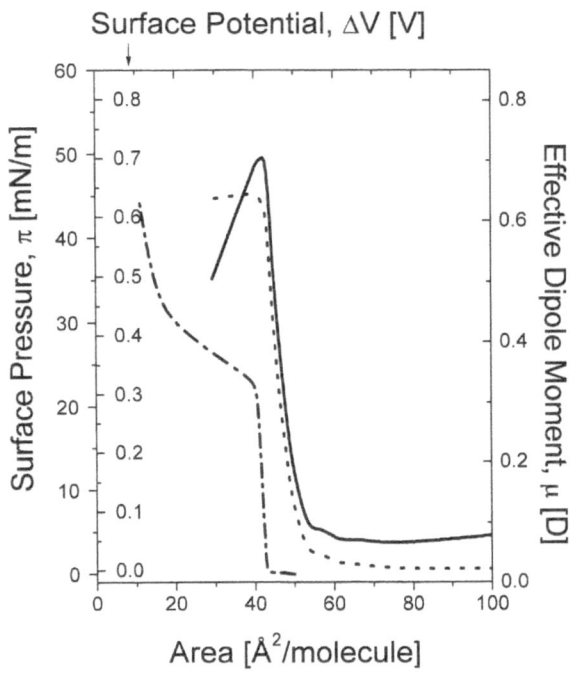

Fig. 3 π (*dashed–dotted line*), ΔV (*dotted line*) and effective dipole moment (μ_n^{exp}) (*solid line*) versus area (A) isotherms of the methyl derivative on a 10^{-3} M HCl aqueous subphase at 20 °C; compression speed 25 mm/min

$$\mu_n = \Delta V \cdot A \cdot \varepsilon_r \cdot \varepsilon_0 , \qquad (1)$$

where A is the area available for a molecule in the monolayer and ε_r and ε_0 are the dielectric constant of the monolayer and the permittivity of free space, respectively. In calculating μ_n, the dielectric constant was taken as 1. The effective dipole moment undergoes changes (a decrease for **c**, **d**, **f** or an increase for **a**, **b**, **d**) along with the surface potential upon compression, reaching extreme values at the same molecular area as $\Delta V_{max/min}$ is attained. Throughout the plateau a drastic decrease in $|\mu_n|$ is seen in all cases. It has been interpreted [6] that when $|\mu_n|$ is a maximum, the molecules are perpendicular to the water surface. The decrease in $|\mu_n|$ during the plateau was attributed to the inclination of the molecules upon further compression, followed by the formation of nonmonomolecular structures [6]. The experimental values of the effective dipole moments, μ_n^{exp}, on acidic subphases are compiled in Table 2. They were obtained using Eq. (1) at the area per molecule corresponding to the termination of the preplateau region. If the same vertical orientation is assumed for the molecules at this area, the experimental values of the effective dipole moments indicate clearly that the resulting contribution from the hydrophobic part of the molecules is different.

For further calculations we applied the Demchak–Fort model, in which the effective dipole moment is written as [1]

Table 1 Maximum/minimum values of electric surface potentials for the compounds investigated spread on pure water or an acidic (10^{-3} M HCl) subphase at 20 °C

Compound	Extreme surface potential, $\Delta V_{max/min}$ (V)	
	Water	10^{-3} M HCl
a	0.450	0.545
b	0.540	0.645
c	−0.220	−0.130
d	−0.420	−0.260
e	0.410	0.560
f	−0.185	−0.035

Table 2 Values of experimental and calculated dipole moments of the compounds investigated. The experimental effective dipole moments (μ_n^{exp}) are calculated for the mean molecular area (column 2) corresponding to the termination of the preplateau region

Compound	Mean molecular area (Å^2/molecule)	Experimental effective dipole moment (μ_n^{exp}, D)	Calculated dipole moment (HyperChem) (μ_{calc}, D)		
			Total (μ_t, D)	Normal (μ_n, D)	Parallel (μ_p, D)
a	43.0	0.465	2.90	2.68	1.12
b	40.5	0.690	3.20	3.00	1.10
c	35.7	−0.111	1.43	−0.90	1.10
d	40.5	−0.280	1.60	−1.20	1.10
e	52.5	0.740	3.09	2.89	0.84
f	37.6	−0.035	1.78	−1.39	1.11

$$\mu_n = \mu_1/\varepsilon_1 + \mu_2/\varepsilon_2 + \mu_3/\varepsilon_3. \qquad (2)$$

In Eq. (2), μ_1 is the normal (to the water surface) component of the dipole moment originating from the reorientation of water molecules in the subphase owing to the presence of the monolayer, μ_2 is the normal dipole moment of the head groups, and μ_3 is the normal dipole moment of the hydrophobic part of the molecule. Each of these dipoles is embedded in a medium with effective dielectric constant, ε_i.

If we assume in Eq. (2) that the contributions from the oriented water molecules (μ_1/ε_1) and from the head – COOH group (μ_2/ε_2) are the same for PTCA and its derivatives, any difference in the experimental effective dipole moments should be ascribed to differences in the contribution from the hydrophobic part of the molecule (μ_3/ε_3). Obviously, a further assumption is that the head group orientation is the same for all compounds, which is reasonable for the termination of the preplateau region.

According to the Demchak–Fort model [1], in order to calculate ε_3, one has to use pairs of Eq. (2) for molecules with different hydrophobic parts and identical hydrophilic groups. Assuming that the contribution of μ_1/ε_1 is the same, one may write

$$\left(\mu_3 - \mu_3'\right)/\varepsilon_3 = \mu_\perp^{exp} - \mu_\perp^{exp'}, \qquad (3)$$

where the prime denotes the second compound in the pair used for comparison. The dipole moment contributions of the hydrophobic part of the molecule (μ_3) were calculated by subtracting the normal component of the dipole moment of the carboxylic group from the normal counterpart of the dipole moment of the free molecule (μ_z), where the latter was calculated using a molecular modelling computer program in the AM1 parameterization [13] for the free molecules in a vacuum (μ) (Table 1). The calculations were performed using the following three methods: HyperChem [14], MOPAC [15] and

GAMESS [16] and gave consistent results. The geometry was optimized using a Digital workstation with the GAMESS program at the semiempirical AM1 level, without any symmetry restrictions. The gradient convergence tolerance was 1×10^{-6} hartree/bohr. Assuming that the configuration of the carboxylic group for the molecules investigated is the same, the difference ($\mu_3 - \mu_3'$) corresponds to ($\mu_z - \mu_z'$). By solving equations of type of Eq. (3) for PTCA (**a**), its methyl derivative (**b**) and DTCA (**e**) in combinations with the halogen derivatives (**c** and **d**), the dielectric permittivity in the vicinity of the hydrophobic groups, ε_3, was found to be 4.8. This mean value is close to that obtained by Demchak and Fort for *p*-terphenyl derivatives ($\varepsilon_3 = 5.3$) [1].

For determination of the local dielectric permittivity in the vicinity of the hydrophilic groups, ε_2, experimental values would be required of effective dipole moments of molecules with the same hydrophobic part and different hydrophilic groups [1]. Since our previous attempts were not successful in forming stable monolayers with head groups other than the carboxylic group attached to the polyphenyl core [4] we may adopt previously quoted values for ε_2, i.e. either 7.6 [1] or 6.4 [17, 18]. Since ε_3 was close to that in Ref. [1] we adopted 7.6 for estimating the theoretical dipole moment using Eq. (2). The group dipole moment for the hydrophilic group (μ_2) is the same for all compounds. The normal component of the carboxylic group (cis conformation), calculated using HyperChem [14] for benzoic acid and then subtracting the C^-H^+ dipole moment (0.4 D [19]), is 1.733 D. The mean value for the contribution from reorientation of the water molecules (μ_1/ε_1) was thus found to be 0.15 D. This value is very close to that obtained for a series of small aromatic molecules ($\mu_1/\varepsilon_1 = 0.1$ D) [20] forming adsorbed (Gibbs) monolayers at the air/water interface.

References

1. Demchak RJ, Fort T (1974) J Colloid Interface Sci 46:191
2. Cadenhead DA, Demchak RJ (1968) J Chem Phys 49:1376
3. Czapkiewicz J, Dynarowicz P, Milart P (1996) Langmuir 12:4966
4. Czapkiewicz J, Dynarowicz-Łaka P, Janicka G, Milart P (1998) Colloids Surf A 135:149
5. Dynarowicz-Łaka P, Czapkiewicz J, Kita K, Milart P, Brocawik E (1999) Prog Colloid Polym Sci 112:15
6. Dynarowicz-Łaka P, Dhanabalan A, Cavalli A, Oliveira ON Jr (2000) J Phys Chem B 104:1701
7. Dynarowicz-Łaka P, Dhanabalan A, Oliveira ON Jr (1999) J Phys Chem B 103:5992

8. Dynarowicz-Łaka P, Kita K, Milart P, Dhanabalan A, Cavalli A, Oliveira ON Jr (2001) J Colloid Interface Sci 239:145
9. Fox HW (1957) J Phys Chem 61:1058
10. Bernett MK, Jarvis NL, Zisman WA (1964) J Phys Chem 68:3521
11. Elbert R, Folda T, Ringsdorf H (1984) J Am Chem Soc 106:7687
12. Davies JT, Rideal EK (1963) Interfacial phenomena. Academic, New York
13. Dewar MJS, Zoebish EG, Healy EG, Stewart JJP (1985) J Am Chem Soc 107:3902
14. Hypercube Inc (1996) HyperChem 5.0, professional version. A molecular visualization and simulation software package. Hypercube Inc, Gainesville, Fl
15. Stewart JJP (1990) J Comput-Aided Mol Des 4:1
16. Schimidt MW, Baldridge KK, Boatz JA, Elbert ST, Gordon MS, Jensen JH, Koseki S, Matsunaga N, Nguyen KA, Su SJ, Windus TL, Dupuis M, Montgomery JA (1993) J Comput Chem 14:1347
17. Oliveira ON Jr, Taylor DM, Lewis TJ, Salvagno S, Stirling CJM (1989) J Chem Soc Faraday Trans 1 85:1009
18. Oliveira ON Jr, Taylor DM, Morgan H (1992) Thin Solid Films 210/211:76
19. Smith JW (1955) Electric dipole moments. Butterworth, London, p 98
20. Paluch M, Dynarowicz P (1987) J Colloid Interface Sci 115:307

Progr Colloid Polym Sci (2001) 118: 11–16
© Springer-Verlag 2001

Klaus Werner Stöckelhuber
Hans Joachim Schulze
Andreas Wenger

Metastable water films on hydrophobic silica surfaces

Abstract The mechanism of the rupture process of liquid films is not fully understood yet, particularly in the case of an asymmetric film between a solid surface and a gas bubble. There are two theoretical approaches describing this problem:

– Growing fluctuation waves (spinodal dewetting) on fluid interfaces under the influence of any kind of attractive force [electrostatic, van der Waals, and maybe a so-called long-range hydrophobic force (LRHF)]. This mechanism was first developed by Scheludko.
– Nucleation inside the film first proposed by Derjaguin.

Metastable wetting films on glass surfaces either hydrophobized by methylation (negatively charged) or with Al^{3+} ions positively charged and hydrophilic, are analyzed by film thinning according to the Reynolds law.

These experiments demonstrate that

– Both mechanisms can be responsible for thin wetting film rupture: in the case of hydrophobic surfaces, the nucleation mechanism; in the case of oppositely charged silica surfaces, the capillary waves mechanism due to the attractive electrostatic double layer force between silica and the air bubble.
– The existence of a LRHF on a hydrophobic surface can be excluded. The apparent interaction can be explained by the presence of gas nuclei formed on heterogeneous sites.

The results provide deeper insight into the mechanisms of wetting film stability, the adhesion process in flotation and droplet coalescence.

Key words Thin liquid wetting films · Long-range hydrophobic force · Thin-film rupture · Nucleation · Capillary waves

K. W. Stöckelhuber (✉) · H. J. Schulze
A. Wenger
Max Planck Research Group Colloids &
Interfaces at the Freiberg University for
Mining and Technology, Chemnitzer
Strasse 40, 09599 Freiberg, Germany
e-mail: werner.stoeckelhuber@ipfdd.de
Tel.: +49-3731-797351
Fax: +49-3731-797407

K. W. Stöckelhuber
Institute for Polymer Research, Hohe
Strasse 6, 01069 Dresden, Germany

Introduction

Dewetting of metastable thin liquid films from a solid surface has been a topic of great interest for more than three decades because of the crucial importance of the film rupture between air bubbles and solid particles in mineral and deinking flotation, in modern technologies such as polymer coatings and some other applications. Also from a fundamental point of view there are many unsolved questions concerning the interpretation of experimental phenomena with respect to the behavior predicted theoretically.

Two different mechanisms can be responsible for the destabilization and rupture of the liquid film, depending on the nature of the solid and liquid, the degree of hydrophobicity, the kind of adsorption layer and its morphological and chemical heterogeneity. The two mechanisms are the growing capillary (Mandelstam) wave mechanism and nucleation.

The first mechanism is based on the instability against thermal fluctuations in the presence of any kind of attractive force which increases the amplitude of the fluctuation. According to the theory first developed by Scheludko [1] and Vrij [2], this instability leads to rupture of the film during its drainage. A particular example recently described in the literature [3] is the dewetting of molten gold films from fused silica substrates which occurs after melting with a laser pulse.

The second mechanism was first proposed by Derjaguin and Gutop [4]. Density fluctuations inside the film in the vicinity of a hydrophobic solid, or tiny gas cavities at defects, could be the reason for the instability, but no kind of an attractive force is necessary. A current example is the rupture of polystyrene layers on silicon wafers [5].

As is well known [6], the van der Waals force in the chosen system, silica (or glass)/water film/air, is repulsive. Because of the repulsive electrostatic disjoining pressure between the negatively charged air bubbles and the negatively charged silica surface, the sum of the interaction forces is also repulsive. Hence, thin water films on silica surfaces must be stable at the equilibrium thickness independently of the ionic strengths in water [7].

Owing to methylation the surface becomes strongly hydrophobic; however neither the charge (or potential, respectively) nor the Hamaker constant of the system is influenced significantly by methylation of the surface hydroxyl groups [8, 9] and, therefore, wetting films should remain stable upon such surfaces. Everybody knows, however, that no stable wetting films can exist on hydrophobic surfaces. In order to explain this discrepancy and to describe the high instability an additional so-called long-range hydrophobic force (LRHF) was postulated during the last decade [10, 11].

Another possibility to change the surface properties of hydrophilic silica surfaces is to change their surface charge to a positive value by adding Al^{3+} cations. In this case, an attractive electrostatic disjoining pressure occurs. Although the surface remains nearly hydrophilic, thin liquid wetting films rupture at a thickness which depends on the range of the electrostatic attractive double-layer force [12].

We are able to demonstrate experimentally for the first time that water films on silica surfaces can also be destabilized by both mechanisms, nucleation and growing capillary waves: When the surface is, in principle, hydrophilic but an attractive electrostatic double-layer force is present, the capillary wave fluctuation rupture occurs. In the case of hydrophobic surface, the nucleation is responsible for the rupture. No LRHF need to be introduced in order to explain the rupture on such strong hydrophobic surfaces.

Experimental

To prove the nonexistence of LRHFs during the time of thinning of the wetting film until its critical rupture thickness, h_{crit}, we had to generate metastable wetting films on hydrophobic surfaces. Formerly, it had been proved that the heterogeneity of the hydrophobic surface has a dramatic influence on the lifetime of the film and its h_{crit}: the larger the heterogeneity, the shorter the lifetime and the larger h_{crit}. The goal was to generate very homogeneous hydrophobic surfaces. We succeeded by gaseous phase silanization of the silanol groups of the glass surface with hexamethyldisilazane (HMDS), where different reaction times lead to different hydrophobicities, expressed by contact angles between 20 and 90°.

Then, we established microscopic wetting films on these modified surfaces using the well-known method of Derjaguin and Scheludko (D-S-film balance) [13, 14]. The thickness was measured with time by microinterferometric means at a wavelength of 470 nm.

If the film is formed at a distance less than 150 nm between the bubble and the surface it is possible to generate a flat, parallel film without a central dimple. In this case the hydrodynamic equation of Reynolds describing the drainage can be used without restrictions. The capillary pressure, P_σ, in the gas bubble was measured separately and was adjusted to 250 Pa in our experiments. Microscopic cover glasses made of soda lime float glass (Marienfeld Superior no. 1 20×20 mm^2) were used as substrates. Their roughness was less than 2 nm. They were cleaned by boiling them in a 70/30 H_2SO_4/H_2O_2 mixture prior to long rinsing under running Milli-Q/Plus water.

Results

The experimentally observed h_{crit} are given in Fig. 1 as function of the lifetime of the thin film at different contact angles (owing to the methylation time). They are compared with the theoretical drainage curve calculated by numerically solution of the Reynolds equation. This equation describes the thickness, $h(t)$, of a flat liquid film during its drainage under laminar flow conditions:

$$\frac{1}{h(t)^2} - \frac{1}{h_0^2} = \frac{4}{3} n \frac{P_\sigma - \Pi_\Sigma[h(t)]}{\eta R_F^2} t \quad ,$$

where h_0 is the reference thickness where the time measurement is started (in our case 87.8 nm), P_σ is the driving capillary pressure in the air bubble (250 Pa), η is the liquid film viscosity, R_F is the radius of the film (in our experiments 90 ± 15 μm), $n = 4$ is a factor that depends on the mobility of the surfaces (here the air/liquid interface is completely mobile, the solid surface completely rigid), t is the drainage time and $\Pi_\Sigma(h)$ represents the sum of disjoining pressure, which consists of the electrostatic force, van der Waals interaction and, in some calculations, the LRHF part.

The most amazing observations on methylated, hydrophobic surfaces are

1. The rupture takes place along the theoretical drainage curve for a system where only repulsive Derjaguin–Landau–Verwey–Overbeek forces are present, no kind of attractive force.

Fig. 1 Drainage and rupture of metastable flat wetting films on gas-phase methylated glass surface at different advancing contact angles between 20 and 90° realized by different reaction times with hexamethyldisilazane. Arbitrarily chosen start time at the film thickness of 89 nm. Larger rupture thickness are indicated by the *arrow* outside the coordinate system. *Continuous line*: calculated drainage according to the Reynolds equation if only repulsive electrostatic and van der Waals forces are present. *Broken lines*: calculated drainage curves if additionally an attractive long-range hydrophobic force is present (denoted according to different approaches)

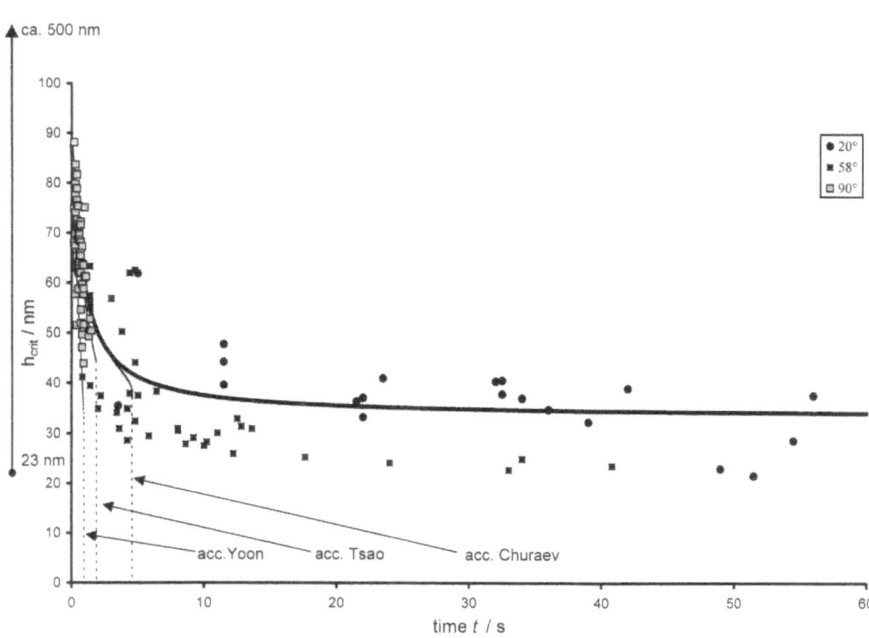

2. The smallest observed rupture thickness corresponds to the equilibrium thickness to which the film could thin if no rupture occurred (in our model system at 34 nm).
3. The largest observed critical thickness is up to several hundred nanometers and is higher, the larger the degree of hydrophobicity.
4. The lifetime depends on the degree of hydrophobicity: the higher the contact angle, the shorter the lifetime.
5. High-speed video frames of hole formation and its expansion (Fig. 2) show that only one formed hole is sufficient for destabilization and dewetting of the whole film. Another remarkable fact is the rather isometric, circular expansion of the newly formed three-phase-contact perimeter.

The behavior of wetting films on recharged glass surfaces (owing to Al^{3+} ions) differs unambiguously from that on the methylated surface:

1. The rupture takes place simultaneously at many places (Fig. 4).
2. The critical thickness is never larger than the range of the attractive electrostatic double-layer force, i.e. not larger than 90 nm under the experimental conditions.
3. The holes do not enlarge. The pinning of the newly formed three-phase contact on the solid surface can be easily visualized by careful receding of the pressed air bubble by means of slow pressure degradation inside the bubble.
4. The distance between the holes is remarkably constant.
5. Although many holes are formed simultaneously, the whole wetting film remains stable for a long time.

6. Such a partially ruptured film could be named a perforated wetting film.

Discussion

On the assumption that attractive LRHF would exist on hydrophobic surfaces, the drainage process should be accelerated considerably and the lifetime of the film could not be longer than a few seconds, depending on the strengths of the attractive force. In Fig. 1 three different approaches for LRHFs [10, 15, 16] and their accelerating influence on the drainage kinetics are given. It is clearly visible that the actual observed lifetimes are much longer. This demonstrates unambiguously that no LRHFs are present; therefore, it is evident that, in contrast to what is widely believed, the capillary wave mechanism cannot be responsible for rupture and nucleation must be the dominating process.

In order to better understand the energetic balance during the process of forming and opening a hole, we use the thermodynamic theory of Sharma and Ruckenstein [17]. This theory does not set any preconditions for the mechanisms of hole formation in a wetting film upon a solid substrate which is characterized by its contact angle. According to this theory, the critical hole size (critical radius of the opening hole) for a given film thickness is reached when the free energy after formation of the hole is equal to the free energy of the initial state of the nonruptured film. Holes with a size smaller than this critical dimension are closing, holes with larger dimensions are opening. The necessary size of the hole for its opening is smaller, the larger the contact angle. Therefore

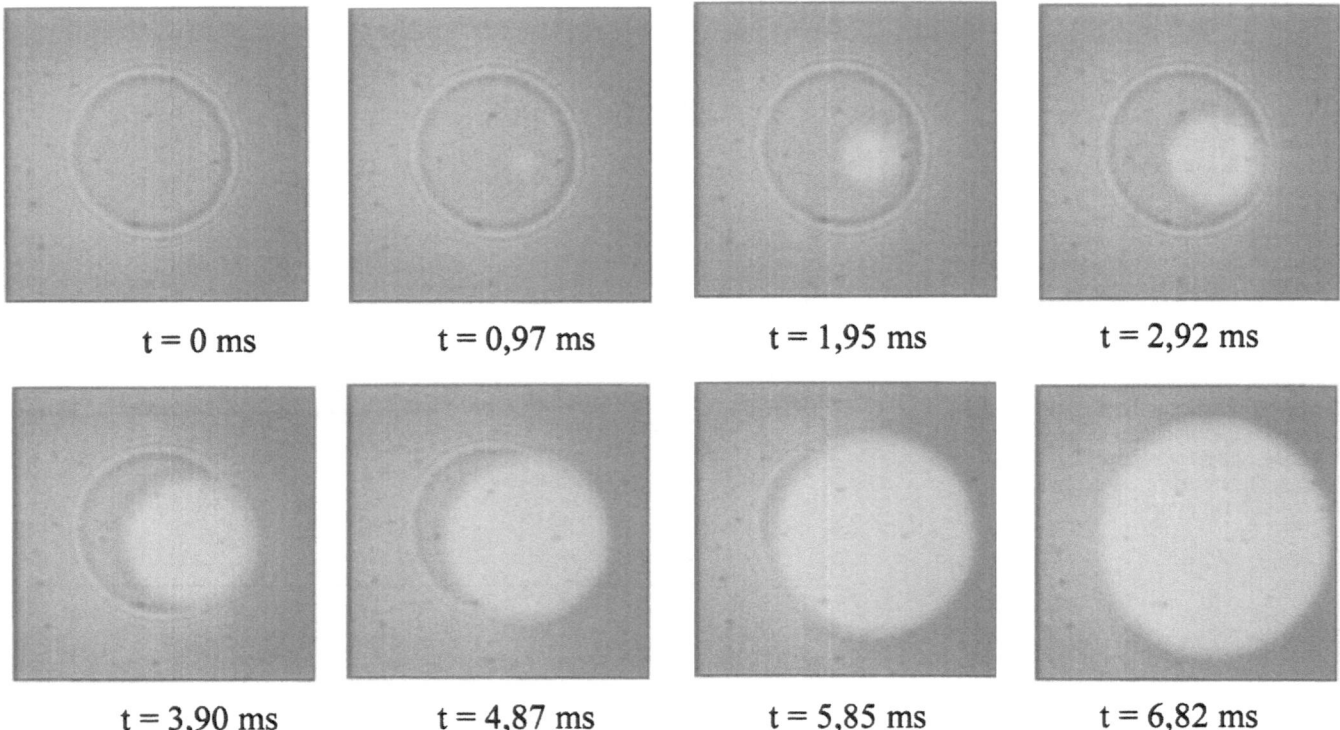

| t = 0 ms | t = 0,97 ms | t = 1,95 ms | t = 2,92 ms |

| t = 3,90 ms | t = 4,87 ms | t = 5,85 ms | t = 6,82 ms |

Fig. 2 High-speed video sequence of the rupture of an aqueous wetting film on methylated glass with a contact angle of 59°. Film diameter: 200 μm. The rupture occurs at 0.97 ms

the probability for rupture of the thin wetting film is higher, the higher the degree of hydrophobicity.

Our calculated values of the critical hole radius cover a range between 125 nm at a contact angle of 85° and 275 nm at 20°. This explains very well the shorter lifetime at high contact angles in comparison to the long lifetime at low contact angles.

We also have evidence for a nucleation mechanism owing to artificial heterogeneities, created either as domains in Langmuir–Blodgett layers [13] or as a striped pattern in skeletonized Langmuir–Blodgett layers [18]. This allows the conclusion that nucleation of holes is obviously based on a heterogeneous process starting at surface inhomogeneities.

The question of what the process of opening of the hole involves still remains unanswered.

No theories have considered the effect of dissolved gas yet, but at a hydrophobic interface the accumulated gas is significant. Effects owing to dissolved gas are explicit in optical cavitation and sonar cavitation experiments, for instance. For this reason it is very probable that the bridging of such nanobubbles or cavities can produce an effect that seems to be equal to the action of a long-range

Fig. 3 Drainage and rupture of an aqueous wetting film of AlCl$_3$ solution on glass at different KCl concentrations. *Lines*: calculated curves according to the Reynolds law

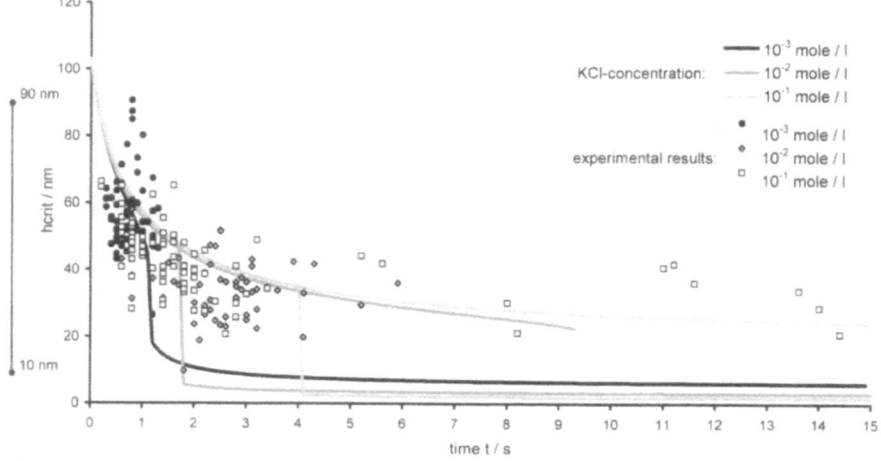

Fig. 4 Ruptured wetting film of electrolyte solution with 10^{-2} KCl and 10^{-4} AlCl$_3$ on oppositely charged glass surfaces. The pinning of the three-phase contact line during removing the bubble on multiple holes indicates the wavelength of the critical fluctuation

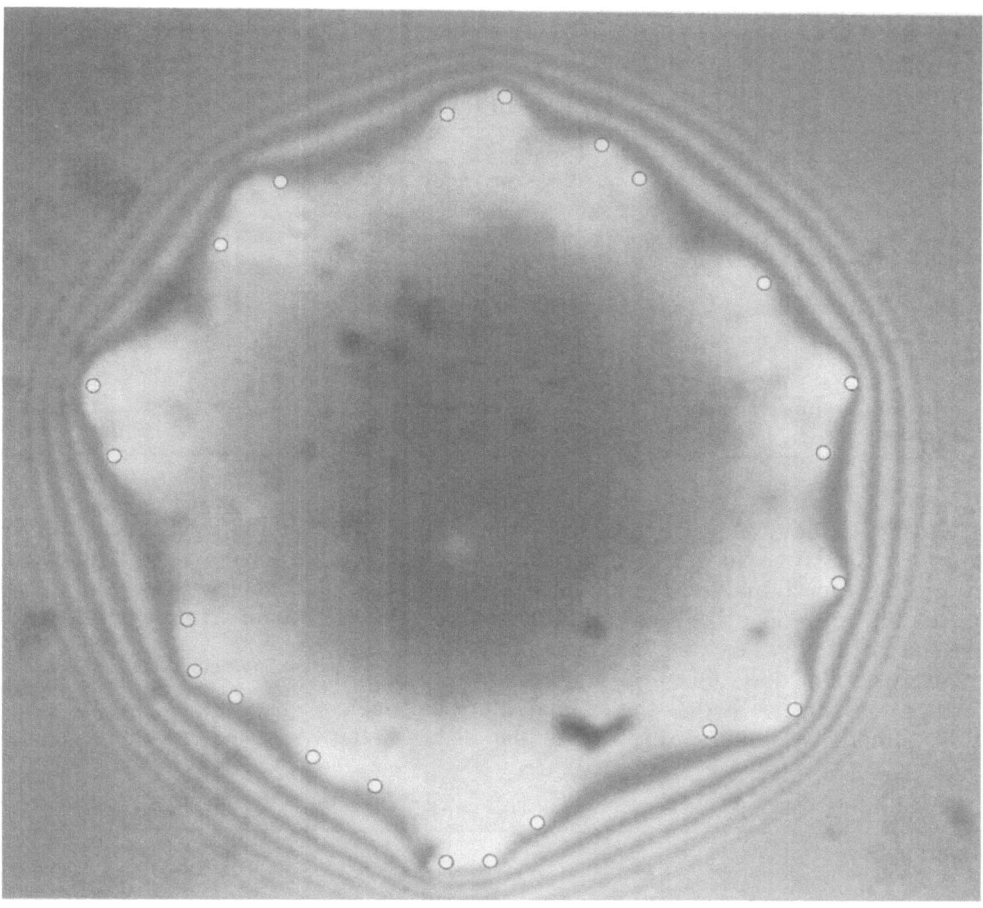

attractive force. Similar conclusions were drawn in Refs. [21, 22]. If the presence of tiny (nano) gas bubbles or cavities is assumed to be the reason, then the observed rupture thickness should be of the same order as the critical thickness of film rupture. Recently such nanobubbles were observed on atomically flat mica surfaces [19] and on silicon wafers [22]. They were stable and did not move around during imaging with the atomic force microscopy technique.

The rupture on oppositely charged glass surfaces is remarkably different from that on a methylated surface. The observed critical thickness values scatter regularly around the calculated drainage curves for different electrolyte concentrations. However, the most amazing observation is the formation of a constant hole distance at the thinnest part of the film. We believe this is good evidence for the existence of a dominating fluctuation wave. In this case thermal fluctuations grow as a consequence of attractive electrostatic forces, leading to hole formations as soon as the amplitude of the dominating wave reaches the size of the film thickness. If one assumes that every trough of a wave leads with large probability to the formation of a hole, the distance between them should be scaled by the wavelength.

According to the theory [20], this critical wavelength is inversely proportional to the square root of the first derivative of the attractive disjoining pressure with respect to the film thickness and is given by

$$\lambda_c = \frac{2\pi}{\sqrt{\frac{1}{\sigma}\frac{\partial \Pi}{\partial h}}} \quad .$$

The evaluation of this relation for the given disjoining pressure leads to a theoretical critical wavelength, λ_c, of approximately 25 μm, which is of the order as the measured distance of the holes. We take that as very good evidence for the existence of the capillary wave mechanism.

Conclusions

Our experiments with thin water films on glass surfaces which were either homogeneously hydrophobized by gas-phase methylation with hexamethyldisilazane or oppositely charged with aluminium chloride solution have detected for the first time both mechanisms of destabilization of wetting films: nucleation and spinodal

Table 1 Comparison of different substrates

	Glass surface	
	Methylated	Oppositely charged by Al^{3+} ions
Interfacial forces (DLVO)	Repulsive	Attractive
Contact angle	20–90° (owing to degree of methylation)	≈20°
Thickness of rupture	25–<500 nm	10–90 nm
Thickness of rupture regarding drainage	In good agreement with the Reynolds curve	Large scattering around the Reynolds curve
Lifetime of wetting film	2 (90°)–<60 s (20°), dependent on contact angle	1 (10^{-3} m KCl)–15 s (10^{-1} m KCl), dependent on range of surface forces
Number of holes	1	Several holes, hole distance in good accordance with critical capillary wavelength
Kinetics of dewetting	Very fast (milliseconds)	Slow (seconds)
Conclusion	Nucleation mechanism	Capillary wave mechanism

dewetting. In the case of strong hydrophobic surfaces the so-called LRHF could not be observed. Its existence is even rather questionable not only in the system investigated, but also on other hydrophobic surfaces. If the existence of tiny gas cavities is assumed to be the reason for wetting film destabilization and its rupture, then a further consequence is that the estimated critical thickness should be commensurable to the size of the nanobubbles, which are in a range between 30 nm and a few hundred nanometers.

The results and conclusions for both systems are summarized again in Table 1.

Acknowledgements Financial support of the Deutsche Forschungsgemeinschaft (SFB 285) is gratefully acknowledged. We thank Emil Manev, for useful discussions, and the Deutsche Akademische Austauschdienst for a grant to Emil Manev.

References

1. Scheludko A (1967) Adv Colloid Interface Sci 1:391
2. Vrij A (1966) Discuss Faraday Soc 42:23
3. Bischof J, Scherer D, Herminghaus S, Leiderer P (1996) Phys Rev Lett 77:1536
4. Derjaguin BV, Gutop JV (1963) Dokl Akad Nauk SSSR 153:859
5. Jacobs K, Herminghaus S (1989) Langmuir 14:965
6. Israelachvili J (1992) Intermolecular and surface forces. Academic, London, p 190
7. Schulze HJ (1984) Physico-chemical elementary processes in flotation. Elsevier, Amsterdam
8. Laskowski J, Kitchener JA (1969) J Colloid Interface Sci 29:670
9. Derjaguin BV, Curaev NV, Muller VM (1985) Surface forces (in Russian). Isdatelstvo Nauka, Moskva
10. Yoon RH (2000) Int J Miner Proc 58:129
11. Pugh RJ, Rutland MW (1995) Prog Colloid Polym Sci 98:284
12. Schulze HJ (1975) Colloid Polym Sci 253:730
13. Mahnke J, Schulze HJ, Stöckelhuber KW, Radoev B (1999) Colloids Surf A 157:1
14. Stöckelhuber W, Schulze HJ, Wenger A (2000) Chem Ing Technol 72 No. 10, 1216
15. Curaev NV (1995) Adv Colloid Interface Sci 58:87
16. Tsao YH, Evans DF, Wennerström H (1993) Langmuir 9:779
17. Sharma A, Ruckenstein E (1990) J Colloid Interface Sci 137:433
18. Mahnke J, Vollhardt D, Stöckelhuber W, Meine K, Schulze HJT (1999) Langmuir 15:8220
19. Jun H, Xu-dong X.Z.-Q, Quyang S-F Imaging nano bubbles in liquids with AFM. Preprint of the Department of Physiology and Biophysics, Fudan University, Shanghai, 200433 China
20. Edwards DA, Brenner H, Wasan DT (1991) Interfacial transport processes and rheology. Butterworth, Boston, p 317
21. Ninham BW (1999) Adv Colloid Interface Sci 83:1
22. Ishida N, Inoue T, Miyahara M, Higashitani K (2000) Langmuir 16:6377

Progr Colloid Polym Sci (2001) 118: 17–21
© Springer-Verlag 2001

V. N. Stathopoulos
P. J. Pomonis

Low–temperature synthesis of spinels MAl_2O_4 (M=Mg, Co, Ni, Cu, Zn) prepared by a sol–gel method

V. N. Stathopoulos · P. J. Pomonis (✉)
Department of Chemistry
University of Ioannina, 451 10 Ioannina
Greece
e-mail: ppomonis@cc.uoi.gr
Tel.: +30-651-98350
Fax: +30-651-98795

Abstract High-surface-area spinels of the general formula MAl_2O_4, where M = Mg, Co, Ni, Cu and Zn have been successfully prepared at low temperature (600 °C) from precursor solutions containing the nitrate salts and the surfactant cetyltrimethylammonium bromide as the gelating agent. Thermal analysis (thermogravimetry/differential thermogravimetry/differential thermal analysis) of the precursors dried at 100 °C showed an exothemal decomposition around 250–260 °C and no mass loss above 600 °C under airflow. The solids were heated at 600, 800 and 1000 °C and at each step the X-ray diffraction spectra were obtained in order to check the development of the spinel MAl_2O_4 crystal phase. For M = Mg, Co and Zn samples treated at 600 °C, the $MgAl_2O_4$, $CoAl_2O_4$ and $ZnAl_2O_4$ crystal phases are formed. Nitrogen porosimetry for the MAl_2O_4 samples, heated at 600 °C, revealed mesoporous solids of specific surface areas from 106 to190 $m^2\ g^{-1}$ depending on the M cation. At 1000 °C the $MgAl_2O_4$ spinel possesses a specific surface area of 57 $m^2\ g^{-1}$ and $ZnAl_2O_4$ has a specific surface area of 75 $m^2\ g^{-1}$.

Key words Mesoporous spinels · High surface area · Cetyltrimethylammonium bromide

Introduction

Naturally occuring spinel minerals are found as minor constituents of both igneous and metamorphic rocks. The prototypic mineral after which the structure is named has the ideal formula $MgAl_2O_4$. The general formula for a spinel is AB_2X_4 where A is usually a 2+ and B a 3+ cation, while X stands for oxygen; however this is an extreme stoichiometry as a great number of cations of various charge and size can occupy the tetrahedral and octahedral interstices that are formed in the close-packed face-centered-cubic configuration of the X anions. Typical sizes of cations involved in the spinel structure are $0.06 < A\,(nm) < 0.1$ and $0.055 < B\,(nm) < 0.1$. In the normal distribution in a binary AB_2X_4 spinel, the twice as abutant B cations are located on half of the octahedral sites, while the A cations are found on one-eighth of the tetrahedral sites, always in an ordered manner.

Spinel aluminates have gained much interest as supports in heterogeneous catalysis or as catalysts themselves (e.g. $MgAl_2O_4$ as sulfur transfer catalysts) [1–8]. This is mainly because of their inherent properties, such as their chemical inertness, high thermal stability and mechanical resistance, and their adequate surface acidity compared to the conventional carriers. Furthermore noble metals supported over aluminate spinels exhibit higher sintering resistance than, for example, Pt/Al_2O_3 and Pt/SiO_2 systems [1]. However, in such applications the textural property of the solids needed is the high surface area and open porosity for efficient mass transfer. Towards this aim various preparation methods have been used, different from the conventional solid-state reaction that result in nonporous solids because of the sever calcination treatment that is required. Coprecipitation [1, 2, 5, 9, 10] and sol–gel [9, 11, 12] methods have been applied. The latter method is more successful, achieving spinel aluminates of increased dispersion. Despite the potential applicability of such solids as catalytic supports, reports of the synthesis of high-surface-area solids are limited. Furthermore, among studies focused on low-temperature synthesis of high-surface-area aluminates (mainly of Mg and Zn) no particular attention has been given to the thermal stability and the textural properties, like specific surface area and porosity, at elevated calcination temperatures. This is a very crucial point since most of the catalytic applications of such supports refer to high temperature processes.

The present study reports a successful, simple sol–gel method for preparing high-surface-area aluminates with varying A site cations (MAl_2O_4, M: Mg, Co, Ni, Cu, Zn) at low temperature, for example, 600 °C. The structural properties of the materials were investigated by means of thermal analysis [thermogravimetry (TG) differential TG (DTG) differential thermal analysis (DTA)], N_2 adsoprtion at 77 K (Brunauer–Emmett–Teller, BET, method) and powder X-ray diffraction (XRD).

Experimental

The preparation of the spinels took place as follows. Calculated amounts of the metals, in the form of nitrate salts (analytical grade), were dissolved in water and mixed with an equal volume of surfactant cetyltrimethylammonium bromide (CTAB) solution. The final solution had a concentration of 0.16 M metal cations and the ratio (moles surfactant):(gram atom of cations) was unity. Immediately after mixing, a transparent viscous gel was formed, and this dried at 100 °C. The dried precursors were treated thermally under ambient air up to 280 °C for 3 h, followed by a 2 °C min^{-1} increase in temperature up to 600, 800 and 1000 °C and were kept at this temperature for 4 h.

The solids obtained were examined for their specific surface area (S_p) and crystal structure (XRD) while small portions (about 100 mg) of the dried precursors were tested for their thermal behaviour by TG/DTG/DTA techniques. Analysis took place in a Chyo-TRDA-3H thermal balance with simultaneous recording of temperature, TG, DTG and DTA. In all cases Al_2O_3 was used as a blank and the analysis took place at a heating rate 5 °C min^{-1} between room temperature and 1000 °C. Phase analysis and crystallite size determination were performed for all the samples after heating at 600, 800 and 1000 °C, using a Siemens Diffract 500 system employing Cu Kα radiation (1.5418 Å).

The S_p of the solids was determined using a Fisons Sorptomatic 1900 volumetric adsorption–desorption apparatus, using N_2 as the adsorbent at 77 K, by applying the BET equation. Prior to the determination of the adsorption–desorption isotherms, the samples were degassed at 250 °C under a 5×10^{-2}-mbar vacuum for 10 h.

The solids obtained are summarized in Table 1 together with some of their properties.

Results and discussion

The thermal behaviour of the solids is shown in Fig. 1 in terms of TG, DTG and DTA. From Fig. 1, it can be seen that the dried precursors of the spinels, containing the metallic cations, nitrate groups as well as CTAB, upon heating lose weight by an exothermal effect around 250–260 °C. The reaction is quite fast and with respect to the high nitrate loading, this effect is attributed to their decomposition. At the same time the carbonaceous species introduced by the CTAB decompose, with a gradual mass loss up to 600 °C. Above 600 °C no reaction is observed. Thus, the solids were calcined at 600, 800 and 1000 °C for 4 h. It is notable that the cations Mg, Co, Ni, Cu and Zr have a different catalytic effect on the fast decomposition of nitrates, Cu, Ni and Co decrease it, while in the samples containing Al and Zn

Table 1 Specific surface areas (S_p), pore diameters (D_p) and the crystal phases of M-6, M-8 and M-10 solids (M = Mg, Co, Ni, Cu, Zn) prepared with cetyltrimethylammonium bromide as surfactant and heated at the temperatures indicated

Sample	Composition	Calcination temperature (°C)	S_p (m^2 g^{-1})	D_p^a (nm)	Crystal phases (X-ray diffraction)
Mg-6	Mg/Al = 1/2	600	190	8.0	$MgAl_2O_4$
Mg-8	Mg/Al = 1/2	800	139	8.2	$MgAl_2O_4$
Mg-10	Mg/Al = 1/2	1000	57	16.8	$MgAl_2O_4$
Ni-6	Ni/Al = 1/2	600	157	4.0	NiO, $NiAl_2O_4$[b]
Ni-8	Ni/Al = 1/2	800	108	4.4	NiO, $NiAl_2O_4$[b]
Ni-10	Ni/Al = 1/2	1000	32	–	$NiAl_2O_4$, NiO
Cu-6	Cu/Al = 1/2	600	112	5.7	CuO, $CuAl_2O_4$[b]
Cu-8	Cu/Al = 1/2	800	21	–	$CuAl_2O_4$, CuO[b]
Cu-10	Cu/Al = 1/2	1000	5	–	$CuAl_2O_4$, CuO[b]
Co-6	Co/Al = 1/2	600	106	5.2	$CoAl_2O_4$
Co-8	Co/Al = 1/2	800	65	5.5	$CoAl_2O_4$
Co-10	Co/Al = 1/2	1000	10	–	$CoAl_2O_4$
Zn-6	Zn/Al = 1/2	600	169	5.9	$ZnAl_2O_4$
Zn-8	Zn/Al = 1/2	800	144	6.6	$ZnAl_2O_4$
Zn-10	Zn/Al = 1/2	1000	75	6.7	$ZnAl_2O_4$

[a] Maximum in the pore size distribution
[b] In traces

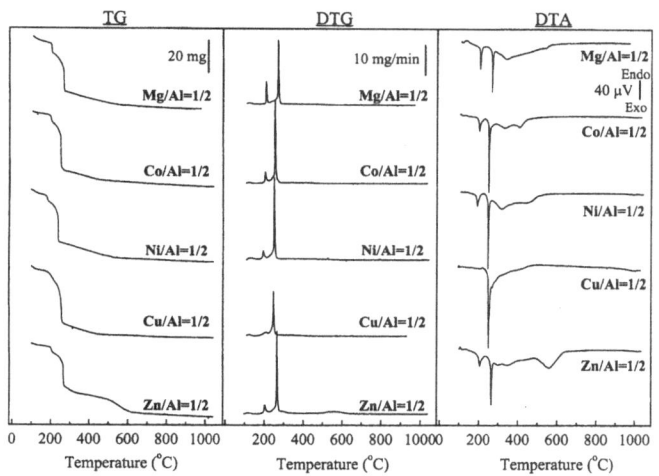

Fig. 1 Thermogravimetry (*TG*)/differential TG (*DTG*)/differential thermal analysis (*DTA*) curves for the *M*/Al–cetyltrimethylammonium bromide precursors (*M* = Mg, Co, Ni, Cu, Zn)

the decomposition takes place at higher temperatures (see DTG) and in a smoother mode (see DTA).

The development of the crystal phases was followed by XRD at each calcination temperature, as shown in Fig. 2. At 600 °C the Mg-, Co- and Zn-containing samples have already developed $MgAl_2O_4$, $CoAl_2O_4$ and $ZnAl_2O_4$ spinel phases, respectively. At the same time, these solids exhibit high surface areas (Table 1) and a porous network was built in the mesopore region, as revealed by the N_2 adsorption studies and the corresponding pore size distribution (Fig. 3). The Cu and Ni samples treated at 600 °C are also mesoporous solids of high surface area but the latter develops the $NiAl_2O_4$ phase only after severe calcination at 1000 °C. Still, at this temperature the spinel phase coexists with significant amounts of the NiO phase, while the Cu-containing solid exhibits significant $CuAl_2O_4$ crystallization at 800 °C. At

Fig. 2 X-ray diffraction patterns of *M*-6, *M*-8 and *M*-10 solids (*M* = Mg, Co, Ni, Cu, Zn) heated at the temperatures indicated (spinel crystal phase, *circles*)

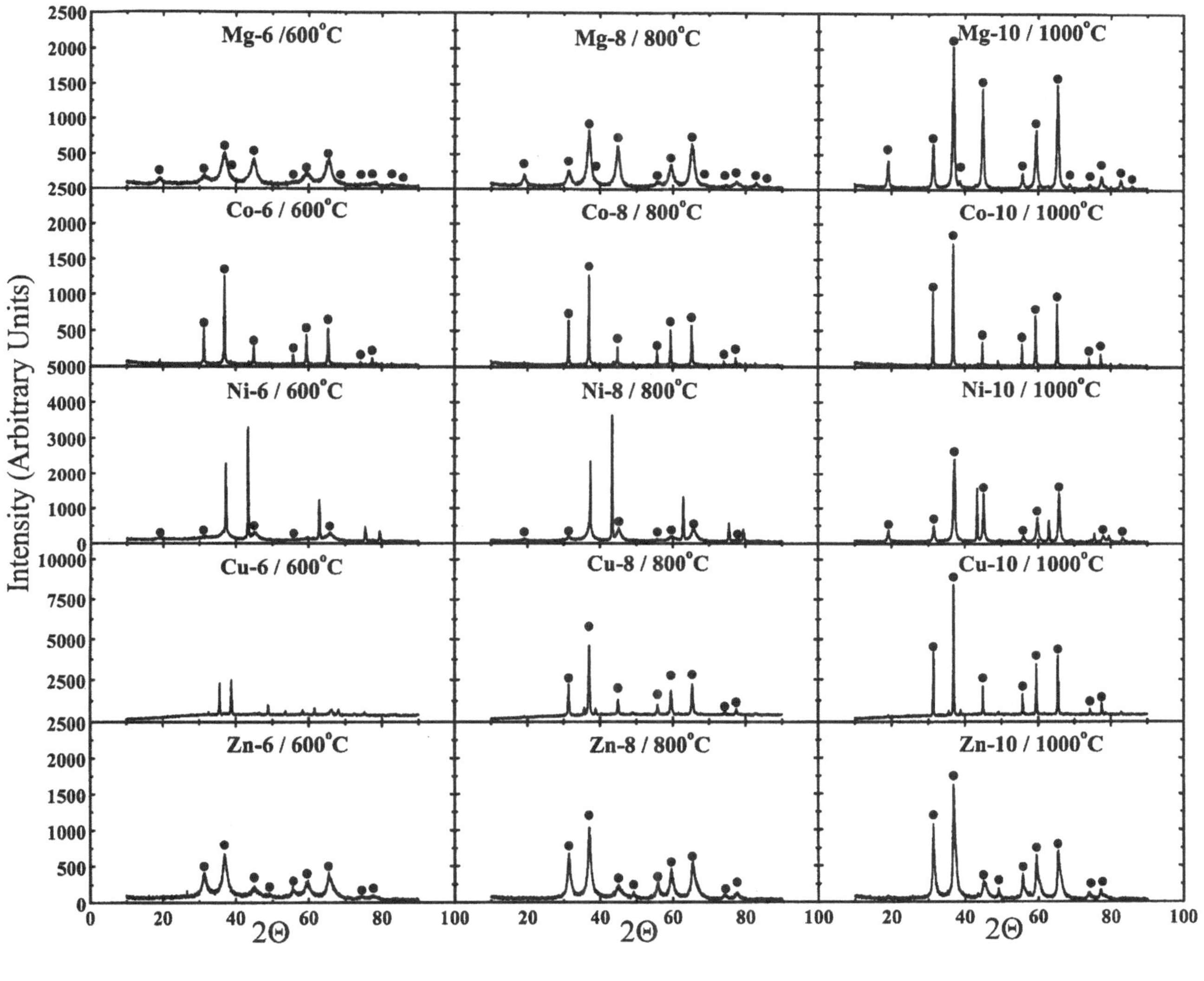

Fig. 3 Adsorption–desorption isotherms (N$_2$ 77 K) and the corresponding pore size distributions of *M*-6, *M*-8 and *M*-10 solids (*M* = Mg, Co, Ni, Cu, Zn) heated at the temperatures indicate

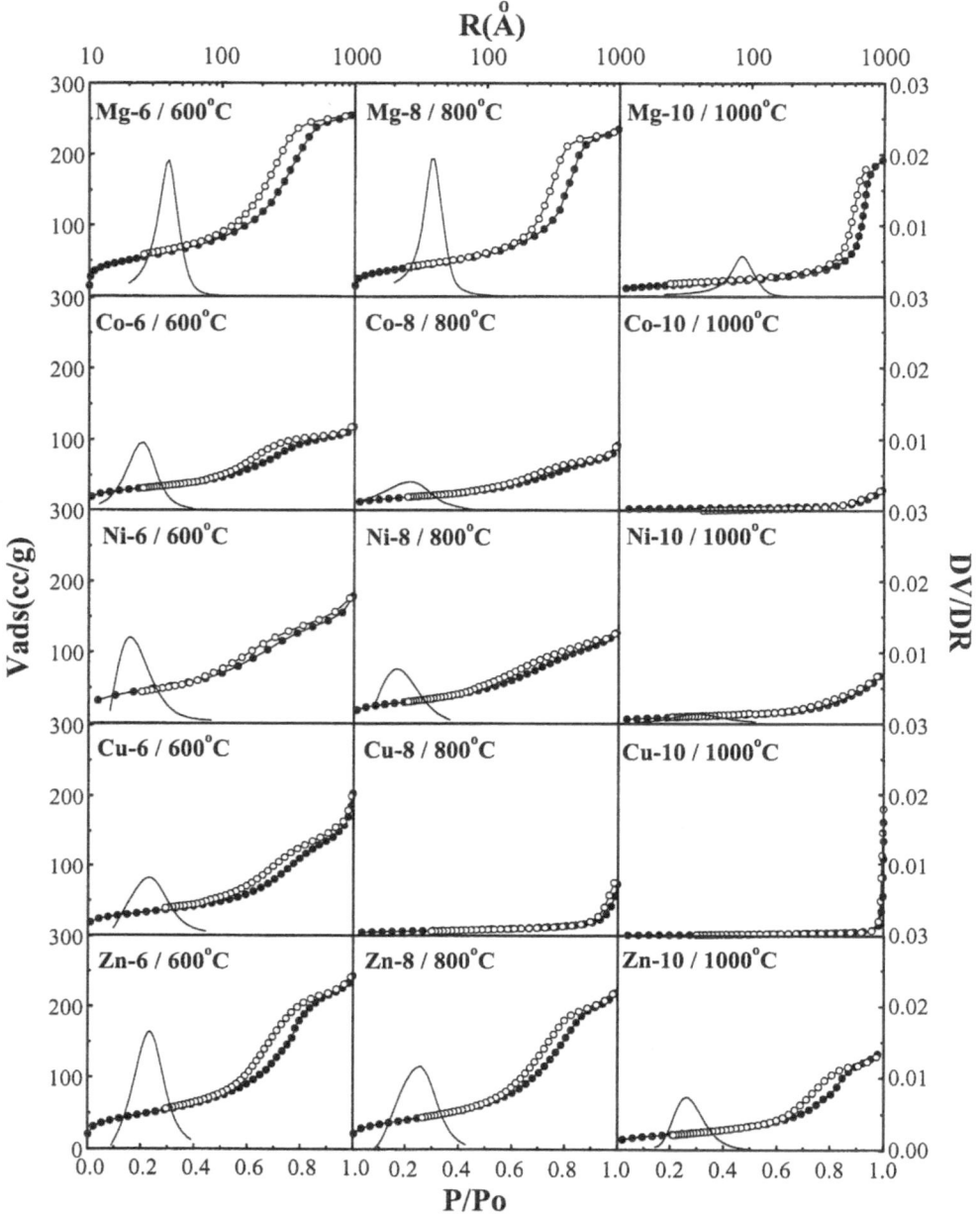

the same time it suffers sintering, which diminishes its S_p (Table 1). From Fig. 3 and the results shown in Table 1 it is clear that all the solids suffer sintering but each to a different degree. Mg and Zn aluminates are the most thermally stable, maintaining their internal porosity to a satisfactory degree after calcination at 1000 °C, 57 m^2 g^{-1} and 75 m^2 g^{-1}, respectively. A comparison is made in Tables 2 and 3 between the Mg and Zn aluminates prepared by the present method and same aluminates prepared by different methods reported in literature. The comparison is made with the best cases in literature by means of S_p and calcination temperature.

For MgAl$_2$O$_4$ the best result concerns the preparation at 600 °C using an alkoxide sol–gel method of mesoporous spinel with a mean pore diameter $D_p = 14$ nm and a surface area of 250 m^2 g^{-1} [12]. Unfortunately there is no data available concerning its thermal stability. This result is followed by the spinel prepared in the present study (Mg-6) treated at the same temperature, with $S_p = 190$ m^2 g^{-1} and $D_p = 8.0$ nm. Among MgAl$_2$O$_4$ solids treated or prepared at higher temperature (e.g. 800 °C), which is more realistic if such solids are to be used as catalytic supports, the Mg-8 sample is the best case together with a commercial one with

Table 2 Comparison of textural properties of MgAl$_2$O$_4$ solids. Coprecipitation (*C-P*) commercial spinel (*Com*) sol–gel (alkoxides) (*S-G/a*) sol–gel (surfactant) (*S-G/s*)

Calcination temperature (°C)	Method	S_p (m²/g)	Refs.
900	C-P	115	[2]
1100		43	
800	C-P	129	[5]
800	Com	140	[4]
1000		36	
600	S-G/a	250	[12]
600	S-G/s	190	This work
800		139	This work
1000		57	This work

Table 3 Comparison of textural properties of ZnAl$_2$O$_4$ solids. Wet-mixing and solid-state reaction (*WMs-s*)

Calcination temperature (°C)	Method	S_p (m²/g)	Refs.
800	C-P	20	[1]
800	C-P	20	[9]
800	S-G/a	50	
1000	WMs-s	22	
300	C-P	340	[10]
500	S-G/a	126	[11]
600	S-G/s	169	This work
800		144	This work
1000		75	This work

$S_p = 140$ m^2 g^{-1} [4]. However, the surface area of the latter drops to 36 m^2 g^{-1} at 1000 °C, while for Mg-10 it is 57 m^2 g^{-1}. Notable is the case of a MgAl$_2$O$_4$ solid prepared by a coprecipitation method with significant surface area at 1100 °C of 43 m^2 g^{-1} [2].

In the case of ZnAl$_2$O$_4$ a very high surface area solid is reported prepared by precipitation and hydrothermal treatment at 300 °C, resulting in a microporous spinel of 340 m^2 g^{-1} and $D_p = 1.1$ nm [10]. This is indeed a very low temperature synthesis but no surface area was checked after higher temperature treatments that are required for catalytic support materials. A Zn aluminate was formed at low temperature (500 °C) by alkoxides with $S_p = 126$ m^2 g^{-1} and this value is lower than the 169 m^2 g^{-1} of the Zn-6 sample prepared in the present study at 600 °C. This is a remarkable sample since even at 800 °C the spinel formed has a surface area of 144 m^2 g^{-1} and at 1000 °C the surface area remains at 75 m^2 g^{-1}, which is the highest surface area reported so far. The next best case is a material of 50 m^2 g^{-1} at 800 °C and only 20 m^2 g^{-1} at 1000 °C prepared by a sol–gel method and wet-mixing, respectively. Taking into consideration the open porosity that allows easy diffusion of molecules this is a promising support for high-temperature catalytic processes such as light hydrocarbon combustion, the oxidation/reduction of pollutants from stationary or mobile sources or other similar applications.

Conclusions

A simple sol–gel templating method was developed and was successfully applied resulting in low-temperature formation (at 600 °C) of MgAl$_2$O$_4$, CoAl$_2$O$_4$ and ZnAl$_2$O$_4$ high-surface-area mesoporous spinels. The method was not successful for low-temperature CuAl$_2$O$_4$ or NiAl$_2$O$_4$ formation.

The porosity of Mg and Zn aluminates is still maintained after heat treatment at 1000 °C. In the case of ZnAl$_2$O$_4$ material, 75 m^2 g^{-1} at 1000 °C is one of the highest surface areas reported in the literature for this spinel.

The shape of the adsorption desorption isotherms and the pore size distribution revealed a random mesoporous network thermally stable up to 1000 °C for Mg and Zn aluminates.

References

1. Aquilar-Rios G, Valenzuela MA, Armedariz H, Salas P, Dominquez JM, Schifter I (1992) Appl Catal A 90:25
2. Marti PE, Maciejewski M, Baiker A (1994) Appl Catal B 4:225
3. Sadykov VA, Pavlova SN, Saputina NF, Zolotarskii IA, Pakhomov NA, Moroz EM, Kuzmin VA, Kalinkin AV (2000) Catal Today 61:93
4. Waqif M, Saur O, Lavalley JC, Wang Y, Morrow BA (1991) Appl Catal 71:319
5. Wang J-A, Li C-L (2000) Appl Surf Sci 161:406
6. Xanthopoulou G (1999) Appl Catal A 182:285
7. Cesteros Y, Salagre P, Medina F, Sueiras JE (2000) Appl Catal B 25:213
8. Le Peltier F, Chaumette P, Saussey J, Bettahar MM, Lavalley JC (1997) Mol Catal A 122:131
9. Valenzuela MA, Jacobs J-P, Bosch P, Reijne S, Zapata B, Brongersma HH (1997) Appl Catal A 148:315
10. Zawadzki M, Wrzyszcz J (2000) Mater Res Bull 35:109
11. Otero Arean C, Sintes Sintes B, Turnes Palormino G, Mas Carbonell C, Escalona Platero E, Parra Soto JB (1997) Microporous Mater 8:187
12. Alvarez Lopez MR, Toralvo Fernadez MJ, Mas Carbonell C, Otero Arean C (1993) J Mater Sci Lett 12:1619

Progr Colloid Polym Sci (2001) 118: 22–26
© Springer-Verlag 2001

Barbara Gzyl
Maria Paluch

Properties of insoluble mixed monolayers of lipids at the water/air interface

B. Gzyl (✉) · M. Paluch
Faculty of Chemistry
Jagiellonian University
Kraków, Poland

Abstract A study of monolayer mixing behaviour in binary D,L-dipalmitoyl phosphatidylcholine/ 3-monopalmitoyl glycerol mixtures was undertaken. For this purpose, the isotherms of surface pressure versus molecular area were acquired at four different temperatures and a surface thermodynamic analysis was applied to these isotherms.

Key words Monolayers · Lipid membrane · Monopalmitoyl glycerol · Dipalmitoyl phosphatidylcholine

Introduction

The study of mixed monolayers is of particular importance because it makes it possible to gain knowledge about the interactions between the monolayer compounds. It provides information about the molecular orientation of the amphiphilic molecules at the interfaces and about their compatibility [1]. It also enables the prediction of the properties of more complex aggregates by the study of molecular interactions in such simple systems. These two-dimensional systems are of particular importance both from the point of view of applications and as useful models of biological membranes. In particular, the two-dimensional miscibility among different components is important in defining the interactions in membrane models [2–4]. It is well known that biological membranes are made up of bilayers of lipidic compounds in which the other components, such as proteins and enzymes, are immersed or bound to the two interfaces.

The aim of this work was to study molecular interactions and characteristics of the monolayers formed by lipids. We studied the behaviour of two lipids: D,L-dipalmitoyl phosphatidylcholine (DPPC) and 3-monopalmitoyl glycerol (PG) and also their mixtures, at different molar ratios, at the water–air interface under diverse temperature conditions. For this purpose the isotherms of surface pressure versus molecular area were recorded using a Langmuir film balance. Then, a classical surface chemistry thermodynamic analysis was performed on these isotherms, which involved calculating the excess free energies of mixing to determine the miscibility properties of these two compounds.

Experimental

DPPC and PG were obtained from Sigma-Aldrich. The spreading solvent was chloroform (distilled before using) from POCH, Poland. Water was distilled four times. Separate stock solutions of DPPC and PG were prepared in chloroform. Solutions of the mixtures were prepared by mixing precisely measured volumes of DPPC and PG.

Surface pressure–area (π–A) isotherm measurements were carried out using a KSV 1000 system (KSV Instruments, Helsinki) with an accuracy of 0.01 mN/m and 0.01 Å^2 per molecule for the surface pressure and area, respectively. The subphase surface was cleaned repeatedly by sweeping the barriers slowly between the maximum and minimum area positions and aspirating the surface until no change in surface pressure was detectable between the "open" and "closed" positions. The experiments were performed at four temperatures: 20, 25, 30 and 35 °C. The subphase temperature was controlled thermostatically to within 0.1 °C by circulating water. For the π–A isotherm experiments, precisely measured volumes of the respective solution were spread on the water surface, using an Hamilton microlitre syringe. As a standard procedure, the monolayers were rested for 20 min before compression to allow sufficient solvent evaporation. A speed of compression of 6 mm/min was used. Below this compression rate no difference in the π–A isotherm was observed – this was therefore considered to be slow enough that the π–A isotherm obtained represented the "equilibrium" isotherm. Each isotherm was measured at least twice. The same procedures were adopted for the pure components and for their mixtures.

Results

The π–A isotherms of the system DPPC and PG at 20, 25, 30 and 35 °C are shown in Fig. 1. It can be seen that the shape of the isotherms is affected by the temperature of the subphase. With increasing temperature, the isotherms become progressively more expanded. All the curves of the mixed systems are included between the pure systems in a regular sequence of mole fraction of one component in the binary mixture. The addition of PG shifts the isotherms of pure DPPC to progressively lower mean molecular areas. This contraction is due to the lower area occupied by the PG molecule.

Information on the mutual miscibility of the two components in the two-dimensional state at the water–air interface may be obtained from the change in surface area as a function of molar ratio at constant surface pressure. If the two components are immiscible or if they behave like an ideal mixture, the following relationship is valid [5]:

$$A_{12} = x_1 A_1 + x_2 A_2 , \tag{1}$$

where A_{12} is the molecular area in the mixed monolayer at the fixed surface pressure, π, A_1 and A_2 are the molecular areas in the pure component monolayer at the same π and x_1 and x_2 are the molar or weight fractions of the pure components in the mixture such that $x_1 + x_2 = 1$.

The molecular areas obtained from the spreading isotherms as a function of monolayer composition are reported in Fig. 2. It is readily evident from Fig. 2 that we observed two-dimensional miscibility. Moreover, negative deviations from ideality indicate that the compounds in the two-dimensional state experience mainly attractive interactions.

The miscibility of the two components at all molar ratios was confirmed by applying the two-dimensional phase rule, since the collapse pressure, π_{coll}, varied with the composition of the mixtures, as can be seen from Fig. 1.

Once the miscibility of DPPC and PG was ascertained, we considered a thermodynamic analysis to be possible and useful; hence, we determined the excess free energy of mixing, ΔG^E_{mix}, following the Goodrich method [6] by integrating the π–A isotherms up to π lower than the discontinuity surface pressure.

$$\Delta G^E_{mix} = N_A \int_{\pi_1}^{\pi_2} (A_{12} - x_1 A_1 - x_2 A_2) \mathrm{d}\pi , \tag{2}$$

where π_1 and π_2 are two fixed surface pressures and N_A is the Avogadro number.

The enthalpic and entropic contributions to the excess Gibbs energy of mixing were calculated according to the Bacon and Barnes method [7], i.e.

$$\Delta S^E_{mix} = -\left(\frac{\mathrm{d}\Delta G^E_{mix}}{\mathrm{d}T}\right)_\pi - \Delta A^E_{mix}\left(\frac{\mathrm{d}\gamma}{\mathrm{d}T}\right) , \tag{3}$$

Fig. 1 π–A isotherms of D,L-dipalmitoyl phosphatidylcholine (*DPPC*)/3-monopalmitoyl glycerol (*PG*) system on water at 20, 25, 30 and 35 °C

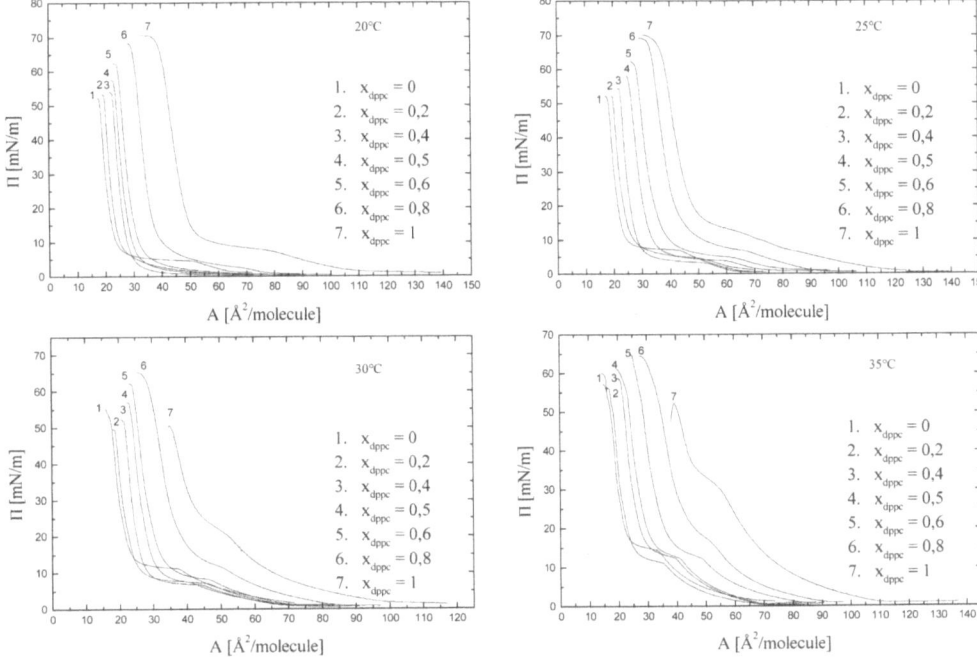

Fig. 2 Surface areas as a function of molar fractions of DPPC, at $\pi = 5$, 20 and 40 mN/m

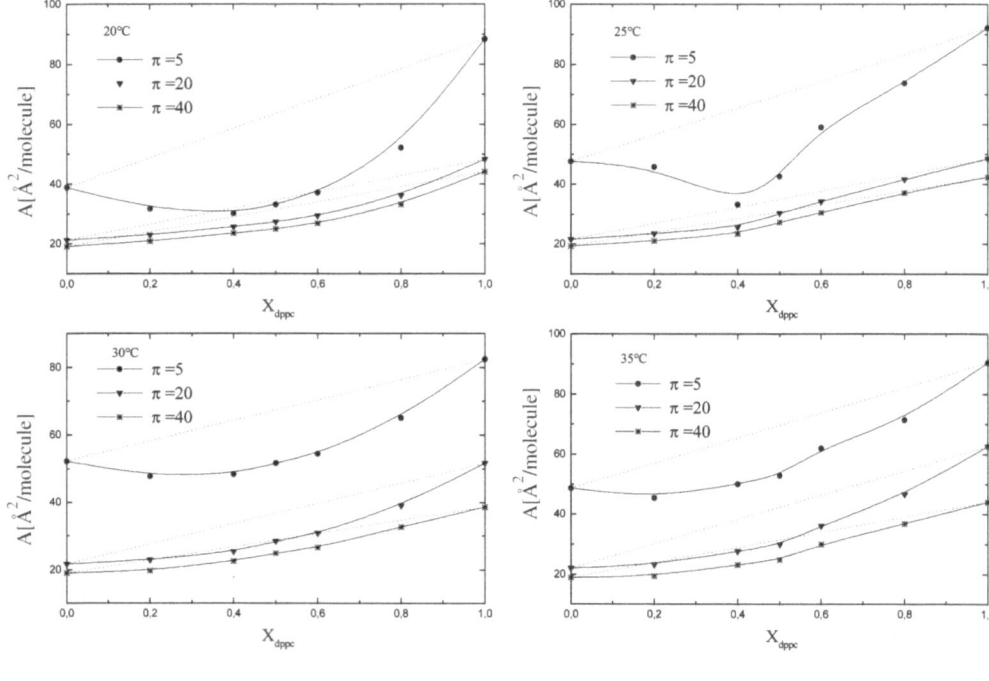

Fig. 3 Excess free energy of mixing of the DPPC/PG system as a function of molar fractions of DPPC, at $\pi = 5$, 20 and 40 mN/m

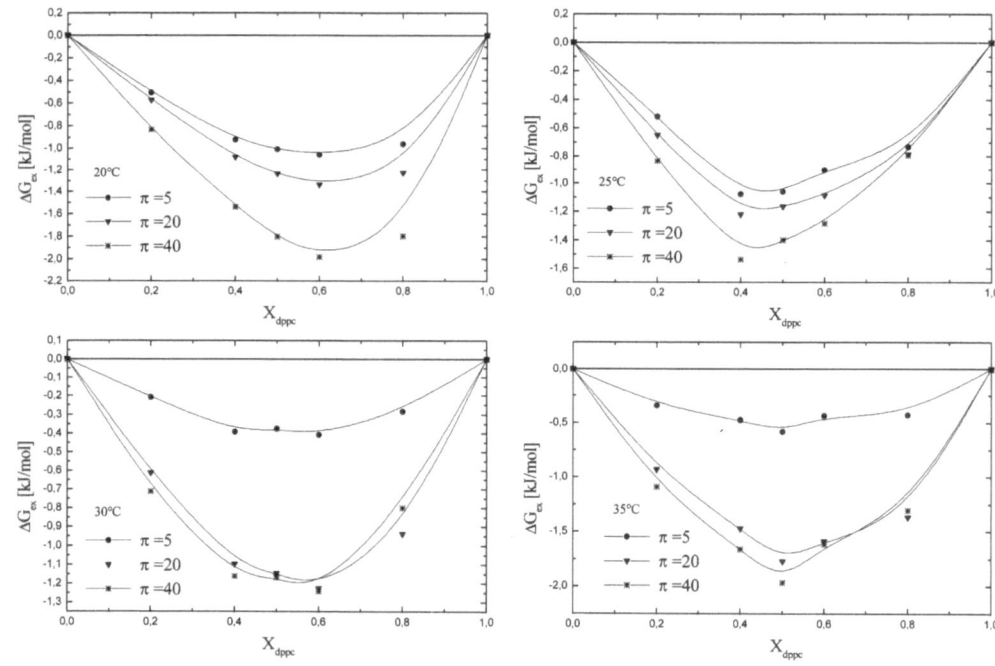

where

$$\Delta A_{mix}^{E} = A_{12} - x_1 A_1 - x_2 A_2,$$
$$\frac{d\gamma}{dT} = -0.154 \text{ mN/m/K},$$
$$\Delta H_{mix}^{E} = \Delta G_{mix}^{E} + T\Delta S_{mix}^{E}. \tag{4}$$

The plot of ΔG_{mix}^{E} versus the molar ratio (Fig. 3) shows that the free energy of mixing differs from zero,

thus indicating a reciprocal nonideal miscibility and the existence of interactions between the components. ΔG_{mix}^{E} always showed negative values, meaning that the monolayers of the mixtures were thermodynamically stable and that the attractive interactions between the components prevailed. Moreover the presence of minima of the ΔG_{mix}^{E} function in the range of molar ratios of DPPC:PG from 2:3 to 3:2 offer definitive proof for the higher thermodynamic stability of the approximately equimolar mixtures.

Detailed results of the calculation of the excess entropy and enthalpy of mixing, ΔS^E_{mix} and ΔH^E_{mix}, for the binary mixture studied, at four temperatures, are presented in Figs. 4 and 5. The trends of ΔS^E_{mix} and ΔH^E_{mix} versuss the molar ratio exhibit negative minima at lower temperatures and positive maxima at higher temperatures: that means that the two-dimensional miscibility between the components is due to enthalpic

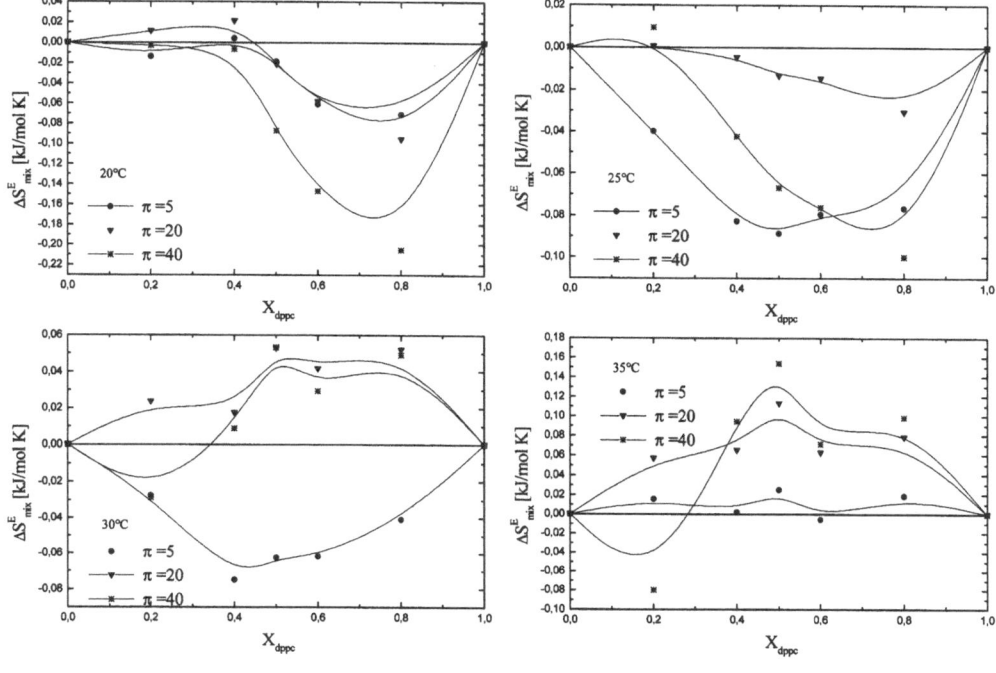

Fig. 4 Excess entropy of mixing of the DPPC/PG system as a function of molar fractions of DPPC, at $\pi = 5$, 20 and 40 mN/m

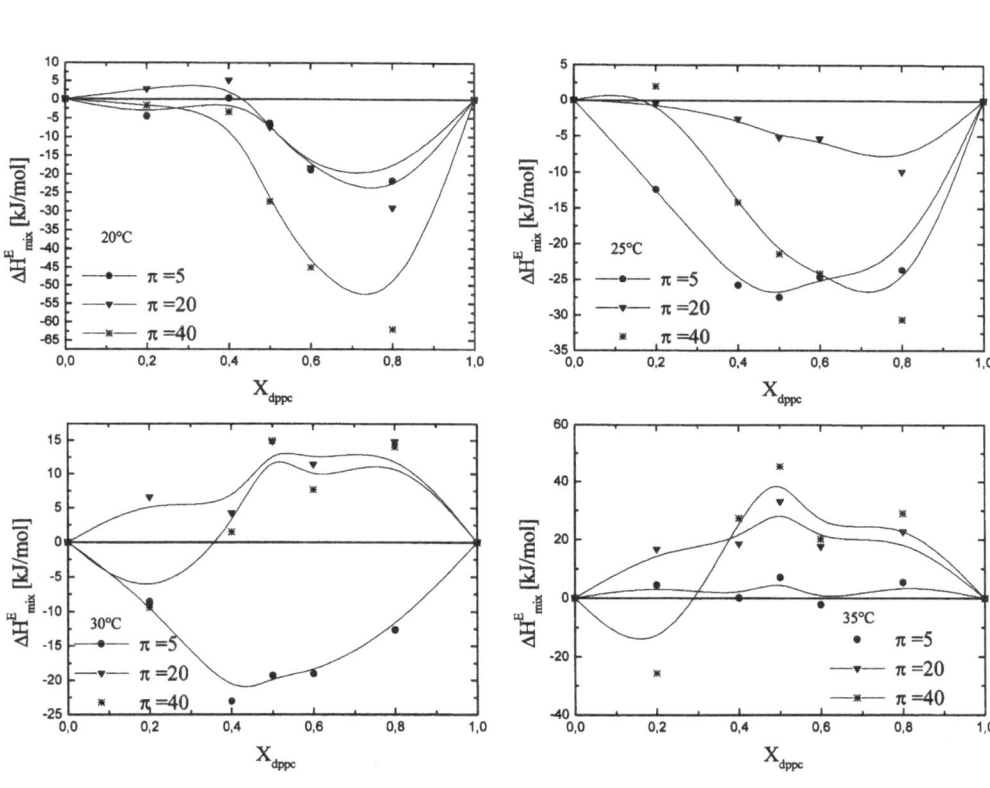

Fig. 5 Excess enthalpy of mixing of the DPPC/PG system as a function of molar fractions of DPPC, at $\pi = 5$, 20 and 40 mN/m

factors at lower temperatures and to entropic factors at higher temperatures. In particular positive values of ΔS^E_{mix} at higher temperatures mean that the higher temperature introduces disorder in the monolayer.

Conclusion

A study of monolayer mixing behaviour in binary lipid mixtures of DPPC and PG was undertaken. The two-dimensional miscibility between these two components was verified in the 20–35 °C temperature range at all surface pressures. We observed negative deviations from ideality for the surface free energy of mixing as a function of the composition. These results indicate that the monolayers of mixtures (in particular of the equimolar composition) were thermodynamically more stable than the monolayers of the pure compounds. The negative deviations from ideality for the mean molecular areas and for the surface free energy of mixing as a function of molar fractions means that the compounds form aggregates of a more closely packed structure and that there are mainly attractive interactions in the monolayer, favoured for enthalpic reasons at lower temperatures and for entropic reasons at higher temperatures.

References

1. Cadenhead DA (1985) Strucrures and properties of cell membranes. CRC, Boca Raton, p 23
2. Gabrielli G, Caminati G, Puggelli M, Gilardoni A, Margheri E (1993) J Phys IV 3:279
3. Birdi KS (1989) Lipid and biopolymer monolayers at liquid interfaces. Plenum, New York
4. Fendler JH (1982) Membrane mimetic chemistry. Wiley, London
5. Gaines GL (1996) Insoluble monolayers at liquid–gas interfaces. Wiley, New York
6. Goodrich FC (1957) in: Schulman JH (ed) Proceedings of the 2nd International Congress on Surface Activity, vol 1. Butterworhs, London, p 85
7. Bacon KJ, Barnes GT (1978) J Colloid Interface Sci 67:70

Progr Colloid Polym Sci (2001) 118:27–29
© Springer-Verlag 2001

Jinman Huang
Vlasoula Bekiari
Panagiotis Lianos

Enhancement of weak radiative transitions of Eu^{3+} in thin surfactant films in the presence of poly(methyl methacrylate)

J. Huang · V. Bekiari · P. Lianos (✉)
Engineering Science Department
University of Patras, 26500 Patras, Greece
e-mail: lianos@upatras.gr
Tel.: +30-61-997587; Fax: +30-61-997803

Abstract Europium ions were dispersed in reverse micellar solutions of bis(2-ethylhexyl) sulfosuccinate (AOT) in chloroform and then thin films were obtained by the dip-coating method. Stabilization of the films was achieved in the presence of poly(methyl methacrylate) (PMMA). The photophysical behavior of Eu^{3+} in these composite organic films was studied. It was found that the combination of AOT with PMMA results in efficient dispersion of Eu^{3+}, a decrease of concentration quenching and enhancement of weak radiative transitions, particularly, of the emission corresponding to the $^5D_1 \rightarrow {}^7F_1$ transition at 538 nm. The importance of the relative polymer and surfactant concentrations as well as some other factors affecting Eu^{3+} emission are discussed.

Key words Poly(methyl methacrylate) · Bis(2-ethylhexyl) sulfosuccinate sodium salt Film · Europeum · Luminescence

Introduction

The luminescence emission of lanthanide ions is being studied with a lot of interest because of their narrow-band emission spectra and their long decay times. This makes them valuable sources of visible and near-IR radiation, some of them being particularly important for lasers and optical communication devices [1]. However, the radiative transitions of lanthanide ions are sensitive to their chemical microenvironment. Thus, their luminescence is quenched owing to coupling with vibrations of the host matrix. For this reason, a lot of effort has been made to improve the emission capacity by employing several different approaches [2–8]. Emission from the upper, D_J, to the lower, F_J, energy level in Eu^{3+} is particularly sensitive to the microenvironment, especially, in oxide host matrices. Indeed, strong coupling with OH groups and limited solubility, leading to aggregation of Eu^{3+}, has a destructive effect on most emission bands, allowing only some weak red emission [4]. Improvement of red emission has recently been achieved in the presence of polymer subphases embedded in a silica matrix [5]. In the present work, we employed a combination of a reverse-micelle-forming surfactant with a matrix-forming polymer in thin film configuration to obtain an enhancement of Eu^{3+} weak transitions, particularly, of the $^5D_1 \rightarrow {}^7F_1$ emission band at 538 nm.

Materials and methods

Poly(methyl methacrylate) (PMMA, $M_w = 1.2 \times 10^5$, Aldrich), bis(2-ethylhexyl)sulfosuccinate sodium salt (AOT, Fluka), europium(III) chloride hexahydrate (Aldrich) and chloroform (Aldrich) were used as received. Millipore water was used in all the experiments. Different amounts of AOT were dissolved in chloroform in the presence of PMMA, then a certain amount of EuCl$_3$ aqueous solution was added. At the beginning the solution was turbid, but after 2 h of stirring it became clear. Films were deposited on clean glass slides under ambient conditions by the dip-coating method. The withdrawal speed was 44 mm/min in all cases. The films were dried in air.

Fluorescence spectra were measured with a home-assembled apparatus, consisting of a 150-W xenon lamp and computer-driven monochromators and detection system. All parts and software were purchased from Oriel Instruments. All measurements were carried out at room temperature under ambient conditions.

Results and discussion

Reverse micelles of AOT in chloroform [9] were made using 0.2 M AOT and 1 M water. This molar ratio of water/surfactant (i.e. $w = 5$) was first kept constant. It is well established that for w up to about 10, all the water present in solution is used up to hydrate AOT polar groups and it is attached on the surfactant by strong forces. PMMA was introduced into the solution before water addition as a stabilizer of the ensuing films on glass supports. Eu^{3+} was introduced by solubilization in the water used to make the original solution. Even though clear reverse micellar solutions can be readily obtained with most solvents in the absence of europium, in its presence Eu^{3+} apparently forms large complexes with the oppositely charged AOT, with a tendency to precipitate. Nevertheless, persistent stirring when PMMA is present results in the clusters redissolving and produces a transparent microheterogeneous system. Films were then obtained by simply dipping substrates in the clear solutions. The data presented later refer to films.

When excited at 396 nm, two emission peaks were observed, situated around 538 and 645 nm, corresponding to the $^5D_1 \rightarrow {}^7F_1$ and $^5D_0 \rightarrow {}^7F_3$ transitions respectively [3]. Both peaks are shown in Fig. 1. The excitation spectrum (not shown) was identical for both peaks, with a maximum at 396 nm. We observed that by increasing the polymer concentration, a dramatic enhancement of the $^5D_1 \rightarrow {}^7F_1$ emission is obtained (Fig. 1). In contrast, by increasing the Eu^{3+} concentration, the ratio of the $^5D_1 \rightarrow {}^7F_1$ to the $^5D_0 \rightarrow {}^7F_3$ emission intensity decreases. It is known that the Eu^{3+} concentration significantly affects the upper ^5D_J luminescence emission [3, 10, 11] owing to concentration-quenching. Good dispersion of these ions in a matrix enhances the emission. Indeed, as seen in Fig. 2, where the relative intensities of the $^5D_1 \rightarrow {}^7F_1$ and $^5D_0 \rightarrow {}^7F_3$ transitions versus the PMMA or Eu^{3+} concentration are compared, higher polymer concentration or lower Eu^{3+} concentration gives higher 538 nm intensities. Figure 2 shows that the highest Eu^{3+} concentration above which a large decrease in the 538-nm band intensity is observed corresponds to about 10^{-3} M (as measured in the original solution). This value coincides with that obtained by Dejneka et al. [3] in fluoride glass and it corresponds to an average Eu–Eu separation of about 40 Å. It is obvious that the presence of PMMA has a double beneficial effect: it stabilizes the surfactant films and it induces a dispersion of Eu^{3+} ions, thus preventing concentration-quenching. At high PMMA concentration, where the dispersion of ions is most efficient, the decrease in concentration-quenching allows the appearance of more transitions and of a rich fine structure, as seen in Fig. 3. Combination of PMMA with AOT is very successful for dispersing lanthanide ions. Complexation of this anionic surfactant with cations and facile dissolution in the polymer hydrophobic matrix allows effective dispersion.

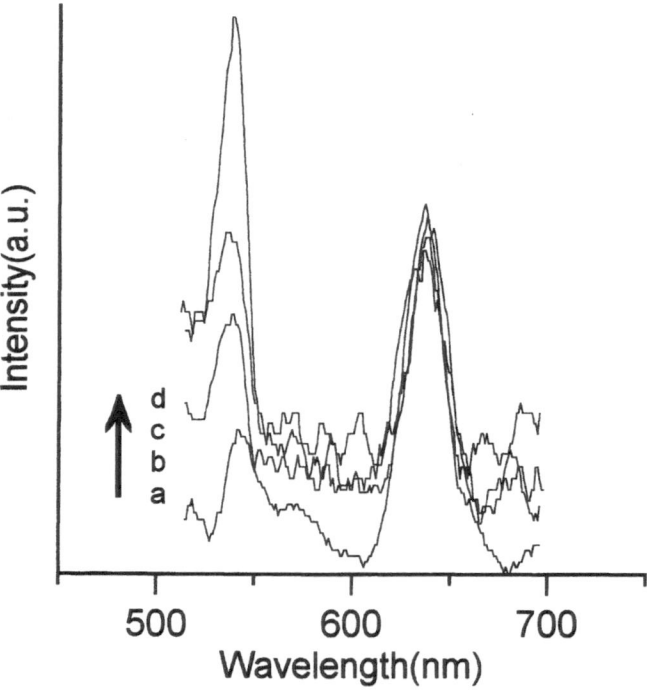

Fig. 1 Effect of poly(methyl methacrylate) (*PMMA*) on the fluorescence spectra of Eu^{3+}-doped thin films. The PMMA concentration in chloroform was *a* 1 g/l, *b* 3 g/l, *c* 5 g/l and *d* 12 g/l. The molar ratio of H₂O/ bis(2-ethylhexyl)sulfosuccinate sodium salt (*AOT*) was 5 and the Eu^{3+} concentration was 1.8×10^{-3} M. The spectra were normalized to the emission peak at 645 nm

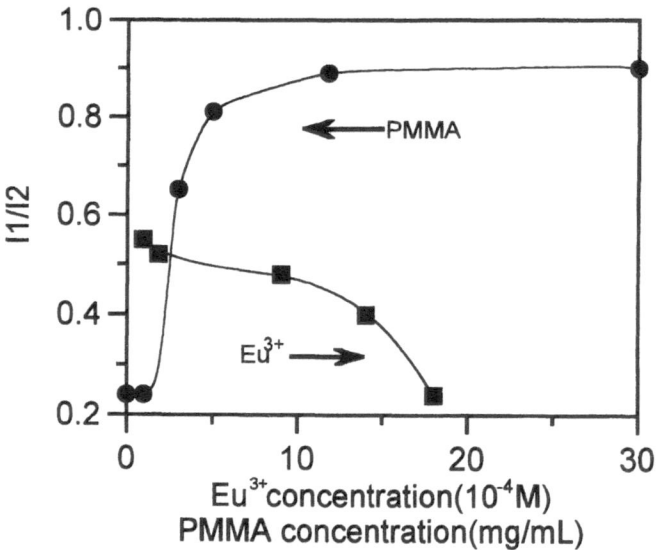

Fig. 2 Relative D_1–F_1/D_0–F_3 emission intensity as a function of PMMA and Eu^{3+} concentration. The molar ratio of H₂O/AOT was 5

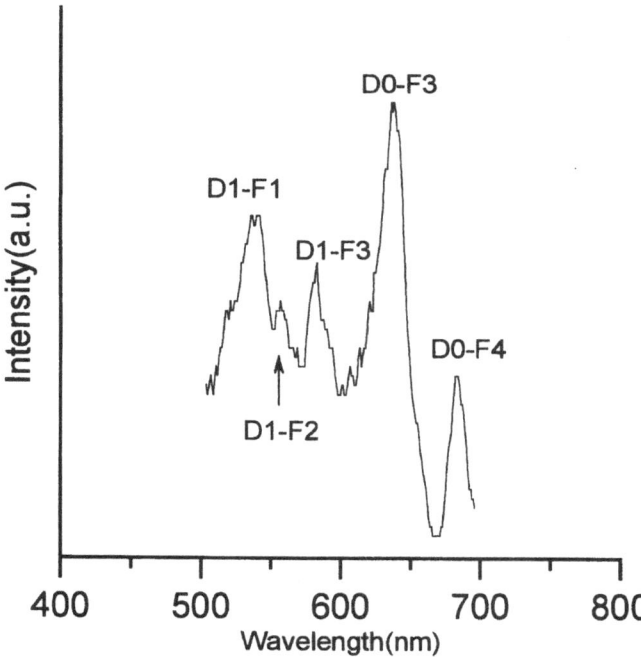

Fig. 3 Fluorescence spectrum of Eu^{3+}-doped thin films with high PMMA concentration (30 g/l). The molar ratio of H_2O/AOT was 5

The size of the reverse micelles did not have any effect on the luminescence emission. By varying, for example, the water/surfactant molar ratio within the range $3 < w < 10$, i.e. below the water-pool-forming regime, no effect on the luminescence emission was observed. An increase in w presumably means an increase in the size of the reverse micelles [12]. Of course, the size of the reverse micelles is expected to be substantially modified in the presence of the polymer chains, while strong complexation with ions at this low water regime is expected to further affect the size of the micelles. At higher water content, no coordination between polymer and surfactant is observed and their mixture remains turbid. Apparently, no reverse micelles, in the strict meaning of the term, are preserved in the final component mixture. The solution, most probably contains polymer chains with Eu^{3+} ions and hydrated surfactant organized in aggregates associated with the polymer chains. This structure is transferred into the film, possibly, without important modifications.

Conclusion

Complex formation between Eu^{3+} ions and the hydrated anionic surfactant AOT allows a fine dispersion of the ions in a PMMA/AOT thin-film matrix. Such a dispersion is facilitated by hydrophobic interaction between the polymer chains and the AOT/Eu^{3+} complex. As a result of dispersion, concentration-quenching is decreased, the $^5D_1 \rightarrow {}^7F_1$ transition is greatly enhanced and fine structure in the emission spectrum is observed.

Acknowledgements We acknowledge financial aid from the program ΠΕΝΕΔ of the Greek General Secretariat of Research and Technology.

References

1. Dejneka M, Snitzer E, Riman RE (1996) J Non-Cryst Solids 202:23
2. Qian DJ, Yang KZ, Nakahara H, Fukuda K (1997) Langmuir 13:5925
3. Dejneka M, Snitzer E, Riman RE (1995) J Lumin 65:227
4. Takada N, Sugiyama J, Minami N, Hieda S (1997) Mol Cryst Liq Cryst 295:71
5. Bekiari V, Pistolis G, Lianos P (1998) J Non-Cryst Solids 226:200
6. Reisfeld R, Greenberg E, Brown RN, Drexhage MG, Jørgensen CK (1983) J Chem Phys Lett 95:91
7. Adam JL, Poncon V, Lucas J, Boulon G (1987) J Non-Cryst Solids 91:191
8. Todoroki S, Hirao K, Soga N (1992) J Non-Cryst Solids 143:46
9. Goto A, Yoshioka H, Kishimoto H, Fujita T (1992) Langmuir 8:441
10. Reisfeild R, Lieblich N (1973) J Phys Chem 34:1467
11. Samek L, Wasylak J, Soga N (1992) J Non-Cryst Solids 140:243
12. Maritra A (1984) J Phys Chem 88:5122

Progr Colloid Polym Sci (2001) 118: 30–33
© Springer-Verlag 2001

D. Vollhardt

Phase transition in monolayers induced by adsorbed amphiphiles

D. Vollhardt
Max-Planck-Institut für Kolloid- und
Grenzflächenforschung
14424 Potsdam/Golm, Germany
e-mail: vollh@mpikg-golm.mpg.de
Tel.: +49-331-5679258
Fax: +49-331-5679202

Abstract A short review is given on recent progress which we achieved in the characterization of condensed monolayer phases induced by the adsorption of amphiphilic species (surfactants, proteins). The combination of surface pressure adsorption kinetics, Brewster-angle microscopy and X-ray diffraction at grazing incidence is the highly effective experimental basis of these investigations. At the beginning, a tailored amphiphile was used to be sure that artefacts caused by highly surface active trace components were avoided. So far, a first-order phase transition has been found in adsorbed monolayers of numerous other surfactants and systems. Various types of 2D modifications have been identified. A first-order phase transition can also be induced by the coadsorption of two surfactants and by the penetration of soluble surfactants into gaseous Langmuir monolayers.

Key words Adsorption · Phase transition · Monolayer penetration · Brewster-angle microscopy · Surfactants

Introduction

Rapid progress in the understanding of the molecular organisation of condensed monolayer phases has been made in the last decade. In particular, Brewster-angle microscopy (BAM) [1, 2] and synchrotron X-ray diffraction at grazing incidence (GIXD) [3, 4] provide powerful methods for characterising 2D condensed phases. In 1996, we provided the first direct evidence that a first-order main phase transition can also occur in adsorbed monolayers [5, 6]. The recent progress which we achieved in the characterisation of condensed monolayer phases induced by the adsorption of ampliphilic species (surfactants, proteins) is given in this short review. We provide evidence that condensed monolayer phases can be formed not only by the adsorption of surfactants but also by the coadsorption of two dissolved surface-active species or can be induced by the adsorption of dissolved amphiphiles into Langmuir monolayers in the gaseous state.

Experimental

Materials

The purity of dodecanol (Fluka, puriss.) distilled twice was 99.5% or greater, as analysed by gel permeation chromotography. Sodium dodecyl sulfate (SDS) was prepared and purified as described previously [7]. The purity of the amphiphile N-dodecyl-γ-hydroxybutyric acid amide (DHBAA) synthesised and purified according to Ref. [6] was 99% or greater. Dipalmitoyl phosphatidylcholine (DPPC) (99% or greater purity) and bovine β-lactoglobulin were obtained from Sigma. Chloroform (p.a. grade) purchased from Baker, Holland, was used as a spreading solvent for DPPC. The water was made ultrapure using a Millipore desktop unit.

Methods

The best possibility for studying first-order phase transitions induced by adsorbed amphiphiles is the coupling of the $\pi(t)$ adsorption kinetics (π: surface pressure, t: time), BAM and GIXD [8, 9]. The $\pi(t)$ adsorption kinetics of the amphiphiles dissolved in the aqueous subphase and the $\pi–A$ isotherms of the amphiphiles spread at the surface were recorded with a computer-interfaced film balance using the Wilhelmy method [6]. The penetration experiments of dissolved amphiphiles into Langmuir monolayers were performed by using the sweeping technique [10].

Results and discussion

On the basis of the results of a tailored amphiphile [5, 6] it has been found that the first-order phase transition is thermodynamically indicated by a break point in the continuous course of the $\pi(t)$ adsorption kinetics. The concentration of the dissolved amphiphile and the temperature of the aqueous solution determine largely whether and after what time the phase transition occurs.

A first-order transition can also be found in adsorbed monolayers of other amphiphiles dissolved in water, even for the model surfactant dodecanol. The dotted line in Fig. 1 shows the $\pi(t)$ adsorption kinetics of a 12 μM dodecanol solution at 15 °C. The characteristic break point in the dynamic surface tension curve indicates a main phase transition in the adsorbed monolayer. Up to the break point the surface pressure increases rather fast and the adsorbed material should be homogeneously distributed in a fluid state. After the break point the pressure increase is at first comparatively small. BAM studies have shown that after the phase-transition point condensed-phase domains surrounded by a homogeneous fluidlike phase are formed. The BAM results of dodecanol monolayers are described in more detail elsewhere [11]. Afterwards there follows a period with a steeper pressure increase, finally approximating the equilibrium pressure. In this region the portion of the

condensed phase increases and in the equilibrium state the adsorbed layer consists completely of condensed phase [11].

Fundamental differences in the adsorption properties were found for the model surfactant SDS. The adsorbed layer of highly purified SDS does not show a phase transition even above the critical micelle concentration and at low temperatures so condensed-phase domains cannot be formed. This provides good preconditions for the study of the coadsorption of SDS and dodecanol in trace amounts from aqueous solutions. The solid line in Fig. 1 represents the adsorption kinetics of the 3 mM SDS and 12 μM dodecanol mixed aqueous solution, i.e. the SDS solution contains only 0.4 mol% dodecanol. The adsorption kinetics of the 0.3 mM SDS solution is obviously so fast that the transient recorded begins at higher surface pressures than the corresponding equilibrium pressure. Then dodecanol adsorption increases and the shape of the transient recorded is similar to that of the pure dodecanol solution of the same concentration, but shifted to much higher surface pressure values; however the phase-transition point is reached at an essentially shorter time just as the equilibrium adsorption. The small dodecanol traces in the SDS main component obviously cause a phase transition of first-order. The development of condensed-phase domains during the adsorption kinetics is demonstrated in Fig. 2. Condensed-phase domains, very similar in shape to those observable at the adsorption of pure dodecanol, are formed in the region after the thermodynamic phase-transition point. They grow with time, coalesce with each other and finally form a nearly complete coverage of the surface. This similarity of the condensed-phase patterns to those of pure

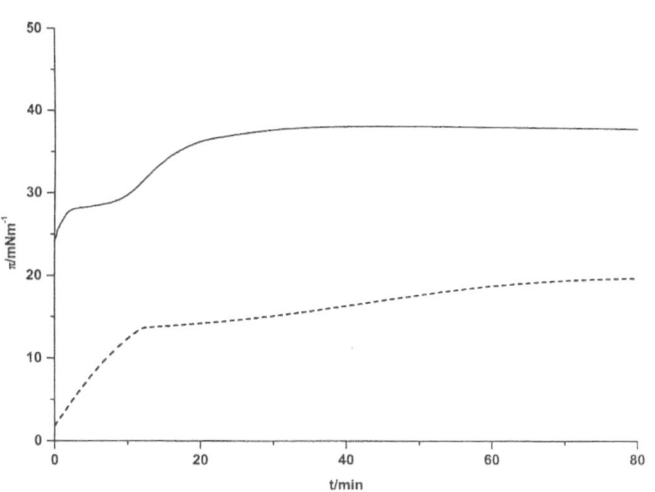

Fig. 1 Dynamic surface tension of aqueous solutions of 12 μM dodecanol solutions (*dotted line*) and mixed 3 mM SDS/12 μM dodecanol solutions; 15 °C

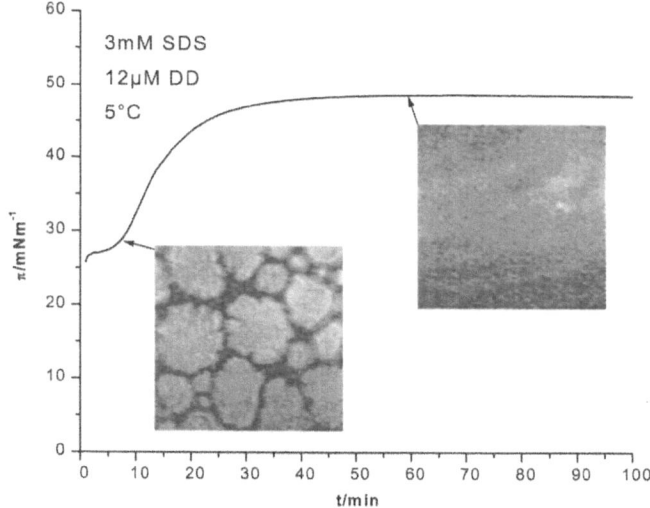

Fig. 2 Typical Brewster-angle microscopy (*BAM*) images for two selected stages during the $\pi(t)$ adsorption kinetics after the phase transition point; 5 °C; image size: 325 μm × 325 μm

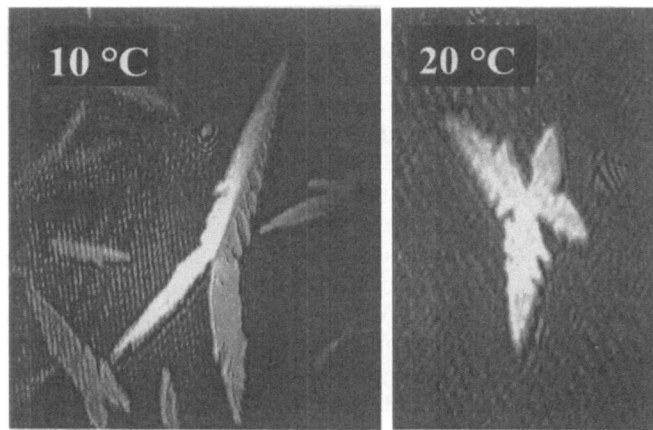

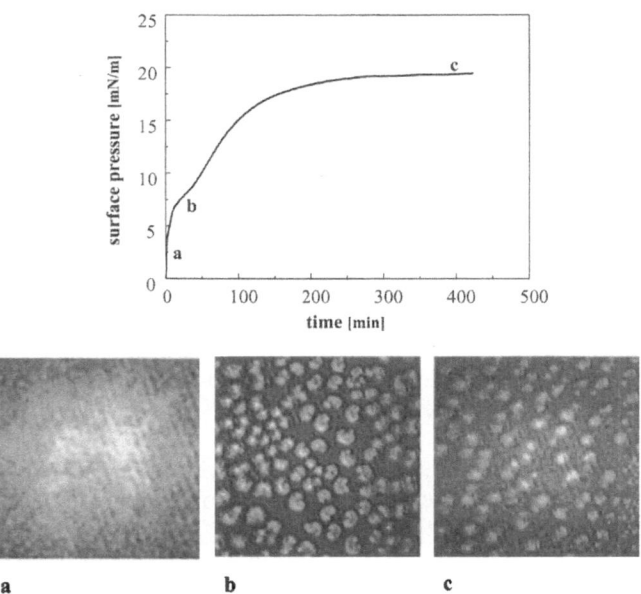

Fig. 3 Domain textures of adsorbed *N*-dodecyl-*γ*-hydroxybutyric acid amide monolayers characteristic for the two modifications formed at $T \leq 10\,°C$ and $T \geq 10\,°C$

Fig. 4 Penetration dynamics for *β*-casein penetration from 10^{-7} M solution at $A = 0.90$ nm^2 per dipalmitoyl phosphatidylcholine (*DPPC*) molecule. The $\pi(t)$ penetration kinetics curve has the first-order phase transition point after 3100 s. The sequence of BAM images **a–c** shows the growth of DPPC domains according to the letters of the $\pi(t)$ curves

dodecanol systems indicates that the condensed phase consists only of a pure dodecanol phase, which increasingly replaces the adsorbed SDS molecules. GIXD measurements very recently performed provide final evidence that the condensed phase indeed consists only of dodecanol. Consequently during the coadsorption kinetics of SDS and dodecanol the properties of the mixed monolayer are increasingly determined by the highly surface active trace component dodecanol.

The BAM experiments have shown that, depending on temperature, the condensed-phase domains of adsorbed monolayers can occur in different modifications. In the case of adsorbed DHBAA monolayers two different types of condensed-phase textures were observed at temperatures above and below 10 °C (Fig. 3). As can be seen, the DHBAA domains have one main growth axis and two additional growth directions at temperatures of 10 °C or lower. The main axis forms an obtuse angle of about 150° with each of the homogeneously reflecting growth directions; these have an acute intersection angle of about 60° with each other. The bisector of the main growth direction subdivides the domain into two sections of different brightness. The analysis of this fact and the defect lines along the bisector suggest an azimuthal tilt direction of the molecules parallel to the homogeneously reflecting growth axes.

At temperatures of 10 °C and above, homogeneously reflecting domain textures are formed having four main growth directions with intersection angles of 60 and 120°. The homogeneous reflectivity indicates the same azimuthal molecule orientation over the whole domain.

First experimental evidence has been provided that the penetration of dissolved surfactants and proteins can induce a first-order main phase transition in lipid monolayers even at zero surface pressure and if the lipid monolayer is in the gaseous state. The penetration

dynamics of bovine *β*-lactoglobulin dissolved in a buffered aqueous substrate into a gaseous DPPC monolayer is presented in Fig. 4. The *β*-lactoglobulin penetration from a 10^{-7} M phosphate buffered solution (pH 7) performed at an area of 0.90 nm^2 per DPPC molecule induces a first-order phase transition in the gaseous DPPC monolayer. This is indicated by the break point in the $\pi(t)$ penetration kinetics curves and visualised by BAM (Fig. 4). The break point is the equivalent of a phase-transition point, as in the time region afterwards condensed-phase domains similar those of pure DPPC are formed. The experimental results indicate that obviously the adsorbed protein induces the condensation of pure DPPC of the gaseous monolayer. GIXD studies of this system corroborate these conclusions [12]. A theoretical model has been developed which supports the experimental findings [13].

Conclusions

Conclusive evidence has been provided that a first-order phase transition and consequently the formation of condensed-phase domains can be induced by the adsorption of amphiphilic species (surfactants, proteins). The combination of $\pi(t)$ adsorption kinetics, BAM and GIXD makes a highly efficient characterisation of the condensed phases formed in adsorbed or penetrated monolayers possible. Depending on the temperature various types of 2D modifications can occur, the molecular ordering of which can be determined as shown for the adsorbed monolayers of DHBAA. A first-order phase transition can also take place by the coadsorption of two surfactants. An interesting system is the coadsorption of dissolved SDS containing only trace amounts of dodecanol. The condensed phase formed after the phase transition consists only of dodecanol and replaces nearly completely SDS in the adsorbed monolayer at equilibrium.

The penetration of dissolved surfactants and proteins can also induce a first-order phase transition in gaseous lipid monolayers. As demonstrated by the penetration dynamics of bovine β-lactoglobulin into a gaseous DPPC monolayer, the condensed phase formed after the phase transition consists solely of DPPC.

References

1. Hönig D, Möbius D (1991) J Phys Chem 95:4590
2. Vollhardt D (1996) Adv Colloid Interface Sci 64:143
3. Als-Nielsen J, Jacquemain D, Kjaer K, Lahav M, Leveiller F, Leiserowitz L (1994) Phys Rep 246:251
4. Kjaer K (1994) Physica B 198:100
5. Melzer V, Vollhardt D (1996) Phys Rev Lett 76:3770
6. Vollhardt D, Melzer V (1997) J Phys Chem 101:3370
7. Czichocki G, Vollhardt D, Seibt H (1981) Tenside Surfactants Deterg 18:320
8. Melzer V, Vollhardt D, Weidemann G, Brezesinski G, Wagner R, Möhwald H (1998) Phys Rev E 57:901
9. Vollhardt D (1999) Adv Colloid Interface Sci 79:19
10. Vollhardt D, Fainerman VB (2000) Adv Colloid Interface Sci 86:103
11. Vollhardt D, Emrich G (2000) Colloids Surf A 161:173
12. Zhao J, Vollhardt D, Brezesinski G, Siegel S, Wu J, Li JB, Miller R (2000) Colloids Surf A 171:175
13. Fainerman VB, Zhao J, Vollhardt D, Makievski AV, Li JB (1999) J Phys Chem 103:8998

Progr Colloid Polym Sci (2001) 118: 34–37
© Springer-Verlag 2001

S. Siegel
M. Kindermann
D. Vollhardt

Molecular recognition under formation of amphiphilic amidinium carboxylates at the air–water interface

S. Siegel · D. Vollhardt (✉)
Max-Planck-Institut für Kolloid- und
Grenzflächenforschung, 14424 Potsdam
Germany
e-mail: vollh@mpikg-golm.mpg.de
Tel.: +49-331-5679258
Fax: +49-0331-5679202

M. Kindermann
Ruhr-Universität Bochum
Organische Chemie 1, 44780 Bochum
Germany

Abstract Amphiphilic "host–guest" assemblies are formed between a "host" monolayer and "guest" molecules dissolved in the aqueous subphase by acid–base interactions. The specific features of the surface films of amphiphilic benzamidinium–benzoate complexes are determined by surface pressure–area isotherms, Brewster-angle microscopy and atomic force microscopy studies. Molecular recognition of the dissolved component by the amphiphilic monolayer causes drastic changes in the properties of the surface film. Not only the area per molecule is considerably increased but also condensed-phase domains of special texture and topography can be formed. Details of a specific substructure reveal that a second layer grows over the primary structures at further compression.

Key words Molecular recognition · Langmuir monolayers · Benzamidinium–carboxylate complexes · Brewster-angle microscopy · Atomic force microscopy

Introduction

Mechanisms of molecular recognition are of general interest not only for the development of systems with new features and functions by specific composition of two (or more) different molecular components but also for understanding specific processes of biological receptors at the surface of supramolecular biological systems. Monolayers at the air–water interface are optimal models to realise the composition of specific molecular components for studying the structural properties of the supramolecular units formed and, thus, the principles of molecular recognition. One approach is based on the regulation of the assembly process by incorporating strong directional interactions at the air–water interface [1, 2]. Amphiphilic "host–guest" assemblies can be formed between a "host" monolayer and "guest" molecules dissolved in the aqueous subphase by formation of complementary hydrogen bonds and/or Coulomb interactions [3].

Here we focus on the specific features of the surface films of amphiphilic benzamidinium–benzoate complexes

mainly on the basis of acid–base interactions. A stable cationic or anionic "host" monolayer was spread on the aqueous surface, while the "guest" component, having the opposite charge, was dissolved in the aqueous subphase.

Materials and methods

The soluble components used were sodium benzoate, sodium phenylacetate, phenylacetamidinium, and methyl benzamidinium purchased from Merck. The spread components were anionic pentadecyl benzoic acid (I) and cationic pentadecyl benzamidinium chloride (II). Pentadecyl benzamidinium chloride was synthesised and purified as described elsewhere [3].

The following combinations were investigated:

1. Pentadecyl benzoic acid monolayers on phenylacet amidinium and benzamidinium solutions.
2. Pentadecyl benzamidinium chloride monolayers on sodium benzoate or sodium phenylacetate solutions.

The long-chain substances were dissolved in a chloroform/ethanol (4:1) mixture and spread on the surface of both pure water and water containing 1 mmol dissolved component with the opposite

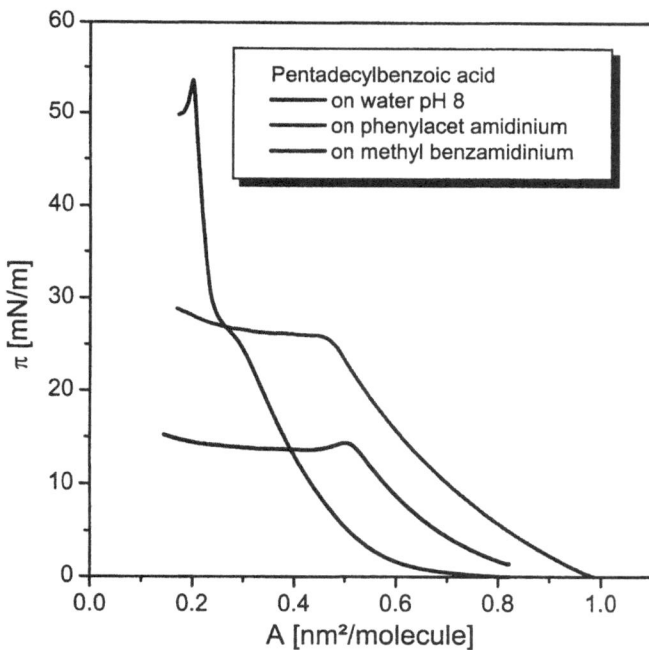

Fig. 1 π–A isotherms of pentadecyl benzoic acid on different aqueous subphases (20 °C)

charge. The monolayers were investigated at 20 °C using a thermostated Langmuir film balance. The surface pressure–area (π–A) isotherms were recorded at a compression speed of 10 Å²/molecule/min.

A BAM 2 Brewster-angle microscope (NFT Göttingen, Germany) was used to image the condensed-phase textures formed on the water surface [4].

After the recognition process, one layer of the surface film was deposited on silicon wafers using the Langmuir–Blodjett technique. These sample were investigated by atomic force microscopy (AFM) using a Nanoscope III (Digital Instruments, Calif.).

The distilled water was made ultrapure by a Milli-Q system.

Results and discussion

The spread pentadecyl benzamidinium chloride forms stable monolayers on pure water. After the spreading, these monolayers are already in the state of the two-phase coexistence between a nontextured condensed state and the surrounding gaseous state. At surface pressures of $\pi > 0$ mN/m a homogeneously reflecting condensed phase completely covers the surface. The anionic pentadecyl benzoic acid monolayers are not stable on water; however, stable monolayers can be formed on water with a higher pH value (pH 8). The effect of the dissolved "guest" components on the spread "host" monolayer changes the properties of the surface film drastically.

In all combinations and independent of the charge of the spread monolayer, the area per molecule is much larger than that obtained on pure water.

Fig. 2 Brewster-angle microscopy (*BAM*) images of the condensed-phase domains of a pentadecyl benzoic acid monolayer spread on 1 mM aqueous solutions of methyl benzamidinium (*top*) and phenylacet amidinium (*bottom*). All the images are on the same scale

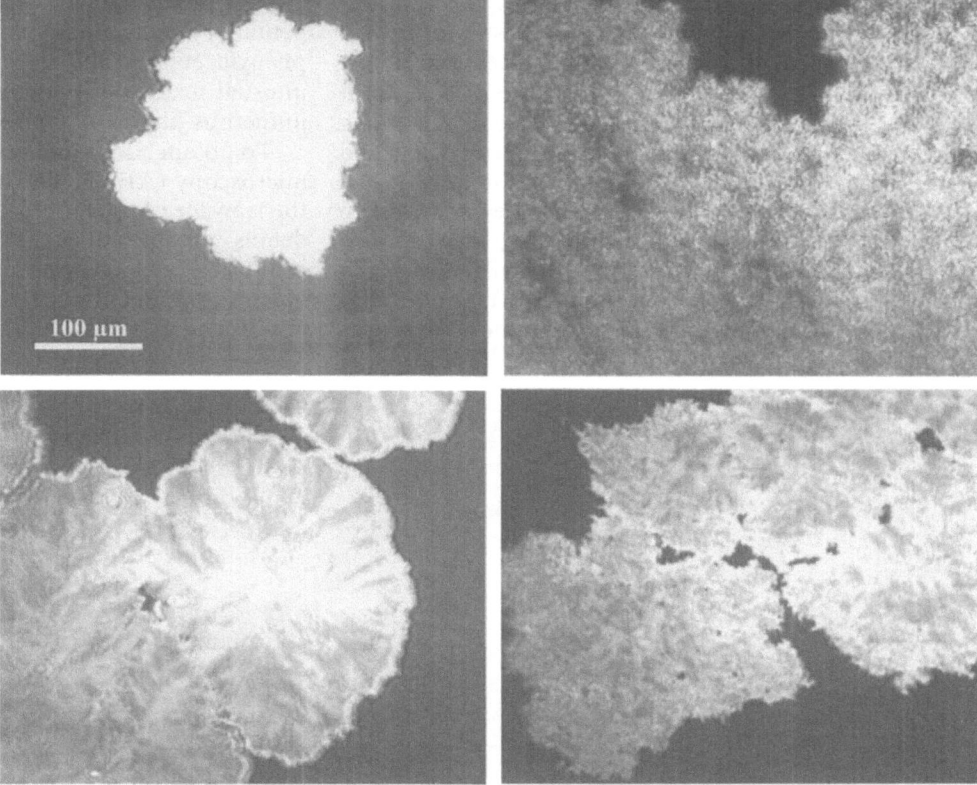

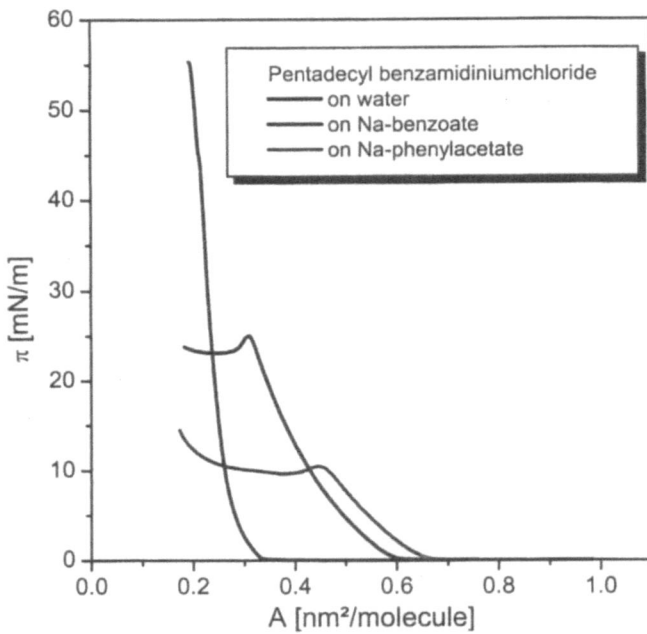

Fig. 3 π–A isotherms of pentadecyl benzamidinium chloride on different subphases (20 °C)

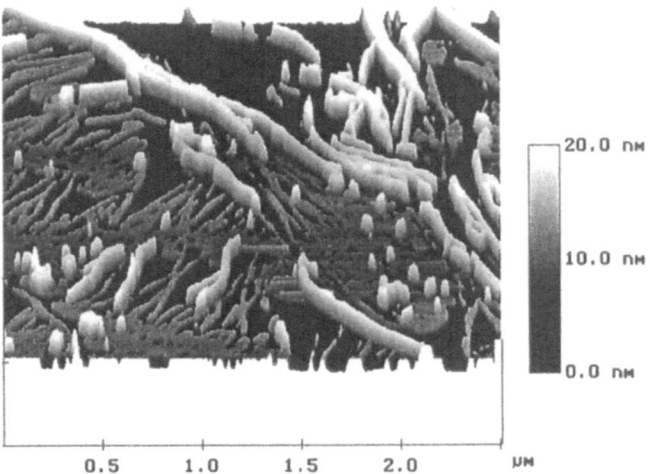

Fig. 5 Atomic force microscopy images of pentadecyl benzamidinium chloride domains transferred onto silicon wafers

This can be clearly seen in Fig. 1, which shows the π–A isotherms of the pentadecyl benzoic acid monolayers spread on pure water and on 1 mM aqueous solutions of methyl benzamidinium and phenylacet amidinium. The π–A isotherms obtained for the dissolved substances show a plateau region which is characteristic for a two-phase coexistence, at a definite surface pressure for the accessible temperature region. A first-order phase transition occurs at the beginning of the plateau region, so the best preconditions are given for the formation and growth of condensed-phase domains. Consequently the Brewster-angle microscopy (BAM) images were taken within this pressure plateau region. Regularly shaped domains of different morphology were observed.

The domains which were obtained after molecular recognition if a pentadecyl benzoic acid monolayer was spread on 1 mM aqueous solutions of methyl benzamidinium and phenylacet amidinium are shown in Fig. 2. It is obvious that the molecules are tilted and a long-range orientational order exists.

The π–A isotherms of the pentadecyl benzamidinium chloride monolayer spread on water and 1 mM aqueous solution of sodium benzoate and sodium phenylacetate (Fig. 3) show similar behaviour as those obtained in the case of opposite charge conditions (Fig. 1). The BAM images in Fig. 4 show the domains for the system of pentadecyl benzamidinium chloride monolayer on 1 mM phenylacetate solution. It is interesting to note the unusual texture of the round domains which consist of numerous filigree branches.

To obtain additional texture information atomic force microscopy (AFM) measurements were performed after the transfer of the monolayers onto silicon wafers. More details on the thickness and the texture at higher resolution are revealed in Fig. 5. Depending on the system components, different subtextures of the domains are formed. Amphiphile I develops compact domains on

Fig. 4 BAM images of the condensed-phase domains of pentadecyl benzamidinium chloride monolayers spread on a 1 mM solution of phenyl-lacetate

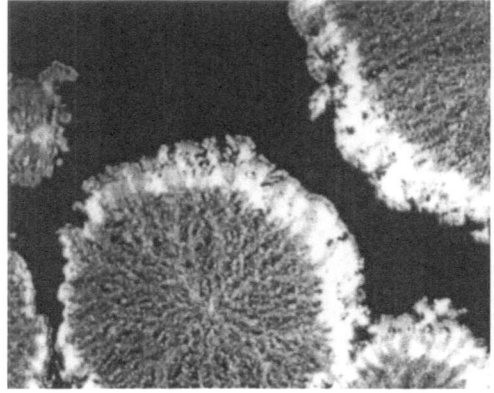
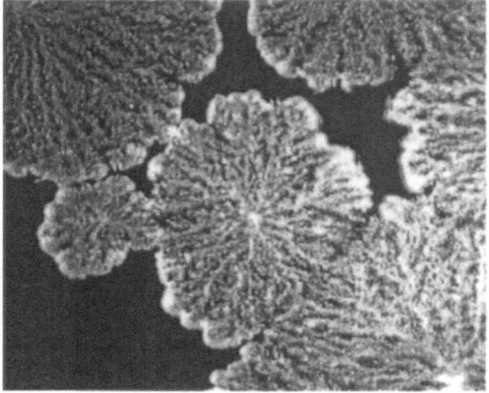

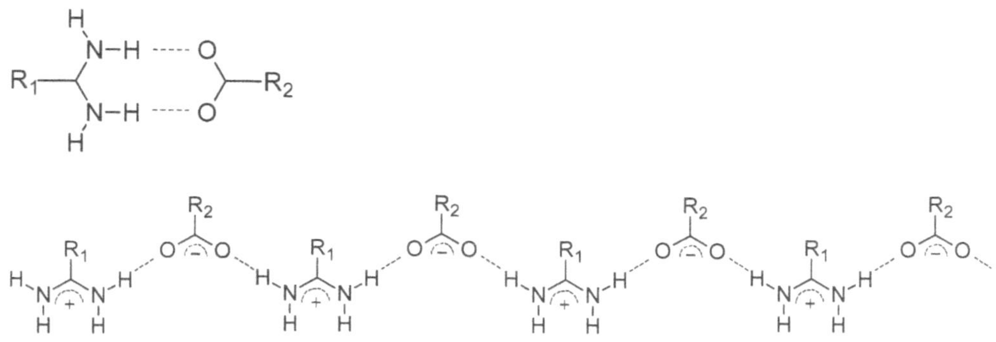

Fig. 6 General schema of the interaction in the amidinium–carboxylate system: *top* as a supramolecular pair; *bottom* as an interdigitated assembly

both subphases, whereas amphiphile II on sodium phenylacetate forms filigree strings. A second and third layer of the same texture grow over the first condensed-phase monolayer. According to the results of the π–A isotherms and the BAM and AFM measurements it can be concluded that the domains consist of well-defined substructures with a defined thickness of 1 or 2, or sometimes of 3, molecule layers.

There are two possibilities of interactions and structure formation which can be expected owing to recognition for the amidinium–carboxylate system (Fig. 6). The preferential formation of bilayers by growing over the filigree monolayer stings indicates interdigitated benzoate–benzamidinium interaction.

Conclusions

Molecular recognition systems can be formed on the basis of acid–base interaction by specific composition of an amphiphilic monolayer component and a dissolved species. The specific features of the surface films of amphiphilic benzamidinium–benzoate complexes were determined by π–A isotherms, BAM and AFM studies. The amphiphilic "host" monolayer can be the acid or the basic component. Molecular recognition of the dissolved component, with opposite charge, changes the properties of the surface films drastically. On the other hand, independent of whether a cationic or anionic "host" monolayer is used, the features of the amphiphilic benzamidinium–benzoate films are more or less similar to each other.

If the two amphiphilic compounds are spread on pure water, stable but nontextured condensed monolayers are formed. The recognition of the guest component leads to drastic changes in the film properties. The π–A isotherms show that the area per molecule is essentially enlarged by recognition of the "guest" components. In the two-phase coexistence region condensed-phase domains of a special texture and topography begin to grow. Particularly supported by AFM studies, a specific substructure can be observed, which reveals that a second molecular layer grows over the primary structures at further compression. All the results suggest interdigitated benzoate–benzamidinium interaction.

References

1. Lehn J-M (1990) Angew Chem 102:1347
2. Desiraju GR (1995) Angew Chem 107:2541
3. Siegel S, Kindermann M, Regenbrecht M, Vollhardt D, von Kiedrowski G (2000) Prog Colloid Polym Sci 115:233
4. Vollhardt D (1996) Adv Colloid Interface Sci 64:143

Progr Colloid Polym Sci (2001) 118: 38–41
© Springer-Verlag 2001

S. Siegel
D. Vollhardt

Equilibrium and dynamics of dissolved homologue penetration into an *N*-tetradecyl-γ-hydroxybutyric acid amide monolayer forming crystalline condensed phases

S. Siegel · D. Vollhardt (✉)
Max-Planck-Institut für Kolloid- und
Grenzflächenforschung
14424 Potsdam/Golm, Germany
e-mail: vollh@mpikg-golm.mpg.de
Tel.: +49-331-5679258
Fax: +49-331-5679202

Abstract Equilibrium penetration and penetration kinetics of the dissolved homologue *N*-decyl-γ-hydroxybutyric acid amide into Langmuir monolayers of *N*-tetradecyl-γ-hydroxybutyric acid amide were studied by surface pressure measurements and Brewster-angle microscopy. If penetration of the dissolved component takes place into the fluid monolayer a first-order phase transition is induced. Under all conditions the condensed phase formed consists only of the longer-chain homologue. The single components were also characterised.

Key words Monolayer penetration · Phase transition · Brewster-angle microscopy · Surfactants · Adsorption

Introduction

Equilibrium and dynamic behaviour of mixed monolayers of soluble and insoluble amphiphiles at fluid/liquid interfaces play an important role in various technological and biological processes. However, even for very simple systems, the thermodynamic analysis is difficult and much more complicated if a condensed phase exists or is induced by the penetration of dissolved surface-active species [1].

On the other hand, in the last decade rapid progress has been made in the understanding of the molecular organisation of condensed monolayer phases [2, 3]. The application of sensitive optical microscopy, particularly Brewster-angle microscopy (BAM), has revealed a variety of textures of condensed phases formed not only in Langmuir monolayers [3] but also in adsorbed monolayers (Gibbs monolayers) [4].

Recent theoretical and experimental progress has led to a better understanding of penetration systems at the air–water interface in which a dissolved amphiphile (surfactant, protein) penetrates into a Langmuir monolayer. The first application of the highly sensitive, direct experimental techniques for penetration experiments has provided new, interesting information on the effect of the dissolved species on the state of the condensed phase of different penetration systems [5–8].

Although a rigorous thermodynamic analysis of penetration systems is unavailable owing to their complexity, some model assumptions resulted in reasonable solutions. Based on the interesting experimental results, new theoretical models describing the equilibrium behaviour of the insoluble monolayers which undergo the 2D aggregation in the monolayer and corresponding equations of state and adsorption isotherms are now available [7, 9, 10].

The present experimental study focuses on a penetration system where a soluble homologue penetrates into a Langmuir monolayer of an amphiphile which can form a highly crystalline monolayer phase.

Materials and methods

N-Tetradecyl-γ-hydroxybutyric acid amide (14-HBAA) and *N*-decyl-γ-hydroxybutyric acid amide (10-HBAA) [C_nH_{2n+1} − NH − CO − $(CH_2)_2$ − CH_2OH, $n = 10, 14$] were synthesised by reaction of butyrolactone with the corresponding alkylamine dissolved in dioxane at 100 °C. The chemical purity of 99% or above obtained by distillation and crystallisation in acetone was checked by elemental analysis and high-pressure liquid chromatography. The spreading solvent was chloroform p.a. (Merck). The distilled water was made ultrapure by a Milli-Q system.

10-HBAA was used as a soluble surfactant. The insoluble monolayer was formed by the homologue 14-HBAA.

The penetration experiments were performed with a circular film balance with two compartments. On one compartment, containing pure water, the monolayer was prepared and compressed. Afterwards, the monolayer was transferred between two barriers onto the second segment, which contained the surfactant solution. The main advantage of this technique is a homogenous distribution of the dissolved 10-HBAA within the subphase.

The film balance was coupled with a BAM 1+ Brewster angle microscope (NFT, Göttingen, Germany). The microscope provides undistorted images with a lateral resolution of approximately 4 μm. All the experiments were performed at 20 °C. The water used for the experiments was made ultrapure by a Millipore desktop system.

Results and discussion

Characterisation of the pure components

The 14-HBAA monolayers were prepared by spreading a 10^{-3} M CHCl$_3$ solution and one π–A isotherm for 20 °C

was recorded with a compression speed of 6 Å^2/molecule/min. The π–A isotherm (Fig. 1) exhibits a distinct plateau region, indicating a first-order phase transition. The BAM images show the structure of the condensed-phase domains which are formed within the plateau region of the isotherm (Fig. 2). The condensed phase appears as highly crystalline rigid needles owing to the strong hydrogen bonds within the headgroup region of the monolayer.

The shorter-chain homologue 10-HBAA can be dissolved in water and adsorbs at the water surface. The surface tension – concentration (σ–log c) adsorption isotherm of 10-HBAA was measured up to the critical micelle concentration at 20 °C (Fig. 3) and the adsorbed monolayers did not form condensed-phase domains.

Equilibrium penetration

After the characterisation of the single components the penetration of the 10-HBAA component dissolved into

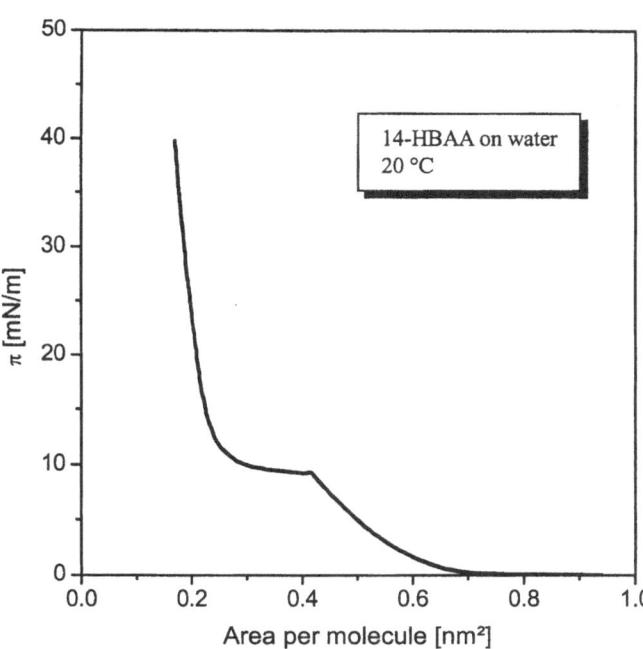

Fig. 1 Surface tension–concentration isotherm of N-decyl-γ-hydroxybutyric acid (10-HBAA) (20 °C)

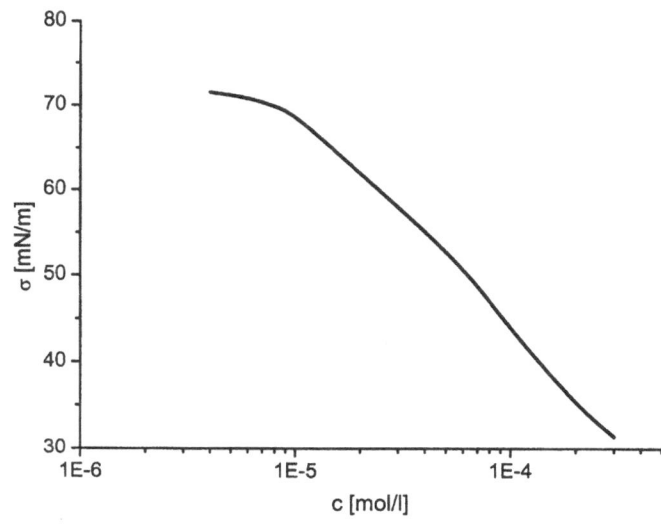

Fig. 3 π–A isotherm of N-tetradecyl-γ-hydroxybutyric acid (14-HBAA) on water (20 °C)

Fig. 2 Brewster-angle microscopy images of condensed-phase domains of 14-HBAA on water (20 °C)

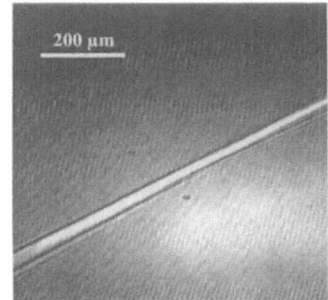

the 14-HBAA monolayer was studied. The 14-HBAA monolayer was transferred onto the surfactant solution, then expanded to an area of more than 1 nm²/molecule and the equilibrium penetration isotherm (Fig. 4) was recorded with a low compression speed (2.5 Å²/molecule/min).

In comparison to the π–A isotherm obtained for pure water as a subphase, the penetration isotherm is shifted to higher pressures and to greater areas. The plateau of the phase transition is nearly diminished and BAM investigations were helpful to find out the region of the phase transition. As can be seen in Fig. 5, the domains have a similar shape as those on water. Small deviations may be caused by the lower surface tension of the expanded phase, while the composition of the condensed domain is not influenced by the surfactant. The condensed phase consists obviously only of pure 14-HBAA, so it can be assumed that the surrounding fluid phase is enriched by 10-HBAA. Minor differences in the domain shape, for example, the needles are thicker and more branched in penetrated monolayers, may result from different line tensions.

Penetration kinetics

The spread 14-HBAA monolayer was compressed to 0.45 nm²/molecule, just before the phase transition point. Under these conditions, only the fluid monolayer phase exists and condensed-phase domains cannot be formed. After sweeping this monolayer onto the surfactant solution the penetration kinetics of the dissolved

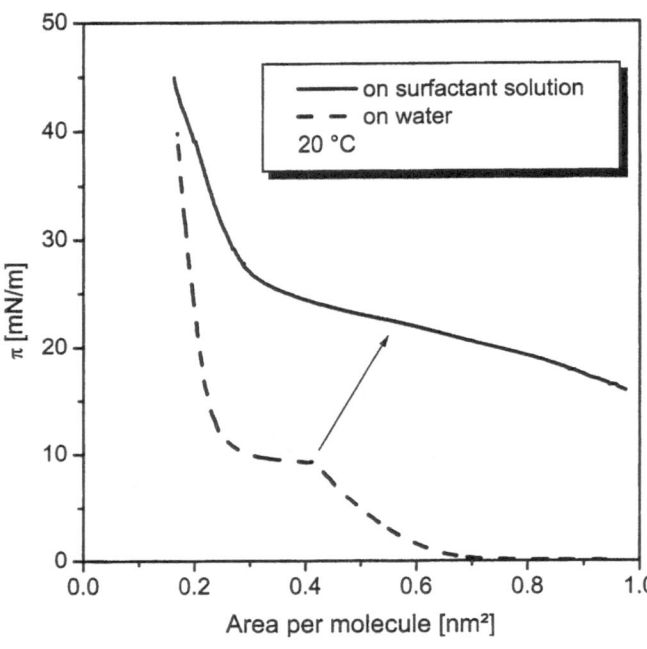

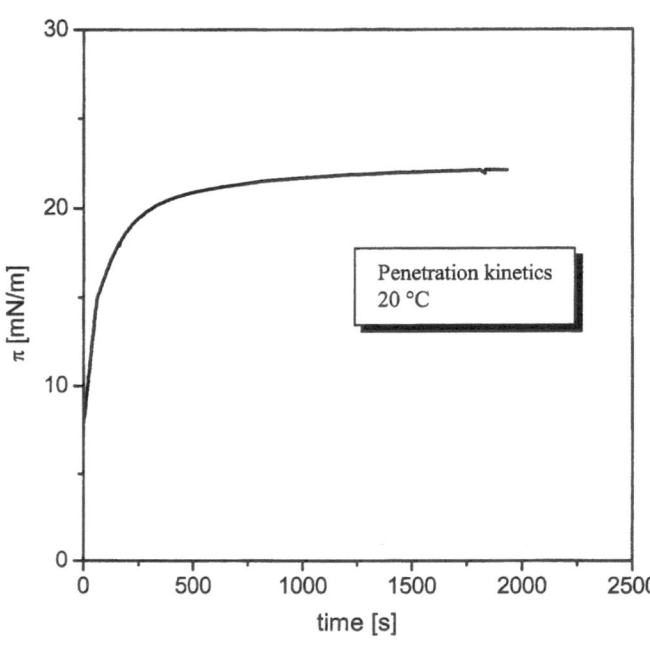

Fig. 4 π–A isotherm of 14-HBAA on a 5×10^{-5} M solution of 10-HBAA. The isotherm on water is given for comparison, and the beginning of the phase transition is indicated

Fig. 6 Adsorption kinetics of 10-HBAA into a monolayer of 14-HBAA. The monolayer was compressed to 0.45 nm²/molecule ($\pi \approx 7$ mN/m) and transferred onto the surfactant solution

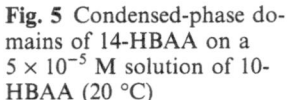

Fig. 5 Condensed-phase domains of 14-HBAA on a 5×10^{-5} M solution of 10-HBAA (20 °C)

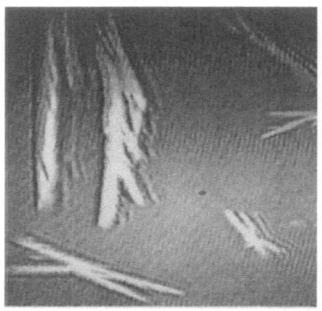

Fig. 7 Condensed-phase domains induced by penetration of 10-HBAA into fluid 14-HBAA monolayers. The phase transition is induced by the penetration of 10-HBAA

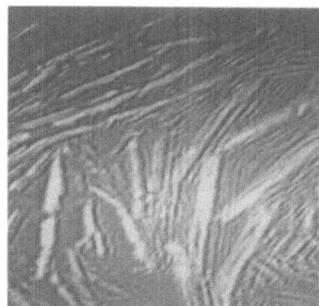 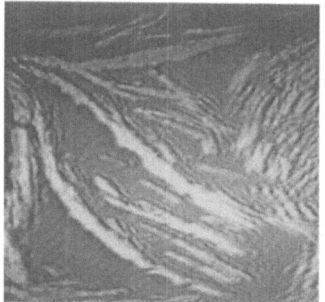 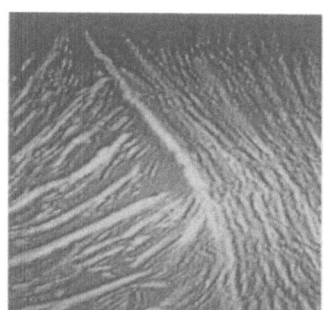

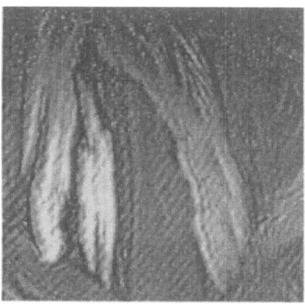 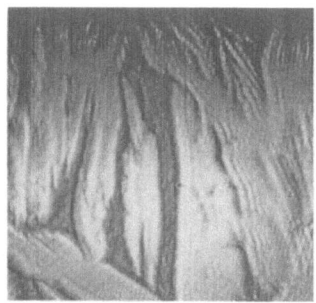

Fig. 8 Using an analyser, the contrast in the image can be changed. This indicates tilted molecules with different azimutal orientation

surfactant can be observed, and the surface pressure increases with time (Fig. 6). The not well developed kink in the penetration kinetics curve after approximately 1 min indicates a first-order phase transition. The BAM images obtained after this time, show the existence of condensed-phase domains (Fig. 7), again similar to those of 14-HBAA monolayers on water. Consequently although the phase transition is induced by the penetration of 10-HBAA molecules into the fluid 14-HBAA monolayer the condensed phase consists of 14-HBAA.

Conclusions

The transition from the fluid to the condensed phase of the 14-HBAA monolayer is shifted to greater areas and higher surface pressures owing to the penetration of the dissolved 10-HBAA surfactant into the monolayer; this induces the first-order phase transition. Although the insoluble and soluble components are miscible in the fluid monolayer phase and both components are only different in the chain length, the condensed-phase domains are formed only by the insoluble homologue.

References

1. Siegel S, Vollhardt D (1993) Colloids Surf A 76:197
2. Möhwald H (1993) Rep Prog Phys 56:653
3. Vollhardt D (1996) Adv Colloid Interface Sci 64:143
4. Vollhardt D (1999) Adv Colloid Interface Sci 79:19
5. Sundaram S, Ferri JK, Vollhardt D, Stebe KJ (1998) Langmuir 14:1208
6. Fainerman VB, Makievski AV, Vollhardt D, Siegel S, Miller R (1999) J Phys Chem B 103:330
7. Fainerman VB, Zhao J, Vollhardt D, Makievski AV, Li JB (1999) J Phys Chem 103:8988
8. Zhao J, Vollhardt D, Brezesinski G, Siegel S, Wu J, Li JB, Miller R (2000) Colloids Surf A 171:175
9. Fainerman VB, Vollhardt D (1999) Langmuir 15:1784
10. Vollhardt D, Fainerman VB (2000) Adv Colloid Interface Sci 86:103

Progr Colloid Polym Sci (2001) 118:42–47
© Springer-Verlag 2001

Chrystel Faure
Emmanuel Belamie

Using multilamellar vesicles to incorporate glucose oxidase into a polypyrrole film

C. Faure (✉)
Centre de Recherche Paul Pascal (CNRS)
Av. A. Schweitzer, 33600 Pessac, France
e-mail: faure@crpp.u-bordeaux.fr
Tel.: +33-5-56845614
Fax: +33-5-56845600

E. Belamie
Department of Physics, Brandeis
University, Waltham, MA 02454, USA

Abstract We report the immobilization of glucose oxidase into a polypyrrole layer owing to multilamellar vesicles. Multilamellar vesicles consist of a mixture of lipids and surfactants and are used to encapsulate glucose oxidase and vectorize it towards an electrode where polypyrrole is electrosynthesized. We show by cyclic voltammetry experiments performed on vesicles free of glucose oxidase that they are likely to interact with pyrrole oligomers during synthesis, leading to their incorporation into the film. Their engulfment into the film is evidenced by scanning electron microscopy and Auger analysis. Vesicles in which glucose oxidase is encapsulated do not seem to be destroyed during film synthesis since the enzymatic activity of the mixed film can be triggered by adding a surfactant able to dissolve the vesicles. The advantage of using vesicles for inserting an enzyme into a polymer film is demonstrated in the special case where a pulsed field is applied for film synthesis. We indeed measure a higher enzymatic activity when glucose oxidase is incorporated in the polypyrrole film using multilamellar vesicles.

Key words Polypyrrole ·
Multilamellar vesicles · Glucose
oxidase · Electric field · Biosensor

Introduction

The making of biosensors requires the immobilization of the highly selective molecule, usually an enzyme, on a solid surface. One of the most attractive process is electrochemical immobilization, mainly because of its simplicity and rapidity. This method involves the electrochemical oxidation of a suitable monomer from a supporting electrolyte containing enzyme to form a polymer film on the electrode surface. However, several drawbacks are inherent in this method: enzymes can be denaturated by the electrochemical environment generated at the electrode surface during the electropolymerization step (change in pH, ionic force, etc.) so that the enzymes are incompatible with certain polymers [1–3]. Moreover, incorporation of the enzyme into the growing film requires electrostatic affinity between the polymer and the enzyme as well as between the electrode and the enzyme, so, for instance, positively charged proteins cannot be efficiently entrapped in polypyrrole (PPy) films [1]. Finally, the low amount of the entrapped enzyme significantly reduces the long-term stability of these biosensors. We believe that one way to solve these problems is to encapsulate the enzyme into multilamellar vesicles (MLV). Vesicles should protect the enzyme from its environment, avoiding both its denaturation and the use of concentrated buffers. Furthermore, the effective charge of the enzyme would not be of importance as long as the appropriate charge is conferred to the vesicles for electrostatic attraction and, finally, the vesicles should allow the concentration of the enzyme in the film to increase because of their greater attraction to an electrode when submitted to an electrical field.

The aim of this article is to show that the incorporation of MLV, with or without encapsulated glucose oxidase (GOD), into a PPy layer is possible by an electrochemical process. We show that the enzyme activity incorporated into the film can be triggered by the destruction of the embedded vesicles and that, for certain conditions, vesicles allow the enzyme concentration in the polymer film to rise.

Materials and methods

Chemicals and solutions

The monomer, pyrrole (Aldrich, 99%), was distilled and stored under nitrogen in the dark prior to use. NaCl, Brij30 and T35 (sodium octadecyl sulfate) were purchased from Aldrich. PC-90 was provided by Rhone-Poulenc Rorer.

The GOD from *Aspergillus niger* (E.C.1.1.3.4, type VII-S) was obtained from Sigma. Peroxidase (POD) from horse radish (grade II) as well as 2,2'-azinobis(3-ethylbenzthiazoline sulfonic acid) (ABTS) were from Boehringer Mannheim. $D(+)$glucose was purchased from Aldrich. A potassium phosphate buffer (0.1 M, pH 7) was obtained by adjusting the pH of a solution of K_2HPO_4 $3H_2O$ (2.28 g/100 ml water) to 7.0 by adding a solution of KH_2PO_4 (1.36 g/100 ml water). Solutions of $D(+)$glucose (1.1 M) and POD (2500 U/ml) were made in water, whereas those of ABTS (2 mM) and GOD were made in phosphate buffer. $D(+)$glucose solution was allowed to mutarotate at room temperature for at least 24 h before use.

Preparation of the MLV dispersions

MLV were obtained mixing the same amount of surfactants [Brij30 (5 wt%), T35 (25 wt%), PC-90 (70 wt%)] and solution [water or GOD in NaCl (10^{-3} M) for GOD encapsulation]. The surfactants were first mixed together, then the required mass of solution was added to this mixture. The paste obtained was mixed until it appeared homogeneous. The dispersion of MLV was performed under agitation using a vortex stirrer in water, NaCl (10^{-3} M) or pyrrole. The formation of MLV was checked by phase-contrast microscopy before each experiment. The diameter of the MLV ranged from 0.1 to 2 μm, with a population peak of around 0.3 μm.

Electrochemical cells

Thin-gap cell

Scanning electron microscopy (SEM) analysis was performed on films synthesized in a thin-gap cell [4, 5]. This electrochemical cell consists of two glass plates covered with a gold film separated by a square silicone joint forming a reservoir. A thin tube is inserted into the reservoir in order to inject pyrrole solution close to the anode. A potential drop, $\Delta\Phi$ ($\Phi_{anode}-\Phi_{cathode}$) is applied between both electrodes using a HP3324A instrument (Hewlett-Packard) as the signal generator.

Three-electrode cell

Cyclic voltammetry was carried out using a three-electrode-cell geometry. The working electrode and the counter electrode were Pt wires (1-mm diameter and separated by 3 mm) and a saturated calomel electrode (SCE) was used as the reference electrode. They were immersed into a 30-ml beaker. Cyclic voltammetry was performed with a PGstat 20 potentiostat in connection with the data acquisition software Autolab GPES 4.4.

SEM and Auger electron analyses

Images of the surface of the anode were obtained with a scanning electron microscope (JEOL 840 A) under a tension of 15 kV. The samples were sputtered with Au/Pd under vacuum before analysis.

Auger electron analysis was made using a Fisons Instruments analyzer.

GOD activity measurement

The activity of GOD was assayed using a spectrophotometric procedure provided by Boehringer Mannheim. This method is a modification of a previously described method where o-dianisine was used as the leuco dye instead of ABTS [6]. The assays were based on the POD catalyzed oxidation of ABTS by the H_2O_2, product of the O_2 reduction by GOD. A magnetic stirrer bar rotating at 500 rpm ensured homogeneity of the solution. The temperature of the spectrophotometer cell was kept constant at 25 °C using flowing water. All the experiments were performed using a PerkinElmer UV–vis spectrophotometer saturated with O_2. The 1.0-cm light-path cell was filled with 1.8 ml ABTS (2 mM), 0.326 ml glucose (1.1 M) and 6.6 μl POD (2500 U/ml). The electrode (Pt wire) on which the mixed film was synthesized, was dipped in the cell and the absorbance was measured at 405 nm for 1–3 min.

Results and discussion

Ex situ analysis of the mixed MLV–PPy film

When submitted to a direct current (d.c.) field, charged particles dispersed in water are attracted to the oppositely charged electrode by electrophoretic transport. This well-known property still applies for MLV [4, 5, 7]. For the negatively charged MLV we used (T35 is an anionic surfactant), we found that the density of the MLV close to the anode sharply increased within about 5 min when $\Delta\Phi = 2$ V. We took benefit of this property to preassemble MLV on the electrode before introducing pyrrole into the electrochemical cell. In so doing, we wanted to optimize the concentration of MLV in the PPy layer. Several seconds after the injection of pyrrole close to the anode, some dark areas corresponding to polymer domains are visible on the anode surface and within 2 min the anode is totally covered by a dark layer [7]. We stopped the electrosynthesis of PPy 5 min after the introduction of monomer into the cell in order to get a "flat" film, thin enough for the MLV to be distinguished. This is illustrated in Fig. 1, which displays a SEM image of the anode surface, after opening the thin-gap cell and rinsing the anode with water.

The surface of the sample exhibits several bumps separated by flat domains. The bumps are ellipsoidal and their size axis ranges from about (0.25 × 0.5 μm^2) to about (0.75 × 2 μm^2). When the experiment is repeated without MLV, the film displays a very fine granulation (not shown) which is similar to that of the flat domains seen in Fig. 1. We therefore believe that the bumps are due to MLV embedded in the film. The ellipsoidal shape

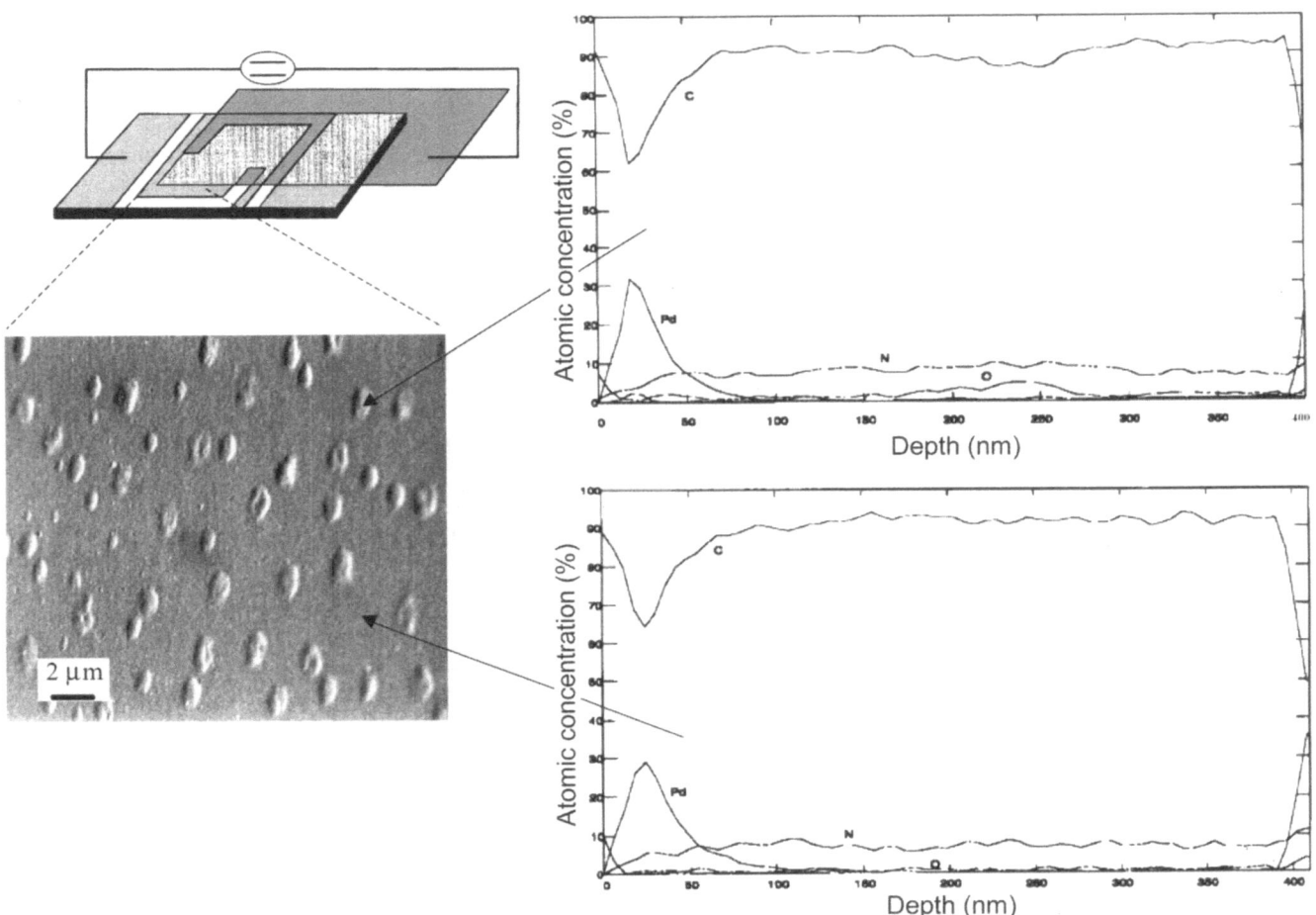

Fig. 1 Scanning electron microscopy image of the polypyrrole (*PPy*)–multilamellar vesicle (*MLV*) film synthesized on the anode of the electrochemical cell depicted and the corresponding Auger electron spectra. The mixed film is obtained by applying a voltage drop of 2 V on an aqueous dispersion of MLV. After 20 min, a solution of pyrrole (0.5 M) is injected into the cell and the direct current field is still applied for some additional 5 min. Auger spectra: (*top*) bump analysis, (*bottom*) "flat" area analysis

of these bumps can be explained by the inclination of the sample (30°) while being shadowed with Pd/Au for SEM analysis. The size of the bumps is slightly greater than the size of the MLV: this can be imputed to a flattening of the vesicles under vacuum or/and sample inclination.

Auger electron spectroscopy was performed on that sample. Two depth profiles were recorded: one (top spectrum) corresponds to an area where a bump was visible, the other (bottom spectrum) to a flat area. This technique gives information on the relative atomic composition of the film for its whole thickness. Let us note, however, that not all the elements (e.g. hydrogen, phosphorus, etc.) can be detected by this technique and that the depth values (0 corresponds to the air–film interface) are only indicative since the apparatus could not be calibrated with our material. Both spectra are very similar. They both present a strong depletion in the carbon relative concentration in the first 100 nm while the Pd peak rises. This corresponds to the Pd/Au conductive layer covering the sample for purpose of SEM analysis. For both spectra, the strongest concentrations are those of carbon and nitrogen (90 and 10%, respectively); however, between 150 and 275 nm in depth, they differ in that oxygen (about 5%) is detected in the bump, whereas for the flat area analysis, the oxygen concentration is close to 0 for all the film thickness. As oxygen can only come from the surfactants and lipids which comprise the MLV (pyrrole being devoid of oxygen), these results suggest that the MLV are located under the bumps, sandwiched between two PPy layers. The low concentration of oxygen is not surprising since its ratio in the mixture of lipids and surfactants is very small (less than 10%). The carbon and nitrogen atoms are indicative of the PPy film which covers all the anode, recovering also the bumps. Their ratio (90/10) is not very far from that of pure PPy (80/20).

Cyclic voltammetry

PPy electrosynthesis is believed to begin with the oxidation of pyrrole on the anode surface followed by the formation of oligomers which are themselves oxidized to give a free-standing film. Oxidation of the repeating units produces polycationic chains, so anions, called dopants, must be inserted for the synthesis to go on. In contrast, when PPy is reduced (by decreasing the working electrode potential) the dopant must be ejected from the film [8, 9]. This is what happens for mobile anions, but when the dopant is large, such as particles or micelles, the dopant remains in the film and the charge compensation is ensured by uptake or ejection of cations from or to the reactive medium [8, 10]. To gain insight into the electrochemical mechanism, one can measure the current intensity which goes through the electrochemical cell as a function of the working electrode potential. An increase or decrease in the current intensity corresponds to a redox mechanism.

The cyclic voltammograms are reported in Fig. 2. The classical cyclic voltammogram of PPy is shown in the Fig. 2a (dotted line) when chloride ions act as dopants. One can see a sharp current intensity increase at 0.75 V/ SCE, which is usually attributed to the monomer oxidation and two waves, 1 and 1′, which correspond to the oxidation ($E = 0.4$ V/SCE) and the reduction ($E = 0.2$ V/SCE) of oligomers, respectively. When MLV are added to pyrrole in NaCl solution (plain line), the current intensity wall (0.75 V/SCE) as well as both waves (1 and 1′) are still detected. However, at potentials below 0 V/SCE, two additional waves are observed: 2 and 2′, respectively, located at -0.2 and -0.6 V/SCE. These two waves suggest that another dopant larger than the Cl⁻ ions is present in the reactive medium. We believe that the negatively charged MLV interact with oligomers during the film synthesis, playing the role of dopant [7]. This interpretation is supported by previous studies of PPy synthesis, notably in the presence of sodium dodecyl sulfate (SDS) [11, 12]. Zhong and Doblhoker [11] performed cyclic voltammetry in KCl and showed that an additional redox couple appears near -0.6 V/SCE when SDS is added to the pyrrole solution. John et al. [12] analyzed a dispersion of SDS micelles in pyrrole and concluded that dodecyl sulfate is incorporated into the polymer during its synthesis. Similarly, a reduction wave was shown to emerge during cycling of pyrrole in propylene carbonate [13]. The two new waves, 2 and 2′, are thus likely to result from the interaction of MLV with the oligomers. However, one can wonder whether oligomers interact rather with the whole vesicle or with the anionic surfactant (which could be present in solution).

When the synthesis is performed in a pyrrole solution devoid of MLV but containing the negatively charged

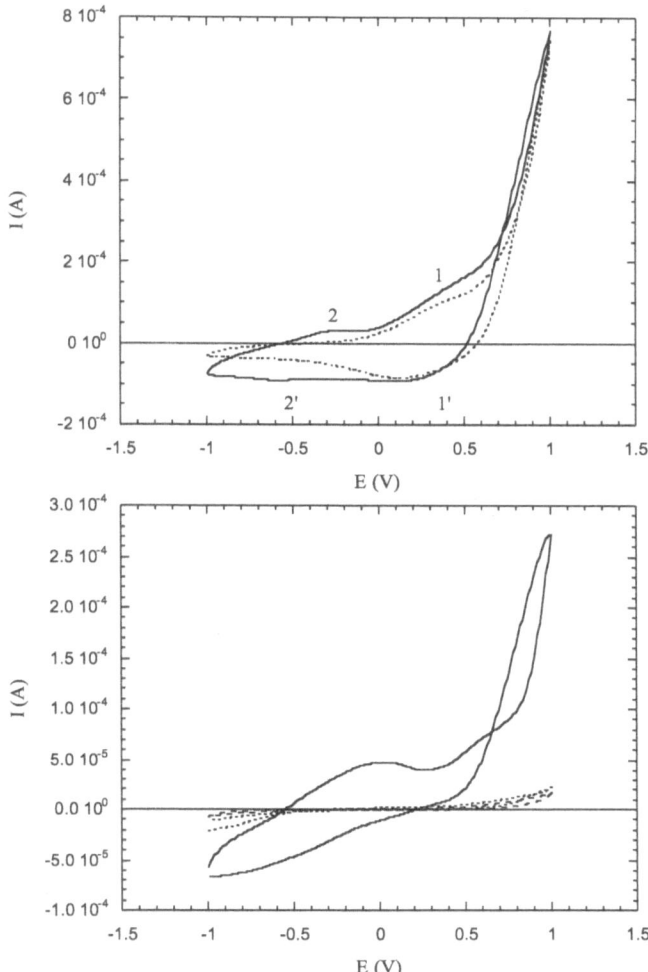

Fig. 2a, b Cyclic voltammograms recorded in a three-electrode cell. The potential, E, is measured versus a saturated calomel electrode. **a** Pyrrole (0.5 M) in NaCl (10^{-3} M):(- - - -) without MLV, (——) with MLV (1 g/l). Scanning rate:100 mV/s. **b** Pyrrole (0.5 M):(- - - -) without MLV, (——) with MLV (1 g/l), (....) in a solution of T35 (concentration below the critical micelle concentration). Scanning rate:10 mV/s

surfactant T35 (used in the preparation of MLV) at a concentration below its critical micelle concentration no current intensity goes through the cell, i.e. no polymerization occurs (Fig. 2b, dotted lines) as when the synthesis is carried out in pure water (dashed line). In contrast, when MLV are dispersed in pyrrole without any dopant (plain line), the PPy film grows, as evidenced by the cyclic voltammogram. All these data indicate that MLV act as dopants for the PPy synthesis.

Another way to assess the entrapment of MLV into the PPy film is to resort to an encapsulated probe. This is what is described in the next section for the encapsulation of GOD into MLV.

GOD immobilization into PPy

GOD is an enzyme which catalyzes the oxidation of glucose in gluconolactone and peroxide in presence of O_2. To follow its ability to catalyze this reaction, i.e. its activity, one can measure the rate of formation of peroxide. This can be achieved using a second enzymatic reaction, the oxidation of a leuco dye (ABTS) by peroxide in the presence of POD. The formation of the reduced leuco dye was followed by UV–vis spectroscopy at 405 nm. If GOD is present in the measuring medium, the plot of the absorbance at 405 nm versus time gives a straight line whose slope increases with GOD concentration.

Before encapsulating GOD into MLV, we checked that the GOD activity was not inhibited by MLV surfactants. We found that the effect of T35 on the GOD activity was negligible. We also checked that TX-100 does not significantly modify the GOD activity. Finally, we compared the activity of a solution of GOD with that of a dispersion of MLV in which GOD is encapsulated (GOD/MLV) for the same global enzyme concentration. We found that encapsulation of GOD into MLV results in a sharp decrease of its activity certainly because of steric hindrance but also because of limitations in the diffusion of glucose and leuco dye through the MLV bilayers. However, we concluded that the enzyme is not denatured by encapsulation since the destruction of MLV by addition of TX-100 leads to the recovery of the enzyme activity.

We thus made a film from a dispersion of GOD/MLV in pyrrole and NaCl by applying 1 V/SCE to the working Pt electrode for 1 h. After several minutes, the Pt wire was covered by a black deposit whose thickness reached about 1 mm within 30 min. At the end of the synthesis, the deposit was wrapped in a white jellylike sheath which is likely to consist of MLV in close contact. We washed the electrode with Milli-Q water and phosphate buffer in order to remove the sheath and MLV which could be "adsorbed" to the film surface. We then measured the activity of the film before and after treatment with TX-100. The corresponding curves are shown in Fig. 3. The activity of the crude film is very low (straight line slope $= 1.7 \times 10^{-4}$ s^{-1}): this activity can be attributed either to GOD which could remain adsorbed at the film surface and/or to enzyme trapped in the film but not encapsulated into the MLV. Indeed, let us note that not all the enzyme is encapsulated into MLV during the preparation of the samples and that free enzyme remains in the dispersion medium. After 5 min of incubation with TX-100 in order to solubilize the MLV, the film activity displays a twofold increase (straight line slope $= 3.4 \times 10^{-4}$ s^{-1}). This increase is even more pronounced after 1 h of treatment since the value of the straight line slope

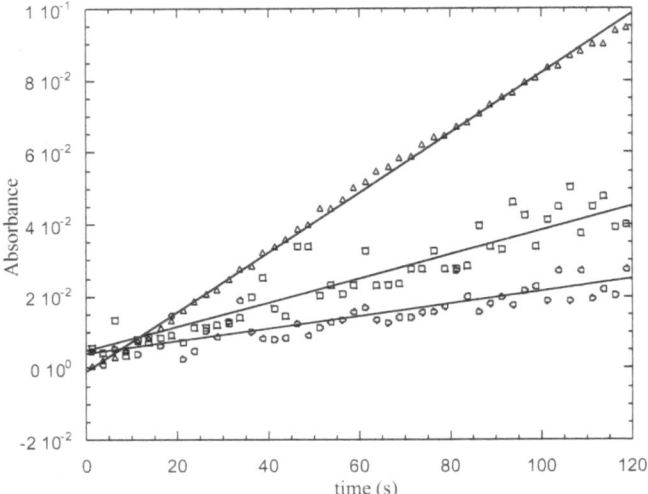

Fig. 3 Measurement of glucose oxidase (*GOD*) activity of a PPy–GOD/MLV film synthesized by applying 1 V/SCE for 1 h. The film is obtained from a dispersion of MLV (10 g/l) in pyrrole (0.5 M) and NaCl (10^{-3} M). In the MLV, GOD (10 g/l) is encapsulated (*circles*). After rinsing the electrode with phosphate buffer, (*squares*), after washing the electrode with a solution of TX-100 (20 vol %) and rinsing it with the phosphate buffer (*triangles*) after 1 h of incubation of the electrode in a solution of TX-100 (1 vol %) and rinsing it with the phosphate buffer

reaches 8.3×10^{-4} s^{-1}. We interpret the rise of the film activity after the action of TX-100 as the release of GOD from MLV into the PPy film. This suggests that not only the enzymes are well immobilized into the PPy layer but also that some of them are still encapsulated into MLV or, at least, interact with the MLV molecular components. This confirms that the MLV were incorporated into the film and, thus, that they are able to vectorize GOD into the film.

Another interesting point is to compare the activity of mixed GOD/PPy films obtained with and without the MLV for the same experimental conditions, i.e. for the same electrical treatment and for the same global concentration of enzyme. Our first experiments in that sense were performed by applying an alternating current (a.c.) field coupled with a d.c. component to get the mixed film. This process was chosen because we previously showed that MLV self-organize on a planar electrode to form a dense layer of particles when they are submitted to an a.c. field [4], whereas GOD should not be very sensitive to an a.c. field because of its smaller charge density. The idea was to induce a greater attraction of the MLV than free enzyme by applying an a.c. field and then to apply a d.c. component to start the PPy synthesis.

Our first results are reported in Fig. 4. One can see that the film activity is indeed higher when the MLV are employed to encapsulate GOD: the slope of the straight line is 3.6×10^{-4} s^{-1} compared with 3.2×10^{-5} s^{-1} when free GOD is used. The curves also show that GOD is well

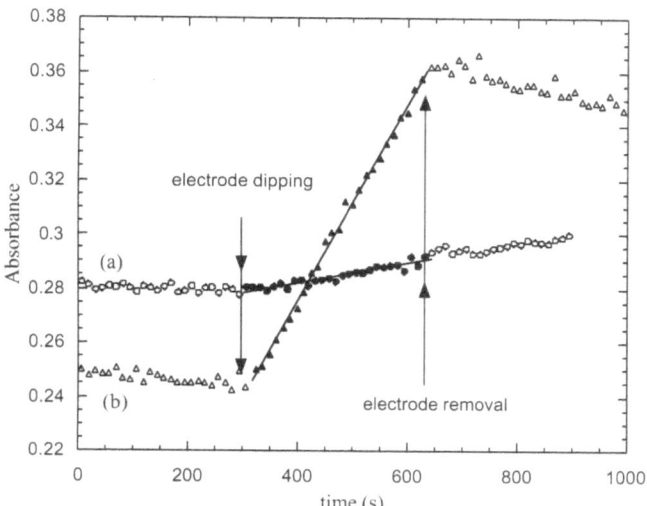

Fig. 4 Measurement of GOD activity of a PPy film synthesized with *a* free GOD (200 g/l) in pyrrole (0.5 M) and NaCl (10^{-3} M) and *b* GOD (200 g/l) in MLV (30 g/l) in pyrrole (0.5 M) and NaCl (10^{-3} M). For both films, an alternating current field was applied for 2 min (frequency:1 kHz, voltage drop ranging from 2 to 8 V with a 2 V step each 30 s) and, then, a direct current component was added (0.3 V for 1 h). The film represented by *a* was incubated in the phosphate buffer for one night. The film represented in b was incubated in a solution of TX-100 (5 vol%) for one night and incubated in the phosphate buffer for 2 h

trapped into the film and does not go out of the film since when the electrode is removed from the spectrophotometric cell, the activity is close to 0 (see curve b).

Conclusion

In this article we presented evidence for the embedment of MLV into a PPy matrix. Direct ex situ characterization of the composite material was achieved by SEM observations and Auger electron spectroscopy. When engulfed in the polymer layer, the MLV appear as bumps on the relatively flat surface and their presence is indicated by slight differences in the atomic concentrations measured through a bump or a flat area.

The rise of an additional redox couple in the cyclic voltammograms reveals a specific interaction of the MLV with the growing PPy. This new redox system reflects the doping of PPy by large, negatively charged entities. We believe these are MLV since the doping effect of T35 (the MLV surfactant) was ruled out by cyclic voltammetry experiments. This electrostatic interaction between negatively charged MLV and positively charged oligomers is likely to result in a coprecipitation to give the mixed film [7].

The incorporation of MLV into the PPy layer is confirmed using a biological probe, GOD. The triggering of GOD activity in a film obtained with MLV in which GOD is encapsulated, demonstrates that not only GOD has been immobilized in the PPy layer but also that GOD remains encapsulated or, at least, surrounded by MLV surfactants once in the film. This is indirect proof that MLV are not completely destroyed while being incorporated in the polymer matrix.

The advantages to resort to MLV to introduce an enzyme into a polymer film are numerous. First, we showed that the enzyme is somehow isolated, protected from its environment, since one needs to add TX-100 to trigger enzyme activity. Moreover, we believe that using MLV should allow the concentration of enzyme into the polymer layer to increase. We showed that this is the case when a combination of a.c. and d.c. fields is used, but we think this should also be valid for more conventional conditions, i.e. when the synthesis is performed with a d.c. field. This study is in progress. Finally, another advantage brought by MLV is that they should vectorize any enzyme into a film, for example, a positively charged enzyme which could not be inserted into a PPy film by applying an electric field because of "electrostatic incompatibility". However, if this enzyme is encapsulated in negatively charged MLV the making of the mixed film is conceivable. This study is in progress.

Acknowledgements We thank S. Ugazio and R. Laversanne for fruitful and stimulating discussions. One of us (E.B.) was supported by the European Community under a Feder grant (no. 13 02 01 Z0028) and CAPSULIS (Pessac, France). We thankfully acknowledge support from the Centre National des Etudes Spatiales (CNES) under grant 793/98/CNES/7315.

References

1. Bartlett PN, Cooper JM (1993) J Electroanal Chem 362:1
2. Palmisano F, Guerrieri A, Quinto M, Zambonin PG (1994) Anal Chem 34:1005
3. Pantano P, Kuhr WG (1995) Electroanalysis 7:405
4. Faure C, Decoster N, Argoul F (1998) Eur Phys J B 5:87
5. Faure C, Ravaine S, Argoul F (2000) J Electrochem Soc 147:575
6. Bergmeyer HU (1983) In:Bergmeyer HU (ed) Methods of enzymatic analysis, vol 1, 3rd edn. p 201
7. Belamie E, Argoul F, Faure C (2001) J Electrochem Soc 148:C301
8. Naoi K, Lien M, Smyrl W (1989) J Electroanal Chem 272:273
9. Suarez M, Compton R (1999) J Electroanal Chem 462:211
10. Pernaut J, Peres R, Juliano V, De Paoli M (1989) J Electroanal Chem 274:225
11. Zhong C, Doblhofer K (1990) Electrochim Acta 35:1971
12. John R, John M, Wallace G, Zhao H (1992) In:Mackay R, Texter J (eds) Electrochemistry in colloids and dispersions. VCH, New York, pp 225–234
13. Smela E, Zuccarello G, Kariis H, Liedberg B (1998) Langmuir 14:2970

Progr Colloid Polym Sci (2001) 118: 48–52
© Springer-Verlag 2001

G. Bokias
I. Iliopoulos
D. Hourdet
G. Staikos

Association of hydrophobically modified poly(sodium acrylate) with cationic copolymers based on N-isopropylacrylamide

G. Bokias (✉)
Department of Chemical Engineering
University of Patras
26500 Patras, Greece
e-mail: bokias@chemeng.upatras.gr
Tel.: +30-61-997501
Fax: +30-61-997266

I. Iliopoulos · D. Hourdet
Laboratoire de Physico-chimie
Macromoléculaire
UMR-7615, ESPCI-CNRS-UPMC
10 rue Vauquelin
75231 Paris cédex 05, France

G. Staikos
Department of Chemical Engineering
University of Patras and
Institute of Chemical Engineering
and High Temperature Processes
ICE/HT-FORTH
P.O. Box 1414
26500 Patras, Greece

Abstract We have studied the behaviour of aqueous mixtures of a hydrophobically modified poly(sodium acrylate) derivative (PANa3C18, containing 3 mol% octadecyl groups) with a cationic copolymer of N-isopropylacrylamide (PNIPAM10, containing 10 mol% cationic groups). Rheology and pyrene fluorescence probing were used to this end. The PANa3C18/PNIPAM10 mixtures were compared to mixtures of PANa3C18 with PNIPAM. The combination of both electrostatic attractions and hydrophobic interactions makes the PANa3C18/PNIPAM10 mixtures more effective thickeners than the PANa3C18/PNIPAM ones. In the latter case, only hydrophobic interactions are effective.

Key words Poly(N-isopropylacrylamide) · Hydrophobically modified poly(sodium acrylate) · Polyelectrolyte complexes · Hydrophobic aggregates · Rheology

Introduction

Hydrophobically modified water-soluble polymers (HMWSP) consist of a hydrophilic backbone onto which a small number of strongly hydrophobic groups are randomly anchored. These products are characterised by a pronounced thickening efficiency [1]. The strongly hydrophobic groups are, most often, long alkyl chains, such as dodecyl or octadecyl. In aqueous solution and above a threshold polymer concentration, c_0, the alkyl groups form interchain hydrophobic aggregates that stabilise a transient network. As a result, a viscosity enhancement is observed. The thickening properties of such systems can be further improved by mixing two HMWSP, especially when they are oppositely charged. A known example is the mixture of hydrophobically modified poly(sodium acrylate) (HMPA) with hydrophobically modified positively charged cellulose derivatives [2, 3]. In that case, both electrostatic attractions and hydrophobic interactions contribute to the stabilisation of the transient polymer network.

An important improvement of the thickening properties of hydrophobically HMPA derivatives is also observed when mixing them with poly(N-isopropylacrylamide) (PNIPAM) [4]. PNIPAM is a moderately hydrophobic nonionic polymer. Its aqueous solutions exhibit lower critical solution temperature behaviour, i.e., they phase-separate when the temperature exceeds about 33 °C [5]. It is well known that PNIPAM associates with anionic surfactants, for instance, sodium dodecyl sulphate (SDS), and forms hydrophobic mixed aggregates [6]. The formation of similar hydrophobic mixed aggregates is suggested for mixtures of PNIPAM with HMPA. These mixed hydrophobic aggregates act as cross-links between the PNIPAM and HMPA chains and the solution viscosity increases [4]. Moreover, the hydrophobic character of PNIPAM is enhanced with increasing temperature, leading to increased stability of the mixed aggregates and to interesting thermothickening behaviour (i.e., the viscosity increases with increasing temperature) [4].

The introduction of positive charges to the PNIPAM chain is expected to influence dramatically the behaviour of its mixtures with HMPA. To study this effect we use a HMPA containing 3 mol% octadecyl groups (PANa3C18, Scheme 1) and NIPAM-based copolymers containing 10 mol% cationic units. We present the main features of the association between PANa3C18 and PNIPAM10 at 25 °C and we compare with the behaviour of PANa3C18/PNIPAM mixtures. A more detailed report, which includes the effects of temperature and of charge density of the NIPAM-based copolymers, was presented elsewhere [7].

Materials and methods

The HMPA derivative (PANa3C18) bears 3 mol% octadecyl groups, randomly anchored on the polymer backbone. Its molar mass is 150,000 [8]. PNIPAM ($M_w = 6.5 \times 10^5$) and two PNIPAM10 samples ($M_w = 1.5 \times 10^5$ and 1.1×10^6) were prepared by (co)polymerisation in water at 28 °C, using N, N-(dimethylaminopropyl)methacrylamide as a comonomer and the reodx couple ammonium persulphate–sodium metabisulphite for initation [9, 10]. A synopsis of their characterisation is given in Table 1. The copolymers were recovered in their uncharged basic form. For the purposes of this study, they were quantitatively neutralised with HCl and transformed to the charged chloride salt form (Scheme 1). The water used was purified with a Seralpur Pro90 apparatus, combining an inverse-osmosis membrane and ion-exchange resins.

A controlled-stress rheometer, Rheometrics SR-200, equipped with a cone/plate geometry (diameter = 25 mm, angle = 2°) was used for the rheology measurements.

Steady-state fluorescence spectra were recorded using a Perkin-Elmer LS50B luminescence spectrometer. Pyrene was used as a fluorescence probe at a concentration of 6×10^{-7} M. The excitation wavelength was 334 nm and the intensities at 373 and 384 nm were used to calculate the intensity ratio of the first to the third vibronic band of the fluorescence emission spectrum of pyrene, I_1/I_3. With our experimental setup, the I_1/I_3 value measured in pure water is 1.75.

All the solutions and mixtures were prepared at least 24 h before the measurements. The experiments were carried out at 25 °C.

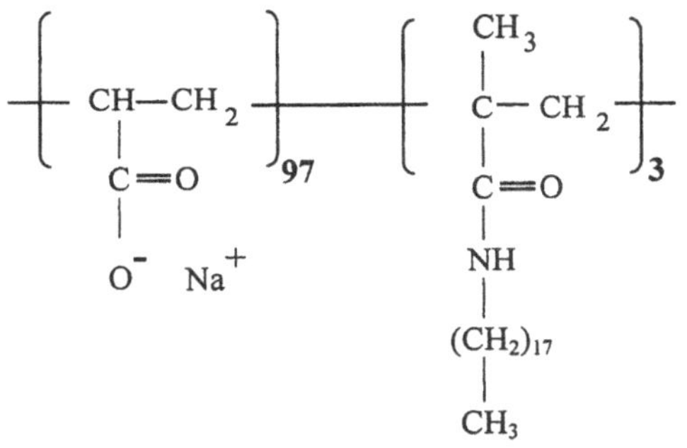

Results and discussion

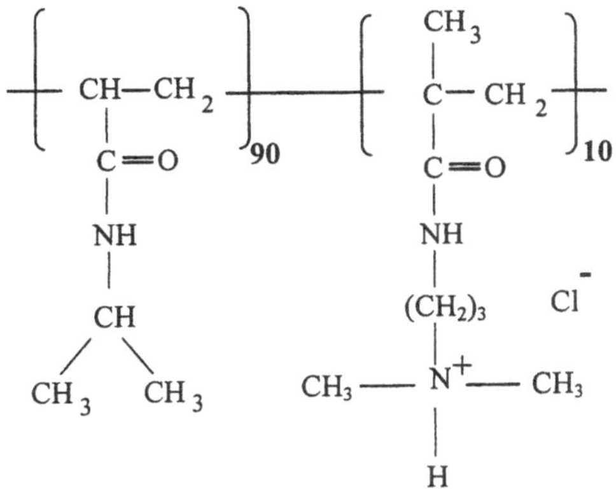

Rheology

The viscosity profile as a function of the shear rate is shown in Fig. 1 for aqueous mixtures of PANa3C18 with PNIPAM or PNIPAM10. The concentration of the NIPAM-based polymers is 1×10^{-3} g/cm^3, while the concentration of PANa3C18 is 1×10^{-2} g/cm^3, slightly lower than c_0 [8]. At this concentration, the viscosity of the solutions containing only one of the polymers is very low: about 0.01 Pas for PANa3C18 (Fig. 1) and even lower for the NIPAM-based polymers (about 3×10^{-3} Pas for the high-molar-mass PNIPAM10; not shown).

The viscosity of the PANa3C18/PNIPAM aqueous mixture is about 1 order of the magnitude higher than

PANa3C18

PNIPAM10

Scheme 1

Table 1 Characteristics of the N-isopropylacrylamide (NIPAM)-based copolymers used in this study

Polymer	N,N-(Dimethylaminopropyl) methacrylamide content (mol%)[a]	$[\eta]$ (cm^3/g)[b]	$M_w \times 10^{-5}$
PNIPAM	–	175	6.5
PNIPAM10 (high M_w)	10	232	11
PNIPAM10 (low M_w)	10	70	1.5

[a] Average values from potentiometric titration and ^{13}C NMR data
[b] $[\eta]$ was measured in a 0.5 M LiNO$_3$ aqueous solution at 20 °C. The copolymers were in the uncharged basic form. M_w was estimated from the relation $[\eta] = 0.047 M_w^{0.61}$ established for the homopolymer PNIPAM under the same conditions [9]

that of the pure PANa3C18 aqueous solution. The viscosity of this mixture remains Newtonian over a large range of shear rate and a slight shear thinning is observed above 200 s^{-1}. This viscosity enhancement is attributed to the hydrophobic interactions between the octadecyl groups of PANa3C18 and the PNIPAM backbone. The PNIPAM chains are the nuclei onto which octadecyl groups of several PANa3C18 chains are "adsorbed", thus cross-linking the system. In this case, the transient network is stabilised only by hydrophobic interactions.

Compared to the PANa3C18/PNIPAM mixture, the thickening properties of the PANa3C18/PNIPAM10 system are remarkably more pronounced. At low shear rate, even the mixture with the low-molar-mass PNI-PAM10 exhibits a viscosity increase of almost 4 orders of magnitude. Obviously, the electrostatic attractions between the negatively charged PANa3C18 backbone and the cationic groups of PNIPAM10 substantially strengthen the interpolymer association and stabilise the transient network; however, the viscosity profile as a function of shear rate is now remarkably different. The Newtonian plateau is restricted to very low shear rates,

followed by a smooth shear-thickening and finally by a strong shear-thinning region. This behaviour is usual for HMWSP [1] and it is attributed to the rearrangement of the transient network under gentle shearing and to the breaking of the interchain aggregates at higher shear rates.

As expected, the viscosity of the mixtures is strongly dependent on the polymer concentration. The viscosity increases monotonically with PANa3C18 concentration when the PNIPAM10 concentration is kept constant (data not shown). For instance, the Newtonian viscosity increases from 10 to 500 Pas when the PANa3C18 concentration increases from 2.5×10^{-3} to 1×10^{-2} g/cm^3, in mixtures containing 1×10^{-3} g/cm^3 PNIPAM10 (high M_w).

The variation of the viscosity of PANa3C18/PNI-PAM10 (high M_w) mixtures as a function of PNIPAM10 concentration, at constant PANa3C18 concentration (2.5×10^{-3} g/cm^3), is presented in Fig. 2. Each curve corresponds to a constant shear rate, ranging from 0.1 to 100 s^{-1}. The viscosity of the mixtures increases with increasing PNIPAM10 concentration, passes a maximum and decreases for higher PNIPAM10 concentrations. When the shear rate is low (0.1 s^{-1}), the viscosity gain at

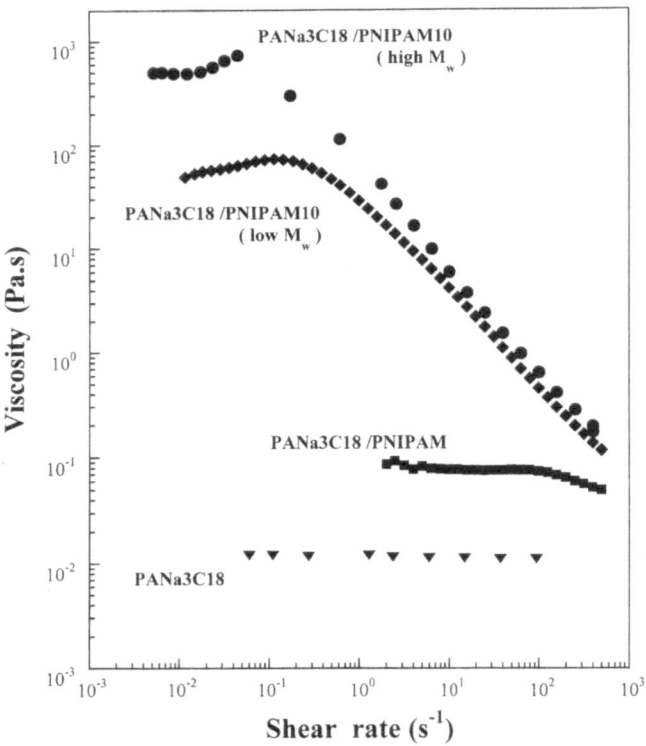

Fig. 1 Viscosity as a function of shear rate for aqueous mixtures containing 1×10^{-2} g/cm^3 hydrophobically modified poly(sodium acrylate) derivative containing 3 mol% octadecyl groups (*PANa3C18*) and 1×10^{-3} g/cm^3 poly(*N*-isopropylacrylamide) (*PNIPAM*) (■), PNIPAM containing 10 mol% cationic groups (*PNIPAM10*) (low M_w) (◆) or PNIPAM10 (high M_w) (●). (▼): pure PANa3C18 aqueous solution 25 °C

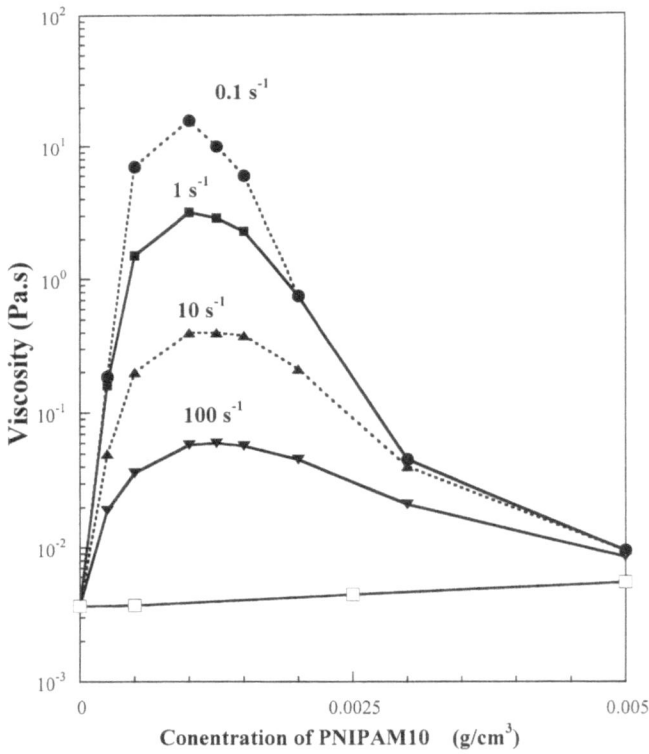

Fig. 2 Viscosity as a function of PNIPAM10 (high M_w) concentration for PANa3C18/PNIPAM10 aqueous mixtures, at various shear rates. 0.1 s^{-1}: (●); 1 s^{-1}: (■); 10 s^{-1}: (▲); 100 s^{-1}: (▼). (□): Newtonian viscosity of PANa3C18/PNIPAM mixtures. 25 °C. The concentration of PANa3C18 is 2.5×10^{-3} g/cm^3

the maximum is about 3 decades, compared to the viscosity of each polymer at the same concentration. Such bell-shaped curves, also observed for mixtures of HMWSP with proteins [11, 12], are qualitatively explained by the change in the cross-linking density with the mixture composition. Initially, addition of PNIPAM10 increases the number of cross-links and strengthens the transient network formed: each PNIPAM chain associates with many PANa3C18 chains and vice versa. At higher PNIPAM10 concentrations, association of PNIPAM10 with individual PANa3C18 chains is favoured and the number of effective interpolymer cross-links decreases, making the network looser. Furthermore, these interpolymer hydrophobic aggregates are reversible and they can easily rearrange when a shear stress is applied to the system. Indeed, the viscosity gain decreases gradually with increasing shear rate, so at a shear rate of $100 \ s^{-1}$ the viscosity enhancement is much lower (Fig. 2).

For the sake of comparison, we have also presented in Fig. 2 the variation of the Newtonian viscosity as a function of PNIPAM concentration for the PANa3C18/PNIPAM system. It is obvious that the PNIPAM concentration has little effect on the viscosity of such mixtures. This comparison confirms once more that the combination of electrostatic attractions and hydrophobic interactions is essential for the thickening properties of these mixed systems.

Fluorescence probing

The fluorescence emission spectrum of pyrene is sensitive to the polarity of the microenvironment experienced by the probe [13]. In practice, the intensity ratio of the first to the third vibrational bands, I_1/I_3, of the fluorescence emission spectrum of pyrene is used to obtain information on the existence of hydrophobic microdomains in aqueous solution. For instance, the value of the I_1/I_3 ratio is about 1.8 in water and decreases to about 1.2 in the presence of SDS micelles.

We applied this method to our systems keeping the PANa3C18 concentration rather low, $5 \times 10^{-4} \ g/cm^3$. The variation of I_1/I_3 as a function of the PNIPAM or PNIPAM10 concentration is presented in Fig. 3. Note that at these low concentrations, the aqueous solutions containing only one of the polymers exhibit I_1/I_3 values in the range 1.70–1.75, i.e., very close to the value measured in pure water (1.75). For the PANa3C18/PNIPAM mixtures, I_1/I_3 decreases very smoothly with PNIPAM concentration, indicating no significant formation of hydrophobic aggregates. In contrast, on adding PNIPAM10 to the PANa3C18 solution, I_1/I_3 decreases substantially. Thus owing to the electrostatic attractions between the two oppositely charged polymers, hydrophobic microdomains are formed by the first addition of PNIPAM10 to the PANa3C18 solution. This trend was

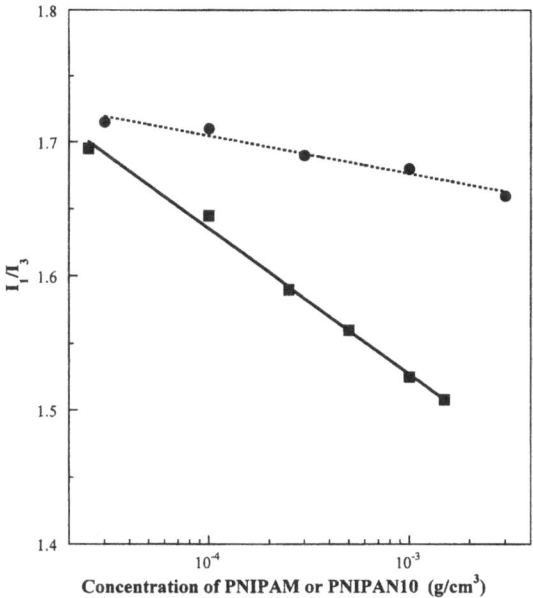

Fig. 3 Variation of I_1/I_3 with the concentration of PNIPAM (●) or PNIPAM10 (high M_w) (■), in the presence of $5 \times 10^{-4} \ g/cm^3$ PANA3C18 25 °C

also confirmed for mixtures of PANa3C18 with NIPAM-based copolymers containing various molar contents of charged groups, ranging from 5 to 25 mol%.

Conclusion

We investigated the association of a HMPA bearing 3 mol% octadecyl groups (PANa3C18), with NIPAM-based copolymers, containing 10 mol% cationic groups (PNIPAM10) in aqueous solution. The behaviour of these mixtures was compared to the behaviour of mixtures of PANa3C18 with the homopolymer PNIPAM. In the latter case, hydrophobic interactions between the octadecyl groups and the PNIPAM chains are the only driving force for interpolymer association. In contrast, in PANa3C18/PNIPAM10 mixtures, both electrostatic attractions between the two oppositely charged polymers and hydrophobic interactions between the octadecyl groups and the NIPAM sequences favour interpolymer association. This leads to important differences in the behaviour of the two systems. Namely, interpolymer association persists even at very low polymer concentrations for PANa3C18/PNIPAM10 mixtures and their thickening properties are very pronounced compared to the thickening properties of PANa3C18/PNIPAM mixtures under similar conditions.

Acknowledgements This work was supported financially by the Greek General Secretariat for Research and Technology (ΠΕΝΕΔ 99EΔ98).

References

1. McCormick CL, Bock J, Schulz DN (1987) In: Mark HF, Bikales NM, Overberger CG, Menges G (eds) Encyclopedia of polymer science and engineering, 2nd edn, vol 17. Wiley-Interscience, New York, pp 772–779
2. Thuresson K, Nilsson S, Lindman B (1996) Langmuir 12:530
3. Tsianou M, Kjøniksen A-L, Thuresson K, Nyström B (1999) Macromolecules 32:2974
4. Bokias G, Hourdet D, Iliopoulus I, Staikos G, Audebert R (1997) Macromolecules 30:8294
5. Schild HG (1992) Prog Polym Sci 17:163
6. Schild HG, Tirrell DA (1991) Langmuir 7:665
7. Bokias G, Mylonas Y (2001) Macromolecules 34:885
8. Wang TK, Iliopoulos I, Audebert R (1988) Polym Bull 20:577
9. Bokias G, Durand A, Hourdet D (1998) Macromol Chem Phys 199:1387
10. Bokias G, Hourdet D (2001) Polymer 42:6329
11. Petit F, Audebert R, Iliopoulos I (1995) Colloid Polym Sci 273:777
12. Borrega R, Tribet C, Audebert R (1999) Macromolecules 32:7798
13. Zana R(1987) In: Zana R (ed) Surfactant solutions new methods of investigation. Surfactant science series, vol 22. Dekkes New York, Chapter 5

Progr Colloid Polym Sci (2001) 118: 53–56
© Springer-Verlag 2001

C. Rodriguez
Md. H. Uddin
H. Furukawa
A. Harashima
H. Kunieda

Effect of mixing lipophilic and hydrophilic silicone surfactant systems

C. Rodriguez · Md. H. Uddin
H. Kunieda (✉)
Department of Artificial Environments and
Systems. Graduate School of Engineering
Yokohama National University
Tokiwadai 79-5. Hodogaya-ku
Yokohama 240-8501, Japan
e-mail: kunieda@ynu.ac.jp
Tel.: +81-45-3394190
Fax: +81-45-3394190

H. Furukawa · A. Harashima
Dow Corning Toray Silicone Co. Ltd
Chigusa-Kaigan 2-2, Ichihara 299-0108
Japan

Abstract The phase behavior of a hydrophilic A–B-type silicone surfactant, $(CH_3)_3SiO-[(CH_3)_2SiO]_{3.8}-(CH_3)_2SiCH_2CH_2CH_2-O-(CH_2CH_2O)_{51.6}H$, $Si_{5.8}C_3EO_{51.6}$, was investigated by phase study and small-angle X-ray scattering $Si_{5.8}C_3EO_{51.6}$ forms a micellar cubic phase and a hexagonal phase in aqueous mixtures. The structure of the cubic phase seems to be face-centered type. When lipophilic surfactant ($Si_{14}C_3EO_{7.8}$ or $Si_{25}C_3EO_{7.8}$) is added to $Si_{5.8}C_3EO_{51.6}$/water systems, a transition from the hexagonal phase to the lamellar phase takes place, owing to a change in the hydrophile–lipophile balance of the system. The change in the surface area per surfactant molecule is larger as the polydimethylsiloxane chain is longer, even if the EO number remains constant. This fact is attributed to the coiling of the long lipophilic chain in order to reduce the entropy loss.

Key words Phase behavior · Small-angle X-ray scattering · Cubic phase · Surfactant mixtures

Introduction

There is a strong relation between the hydrophile–lipophile property of surfactant and the type of self-organized structures in water. In the case of poly (oxyethylene) (PEO) type nonionic surfactants, the surfactant layer curvature is successively changed from negative to positive with increasing EO chain length and the types of liquid crystals also change accordingly [1]. In nonionic surfactants the length of the hydrophobic part of the surfactant is usually limited, therefore, with increasing EO chain length the molecular size as well as the Griffin hydrophile–lipophile balance (HLB) value increase [2]. In order to know the effect of the HLB on the surfactant layer curvature, it would be necessary to increase the lipophilic chain length while decreasing the EO chain length, keeping the molecular size constant. However, the lipophilic parts of conventional hydrocarbon surfactants are usually only in the range C_8–C_{18}, and we cannot change the size of the lipophilic part significantly.

When homologous surfactants are mixed, usually mixed micelles are formed. For example, nonionic surfactants having the same lipophilic group form a mixed micelle in water; however, there is almost no report on the mixing of two surfactants with similar molecular sizes but very different HLB values.

In this context, we studied the effect of added lipophilic A–B-type silicone surfactant on the surfactant layer curvature in liquid crystals formed in a water–hydrophilic A–B-type silicone surfactant.

Experimental

Materials

The series of surfactants with the general formula $(CH_3)_3 SiO-[(CH_3)_2SiO]_{m-2}(CH_3)_2SiCH_2CH_2CH_2-O-(CH_2CH_2O)_nH$ ($Si_m C_3EO_n$) were obtained from Dow Corning-Toray Co, Japan. m is the total number of silicon atoms and n is the average number of ethylene oxide units. Their purities are 99% for $Si_{5.8}C_3EO_{51.6}$, 99.9% for $Si_{14}C_3EO_{7.8}$ and 96.2% for $Si_{25}C_3EO_{7.8}$. The main impurity is unreacted polyether, $CH_2=CHCH_2-O-(CH_2CH_2O)_nH$, which is soluble in water and is considered to have a negligible effect on the phase behavior of concentrated solutions.

Structural characterization by small-angle X-ray scattering

Interlayer spacings were measured using small-angle X-ray scattering (SAXS) performed with a small-angle scattering goniometer with a 15-kW Rigaku Denki rotating anode generator (RINT-2500) at 25 °C.

For the lamellar and hexagonal phases, the peak ratios are 1:2:3:4... and $1:\sqrt{3}:2:\sqrt{7}:3...$, respectively. The half-thickness of the lipophilic part in the lamellar structure, r_L, and the radius of the lipophilic cylinder in the hexagonal phase, r_H, were calculated by the following Equations. [1]

$$r_L = d\phi_L/2 , \tag{1}$$

$$r_H = d\left(\frac{2}{\sqrt{3}\pi}\phi_L\right)^{1/2} , \tag{2}$$

where d is the interlayer spacing and ϕ_L is the volume fraction of the lipophilic part. The effective cross-sectional areas per surfactant molecule for the lamellar phase, a_{SL}, and for the hexagonal phase, a_{SH}, are given by

$$a_{SL} = v_L/(N_A r_L) , \tag{3}$$

$$a_{SH} = 2v_L/(N_A r_H) , \tag{4}$$

where v_L is the molar volume of the lipophilic moiety in the surfactant and N_A is the Avogrado constant. The values of v_L/N_A are 0.84 nm^3/molecule for Si$_{5.8}$C$_3$-, 1.89 nm^3/molecule for Si$_{14}$C$_3$- and 3.29 nm^3/molecule for Si$_{25}$C$_3$- [6].

For the calculation of the parameters in the micellar cubic phase, it was assumed that spherical micelles are packed in a cubic array, so the following relations can be derived [3]

$$r_I = d\sqrt{h^2+k^2+l^2}\left(\frac{3}{4n_m\pi}\phi_L\right)^{1/3} , \tag{5}$$

$$a_{SI} = 3v_L/(N_A r_I) , \tag{6}$$

where h, k, l are the Miller indices corresponding to the diffracting planes, and n_m the number of micelles per unit cell.

Phase diagrams

The surfactants were first melted and mixed and then water was added. Homogeneity was attained using a vortex mixer and repeated centrifugation through a narrow constriction in the sealed sample tube. The phase change was detected by direct visual inspection of the samples and with crossed polarizers for birefringence. The types of liquid crystals were identified by a video enhanced microscope (Nikon X2F-NTF-21) and SAXS.

Results and discussion

Phase diagram of Si$_{5.8}$C$_3$EO$_{51.6}$ water system

Si$_{5.8}$C$_3$EO$_{51.6}$ is a very hydrophilic surfactant and forms successively an aqueous micellar solution, a discontinuous cubic phase, and a normal hexagonal phase with increasing concentration, as shown in Fig. 1. The two-phase regions between the liquid-crystalline phases and the isotropic solution seemed to be very narrow and they are not shown in the diagram. Although the hydrophobic moiety is larger than a stearyl group, the Krafft

temperature does not appear and the surfactant is in a liquid state in a wide range of composition owing to the flexible polydimethylsiloxane chain. The maximum melting point of the cubic phase (104 °C) is about 20 °C higher than that of a poly(oxyethylene) oleyl ether with a similar EO chain [4], indicating that the lipophilic moiety has an effect on the stability of the mesophases. The cloud point phenomenon occurs at very high temperatures (above 110 °C), indicating the highly hydrophilic character of the surfactant.

The interlayer spacing of each liquid crystal was measured by SAXS. The four peaks resolved in the cubic phase showed spacing ratios of $\sqrt{3}:\sqrt{4}:\sqrt{8}:\sqrt{11}$. It is considered that these peaks represent diffractions from the (1 1 1), (2 0 0), (2 2 0), and (3 1 1) planes, respectively, that correspond to a face-centered-cubic structure. This kind of cubic structure has been found in trisiloxane PEO surfactant/water/oil systems [5]. The effective cross-sectional area per surfactant molecule and the radius of the lipophilic part of the micelle in the I$_1$ phase were calculated assuming four micelles per unit cell ($n_m = 4$ in Eq. 5). The results are shown in Fig. 2. Both a_s and r do not significantly change with water content, suggesting a small variation in the hydration of the EO groups. The values for the radius of the lipophilic core of the micelles in the hexagonal phase, r_H are close to the fully extended length of the Si$_{5.8}$C$_3^-$ chain, $l_{max} = 2.1$ nm [6], indicating that the polydimethylsiloxane chains are quite stretched

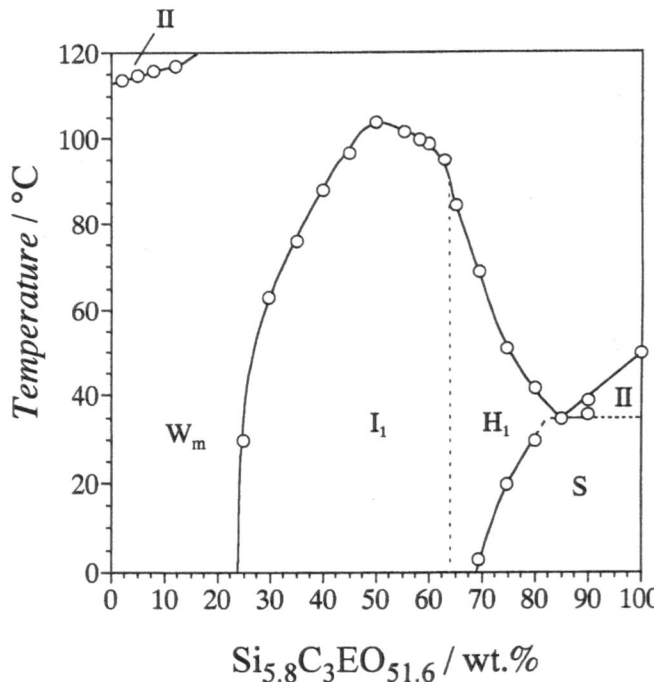

Fig. 1 Phase behavior of Si$_{5.8}$C$_3$EO$_{51.6}$/water systems. W_m: micellar solution; I_1: cubic phase; H_1: hexagonal phase; S: solid region; II: two-phase region

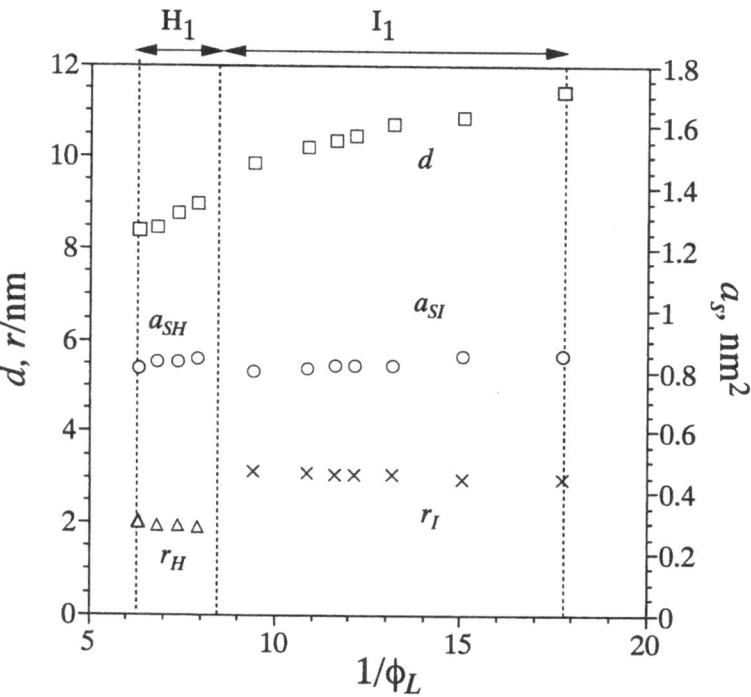

Fig. 2 Small-angle X-ray scattering (SAXS) data for $Si_{5.8}C_3EO_{51.6}$/water systems at 25 °C. Interlayer spacing, d (squares); effective cross-sectional area per surfactant molecule, a_s (circles); radius of lipophilic cylinders in the hexagonal phase, r_I (triangles); radius of the lipophilic core of micelles in the cubic phase, r_I (crosses). ϕ_L is the volume fraction of the lipophilic part in the system

in the H_1 phase. On the other hand, the micellar radius in the I_1 phase, r_I, exceeds l_{max}. One possibility is that the micelles are not completely spherical; another is that some of the headgroups are embedded in the lipophilic chains [7].

Phase change by mixing with $Si_mC_3EO_{7.8}$

When part of the hydrophilic surfactant $Si_{5.8}C_3EO_{51.6}$ is replaced with a lipophilic one with a similar molecular weight, $Si_{14}C_3EO_{7.8}$, a phase transition from H_1 to the

$Si_{14}C_3EO_{7.8} + Si_{5.8}C_3EO_{51.6}$, 30% water

H_1	L_α	M

$Si_{25}C_3EO_{7.8} + Si_{5.8}C_3EO_{51.6}$, 30% water

H_1	L_α	M

0   0.1 0.2 0.3 0.4 0.5 0.6 0.7 0.8 0.9 1

lipophilic/hydrophilic surfactant, molar ratio

Fig. 3 Schematic phase behavior of $Si_{5.8}C_3EO_{51.6}$/$Si_{14}C_3EO_{7.8}$/water and $Si_{5.8}C_3EO_{51.6}$/$Si_{25}C_3EO_{7.8}$/water systems at 25 °C. The water weight fraction is kept at 0.3. M is a multiphase region

lamellar (L_α) phase takes place as shown in Fig. 3. This phase transition can be attributed to a change in the HLB of the surfactant mixture. Further addition of lipophilic surfactant leads to phase separation. Although both are PEO-type silicone surfactants, the lipophilic surfactant cannot participate to form aqueous micelles with hydrophilic surfactant owing to the difference in the lipophilic chain length.

When the number of polydimethylsiloxane groups of the lipophilic surfactant is increased from 14 to 25 keeping the EO number constant, the H_1-L_α transition and the phase separation are shifted to lower lipophilic surfactant molar ratios. This indicates that the lipophilic chain length has an effect on the curvature of the aggregates and also that the surfactant molecules are more incompatible to form aggregates.

Phase separation took the place upon addition of a small amount of lipophilic surfactant to the cubic phase of $Si_{5.8}C_3EO_{51.6}$ system and the one-phase region was narrow; therefore, it was decided to carry out the SAXS study on the hexagonal phase.

Structural change during the phase transition

The interlayer spacings of each liquid crystal were measured in the $Si_{5.8}C_3EO_{51.6}$-$Si_{14}C_3EO_{7.8}$ and $Si_{5.8}C_3EO_{51.6}$-$Si_{25}C_3EO_{7.8}$ systems and the mean values for the effective cross-sectional area, the radius of the cylindrical micelle, and the half-thickness of bilayer were calculated as shown in Fig. 4. a_s decreases with

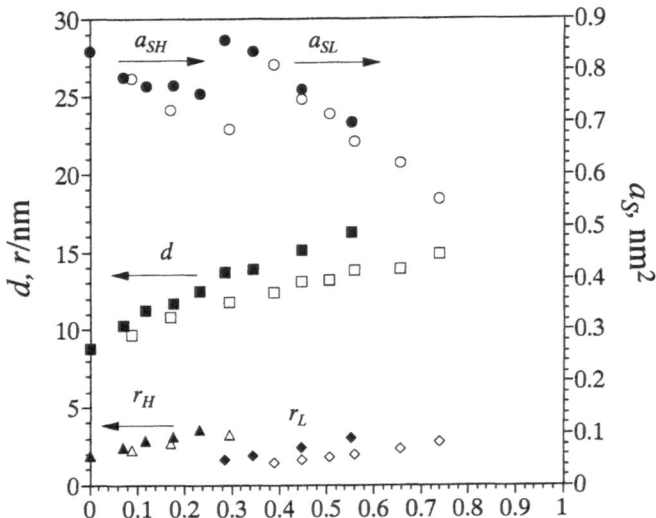

Fig. 4 SAXS data for $Si_{5.8}C_3EO_{51.6}/Si_{14}C_3EO_{7.8}$/water and $Si_{5.8}C_3$ $EO_{51.6}/Si_{25}C_3EO_{7.8}$/water systems at 25 °C. Interlayer spacing, d (*squares*); effective cross-sectional area per surfactant molecule, a_s (*circles*); radius of lipophilic cylinders in the hexagonal phase, r_H (*triangles*); lipophilic thickness in the lamellar phase, r_L (*diamonds*). $Si_{5.8}C_3EO_{51.6}/Si_{14}C_3EO_{7.8}$/water (*open symbols*); $Si_{5.8}C_3EO_{51.6}/Si_{25}$-$C_3EO_{7.8}$/water (*filled symbols*)

increasing $Si_{14}C_3EO_{7.8}$ ratio within the H_1 and L_α phases. This means that the surfactant curvature decreases by adding lipophilic surfactant with almost the same molecular size, but different HLB number. In this process, the hydrophobic part is elongated in the aggregates. Interestingly, the H_1-L_α transition takes place when the radius of the cylindrical micelles, r_H, almost reaches the value for the fully extended length of the $Si_{14}C_3^-$ chain, $l_{max} = 3.9$ nm [6].

When $Si_{14}C_3EO_{7.8}$ is replaced by $Si_{25}C_3EO_{7.8}$, the Si chain length of the lipophilic surfactant is increased keeping the EO chain length constant. Again a_s decreases and r_H and r_L increase with increasing $Si_{25}C_3EO_{7.8}$ ratio within the H_1 and L_α phases. However, r_H and r_L show practically no change when compared with the $Si_{14}C_3EO_{7.8}$ system, even if the lipophilic chain is almost twice as long. This means that the Si chain is coiled in the aggregates. As a matter of fact, the actual Si chain length is less than half of the fully extended length of the $Si_{25}C_3^-$ chain, $l_{max} = 6.7$ nm [6]. It can be seen in Fig. 4 that a_s increases with increasing lipophilic chain length even if the EO chain length is fixed. According to polymer theory [8], a long polymer chain has a short, bulky shape compared with a short polymer chain, because the entropy loss is very large when a long chain is in its extended form. Namely, it is energetically unfavorable for the long lipophilic chain to pack in a cylindrical micelle because the chain has to be elongated. Hence, at a certain point, the phase transition takes place to expand the a_s and to decrease the radius. The bulky conformation of the Si_{25} chain increases the steric repulsive interactions and induces a larger a_s when compared with the Si_{14} chain.

Conclusions

A phase transition from the hexagonal phase to the lamellar phase is induced by adding lipophilic silicone surfactant ($Si_{14}C_3EO_{7.8}$ or $Si_{25}C_3EO_{7.8}$) to a hydrophilic one ($Si_{5.8}C_3EO_{51.6}$). The effective cross-sectional area per surfactant molecule increases with the Si chain length even if the EO number is constant since the lipophilic moiety becomes coiled and bulky, increasing the steric repulsive interactions between the surfactant molecules.

References

1. Kunieda H, Shigeta K, Ozawa K (1997) J Phys Chem B 101:7952
2. Griffin WC (1954) J Soc Cosmet Chem 5:249
3. Rodriguez C, Shigeta K, Kunieda H (2000) J Colloid Interface Sci 223:197
4. Shigeta K, Rodriguez C, Kunieda H (2000) J Dispersion Sci Technol 7:1023
5. Li X, Washenberger RM, Scriven LE, Davis HT (1999) Langmuir 15:2267
6. Kunieda H, Uddin Md.H, Horii M, Furukawa H, Harashima A (2001) Phys Chem B 105:5419
7. Luzzati V, Delacroix H, Gulik A (1996) J Phys II 6:405
8. Kuhn W, Grun FJ (1946) Polym Sci 1:183

Progr Colloid Polym Sci (2001) 118: 57–62
© Springer-Verlag 2001

E. Leontidis
T. Kyprianidou-Leodidou
W. Caseri
K. C. Kyriacou

Surprising effects of polymer–surfactant solutions on inorganic crystallization processes

E. Leontidis (✉)
T. Kyprianidou-Leodidou
Department of Chemistry
University of Cyprus
P.O. Box 20537
1678 Nicosia, Cyprus
e-mail: psleon@ucy.ac.cy
Tel.: +357-2-892185
Fax: +357-2-339063

W. Caseri
Institut für Polymere, ETH Zentrum
8092 Zurich, Switzerland

K. C. Kyriacou
The Cyprus Institute of Neurology and
Genetics, P.O. Box 3462
1683 Nicosia, Cyprus

Abstract Polymer–surfactant solutions, in which the surfactant molecules interact strongly with the polymer chains forming polymer-bound micelles above the critical association concentration, are used as media for inorganic precipitation reactions. The formation of PbS is used as a prototype reaction with unexpected results. Under a wide range of conditions, the PbS crystallites initially produced evolve into a range of metastable structures composed of PbS and lead dodecyl sulfate. X-ray diffraction and transmission electron microscopy serve as valuable tools to study the evolution of the crystallizing system in detail.

The coexistence of three different colloidal particles (polymers, surfactant micelles and inorganic nanocrystals) leads to extremely complex behavior. The present work highlights the significance of coupling colloidal aggregation to ionic equilibria, and introduces polymer–surfactant solutions as a new medium for the study of inorganic crystallization reactions and the production of organic–inorganic nanocomposite materials.

Key words Nanocrystals ·
Polymer–surfactant aggregates ·
Metastable structures ·
Electron microscopy

Introduction

One of the "Holy Grails" of materials science is to achieve perfect control over the size, overall shape and morphology of crystals used in the production of technologically interesting materials [1]. With the revolution of nanotechnology a remarkable array of methods to control the size and monodispersity of small compact crystals has appeared. It has proved much more difficult to tailor the shape and morphology of crystals [1]. Because of a strong interfacial energy component to the free energy of a growing crystal, the preferred shape is almost always compact (spherical, cubic, normal polyhedral). The use of soft colloids (micelles, microemulsions, Langmuir–Blodgett films, lyotropic liquid crystal, polymer solutions and phases) as templates for inorganic crystallization has attracted considerable attention in recent years [2, 3, 4, 5]. Block copolymers [4] and polyelectrolyte–surfactant gels [5] have already been used

with some success in this respect, and morphology control has been reported in some cases. We focus our present discussion on polymer–surfactant systems as novel media for inorganic crystal growth. We use the reaction between Pb^{2+} and S^{2-} ions for the formation of PbS as a prototype reaction to examine the effect of the organization of the solution on the shape of the reaction product. PbS is in itself an interesting material, being a semiconductor with a wide range of potential application [6]. Recent work has shown that the shape and overall morphology of PbS nanocrystals can indeed be controlled by templating phenomena in systems in involving a strong interaction of Pb^{2+} ions and a moiety of the template [7].

This investigation started some years ago, when we examined the size and shape variation of PbS nanocrystals formed in poly(ethylene oxide) (PEO) sodium dodecyl sulfate (SDS) solutions [8]. It was realized that the use of relatively low polymer and surfactant concentrations but a high Pb^{2+} concentration results in coprecipitation of PbS crystals and polymer in the form of a gel [8]. Upon variation of the composition of the crystallizing system, taking into account the known physicochemical characteristics of polymer–surfactant solutions [9], we obtained an organic/inorganic composite material in the form of long rodlike particles, which evolved into a layered structure with time [10]. This complex behavior was subsequently investigated in detail [11]. It was concluded that the surfactant (SDS) forms a precipitate itself, since it interacts strongly with the Pb^{2+} cation, and that under appropriate conditions the organic precipitate may predominate. The presence of many possible precipitates and the interconversion of one to the other gives rise to complex dynamic phenomena and to the appearance of many metastable or even unstable product structures that exist "in transition". The purpose of this work is the further study of some of these structures using a set of methods, among which transmission electron microscopy (TEM) plays a central role.

Why use polymer–surfactant solutions as crystallization media

The micelles formed in the PEO/SDS or poly(vinyl pyrrolidone) PVP/SDS systems have been particularly well characterized in the last 30 years [12] and have become a standard reference system. We decided to use polymer–surfactant solutions as crystallization media for PbS on the basis of the following considerations:

1. Counterions are strongly absorbed on micellar surfaces of the opposite charge; hence, crystallization occurs at a much faster rate at the surface of the micelles. Heterogeneous nucleation should be the main, if not the only, PbS crystal formation process in such a system.
2. The spatial confinement of the micelles should have a strong influence on the crystallization process. By fine-tuning the composition of the reactants and the polymer–surfactant system, we may achieve nucleation of PbS almost exclusively in the domains of the solution occupied by polymer chains. Because of this, the nature of the polymer solution (dilute versus semidilute) is of the utmost importance.
3. Pb^{2+} is known to precipitate the surfactant (SDS); hence, it may be possible to influence the crystallization conditions towards various possible products by tuning the pH and the solution composition.
4. The polymer can act as a stabilizer for the crystal dispersion, preventing the direct precipitation of the product. On the other hand, the polymer can help to connect the crystals into loose flocs through bridging flocculation [13].

Experimental

Materials

PEO was from Polysciences, with a molecular weight of 2×10^5, and was used as received. PVP of molecular weight of 3.6×10^5 was from Sigma and was used as received. SDS (99% purity) was obtained from Sigma or Fluka and was used without further purification. $Na_2S \cdot 9H_2O$, $Pb(NO_3)_2$, $Pb(CH_3COO)_2 \cdot 3H_2O$ and other salts, acids and bases were purchased from Sigma and were used as received.

Procedures for carrying out the crystallization experiments

In a typical experiment, the ratios of the two components in the polymer/SDS system are selected so that the system contains mostly polymer-bound micelles (no free micelles). SDS concentrations between 4 and 8 mmol/l were employed in most experiments. PEO or PVP solutions with polymer concentrations between 3 and 15 g/l were employed. In this concentration range, the polymer solutions range from dilute to semidilute. To these solutions Pb^{2+} salts were added (lead acetate or lead nitrate, to a final total Pb(II) concentration of 3–4 mmol/l) and subsequently S^{2-} ions (sodium sulfide, to a final total S(II) concentration of 3–4 mmol/l). The pH value of the solutions was adjusted with concentrated nitric or acetic acid for acidic solutions and with sodium hydroxide or sodium acetate for basic solutions. After the initial addition of lead ions, the polymer/SDS solution became cloudy because of the association of Pb^{2+} with SDS. Upon addition of S^{2-} the solution immediately acquired a brown-black color. The colloidal particles formed remained in suspension at room temperature (20–25 °C) without stirring, being stabilized by the surfactant and by the polymer chains. Direct dipping of Formvar-covered copper grids provided samples for TEM work. Small amounts of the reaction solution were periodically withdrawn and analyzed by UV spectroscopy. Powder samples for X-ray diffraction (XRD) were obtained mostly by vacuum filtration of the colloidal solutions using filters with pore diameters of 0.2 or 0.5 μm, and in some cases by prolonged centrifugation. The solids were always washed with large amounts of water and dried at atmospheric pressure and room temperature for at least 24 h.

Spectroscopic investigation

UV spectra of the reaction solutions were obtained with a Shimadzu 160A UV–vis spectrophotometer.

X-ray measurements

X-ray powder diffraction spectra were obtained using a Siemens D5400 diffractometer at room temperature with $CuK\alpha$ radiation (1.54 Å).

Electron microscopy

The electron microscopy work was performed at the Institute of Neurology and Genetics, Cyprus, using a JEOL-1010A instrument, with an acceleration voltage of 80 kV.

Investigation of evolution of reaction products

Negatively charged SDS micelles were initially formed on the polymer chains. A Pb(II) salt was subsequently added. As long as the concentration of Pb(II) is less than the stoichiometrically required amount for complete precipitation of lead dodecyl sulfate [Pb(DS)$_2$], the Pb^{2+} ions added are concentrated near the micelles because Pb^{2+} interacts strongly with SDS and also for micellar neutralization (double-layer effect [14]). Addition of Na$_2$S leads to fast nucleation of PbS crystallites mainly along the contour of the polymer chain since Pb^{2+} is found almost exclusively there. The crystals do not precipitate, since their surface is saturated by surfactant molecules, and they are further stabilized by the polymer. However, the crystals aggregate through a bridging flocculation mechanism mediated by the polymer chains, forming characteristic aggregates, which can be either compact or loose and floclike, depending on the polymer concentration. At longer times, organic superstructures start forming either around or from within the initial aggregates. Previous work [11] showed that a slow transformation takes place, with dissolution of the PbS aggregates and incorporation of lead ions into a lamellar superstructure of Pb(DS)$_2$. A typical PbS aggregate formed in a PEO/SDS solution is shown in Fig. 1a and Fig. 1b shows Pb(DS)$_2$ particles developed in a similar solution, the pH of which was reduced to a value of approximately 4.0 by addition of acetic acid. The XRD patterns corresponding to these two aggregates are shown in Fig. 2. Figure 2b reveals the clear layered structure of Pb(DS)$_2$, with a spacing of 31–32 Å, while Fig. 2a, corresponding to the PbS aggregates, also shows signatures of layers with three characteristic spacings, 40.5 ± 2.0 – at higher wavelengths (not shown here) the Pbs signals are very weak! The value of 40–42 Å is close to twice the fully extended length of an SDS molecule, indicating either that the core of the "PbS" aggregates must consist of surfactant layers or that layers form on the outside of the PbS particles. The transition from one of these structures to the other is slow and follows a pathway that depends strongly on the pH and the polymer type and concentration. Figure 3 shows a loose aggregate of PbS particles formed at neutral pH in a dilute PEO solution (0.5% by polymer weight), in which the aggregation process is slow since the individual chains do not interpenetrate and do not allow the crystallites that "decorate" them to mix and aggregate. However, the organic growth is clearly visible in Fig. 3, connecting PbS particles in the form of a ladder. At acidic pH, where Pb(DS)$_2$ predominates, the organic structure envelopes the fewer PbS particles completely, giving rise to microbelike particles (Fig. 4). Other exotic possibilities of transition structures exist, some of which were examined in our previous work [11]. The transformation of PbS to Pb(DS)$_2$ is a generic phenomenon which is observed with polymers other than PEO. The remarkable dendritic growth of the organic structure which originated from small PbS aggregates in a PVP/SDS solution is shown in Fig. 5. PVP systems are generally "slower" than PEO systems, probably because

the strong complexation of Pb^{2+} by PVP leads to the formation of fewer PbS particles. This is seen from the UV spectra in Fig. 6 obtained for reactions in PEO/SDS

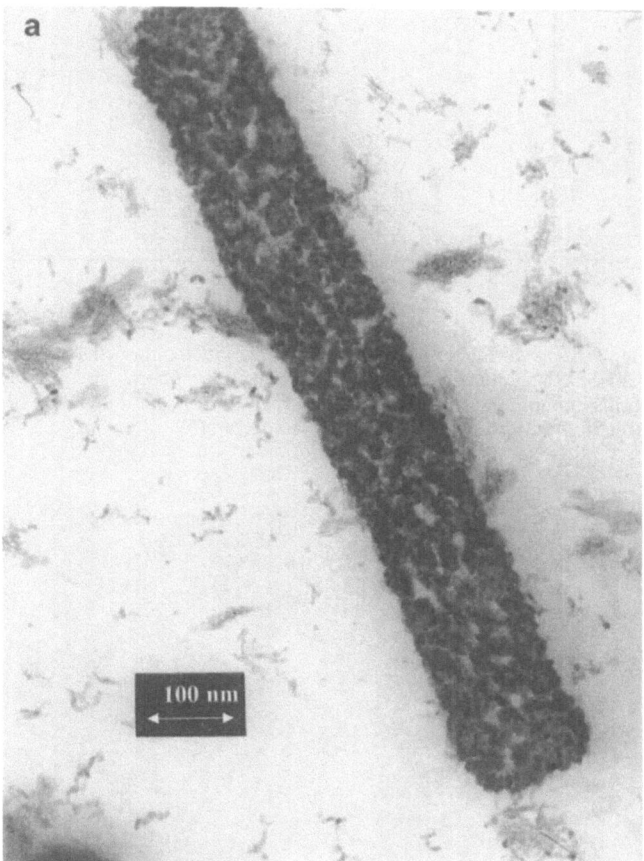

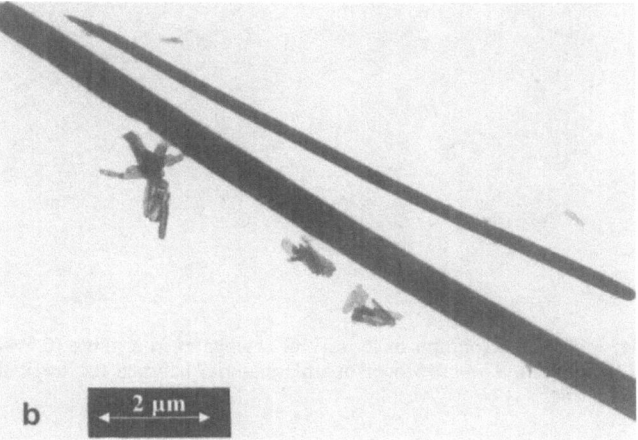

Fig. 1 Transmission electron microscopy (*TEM*) micrographs of **a** a typical PbS aggregate from the precipitation reaction between lead acetate and sodium sulfide in the poly(ethylene oxide) (*PEO*)/sodium dodecyl sulfate (*SDS*) system 5 days after reaction onset and **b** a typical lead dodecyl sulfate [*Pb(DS)$_2$*] rodlike particle formed in the same system 2 weeks after reaction onset

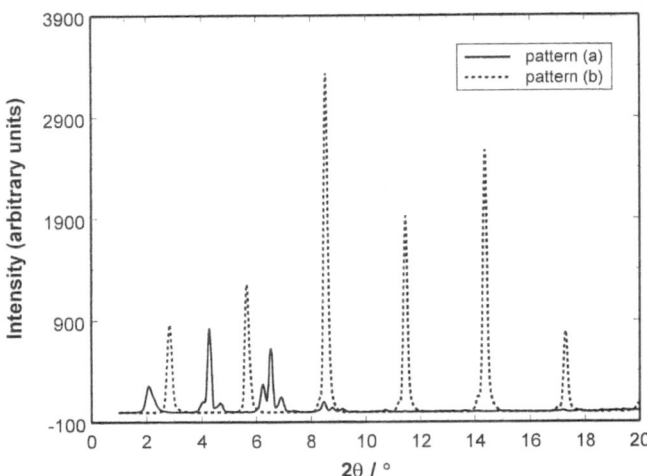

Fig. 2 X-ray diffraction patterns of the product obtained from the PEO/SDS solutions 5 days after reaction onset, presumably corresponding to the structure in Fig. 1a (*a*), and of Pb(DS)$_2$ structures with the characteristic 32-Å spacing (*b*)

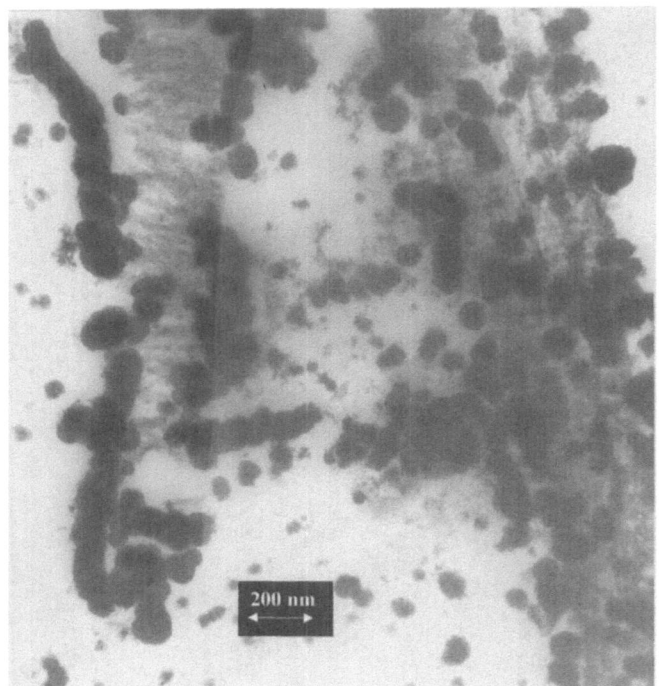

Fig. 3 TEM micrograph of loose PbS aggregates in a dilute (0.5%) PEO solution. The growth of organic material between the original PbS particles is clearly visible

and PVP/SDS solutions with identical compositions 1 day after the onset of the precipitation reaction. PbS absorption is clearly smaller in the PVP case, while the higher peak at 210 nm indicates that a larger percentage of Pb^{2+} was not reacted.

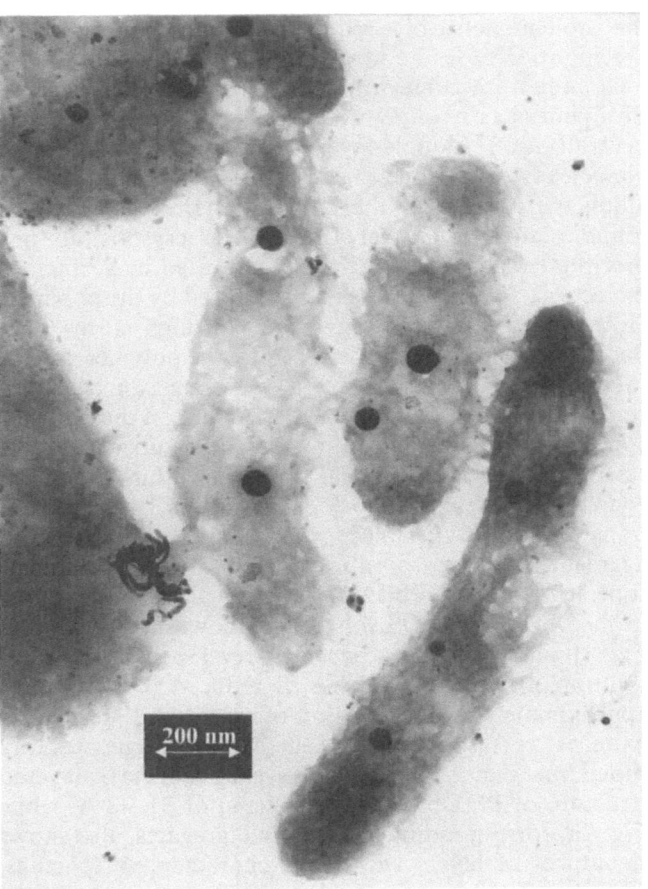

Fig. 4 TEM micrograph of structures obtained in the PEO/SDS system at pH ≈ 4.0, 4 days after reaction onset. The organic structure develops surrounding small PbS particles that act as nucleating centers

We are currently investigating the production of sulfides of Cu^{2+}, Co^{2+} and Cd^{2+} in PEO/SDS solutions. In all these cases, aggregation of primary inorganic particles is typically observed, with rodlike aggregates or tactoids being formed [15]. At least in the case of Cu^{2+} we can safely say that the initial Cu$_x$S is transformed to Cu(DS)$_2$, rather fast, which is very surprising since Cu(DS)$_2$ is a relatively soluble substance.

Summary

It appears that polymer–surfactant solutions interfere with the crystallization of inorganic materials in multiple ways. The surfactant interacts strongly with the cation of the inorganic salt, participates in the crystallization reaction and leads to completely different – and sometimes unexpected – products. It has also been observed before that inorganic reactions in organized surfactant systems may yield unexpected results if the surfactant participates in the reaction [16]. The role of the polymer

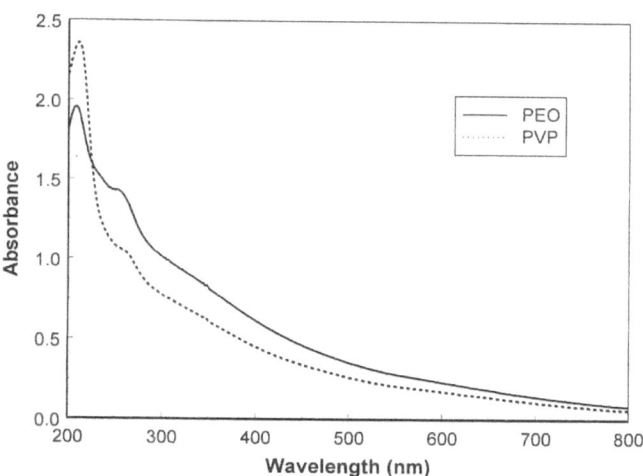

Fig. 6 UV spectra of reaction solutions 1 day after onset for the PEO/SDS and for the PVP/SDS system

Fig. 5 TEM micrograph of dendritic organic growth originating from PbS particles in a poly(vinyl pyrrolidone) (*PVP*)/SDS solution 10 days after reaction onset

type is significant and multifaceted. Strong interaction with the surfactant molecules ensures that the subsequent nucleation and crystallization will be largely localized in the vicinity of polymer chains. Cation complexation by the polymer may lead to slower reactions and even influence the final products, as it affects important species equilibria. Polymer concentration is also important. Apparently, chain interpenetration in semidilute solutions allows a faster bridging flocculation process and a

fast formation of the aggregates that eventually evolve into the layered structures observed in this work. Although dilute solutions provide isolated reaction pockets in the solution, subsequent aggregation of the primary particles and organic structure formation and growth are faster in semidilute solutions. The aggregation of primary nanocrystals is a typical phenomenon in polymer–surfactant solutions, which provide a largely unexplored terrain for inorganic chemistry and materials synthesis.

Acknowledgements This work was partly supported by funds from the ETH and the University of Cyprus.

References

1. (a) Sugimoto T (1987) Adv Colloid Interface Sci 28:65; (b) Ocaña M, Matijević E (1990) J Mater Res 5:1083; (c) Matijević E (1994) Langmuir 10:8; (d) Sugimoto T, Muramatsu A (1996) J Colloid Interface Sci 184:626; (e) Goia DV, Matijević E (1998) New J Chem (1998) 22:1203; (f) Ocaña M, Morales MP (1999) J Colloid Interface Sci 212:317
2. (a) Pileni MP (1993) J Phys Chem 97:6961; (b) Tanori J, Pileni MP (1997) Langmuir 13:639; (c) Guo S, Popovitz-Biro R, Arad T, Hodes G, Leiserowitz L, Lahav M (1998) Adv Mater 10:657
3. (a) Ozin GA (1992) Adv Mater 4:612; (b) Heywood BR, Mann S (1994) Adv Mater 6:9; (c) Fendler JH, Meldrum FC (1995) Adv Mater 7:607; (d) Braun PV, Osenar P, Stupp SI (1996) Nature 380:325; (e) Goltner CG, Antonietti M (1997) Adv Mater 9:431; (f) Antonietti

M, Goltner C (1997) Angew Chem Int Ed Engl 36:910
4. (a) Chan YNC, Craig GSW, Schrock RR, Cohen RE (1992) Chem Mater 4:885; (b) Spatz JP, Roescher A, Sheiko S, Krausch G, Möller M (1995) Adv Mater 7:731; (c) Seregina MV, Bronstein LM, Platonova OA, Chernyshov DM, Valetsky PM, Hartmann J, Wenz E, Antonietti M (1997) Chem Mater 9:923; (d) Möller M, Spatz JP (1997) Curr Opin Colloid Interface Sci 2:177; (e) Moffitt M, Vali H, Eisenberg A (1998) Chem Mater 10:1021; (f) Bronstein L, Chernyshov D, Valetsky P, Tkachenko K, Lemmetyinen H, Hartmann J, Förster S (1999) Langmuir 15:83
5. (a) Antonietti M, Gröhn F, Hartmann J, Bronstein L (1997) Angew Chem Int Ed Engl 36:2080; (b) Bronstein LM, Platonova OA, Yakunin AN, Yanovs-

kaya IM, Valetzky PM, Dembo AT, Makhaeva EE, Mironov AV, Khokhlov AR (1998) Langmuir 14:252
6. (a) Rossetti R, Hull R, Gibson JM, Brus LE (1985) J Chem Phys 83:1406; (b) Wang Y, Suna A, Mahler W, Kasowski R (1987) J Chem Phys 87:7315; (c) Gallardo S, Gutiérrez M, Henglein A, Janata E (1989) Ber Bunsenges Phys Chem 93:1080
7. (a) Schneider T, Haase M, Kornowski A, Naused S, Weller H, Förster S, Antonietti M (1997) Ber Busenges Phys Chem 101:1654; (b) Dutta AK, Ho T, Zhang L, Stroeve P (2000) Chem Mater 12:1042; (c) Pitcher MW, Cates E, Raboin L, Bianconi PA (2000) Chem Mater 12:1738; (d) Wang S, Yang S (2000) Langmuir 16:389
8. Kyprianidou-Leodidou T, Caseri W, Suter UW (1994) J Phys Chem 98:8992

9. (a) Goddard ED (1986) Colloids Surf 19:255; (b) Brackman JC, Egberts JBFN (1993) Chem Soc Rev 85; (c) Jönsson B, Lindman B, Holmberg K, Kronberg B (1998) Surfactants and polymers in aqueous solutions. Wiley, New York

10. Leontidis E, Kyprianidou-Leodidou T, Caseri W, Kyriakou K (1999) Langmuir 15:3381

11. Leontidis E, Kyprianidou-Leodidou T, Robyr P, Krumeich F, Caseri W, Kyriakou K (2001) J Phys Chem B 105:4133

12. (a) Cabane B (1997) J Phys Chem 81:1639; (b) Cabane B, Duplessix R (1982) J Phys (Paris) 43:1529; (c) Cabane B, Duplessix R (1987) J Phys (Paris) 48:651; (d) Jones MN (1967) J Colloid Interface Sci 23:36; (e) Schwuger MJ (1973) J Colloid Interface Sci 43:491; (f) Gao Z, Wasylishen RE, Kwak JCT (1991) J Phys Chem 95:462; (g) Xia J, Dubin PL, Kim Y (1992) J Phys Chem 96:6805; (h) Van Stam J, Almgren M, Lindblad C (1991) Prog Colloid Polym Sci 84:13; (i) Fox GJ, Bloor DM, Holzwarth JF, Wyn-Jones E (1998) Langmuir 14:1026

13. Stoll S, Buffle J (1996) J Colloid Interface Sci 180:548

14. Evans DF, Wennerström H (1999) The colloidal domain, 2nd edn. Wiley-VCH, New York

15. Sonin AS (1998) J Mater Chem 8:2557

16. Yueying G, Xin A (1997) J Dispersion Sci Technol 18:423

Progr Colloid Polym Sci (2001) 118: 63–67
© Springer-Verlag 2001

E. Carretti
L. Dei
C. Miliani
P. Baglioni

Oil-in-water microemulsions to solubilize acrylic copolymers: application in cultural heritage conservation

E. Carretti · L. Dei (✉) · P. Baglioni
Department of Chemistry and Consortium
CSGI, University of Florence, Via Gino
Capponi 50121 Florence, Italy
e-mail: dei@ apple.csgi.unifi.it
Tel.: +39-55-2757576
Fax: +39-55-240865

C. Miliani
Photochemistry Laboratory, Department
of Chemistry, University of Perugia, Via
Elce di Sotto 8, 006123 Perugia, Italy

Abstract Acrylic copolymers have been widely used in the past for the surface coating of porous materials of artistic/architectonic interest. Information on the alteration of the physicochemical properties of the porous materials is quite scarce. In this study we showed that the surface area, the contact angle, the water vapour permeability, and the capillary rise profiles of aerial mortar specimens were strongly modified by coating the surface sample with these copolymers. Therefore, a crucial topic is the development of a suitable method to remove these copolymers from the surfaces of the work of art/architecture. Moreover, fluorescence spectra collected from these acrylates photochemically aged by UV radiation indicated an alteration of the chemical structure enhancing the difficulty in removing the coating layers. We succeeded in developing two different four-component oil-in-water (o/w) microemulsions where the oil phase was p-xylene: the first using Tween 80 as surfactant and 1,2-propandiol as cosurfactant and the second with sodium dodecyl sulfate and 1-pentanol. These o/w microemulsions were shown to be able to solubilize and remove acrylic copolymers from the surface of porous materials constituted of aerial mortar. The microemulsions were tested during the restoration of the wall paintings by Spinello Aretino in the Cappella Guasconi in the San Francesco church, Arezzo. Scanning electron microscopy images, Fourier transform IR spectra and energy-dispersive X-ray data indicated that the removal of the hydrophobic polymeric resins from the painted surface was very satisfying without any negative side effects.

Key words Microemulsion(s) · Acrylic copolymer(s) · Cultural heritage conservation · Porous Materials

Introduction

The principal aim of this study was the development of oil-in-water (o/w) microemulsions effective in the solubilization of acrylic copolymers. These copolymeric substances have been widely used in cultural heritage conservation since 1960. They have mainly been applied as protectives, consolidants and adhesives for different kinds of porous materials [1]; several of these copolymers are constituted of acrylic monomers such as ethyl acrylate (EA), methyl acrylate (MA), methyl methacrylate (MMA), ethyl methacrylate (EMA), n-butyl methylacrylate (nBMA) and isobutyl methylacrylate (iBMA) in different molar ratios. The most critical aspect associated with the use of these materials is ageing [2]: due to both thermal degradation and photochemical reactivity, these substances are subjected to depolymerization and/or cross-linking reactions [3, 4]. The features observed are the change in the polychromy of the surface, the formation of microfractures called "craquelets" and an alteration of the properties of the interface between the work of art and the environment. All these effects suggest that their removal from the painted surfaces is both useful and important. Another consequence of these processes is the drastic loss of solubility of the copolymers [2] owing to photochemical ageing: this leads to several difficulties in their removal from the surfaces of the work of art/architecture. At present the most widely used method is based on the application of solvents such as xylenes or chloro derivatives. This technique has two main negative side effects related both to the toxicity of these compounds and to the spreading of the solubilized materials into the porous support. In order to overcome these contraindications, we tried to develop a new method based on o/w microemulsions. We selected p-xylene [5, 6] as the dispersed phase, since this solvent is very powerful in the solubilization of these acrylic copolymers in bulk [7]. With this method we expected to achieve two different results: reducing the toxicity of the solubilizing agent and inhibiting the copolymer penetration into and spreading over the porous support. Furthermore, the work was also aimed at the characterization of some porous materials (aerial mortar specimens) affected by these copolymers to investigate how the physicochemical properties of the coating support are altered and modified by the copolymers, even simulating photochemical ageing by UV radiation.

Results and discussion

The behavior of some physicochemical parameters measured for aerial mortar samples before and after the coating by three different acrylic copolymers was studied. The porous samples were prepared in the laboratory by putting an aerial mortar on a $5 \times 5 \times 2$-cm brick; the aerial mortar was obtained by mixing pure quartz sand with slaked lime [mixture of $Ca(OH)_2$/water about 0.8 w/w] in a 1:3 ratio (v/v) of slaked lime to quartz. The application of the copolymers was done 90 days after the preparation of the mortar in order to ensure the carbonation of $Ca(OH)_2$ and the consequent hardening. During this period all the samples were maintained at controlled environment conditions (25 ± 1 °C, relative humidity of $52\% \pm 1$). The copolymers were uniformly applied on the surface of the samples using a brush as p-xylene solutions in the case of EMA/MA 70/30 and EA/MMA 60/40 (4% by weight) or water emulsions in the case of nBMA/iBMA 50/50 (4% by weight). The specific amount of the copolymer coating the surface is indicated in the third column of Table 1 for two subsequent applications of each polymeric resin.

The specific area, obtained with the nitrogen absorption method [8], the water vapor permeability, the capillary adsorption coefficient, calculated according to the procedure reported in the literature [9, 10], and the contact angle for a water drop (10 μl) on the mortar surface [11] are given in Table 1. From Table 1 we deduce a significant alteration of each parameter. This infers a drastic change in the physicochemical properties of the work of art/environment interface. In particular, the treatment with the copolymers strongly modified the hydrophilicity/hydrophobicity properties of the surface as indicated by all the parameters reported in Table 1. The main consequence is a loss of both wettability, expressed by contact angle values, and water transpiration (see water vapor permeability and capillary adsorption coefficient values in Table 1). In particular the decrease in water transpiration can dramatically alter the conservation of wall paintings.

The capillary rise absorption profiles [9] for samples 2 and 3 (see Table 1) are shown in Fig. 1. The main effect is the presence of an induction time, shown by the sigmoidal shape of the curve and the decrease of the slope of the straight line portion of the curves. This allowed the decrease in the rate of water absorption caused by the copolymer coating to be quantified. Another feature deduced from Fig. 1 is the decrease in the maximum amount of water absorbed (the value of the flat asymptote at long times). All these data indicated the alteration of the surface properties of the porous materials investigated after the application of the copolymers. Since recent advances in conservation science have shown that a work of art/architecture should be restored according to the rule that the introduction of foreign materials must be avoided, the first task of a restoration procedure consists of solubilizing and removing the copolymers from the surface. As indicated earlier, the main problem associated with this operation is related to the decrease in the solubility of the copolymers as a result of their ageing. The effect of UV radiation on the copolymer coating is shown in Fig. 2. A simulated accelerated ageing was carried out with 20 h of exposure of the mortar sample surface coated by poly(EMA/MA) to UV radiation from a 150-W medium-pressure Hg lamp [3, 12]. The distance between the light source and the sample was about 20 cm. The fluorescence spectra of the sample surface were recorded using a Spex Fluorolog-2 FL 112 spectrofluorimeter equipped with a fiber optic accessory. The fluorescence emission band of the copolymer before (maximum at 279 nm) and after UV exposure (maximum at 460 nm) is shown in Fig. 2. The new band at 460 nm could be ascribed to photooxidation processes that cause the formation of oxygen bridges between the different chains of the copolymer [3].

Two o/w microemulsions were chosen for the extraction of the copolymers. The dispersed phase selected was p-xylene, a solvent very effective in the solubilization of acrylic polymers in bulk. The amount of oil never

Table 1 Behavior of capillary absorption coefficients (*CA*), water vapor permeability (*WP*), specific surface area (*SA*) and contact angle (θ_c) for aerial mortar specimens coated by acrylic copolymers; methyl acrylate (*MA*), methyl methacrylate (*MMA*), ethyl methacrylate (*EMA*), *n*-butyl methylacrylate (*nBMA*) and isobutyl methylacrylate (*iBMA*)

Acrylic copolymers	Samples	Specific amount of copolymer (mg/cm^2)	CA [(g/cm^2) s$^{1/2}$]	WP (g/m^2, 24 h)	SA (m^2/g)	θ_c (degree)
Not coated sample	1	–	0.050	311	2.5	–
EMA/MA	2	1.9	0.015	280	3.0	83
70/30	3	4.1	0.014	241	2.3	109
EA/MMA	4	1.9	0.017	277	2.5	86
60/40	5	4.9	0.030	226	2.7	103
nBMA/iBMA	6	1.9	0.016	272	2.2	85
50/50	7	4.1	0.013	219	2.5	110

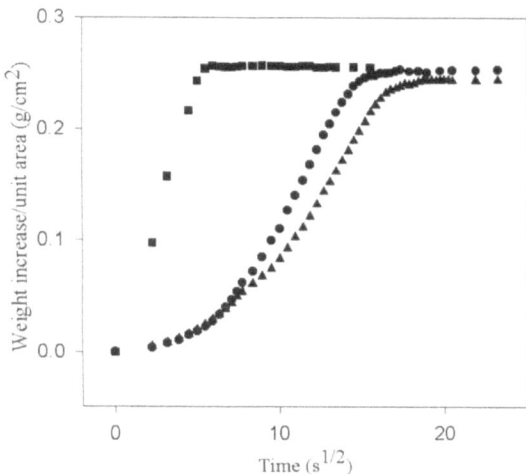

Fig. 1 Capillary rise profiles for aerial mortar specimens coated with different amount of copolymer of ethyl methacrylate and methyl acrylate [*poly(EMA/MA)*]: no coating (*squares*), 1.9 mg/cm^2 coating copolymer (*triangles*), 4.1 mg/cm^2 coating copolymer (*circles*)

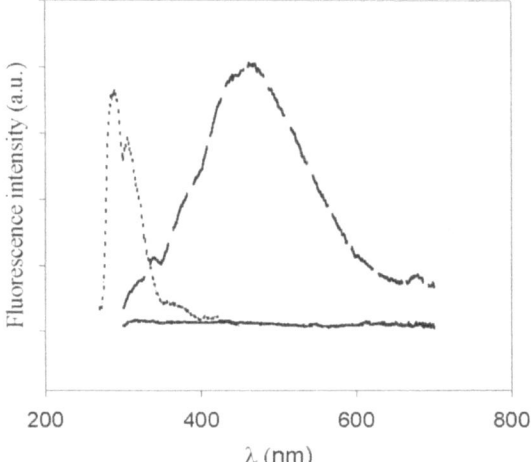

Fig. 2 Fluorescence spectra of the mortar specimen surface: no coating (*continuous line*); poly(EMA/MA) coating (*dotted line*); poly(EMA/MA) coating after UV irradiation (*dashed line*). The spectra were obtained by exciting the sample at 255 nm

exceeded 10% by weight [5–7]. The compositions of the o/w microemulsions are indicated in Table 2.

To verify the efficacy of these systems in copolymer solubilization, we applied the o/w microemulsions on the surface of either aerial mortar samples or frosted glass specimens, both coated by a thin layer of copolymer. To simulate their photochemical ageing the samples were irradiated with the same lamp as described previously for 20 h. Water contact angle measurements were carried out to follow the decrease in the value of this parameter after the application of the microemulsions: we quantified this decrease as 20% [13]. The resulting increase in the hydrophilicity of the surface supported the extraction of the acrylic coating. Scanning electron microscopy (SEM) analysis confirmed that these dispersed systems were highly effective in solubilizing and extracting the copolymer films from the glass surface. The opportunity to test this method in a real case of wall painting conservation occurred on the occasion of a restoration workshop in Arezzo (Guasconi Chapel – San Francesco church). The experiment was carried out on Spinello Aretino's frescoes (fourteenth century) treated with acrylic copolymers on the surface during the restorations carried out about 30 years ago. The presence of these acrylic copolymers was ascertained both by microreflectance Fourier transform IR analysis and by optical microscopy in grazing light. Microemulsions A and B (Table 2) were applied by means of cellulose wood poultice compresses [14], with an application time of about 2.5 hrs during this period the external temperature was 17–18 °C. After this application the treated surface

Table 2 Composition by weight percent of the oil-in-water microemulsions selected for solubilization of acrylic copolymers

Components	Composition Microemulsion A	Components	Composition Microemulsion B
Tween 80	2.7	Sodium dodecyl sulfate	4.1
1,2-Propandiol	1.9	1-Pentanol	7.9
Water	92.7	Water	85.4
p-Xylene	2.7	*p*-Xylene	2.6

was repeatedly washed with deionized water to remove residual traces of the surfactants. Two small samples were taken before and after the application of the microemulsions. The SEM micrographs obtained from these samples are shown in Fig. 3.

The SEM image of the surface of the fresco before the application of microemulsion B is shown in Fig. 3a: the presence of the acrylic copolymer smoothes the natural roughness of a wall painting surface owing to the perfect adhesion of the polymeric film onto the fresco surface.

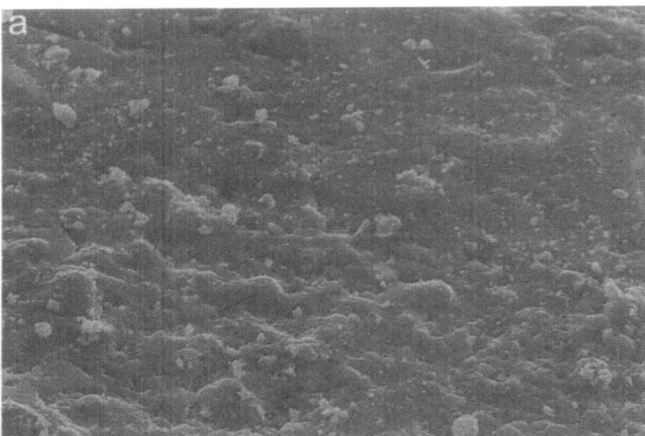

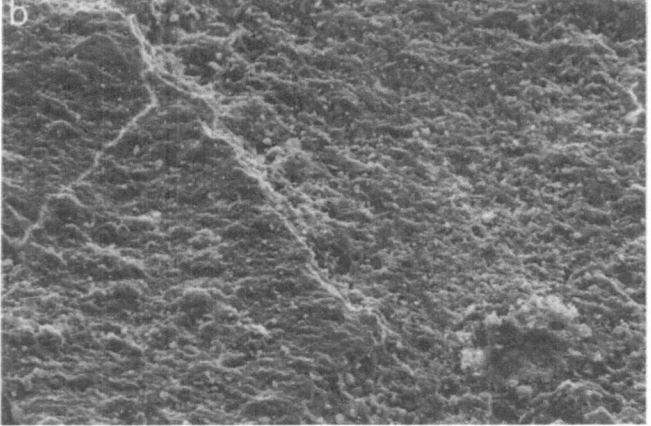

Fig. 3 Scanning electron microscopy micrographs of the painted surface of a microsample from the Spinello Aretino frescoes: before the test (*a*) and after the application of microemulsion B (*b*) (Table 2)

Figure 3b refers to the micrograph collected from the surface of the sample taken from a region of the mural painting where microemulsion B had been applied. The roughness is now perfectly evident, indicating that the o/w microemulsion tested showed good performance in solubilizing and removing acrylic copolymers from painted surfaces. No peaks typical of acrylic copolymers were detected in the Fourier transform IR microreflectance spectra collected from the surface of these samples, suggesting the cleaning of the painting surface. Even microsamples from the inner layers of the mural painting did not show spectroscopic features attributable to acrylic compounds, confirming the complete removal of the coating films.

Conclusions

This study showed that the application of acrylic copolymers as coating agents onto the surfaces of porous materials greatly alters both the physicochemical properties of the porous material/environment interface and the surface morphology. Usually to remove them from surfaces of artistic/architectonic interest solvents such as *p*-xylene, toluene or chloro derivatives are used. For this operation we developed a new method based on the application of some o/w microemulsions. The dispersed phase was constituted of *p*-xylene, while the surfactant and cosurfactant were Tween 80/1,2-propanediol and sodium dodecyl sulfate/1-pentanol, respectively. We verified their efficacy in cleaning aerial mortar and frosted glass specimens in laboratory experiments. We also tested these systems to extract copolymer layers from wall paintings. The results achieved indicated fair efficiency in removing the naturally aged copolymers from the painted surfaces and we think that this method could open interesting research perspectives both in the field of cultural heritage conservation and in the technological application of large interface systems.

Acknowledgements The authors express their gratitude to Anna Maria Maetzke, Soprintendente ai Beni Ambientali Architettonici Artistici e Storici di Arezzo, to P. Sarti Fantoni, Department of Organic Chemistry "U. Schiff" University of Florence and to Giovannoni and S. Lazzeri, restorers of the San Francesco workshop, for their cooperation. Financial support from CNR Progetto Finalizzato "Beni Culturali" and from Consorzio Interuniversitario per lo Sviluppo dei Sistemi a Grande Interfase, CSGI, is gratefully acknowledged.

References

1. De Witte E (1988) In: Proceedings of the symposium Resins in Conservation. Scottish Society for Conservation and Restoration, Edinburgh, 1.1
2. Feller RL (1994) In: Photochemical and thermal aspects. The Getty Conservation Institute, Los Angeles, p 63
3. Ciabah J (1988) In: Proceedings of the symposium Resins in Conservation. Scottish Society for Conservation and Restoration, Edinburgh, 1.1
4. Morimoto K, Suzuki S (1972) Appl Polym Sci 16:2947

5. Alba-Simionesco C, Teixeira J, Angell CA (1989) J Chem Phys 91:395
6. Rance G, Frieberg S (1977) Colloid Interface Sci 60:207
7. Baglioni P, Dei L, Carretti E (1999) Italian Patent FI/99/A/000071 2 April 1999
8. Allen T (1997) In: Scarlett B, Jimbo G (eds) Surface area and pore size distribution, vol 2, 5th edn. Powder technology series. Chapman and Hall, New York, pp 39–148
9. Various authors (1986) CNR-ICR Normal 11/85 CNR-ICR Roma 1–7
10. Various authors (1986) CNR-ICR Normal 21/85 CNR-ICR Roma 1–5
11. Drelich J, Wilbur JL, Miller JD, Whitesides GM (1996) Langmuir 12:1913
12. Biscontin G, Zendri E, Schionato A (1991) Materiali e strutture – Problemi di conservazione 1:95
13. Baglioni P, Dei L, Carretti E (2000) In: Proceedings of the 5th World Surfactants Congress CESIO 2000, Federchimica – assobase. PTIO, Florence, p 517
14. Ferroni E, Dini D (1977) In: Scritti di storia dell'arte in onore di Ugo Procacci. Electa, Florence, p 17

Progr Colloid Polym Sci (2001) 118: 68–72
© Springer-Verlag 2001

Moira Ambrosi
Luigi Dei
Rodorico Giorgi
Chiara Neto
Piero Baglioni

Stable dispersions of Ca(OH)$_2$ in aliphatic alcohols: properties and application in cultural heritage conservation

M. Ambrosi · L. Dei (✉) · R. Giorgi ·
C. Neto · P. Baglioni
Department of Chemistry and Consortium
CSGI, University of Florence via Gino
Capponi, 9, 50121 Florence, Italy
e-mail: dei@apple.csgi.unifi.it
Fax: +39-055-240865

Abstract The kinetic stability of dispersions of Ca(OH)$_2$ particles (1–2 μm) in short-chain aliphatic alcohols was investigated. The alcohols were shown to strongly enhance the kinetic stability with respect to water. Ca(OH)$_2$ crystalline nanoparticles were also synthesised at 60 °C from aqueous supersaturated solutions. The nanoparticles were characterised by scanning electron microscopy and atomic force microscopy techniques. The kinetic stability of the Ca(OH)$_2$ nanoparticle dispersions was higher than that of the micron-sized particles. Dispersions of the nanoparticles in 1-propanol were successfully tested as consolidating on aerial mortar and carbonatic stone specimens. Applications of these dispersions on carbonatic stones presenting flaking and powdering from architectonic sites in Rome and near Padua gave positive results.

Key words Calcium hydroxide · Solid–liquid dispersions · Colloidal particles · Cultural heritage conservation

Introduction

The effect of organic additives on the stability of solid/liquid aqueous dispersions has been studied extensively with the aim of understanding the mechanisms of interaction between the organic molecules and the solid/liquid interface including the charged double layer [1, 2]. In contrast, there are relatively few studies [3] on solid/liquid organic dispersions and the extension of the Derjaguin–Landau–Verwey–Overbeek theory to nonaqueous systems is only qualitative [4]. In particular, dispersions of inorganic solids in organic media have been studied for Al$_2$O$_3$ and Al(OH)$_3$ in short-chain aliphatic alcohols [5] and for several inorganic solids in xylene [6]. In this field we were interested in Ca(OH)$_2$ dispersions for possible application in cultural heritage conservation [7]. One of the aims of this study was to investigate the effect of the continuous phase – water and short-chain aliphatic alcohols – on the kinetic stability of the dispersions.

On the other hand, we were also interested to check the effects of the particle size on the stability of the dispersions and to try to synthesise Ca(OH)$_2$ nanoparticles. Several studies [8, 9] have been made on the synthesis and characterisation of insoluble metal oxides, hydrous oxides, and hydroxides of nanoparticles by precipitation from salt solutions. These compounds have very low solubility products and can be easily prepared in highly supersaturating conditions. Ca(OH)$_2$ has moderate solubility and presents some complications in the preparation of nanoparticles; in fact, achieving a high degree of supersaturation becomes more difficult and the influence of co-ions is also crucial [10]. In the present work we also aimed to synthesise and characterise Ca(OH)$_2$ nanoparticles prepared by hydrolithic methods at high temperature and degree of supersaturation. The physicochemical characterisation was performed by scanning electron microscopy (SEM) analysis and by atomic force microscopy (AFM).

The dispersions studied were interesting as innovative materials for cultural heritage conservation. In fact, aqueous saturated solutions of Ca(OH)$_2$ are used for consolidation of wall paintings and/or carbonatic stones [11], but the low solubility in water makes difficult their application. Moreover, it is to be underlined that inorganic products for consolidation should be preferred for the durability and the high compatibility of the materials with the substrate [12]. The application of $Ca(OH)_2$ as a dispersion could avoid the limitation imposed by the low concentration of $Ca(OH)_2$ in aqueous solutions. We tried to apply the $Ca(OH)_2$/alcohol dispersions as a new consolidating agent in porous material consolidation. We also report the results of tests carried out with these $Ca(OH)_2$/alcohol dispersions in a restoration workshop at Rome and Vigonza near Padua.

Results and discussion

We illustrate and discuss the experimental results in two separate sections: the first concerns the characterisation of the $Ca(OH)_2$ dispersions; the second includes the evaluation of the performances of these dispersions for consolidating porous media with flaking and powdering of the outer surface layers, including tests in the workshops.

$Ca(OH)_2$ dispersions: effect of continuous phase and particle size

The first aspect of the stability of the $Ca(OH)_2$ dispersions we investigated was the influence of the continuous phase. The optical density, i.e. absorbance, at 600 nm of some $Ca(OH)_2$ dispersions is shown as a function of time in Fig. 1 [7, 13]. The $Ca(OH)_2$ particles used to prepare these dispersions were micron-sized (diameter 1–2 μm) [7]. It is known that the behaviour of the optical density can be considered as an estimate of the dispersion kinetic stability, since it is proportional to the turbidity. In particular, a slow decrease in the absorbance as a function of time indicates kinetic stability. The change in the continuous phase from water to short-chain aliphatic alcohols strongly increased the kinetic stability of dispersions. The enhanced stability is in the order 1-propanol > ethanol > 2-propanol (Fig. 1), indicating that the critical factor is the length of the hydrophobic tail. The adsorption of the alcohol molecules onto to the surface of the $Ca(OH)_2$ nanoparticles could be considered the main cause for such enhanced stability, since it plays a fundamental role against particle agglomeration. Particle agglomeration, in fact, is known to occur in water by a mechanism of bridging driven by hydrogen bonding [14]. Further studies – electrokinetic potential measurements – are in progress to determine the exact role of electrostatic repulsive interactions that can be significant even at low surface potential values in a continuous phase characterised by very weak polarity [15]. It is worth noting that the hydrophobic chain conformation also plays an important role. Dispersions

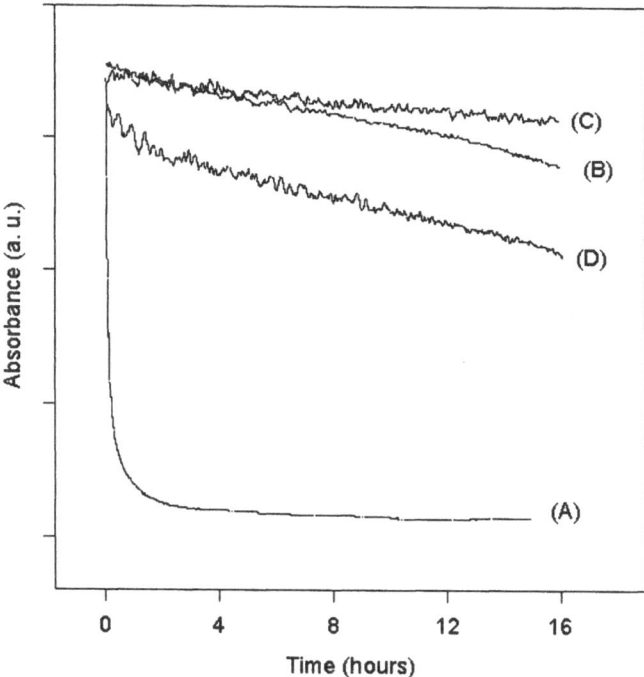

Fig. 1 Behaviour of the absorbance, A, at 600 nm for dispersions of $Ca(OH)_2$ particles (1–2 μm) as a function of the continuous phase: **a** water, **b** ethanol, **c** 1-propanol, **d** 2-propanol

with 1-propanol are more stable than those with 2-propanol, suggesting that the kinetic stability of the dispersions is proportional to the thickness of the hydrophobic layer present on the $Ca(OH)_2$ nanoparticles with adsorbed alcohol. This is confirmed by preliminary results achieved with continuous phases constituted by n-butanol, 2-butanol, and 1-octanol [16]. According to the procedure reported in the literature [7, 17], it is possible to calculate an empirical parameter, ξ, that quantifies the kinetic stability of the dispersions after a certain time. This parameter, determined from the curves reported in Fig. 1 at 16 h, is 87 for 1-propanol, 65 for 2-propanol, 79 for ethanol, and 4 for water.

We have previously shown that the particle dimensions are a factor to be controlled for obtaining high kinetic stability [18]. In particular, 3–4 μm $Ca(OH)_2$ particles had ξ values of 4 in water and 64 in 1-propanol [7], while the nanoparticles constituting the dispersions of Fig. 1, that give the ξ parameters above earlier, were 1–2 μm.

In order to study this aspect further we tried to synthesise finer particles. The approach we followed is based on a hydrolytic method in high temperature and supersaturating conditions [10]. A SEM image of the particles so synthesised is shown in Fig. 2. The tendency was for hexagonal platelets, typical of $Ca(OH)_2$ crystals, and the size was in the range 80–200 nm, indicating that we succeeded in obtaining a considerable fraction of

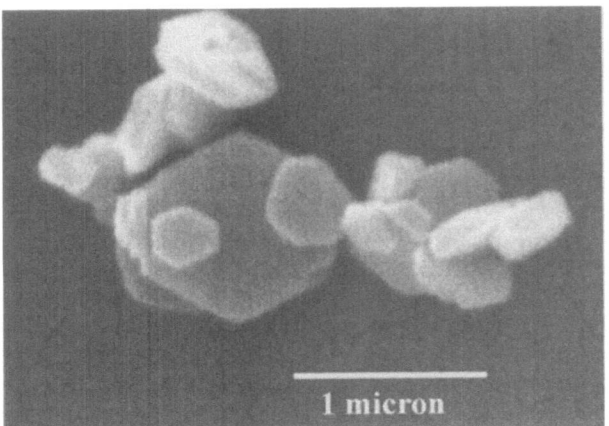

Fig. 2 Scanning electron microscopy micrograph of Ca(OH)$_2$ nanoparticles obtained from the homogeneous phase at 60 °C

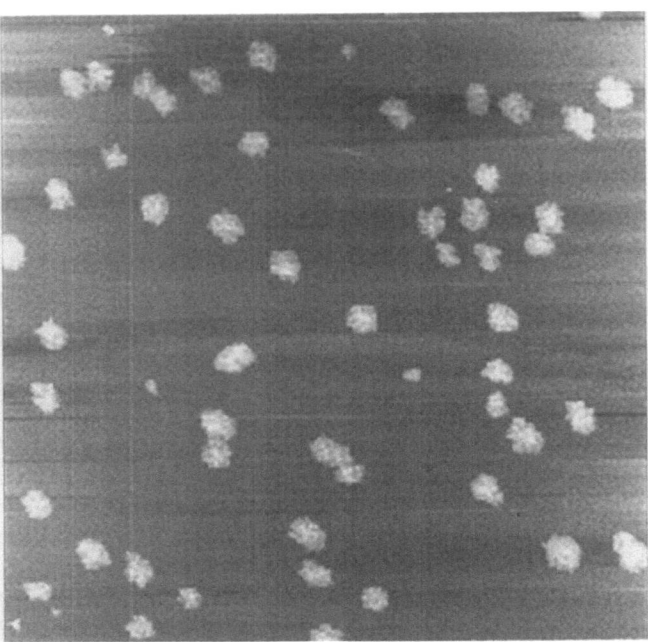

Fig. 3 Atomic force microscopy image taken in the liquid phase (1-propanol) of the Ca(OH)$_2$ nanoparticles obtained at 60 °C: image dimensions 3.3 μm $\times$ 3.3 μm

nanoparticles. The nanoparticles were obtained easily with significant yields, but great care was made to eliminate NaCl adsorbed in the collected Ca(OH)$_2$ particles. ξ determined for the dispersions prepared with the synthesised Ca(OH)$_2$ nanoparticles was 92, in comparison with 87 for the micron-sized particles (see earlier); therefore, we observed a further increase in the kinetic stability.

We also wanted to check the shape and the morphology of the synthesised nanoparticles directly in the alcohol dispersions: to do this, we performed AFM measurements in the liquid phase [19]. The AFM image of the synthesised nanoparticles dispersed in 1-propanol captured in the liquid phase is shown in Fig. 3. The dimensions are in agreement with the SEM data (Fig. 2) and we also observed a fair degree of monodispersity. Moreover, no particle agglomeration was observed. All the results indicated that the best dispersions we could test as new consolidating materials were constituted of synthesised nanoparticles in 1-propanol; therefore, these dispersions were selected to verify potential application as inorganic consolidating agent materials for exfoliated wall paintings and/or carbonated stones.

Ca(OH)$_2$ dispersions: consolidation tests

In the present section we illustrate the results of some experimental tests carried out by means of Ca(OH)$_2$/1-propanol dispersions on laboratory samples and on real systems in the restoration workshops. The aim of this experiment was to check if these dispersions could find interesting application in cultural heritage conservation to consolidate carbonate porous materials affected by flaking and/or powdering of the external surface layers.

The optical microscopy under glazing light was performed on aerial mortar surfaces of laboratory specimens. These samples were prepared using a very low lime (binder)/sand ratio (1/8 by volume) in order to simulate surface flaking. A poorly compact surface under optical microscopy in grazing light is shown in Fig. 4a. In particular, it is to be noticed that the sand grains are not bound to each other and this is due to the low content of the binder. The same sample, after the application of the Ca(OH)$_2$ nanoparticles/1-propanol dispersion, is reported in Fig. 4b. The detectable effect is an increase in the surface compactness and the appearance of intergrain junctions that reduces the voids that were evident before the treatment (Fig. 4a). Mechanical and physicochemical properties confirmed the consolidating power of this application [7]. These dispersions were also tested in some wall paintings workshops with very encouraging results [20].

Since the experiments in the laboratory and in the wall paintings restoration workshops gave fair indications, we tried to test the efficacy of these dispersions on carbonatic stones as a consolidating agent. The results obtained on Pietra di Nanto (a calcareous stone) samples constituting the external walls of the S. Margherita Abbey in Vigonza (Padua) are summarised in Table 1. The Scotch tape test parameter [7], which is inversely proportional to the stone surface compactness, shows a distinct decrease, indicating consolidation of the surface layers; therefore, we succeeded in achieving reinforcement of superficial cohesion. A similar decrease in the values of water absorption was observed (Table 1), suggesting that the

Fig. 4 Optical microscopy images under glazing light of aerial mortar surfaces: **a** decohered surface before application of Ca(OH)$_2$ dispersions, **b** the same surface consolidated by Ca(OH)$_2$/1-propanol dispersions

Table 1 Scotch tape test (*STT*) and water absorption parameters determined on the calcareous stone called Pietra di Nanto submitted to consolidating treatment by Ca(OH)$_2$/1-propanol dispersions

Sample	STT (mgcm^{-2})	Water absorption (%)
East part before the treatment	10.3	7.6
East part after second treatment	2.8	4.7
West part before the treatment	3.5	10.3
West part after second treatment	1.4	0.2

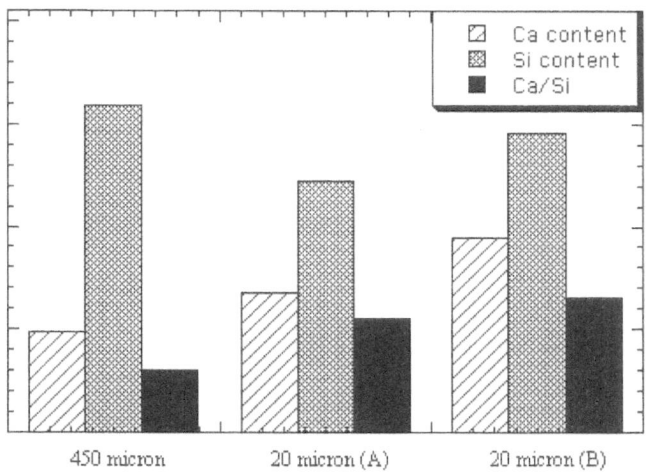

Fig. 5 Histograms of Ca and Si atomic proportion and of the Ca/Si ratio taken by energy-dispersive X-ray microanalysis on a brick sample from the Santa Prisca in Aventino church, Rome, treated with Ca(OH)$_2$/1-propanol dispersions as a function of the depth with respect to the external surface. *A* and *B* refer to two different dispersions [20]

external surfaces increased their cohesion causing less penetration of water into the wall.

Another experiment was made during the restoration of the external walls of the Santa Prisca in Aventino church apse in Rome. The tests were carried out on brick surfaces to be protected by a thin surface layer (about 50–100 μm) [20]. The histograms in Fig. 5 refer to the Ca and Si atom percent detected by energy-dispersive X-ray microanalysis monitored as a function of the depth in microns with respect to the external surface. The Ca/Si ratio as a function of the same depth is also reported in Fig. 5. The main feature from Fig. 5 is the increase in the Ca/Si ratio from the bulk (300 μm from the external surface) up to the surface (25 μm from the external surface), indicating that the application of the dispersion created an outer layer rich in Ca. SEM micrographs confirmed that an external layer about 70-μm thick was present on the brick surfaces treated with the dispersions [20].

Conclusions

This study investigated the effect of both the continuous phase and the particle size on the kinetic stability of Ca(OH)$_2$ dispersions. Alcohol molecules were shown to stabilise the dispersions probably by adsorption on the particle surface and by hindering the particle agglomeration that causes sedimentation. It is worth noting that the hydrophobic chain conformation also played an important role. Dispersions with 1-propanol were shown to be more stable than those with 2-propanol, suggesting that the kinetic stability of the dispersions is proportional to the thickness of the hydrophobic layer present on the Ca(OH)$_2$ nanoparticles with adsorbed alcohol.

The present study also showed that Ca(OH)$_2$ nanoparticles can be synthesised in aqueous media in supersaturating conditions at moderately high temperature. The nanoparticles obtained by this method were hexagonally (prismal) shaped, as evidenced by SEM analysis. AFM liquid imaging experiments on Ca(OH)$_2$ nanoparticles dispersed in 1-propanol enabled the size and morphology of the dispersed phase to be determined in situ.

The experimental results indicated that the dispersions in 1-propanol are very promising as innovative material for consolidation of porous materials. The tests carried out in the laboratory and in situ showed fair results in the application of Ca(OH)$_2$/1-propanol dispersions. The method seems to be useful for various kind of stone materials and can be considered as an innovative tool for cultural heritage conservation.

Acknowledgements The authors wish express their gratitude to G. Monti and E. Norbiato, Soprintendenza ai Beni Ambientali ed Architettonici per il Veneto Orientale, F. Zurli and P.R. David, Soprintendenza ai Beni Ambientali ed Architettonici per la Provincia di Roma, and G. Schonhaut, conservator, for cooperation in the restoration projects. Financial support from CNR, National Council of Researches, Italy, "Progetto Finalizzato Beni Culturali 1996–2000", MURST (60%), and Consorzio Interuniversitario per lo Sviluppo dei Sistemi a Grande Interfase, CSGI, Italy, is gratefully acknowledged.

References

1. Bijsterbosch BH (1987) In: Tadros TF (ed) Solid/liquid dispersions. Academic, New York, p 91
2. Vincent B, Lyklema J (1971) J. Colloid Interface Sci 31:171
3. Lyklema J (1968) Adv Colloid Interface Sci 2:65
4. Parfitt GD, Peacock J (1978) In: Matjievic E (ed) Surface and colloid science, vol 10. Wiley, New York, p 180
5. Romo LA (1966) Discuss Faraday Soc 1652
6. Koelmans H, Overbeek JTG (1954) Discuss Faraday Soc 18:51
7. Giorgi R, Dei L, Baglioni P (2000) Stud Conserv 45:154
8. Hamada S, Kudo Y, Minigawa K (1990) Bull Chem Soc Jpn 63:102
9. Matijevic E, Scheiner PJ (1978) Colloid Interface Sci 63:509
10. Arai Y (1996) In: Scarlett B (ed) Powder technology series. Chapman and Hall, London, pp 150–161
11. Brajer I, Kalsbeek N (1999) Stud Conserv 44:145
12. Mora P, Mora L, Philippot P (1984) In: Rees-Jones SG, Lindstrum D (eds) Conservation of wall paintings. Butterworths, London, p 216
13. Ma C, Xia Y (1992) Colloids Surf A 66:215
14. Gregory J, Tadros TF (1987) In: Tadros TF (ed) Solid/liquid dispersions. Academic, New York, p 163
15. Parfitt GD (1981) In: Parfitt GD (ed) Dispersion of powders in liquids, 3rd edn. Applied Science, London, pp 1–50
16. Giorgi R (1996) Chemistry degree thesis. University of Florence
17. Gabrielli G, Cantale F, Guarini GGT (1996) Colloids Surf A 119:163
18. Ambrosi M, Dei L, Giorgi R, Neto C, Baglioni P (2001) Langmuir 17:4251
19. Neto C, Aloisi G, Baglioni P, Larsson K (1999) J Phys Chem B 103:3896
20. Ambrosi M, Baglioni P, David PR, Dei L, Giorgi R, Lalli C, Mairani A, Matteini M, Rizzi M, Schonhaut G, Lanterna G (2000) In: Guarino A (ed) Proceedings of the 2nd International Congress for the Safeguard of the Cultural Heritage in the Mediterranean Basin, Paris, 5–9 July 1999, Elsevier, Amsterdam, pp 873–877

Progr Colloid Polym Sci (2001) 118: 73–76
© Springer-Verlag 2001

R. Bubeck
P. Leiderer
C. Bechinger

Structure and dynamics of two-dimensional colloidal systems in circular cavities

R. Bubeck · P. Leiderer · C. Bechinger (✉)
University of Constance, Department of
Physics, 78457 Constance, Germany
e-mail: clemens.bechinger@uni-konstanz.de
Tel.: +49-7531-883562
Fax: +49-7531-883127

Abstract The properties of 2D colloidal crystals have been widely investigated over the last 20 years. Recently, it has been recognized that colloids are also useful for the investigation of systems comprised of only a few particles. We study the melting behavior of a finite number ($N < 30$) of paramagnetic colloidal spheres ($\sigma = 4.5$ μm) in 2D circular cavities. By applying a magnetic field, B, the interaction strength between the particles is varied. At high B, i.e. strong dipole interaction, the particles are arranged in a highly ordered shell-like structure. With decreasing B we observe a loss of angular order between adjacent shells. Upon further reduction of the external field, however, different scenarios are observed. For cavities with hard-wall confinement and commensurate particle numbers angular order is restored again. In contrast, the latter effect is absent for soft-confinement potentials. In both cases the system melts completely for small magnetic fields.

Key words Colloids · Two-dimensional · Two-dimensional freezing · Phase transitions · Classical atoms

Introduction

In the last few years there has been considerable progress in the field of localization and cooling of ions and electrons in artificial confining fields. Typical examples for 3D and 2D systems are ions in radio-frequency traps [1], electrons on the surface of liquid He [2], and electrons in quantum dots [3]. With the help of present-day powerful imaging techniques such examples may be promising subjects for the experimental investigation of systems in lateral confinements. Additionally, the structural and dynamical properties of few-body systems are also attractive from the theoretical point of view [4]. Several authors considered 2D systems with finite numbers of ions or electrons in lateral confinements using Monte Carlo (MC) simulations [5–8]. So far, the majority of theoretical studies investigated the behavior of Coulomb particles in a harmonic external potential. At low temperatures and in the case of a small number of particles, the particles are found not to crystallize in a triangular lattice (Wigner crystal), but are arranged in a shell-like structure. Accordingly, it was pointed out that such systems might be a "realization" of a 2D Thomson atom, where the structure as a function of the particle number can be analyzed in terms of a Mendeleev-type table [6, 9]. The melting of laterally confined 2D systems with particle numbers of the order of 100 or smaller is predicted to occur via a two-step process [6–8]. Upon increasing the temperature, first intershell rotation becomes possible where orientational order between adjacent shells is lost. At even higher temperatures, radial diffusion between shells sets in, and this finally destroys the shell structure of the cluster. In recent experiments with magnetic colloidal particles in hard-wall confinements, however, a deviation from this scenario was found: when the effective temperature is increased beyond the point where intershell rotation sets in, angular order is restored again before the system melts completely [10]. In the meantime this reentrant behavior has also been confirmed by MC simulations

[11]. It has been pointed out that the occurrence of this effect is related to the boundary condition of the confinement, which determines to what extent the particles can escape in radial direction upon changing the temperature. This also explains why such behavior was not seen in previous calculations because they concentrated on the behavior of particles in parabolic, i.e. soft, confinements.

Here we present an experimental study of the phase behavior of 2D systems consisting of a small number of particles interacting via a dipole potential in different circular confinements. In the first part we discuss a system of 29 particles in a hard-wall confinement. These results are in excellent agreement with earlier experimental data [10] and recent numerical simulations [11]. In addition we also investigated a system of 26 particles in a soft potential. As shown, the properties in both cases are rather different, which is also in agreement with recent numerical simulations [11].

Experimental

The experiments were performed with superparamagnetic colloidal spheres with a diameter of 4.55 μm. Since the pair interaction potential between the spheres can be conveniently varied by means of a magnetic field such systems provide excellent model systems for the investigation of 2D systems. The advantages of colloidal suspensions are their convenient time (milliseconds) and length scales (microns), which allow the detailed observation of single particle trajectories by means of video microscope [12]. As a substrate for the 2D colloidal system we used fused silica plates onto which a 3–4-μm thick smooth film of poly(methylmethacrylate) (PMMA) was deposited by spin-coating. This was necessary to prevent the particles from sticking to the surface. The lateral confinements were realized by using a transmission electron microscope (TEM) grid made of copper that was pressed into the thin PMMA film, which was heated to about 150 °C. After this process typically several tens of identical circular hard-wall compartments with perpendicular wall were obtained (Fig. 1a). In addition to such hard-wall boundaries we also fabricated compartments with soft-wall potential. This was achieved by adding some additional PMMA solution to a sample after the TEM grid had been pressed onto the substrate. Owing to capillary forces the liquid

preferentially wetted the previously formed hard-wall compartments, where under the influence of the radial forces during spin-coating a radially symmetric meniscus was formed. After the solvent had evaporated this resulted in a soft-wall potential (Fig. 1b).

After insertion of the superparamagnetic colloidal spheres which were suspended in water (DynaBeads 4.55 μm, Dynal, lot no. B20100) and stabilized with 2.0 g/l sodium dodecyl sulfate, the particles sedimented towards the patterned (see earlier) bottom plate, where they were confined by gravity to a 2D colloidal system in each compartment (Fig. 1). In our experiments the number density in different compartments varied statistically between about 25 and 45, depending on the particle concentration. The whole cell was placed in the center of a copper coil, which produced a magnetic field, B, perpendicular to the substrate. The magnetic field induces a magnetic moment, $M(B)$, within the particles which led to a repulsive magnetic dipole pair potential, $V_{i,j} = \mu_0 M(B)^2 / 4\pi r_{ij}^3$, where μ_0 denotes the permeability of vacuum and r_{ij} the distance between particles i and j. The colloidal spheres were imaged with a home-built inverted video microscope onto a charge-coupled-device camera. An image processing system (Visiometrics IPS) with particle recognition facilities was used to determine the particle coordinates for each frame via a particle recognition algorithm. From these coordinates we determined the particle trajectories by using an object tracking software (Visiometrics TRACE).

To describe the effective interaction between the paramagnetic spheres it is useful to introduce the dimensionless plasma parameter, Γ, which can be defined in several ways. In order to avoid confusion and to allow direct comparison with the work of Schweigert et al. [11] we used their definition of the plasma parameter, where $\Gamma = q^2 / (a^3 k_B)$, with $q^2 = \mu_0 M(B)^2 / 4\pi$ and $a = 2R/N^{1/2}$. Here R denotes the radius of the circular hard-wall confinement, k_B the Boltzman constant and T the temperature of the suspension. Within the range of magnetic fields used in this work, M is to a good approximation proportional to B. We want to emphasize that in our previous work we used a slightly different definition for Γ [10].

Accordingly the plasma parameter for $N = 29$ in the present work is a factor of 2.2447 smaller than in the previous one.[1] In our experiments we varied Γ by changing B and kept T constant at 295 K. For each B, we measured the particle trajectories up to several hours with a lateral and temporal resolution of 0.25 μm and 20 ms, respectively. In contrast to previous measurements where the measuring time was only 30 min, here the particle trajectories were taken up to several hours, which largely increased the statistics and allowed us to calculate dynamic properties such as mean square displacements.

Results

Hard walls

A real-space image for the case of $N = 29$ particles in a hard-wall confinement at $\Gamma = 64$ is shown in Fig. 1a. In this case the particles arrange in shells consisting of 3, 9 and 17 particles, respectively. Such a configuration is denoted in the following as (3, 9, 17) and has been shown to correspond to the ground state of such system [10]. In an earlier measurement we presented the structure of such a system as a function of Γ. In contrast to these experiments, where the measurement time was limited to 30 min, here the time was increased by a factor of 20.

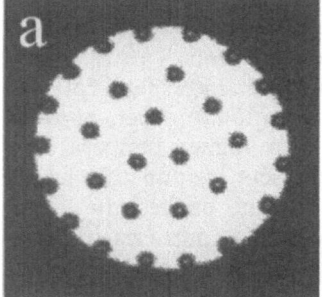
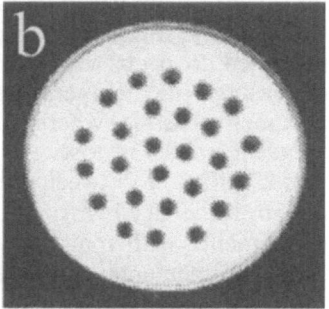

Fig. 1 a Bottom view of the 72-μm openings of a transmission electron microscope mesh, which serve as lateral hard-wall confinements for the particles. **b** An additional poly(methyl methacrylate) film on top of the openings produces a soft potential for the particles

---

[1] In our previous work the plasma parameter was defined as $\Gamma = \langle V_{i,j} \rangle / N k_B T$, where the angled brackets correspond to the sum over all particle pairs i,j

This allowed us to calculate the angular diffusion coefficients of the shells.

To visualize the shell-like behavior of our system, the particle trajectories of the particles for $\Gamma = 17$ are plotted in Fig. 2a. The three particles of the inner ring also distort the symmetry of the second shell, which shows also a pronounced threefold symmetry, whereas the structure of the outer shell is only given by the circular hard-wall confinement. It can also be seen that the outer particles are more localized than the inner particles, which is a consequence of the hard-wall confinement. In addition the diffusion of the inner particles is highly anisotropic. Both findings are in good agreement with numerical calculations [7]. In order to characterize the angular movement of the innermost shell relative to the second shell, we first transformed the data shown in Fig. 2a into a reference frame, which follows the collective angular motion of the second shell. As a result, the angular positions of the particles of the second shell become rather localized, indicating that the particles move in a highly cooperative manner (Fig. 2b).

From the linear slope of the mean angular displacements of the particles in the first shell at long times one directly obtains the angular diffusion coefficient with respect to the second shell. In order to determine to which shell a particle belongs (this becomes increasingly difficult upon decreasing Γ) we used the minima in the radial distribution of particles, which show three maxima (corresponding to the three shells) with pronounced minima in-between, the latter characterizing the radial extension of the shells.

The angular diffusion constant, D_Θ, of the first shell is plotted against Γ in Fig. 3. At high Γ values, the energy barriers for shell rotation are rather small, corresponding to a small D_Θ. With decreasing Γ, however, D_Θ increases by more than a factor of 8 and indicates the high mobility of the second shell versus the first one. For even lower Γ values, D_Θ decreases again until the shell structure finally vanishes and an angular diffusion coefficient cannot be defined anymore. Such a maximum of the angular mobility was observed in earlier measurements [10] and

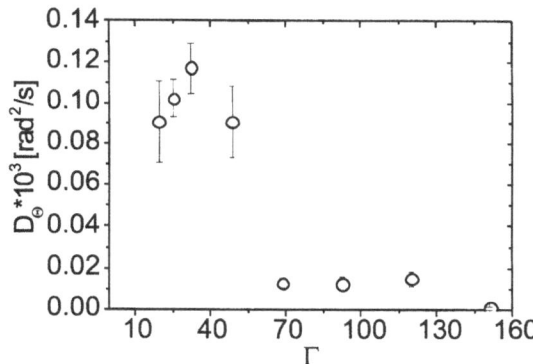

Fig. 3 Angular diffusion constant of the first shell over Γ for the system of 29 particles in a hard-wall confinement

has been explained by radial particle fluctuations, which tend to increase the registration between adjacent shells. With increasing Γ, the angular mobility of the first shell increases at first but is then reduced again by the increasing radial particle fluctuations of the second shell, which finally increase the registration between adjacent shells. This scenario has been supported by recent calculations of Schweigert et al. [11].

Soft walls

It has already been suggested that the occurrence of the previously discussed reentrance behavior is strongly related to the details of the confining potential. In the presence of a circular cavity with hard walls, the particles cannot escape in a radial direction, which seems to be important for the unusual Γ dependence of the angular diffusion coefficient. In contrast, if the confinement is comprised of soft walls, e.g. a parabolic potential, the whole system will expand if the magnetic field is increased and the melting scenario should be rather different [5, 6, 13–15].

Accordingly, we also performed measurements where the confinement was a soft wall. A typical snapshot at a plasma parameter of $\Gamma = 32.6$ with $N = 26$ is plotted in Fig. 1b. In contrast to the hard-wall potential, it can be clearly seen that now the particles are located in the center of the confinement with a configuration (3:9:14). Taking the preparation process of the samples into account (spin-casting of a liquid in a circular confinement), the shape of the confinement should roughly correspond to a rotational parabola. Owing to surface tension effects, however, deviations from a perfect parabolic potential might occur and we assume the potential to be given by $V(r) = \alpha r^\beta$. By comparing our ground-state configuration with the results from MC simulations[2] this suggests β is

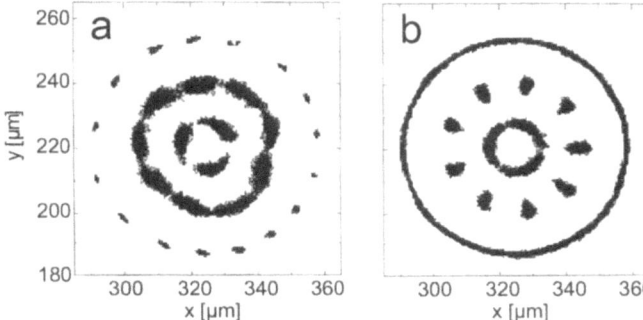

Fig. 2 Particle trajectories over 10 h for $\Gamma = 32$: **a** raw data; **b** after correction of angular movement of the second shell

[2] The transition from ground-state configurations (4:9:13) to (3:9:14) takes place between $r^{2.04}$ and $r^{2.05}$ and that from (3:9:14) to (3:8:15) between $r^{6.57}$ and $r^{6.575}$

between 2.05 and 6.57. In contrast for $\beta = 2.0$, for example, a slightly different arrangement (4:9:13) should be observed. α can be estimated by the dynamical behavior of the particles. When the external magnetic field is turned off the particles begin to diffuse towards the center of the cavity, with their dynamics given essentially by their interaction with the confining potential, i.e. the meniscus of the polymer film. Since this dynamic is highly overdamped, the corresponding equation of motion can be easily solved, which allows us to obtain $\alpha = 0.3 \pm 0.1$ 10^{-9} N/m. Accordingly the outmost particles are only raised about 270 nm (corresponding to 6% of the particle diameter) with respect to the center of the compartments; hence the a priori assumption of a 2D system is largely fulfilled in this situation.

Similarly, to the experiments described earlier we followed the particle trajectories at different Γ values and obtained the angular diffusion constants, which are plotted in Fig. 4. The experimentally accessible Γ range is now restricted to values above 20 since at smaller values it is difficult to define a shell structure and thus allow calculation of D_Θ. The available data, however, cover the region just above complete melting, where the reentrance has been observed in hard-wall potentials.

In contrast to Fig. 3, for soft walls we do not find any indications for a maximum as observed before, but the data points roughly show a linear Γ dependence. In addition the angular diffusion coefficient is higher compared to the hard-wall confinement. Both findings have also been observed in MC simulations [11]. Unfortunately, these studies were performed only with $N = 25$ particles in contrast to $N = 26$ in our case. This difference, however, should not affect the general diffusion behavior, since the configuration of the two inner shells is identical for $N = 25$ and $N = 26$.

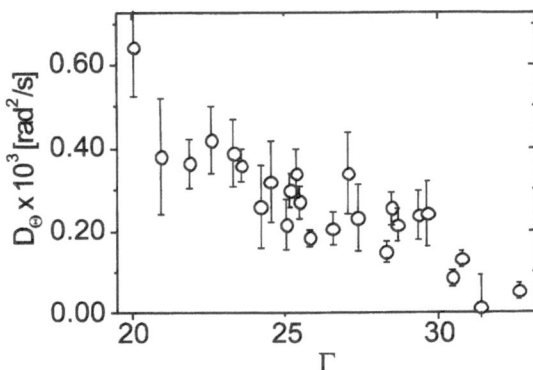

Fig. 4 Angular diffusion constant of the first shell over Γ for 26 particles in a soft-confinement potential

Conclusion

In summary, we have studied the melting process of finite 2D systems in different external cavities. In the case of hard-wall confinements we find in agreement with recent theoretical studies reentrant behavior. This is interpreted in terms of radial particle fluctuations, which lead at lower Γ to a higher coupling of adjacent shells and, therefore, to higher angular order. In contrast, when the external 2D cavity is comprised of soft potentials the system size varies as a function of Γ. Consequently, the coupling of adjacent shells by radial particle fluctuation is smaller than in the hard-wall case. This finally results in a monotonic increase in the angular mobility of the shells with decreasing Γ as has been predicted by recent MC simulations.

Acknowledgement We gratefully acknowledge financial support from the Deutsche Forschungsgemeinschaft (SFB 513).

References

1. Birkl G, Kassner S, Walther H (1992) Europhys News 23:143
2. Leiderer P (1992) J Low Temp Phys 87:247
3. Reed MA, Kirk WP (1989) In Proceedings of the international conference on nanostructure physics and fabrication, Academic, Boston, p 523
4. Date G, Murthy MVN, Vathsan R (1998) J Phys Condens Matter 10:5875
5. Lozovik YE, Mandelshtam VA (1992) Phys Lett A 165:469
6. Peeters FM, Schweigert VA, Bedanov VM (1995) Physica B 212:237
7. Bedanov VM, Peeters FM (1994) Phys Rev B 49:2667
8. Schweigert VA, Peeters FM (1995) Phys Rev B 51:7700
9. Partoens B, Peeters FM (1997) J Phys Condens Matter 9:5383
10. Bubeck R, Bechinger C, Neser S, Leiderer P (1999) Phys Rev Lett 82:3364
11. Schweigert IV, Schweigert VA, Peeters FM (2000) Phys Rev Lett 84:4381
12. Bubeck R, Neser S, Bechinger C, Leiderer P (1998) Prog Colloid Polym Sci 110:41
13. Lozovik YE, Rakoch EA (1998) Phys Lett A 240:311
14. Lozovik YE, Rakoch EA (1998) Phys Solid State 40:1255
15. Belousov AI, Lozovik YE (2000) Eur Phys J D 8:241

Progr Colloid Polym Sci (2001) 118: 77–81
© Springer-Verlag 2001

K. Mangold
R. Bubeck
P. Leiderer
C. Bechinger

Substrate-induced phase transitions in two-dimensional colloidal systems

K. Mangold · R. Bubeck · P. Leiderer
C. Bechinger (✉)
University of Constance
Department of Physics
78457 Constance, Germany
e-mail:clemens.bechinger@uni-konstanz.de
Tel.: +49-7531-883562
Fax: +49-7531-883127

Abstract We study the phase behavior of a 2D colloidal system in the presence of an external periodic 1D potential. As colloidal suspension we use an aqueous suspension of superparamagnetic spheres with a diameter of 4.5 μm. The periodic 1D potential is fabricated by evaporation of thin magnetic nickel lines onto a glass substrate, which is afterwards covered with a protective poly(methyl methacrylate) layer. When the phase behavior of the colloidal system is investigated as a function of the applied external magnetic field, we observe a promoting effect of the underlying periodic 1D potential to the crystallization of the 2D colloidal system.

Key words Colloids · Two-dimensional melting · Two-dimensional freezing · Phase transitions · Periodic potential

Introduction

The study of 2D melting by means of colloidal particles has attracted the interest of many researchers during the last few decades. The basic advantages of the use of such mesoscopic model systems are their convenient length and time scales which allow direct imaging of particle positions by means of optical methods, like video microscopy [1]. Additionally, the relevant interaction potentials in colloidal suspensions are well defined and can be adjusted over a wide range. During the last few years there has been strong experimental support of the basic ideas of 2D melting developed by Kosterlitz, Thouless, Halperin, Nelson and Young (KTHNY) [2, 3] by employing colloidal suspensions [4]. In particular, evidence for a two-stage melting process has been recently given by Zahn et al. [5].

While there exist numerous theoretical and experimental studies on 2D melting on homogeneous substrates, only little is known abut 2D melting in the presence of substrate potentials as typically provided by the atomic corrugation of a crystalline substrate [6]. Such studies, however, are highly demanding since the interplay of a 2D system with a periodic substrate will lead to strong changes in the phase behavior.

Here we demonstrate the effect of a periodic 1D potential on the phase behavior of a 2D colloidal system. The substrate potential is obtained by an array of thin paramagnetic Ni lines, which were deposited onto the substrate. As particles we used superparamagnetic colloidal spheres whose pair interaction potential can be varied over a wide range by an external magnetic field, B. Our results show that the presence of such a substrate potential drastically changes the dynamical and static properties of the colloidal system.

Experimental setup

In order to fabricate periodic 1D potentials we first deposited a Ni film of 3-nm thickness onto a glass substrate by thermal evaporation. During the evaporation process the substrate was masked with a grid, which resulted in equidistant quadratic Ni patches with a side length of 280 μm. The uncoated areas were used to compare the phase behavior of the colloids without the periodic potential. By a subsequent photolithographic process we then obtained periodically aligned Ni lines of 4 μm width and a periodicity of 10 μm. To protect the Ni grid when in contact with the aqueous colloidal suspension and to provide a smooth surface to the particles, a 400-nm-thick poly(methyl methacrylate) (PMMA) film was spin-coated on top of the Ni stripes (Fig. 1). The resulting sample was used as the bottom plate of a sample cell, which contained the aqueous suspensions of superparamagnetic particles (DynaBeads) with a diameter of 4.55 μm [7]. The particles were additionally stabilized with sodium dodecyl sulfate (SDS), which causes a short-range steric repulsion between the spheres and prevents agglomeration in particular when the external field was absent. When the sample cell was filled with colloidal suspension the particles owing to their density (1.5 g/cm^3) immediately started to sediment towards the bottom plate, where they formed a 2D system. The cell was placed in the center of a copper coil, which could induce a magnetic moment, M, in the particles, leading to a magnetic dipole pair interaction given by $V_{mag} = \mu_0 M(B)^2 (\pi \rho)^{3/2} / 4\pi$, where $M(B)$ is the induced magnetic moment of the particles, ρ the single particle density and μ_0 the magnetic permeability. We found for small B that the magnetic moment of a sphere is a linear function of the field. To describe the effective interaction between the paramagnetic spheres it is useful to introduce the dimensionless plasma parameter $\Gamma = V_{mag}/k_B T$, with $k_B T$ being the thermal energy in the system. In the following experiments variation of Γ was only achieved by changing the magnetic field, whereas the temperature was kept constant at room temperature.

To calculate the interaction between the colloidal spheres and the Ni lines we employed a finite-element method where the spheres and Ni lines were divided in small volume segments. Numerical integration of the dipole–dipole interaction over all the volume elements leads then to the potential energy of a single sphere above an array of Ni lines. The magnitude and direction of the moment of the particles is well known by the applied external magnetic field. In contrast, the magnetic moment of the Ni lines is more difficult to determine owing to the limited accuracy of the quartz mass balance method, which was employed to measure the thickness of the Ni layer. It is well known, however, that the direction (in plane or out of plane) of the magnetization of the ferromagnetic Ni lines strongly depends on the film thickness. In addition, a thin layer of antiferromagnetic nickel oxide will probably form on top of Ni immediately after the evaporation chamber is opened.

Characterization of the substrate

In order to determine how different directions of the magnetization affect the interaction between a colloidal sphere and a Ni line, we calculated the potential energy of a superparamagnetic particle above an array of Ni lines for both in-plane and out-of-plane magnetization. The resulting curves are shown in Fig. 2. When the orientation of the magnetic moment is in plane the minimum of the potential energy is close to the right or left edge of the Ni lines

(depending on the directions of the in-plane magnetization and that of the vertical external B-field vector) as seen in Fig. 2a as open and closed symbols, respectively.

Accordingly, a rotation of the direction of the external B field by 180° should also induce in a change in the equilibrium position of the colloidal particles. Indeed, such behavior has been observed experimentally for Ni grids with about 20-nm thickness (Fig. 3a, b). In contrast, if the potential energy of a superparamagnetic sphere is calculated in front of an array of out-of-plane magnetized Ni lines, the minima are located in the middle of the Ni stripes (Fig. 2b). Such behavior was observed experimentally for Ni grids of 3-nm thickness (Fig. 3c). Upon reversal of the direction of B the position of the particles was not altered. This can be explained by the small coercive force of the Ni which causes the out-of-plane magnetization to be rotated together with B. Altogether, the thickness-dependent changes in the magnetic properties of Ni films are in good agreement with findings of other authors who observed that Ni layers below 20 nm are ferromagnetic with out-of-plane magnetization, whereas above 20 nm in-plane magnetization is observed [8].

Results

In the following we present data on the phase behavior of a 2D colloidal system in the presence of a pattern of Ni

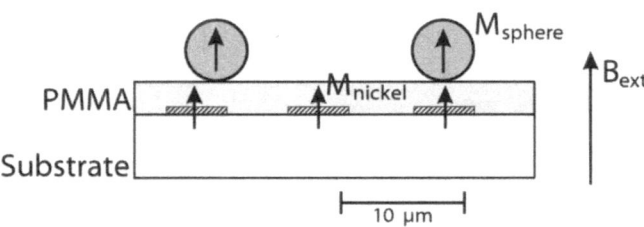

Fig. 1 Schematic side view of the experimental cell. Underneath a protective poly(methyl methacrylate) (*PMMA*) layer an array of Ni lines (*dashed regions*) has been formed on top of a silica substrate by means of photolithography. Also shown are the magnetic moments of the ferromagnetic Ni lines (M_{nickel}) and those of the particles (M_{sphere}), the latter being induced by the external magnetic field (B_{ext})

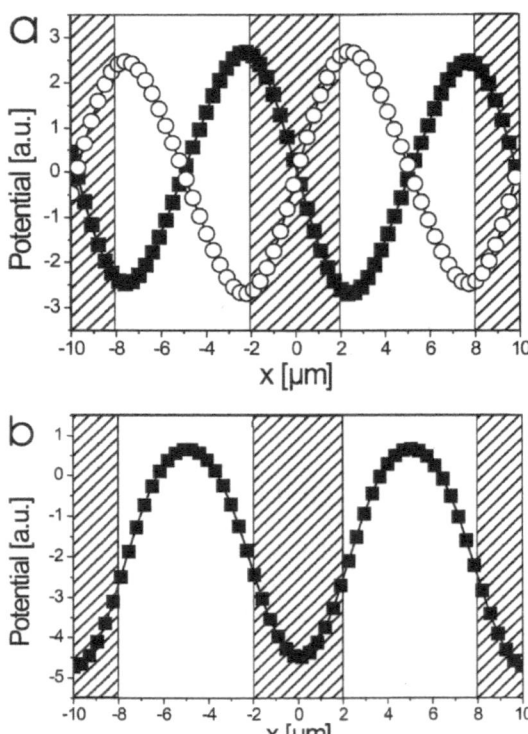

Fig. 2 a Numerical calculation of the potential energy of a superparamagnetic colloidal sphere in front of an array of in-plane-magnetized Ni stripes (*dashed areas*). After reversing the external field the induced magnetic moments of the particles are rotated by 180°, which causes the minima of the curves to be shifted in a horizontal direction. **b** The same calculation for out-of-plane magnetized Ni lines

lines as described earlier. In addition, we compare the results to that of a homogeneous substrate, which allows the influence of the Ni grid to be studied. The particles were observed by means of a homebuilt inverted microscope, which allowed the particles to be imaged onto a charge-coupled-device camera which was connected to a computer. With the help of a particle recognition algorithm we were able to identify the particle trajectories from which dynamical and static quantities were obtained.

The mean square displacements (MSD) for the x-(closed symbols) and y-directions (open symbols) directions are shown in Fig. 4 for the situation without and with Ni lines for different magnetic fields, i.e. Γ. The Ni lines were adjusted in such a way that the x- and y-directions corresponds to the orientation perpendicular and parallel to the lines. At first glance, the curves show similar behavior, i.e. a steep increase at short times which gradually decreases and becomes almost linear at longer times. The initial part of the curves is due to the short-

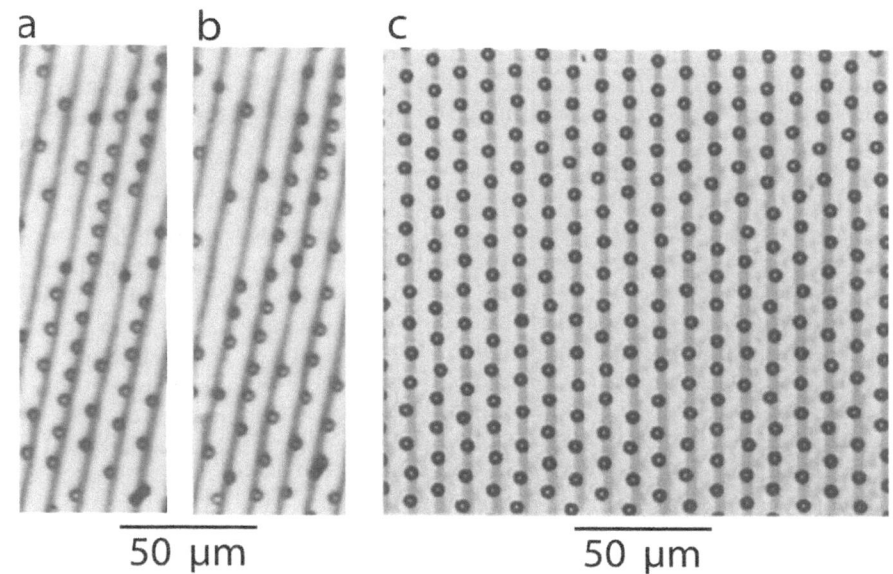

Fig. 3a, b Photos of colloidal particles on a Ni grid of 20-nm thickness, the latter corresponding to the almost vertical *black lines*. Upon reversing the external magnetic field the particles move to the opposite edge, which suggests in-plane-magnetization of the Ni. **c** Same situation for a Ni grid of 3-nm thickness. Upon reversal of the magnetic field no change in the particle positions is observed, which is in agreement with out-of-plane magnetization of the Ni at that thickness

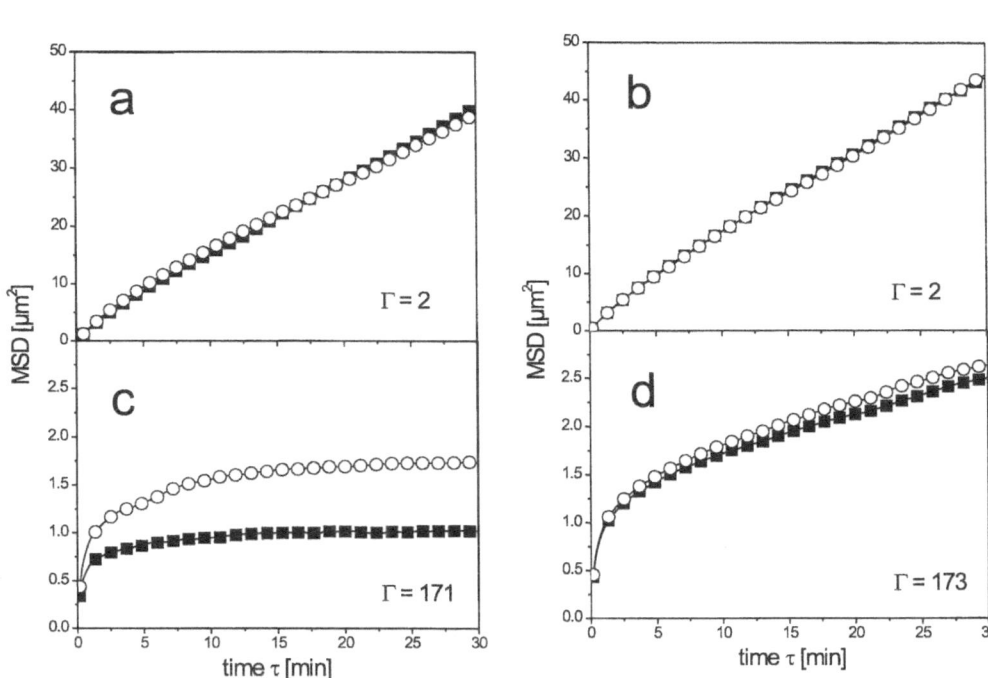

Fig. 4 Mean square displacements (*MSD*) at different plasma parameters for a system with (**a,c**) and without (**b,d**) Ni lines. The *closed symbols* denote the MSD perpendicular to the lines, the *open symbols* indicate the MSD along the lines

time diffusional behavior, which describes the motion of the particle at sufficiently small times (below 30 s) where the influence of the surrounding particles is negligible. From this we can calculate the diffusion constant $D_0 = 0.014$ $\mu m^2/s$. The MSD at longer times is determined by particle–particle interactions, which generally leads to a smaller slope compared to the short-time behavior. At $\Gamma = 2$, the Ni lines show hardly any effect on the MSD as can be seen by the similar behavior of the data in Fig. 4a and b. Apparently, the influence of the Ni grid can be neglected in this case. With increasing Γ, however, the influence of the Ni stripes on the diffusional behavior of the particles becomes significant and manifests itself both in a smaller absolute value of the MSD and also in its different behavior for the x- and y-directions. At $\Gamma = 171$ the MSD perpendicular to the lines is almost constant, indicating the confinement effect of the Ni grid to the lateral movement of the particles, whereas the MSD parallel to the lines still shows a linear increase characteristic for free diffusion (Fig. 4c). At the same magnetic field the particles on the nonpatterned parts of the substrate still perform an isotropic free diffusion as can be seen from Fig. 4d and clearly demonstrates the influence of the Ni stripes on the dynamical properties of the particles.

To study additionally the effect of a periodic substrate potential on the phase behavior of the 2D particle system we calculated the mean square excursions. In analogy to the Lindemann criterion where the MSD of the particles $\left\langle \vec{u}(\vec{R})^2 \right\rangle$ is used as a melting criterion, Bedanov and Gadiyak [9] defined a similar melting criterion for 2D systems where the relative displacements of neighboring particles are considered, i.e.

$$\gamma = \frac{\left\langle \left[\vec{u}(\vec{R}+\vec{a}) - \vec{u}(\vec{R})\right]^2 \right\rangle}{a^2} \ ,$$

with $\vec{a}$ being the lattice vector of the system. It has been shown that, independent of the pair interaction potential, in 2D systems the melting transition occurs at $\gamma = 0.033$. From the experimentally determined particle positions we calculated the melting transition for different magnetic fields. As can be seen in Fig. 5, the freezing transition in the presence of the Ni stripes occurs at substantially smaller Γ values compared to the homogeneous substrate. This can be understood by the additional localization of the particles in the presence of the Ni lines, which support their crystallization. Accordingly, we call this effect magnetic-induced freezing.

The promoting effect of a periodic, 1D substrate potential to the freezing of 2D systems is not limited to the example described here but is also found for other particle–particle and particle–substrate interactions. It has been demonstrated that the presence of an optical interference pattern, which provides a periodic, 1D potential for highly charged dielectric polystyrene spheres, can induce crystallization into a previously disordered 2D colloidal suspension [10]. In addition, recent Monte Carlo simulations suggest a similar behavior to occur also in the case of 2D hard discs [11].

Fig. 5 Relative quadratic excursion, γ, of the colloidal particles as a function of $1/\Gamma$. Above $\gamma^C_M = 0.033$ (*horizontal line*) the system is liquid, below it is solid. The melting point for the system with additional Ni lines, Γ_{C2}, occurs at significantly smaller $1/\Gamma$ values than the melting point for the undisturbed system, Γ_{C1}. The *straight lines* with the *gray areas* are guides for the eye

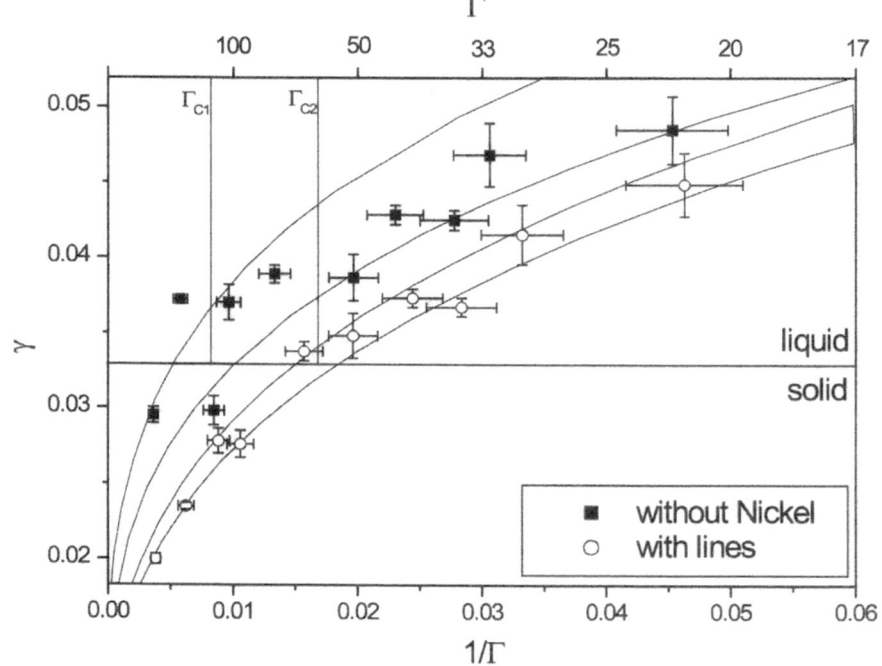

Summary

In summary, we have studied the phase behavior of 2D superparamagnetic colloidal particles in the presence of a periodic, 1D substrate potential. The latter was obtained by photolithographically manufactured Ni lines, which additionally interact with the particles. It has been demonstrated that both the dynamical and the phase properties of the colloidal system are largely affected by the presence of the Ni grid. Owing to the presence of the substrate potential the freezing transition of the particles is shifted to smaller particle–particle interactions compared to a homogeneous substrate. This is in agreement with results of other systems, for instance, light-induced freezing, where a spatial periodic light field causes a colloidal suspension to freeze, and thus demonstrates the generic effect of substrate potentials.

Acknowledgement We gratefully acknowledge financial support from the Deutsche Forschungsgemeinschaft (SFB 513).

References

1. Crocker JC, Grier DG (1996) J Colloid Interface Sci 179:298
2. Kosterlitz JM, Thouless DJ (1973) J Phys C 6:1181
3. Nelson DR, Halperin BI (1979) Phys Rev B 19:2457
4. Kusner RE, Mann JA, Kerins J, Dahm AJ (1994) Phys Rev Lett 73:3113
5. Zahn K, Lenke R, Maret G (1999) Phys Rev Lett 82:2721
6. Strandburg KJ (1988) Rev Mod Phys 60:161
7. Bubeck R, Bechinger C, Neser S, Leiderer P (1999) Phys Rev Lett 82:3364
8. Baberschke K (1996) Appl Phys A 62:417
9. Bedanov VM, Gadiyak GV, Lozovik YE (1985) Phys Lett A 109:289
10. Wei Q-H, Bechinger C, Rudhardt D, Leiderer P (1998) Phys Rev Lett 81:2606
11. Chakrabarti J, Krishnamurthy HR, Sood AK, Sengupta S (1995) Phys Rev Lett 75:2232

Progr Colloid Polym Sci (2001) 118: 82–86
© Springer-Verlag 2001

H.-J. Schöpe
T. Palberg

Crystal nucleation versus vitrification in charged colloidal suspensions

H.-J. Schöpe · T. Palberg (✉)
Institut für Physik der Universität Mainz
Staudinger Weg 7
55099 Mainz, Germany
e-mail: thomas.palberg@uni-mainz.de

Abstract We investigated the solidification behaviour of thoroughly deionised aqueous suspensions of polystyrene latex spheres by various optical scattering methods. We found a dramatic increase in the nucleation rate densities with increasing particle number density. Crystalline and nanocrystalline samples showed two relaxation processes on widely separated time scales. For an index-matched suspension of perfluorinated particles an amorphous state was accessible with the glass-typical signature of frozen long-time relaxation. From our results we propose a route into the amorphous state different to that observed in hard-sphere suspensions. It seems that in charged-sphere systems the increased nucleation rate density triggers the appearance of a Bernal-type glass.

Key words Colloids · Charge spheres · Glasses · Light scattering · Shear modulus

Introduction

Amorphous solids are important materials for technical applications and are still a challenge to fundamental research concerning their formation, structure and dynamics [1]. In general, they are obtained by suppressed crystallisation via geometrical mismatch (mixtures, directed interactions), competition to molecular degrees of freedom (polymers) or phase separation (protein solutions). For one-component materials of spherical interaction (e.g. hard spheres, HS) a purely dynamical glass transition (GT) is predicted [2, 3]. However, many real systems (e.g. metals) exhibit fast crystallisation, even when rapid cooling slows the dynamics and shortens the time available for nucleation and growth. This also applies to colloidal suspensions of charged latex particles, a model system with no substantial temperature dependence of its dynamics [4]. How then can an amorphous state be reached, where conventional strategies do not apply? Somewhat counterintuitively, such a state of short-range order, frozen large-scale motion and finite shear modulus may also be obtained via increased nucleation rates.

Owing to their specific time and length- scales and the analytically tractable HS or Yukawa interactions colloidal suspensions have become fascinating and valuable mesoscopic model systems for many condensed matter problems and for vitrification in particular [5]. Many experimental and theoretical studies were performed on both the amorphous state and the GT [2, 5–11]. In colloidal shear melts [12] an increasing packing fraction, Φ, leads to a slowing of large-scale density fluctuations. For HS, this and low crystallisation enthalpies lead to macroscopically long induction times and slow growth. The addition of a second component [13] or polydispersity [7, 14, 15] further supports vitrification. Also in Yukawa systems vitrification was reported for mixtures [16, 17] and polydisperse platelets [18]; there are only a few reports on charged-sphere glasses [19] or their GT [20, 21]. This is conceivable recalling that Yukawa spheres in general reveal considerably larger crystallisation enthalpies [4]. Solids are readily formed at $\Phi \approx 10^{-4}$, where diffusion is still fast. While a HS-like GT is predicted to occur at about $\Phi \approx 0.1–0.2$ [8], crystallisation is expected to interfere significantly.

Experimental

We studied two samples [moderately polydisperse polystyrene PS120 and less polydisperse poly(tetrafluoroethylene) PTFE180] with increasing particle number density, n. All the suspensions were prepared from diluted and precleaned stock suspensions of approximately 5.15% packing fraction. Thoroughly deionised conditions (residual ion content $c \leq 5 \times 10^{-7}$ mol l^{-1}) were achieved either using an advanced conditioning procedure (PS120) [22] or by introducing an ion-exchange resin into the carefully sealed sample cell (PTFE180).

Quasisimultaneous measurements of static and dynamic light scattering and the shear modulus, G', were performed without the need to transfer the fragile solid to another experiment or sample cell. Details of the novel two-arm light scattering goniometer with counterpropagating illumination and three independent detection schemes are given elsewhere [23, 24].

Results

Our results are shown in Figs. 1, 2, 3 and 4.

Static scattering was recorded immediately after complete solidification as checked via the appearance

Fig. 1 Scattered intensity as a function of scattering vector, q. Plots for different particle number densities n are shifted for clarity. *Left*: PS 120, hydrodynamic diameter from dynamic light scattering $d = 120$ nm, effective charge from conductivity $Z^*_\sigma = 685$, and *a* $n = 0.54$ μm^{-3}, *b* $n = 1.6$ μm^{-3}, *c* $n = 2.6$ μm^{-3}, *d* $n = 3.7$ μm^{-3}, *e* $n = 4.8$ μm^{-3}. *Right*: PTFE180 (which was a kind gift from Clariant, Germany), $d = 180$ nm, $Z^*_\sigma = 520$, and *a* $n = 0.54$ μm^{-3}, *b* $n = 1.6$ μm^{-3}, *c* $n = 26$ μm^{-3} Note that for PS120 turbidity restricts the maximum packing fractions to $\Phi < 10^{-2}$, corresponding to $n < 10$ μm^{-3}. PTFE180 can be studied up to $n \approx 40$ μm^{-3} as its index of refraction is close to that of the suspending water. For low-concentration polycrystalline samples the structure is identified from the sequence of Bragg reflections. For samples showing double-peak structures (PS120, curve *d*) and for the more turbid samples additional measurements were carried out exploiting the different symmetry of the 2D- scattering patterns [24]

of a finite shear modulus, G'. Close to the freezing transition both samples form body-centred-cubic (bcc) polycrystalline materials. PS120 shows a transition to face-centred cubic [24] which is absent for PTFE180. Long range order is lost for PTFE180 at $n = 35$ μm^{-3}. As the elasticity data (sensitive to the local neighbourhood of a particle [23]) are well described by a single theoretical curve, the observed short-range order remains of bcc structure.

Visual inspection shows a rapid decrease in the crystallite size with increasing n. The average linear dimension, L, of the crystallites (assumed to be cube shaped) was estimated from the full width at half-maximum, Δq, using $L = 2\pi K/\Delta q$, with the Scherrer constant $K = 1.155$. For both samples L decreases with increasing n, with minimum values of a few microns only. Following Aastuen et al. [25] the nucleation rate densities J, were determined using $J = 1.158$ v$\rho^{4/3}$, where the crystalline number density is $\rho = 1/L^3$ and the growth velocities are v ≈ 0.1 D_0/d_{NN} (self-diffusion coefficient D_0, nearest-neighbour distance d_{NN}) [4]. The calculated J values are much larger than for HS and keep increasing.

The PTFE180 intermediate scattering functions [26] show a monoexponential relaxation for the fluid and a short- and long-time decay separated by a plateau for the solids. With increasing n its height decreases. For both freshly prepared (shaken) and aged amorphous samples the long-time decay is nearly absent. The incomplete short-time relaxation is interpreted conventionally as "rattling in the cage" of nearest neighbours. Possibly, the long-time relaxation is due to grain boundary diffusion [27], ripening processes or crystal stress relaxation. Our observation is consistent with an increase in the ratio of the cluster surface to

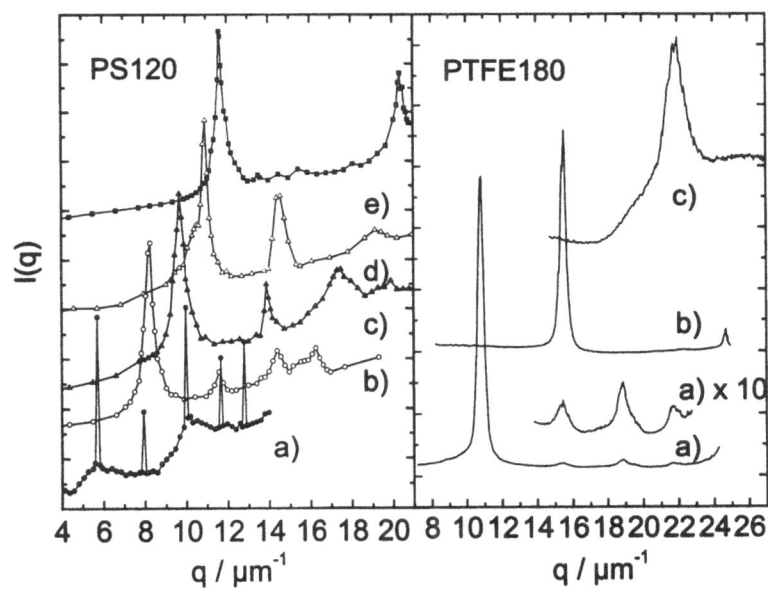

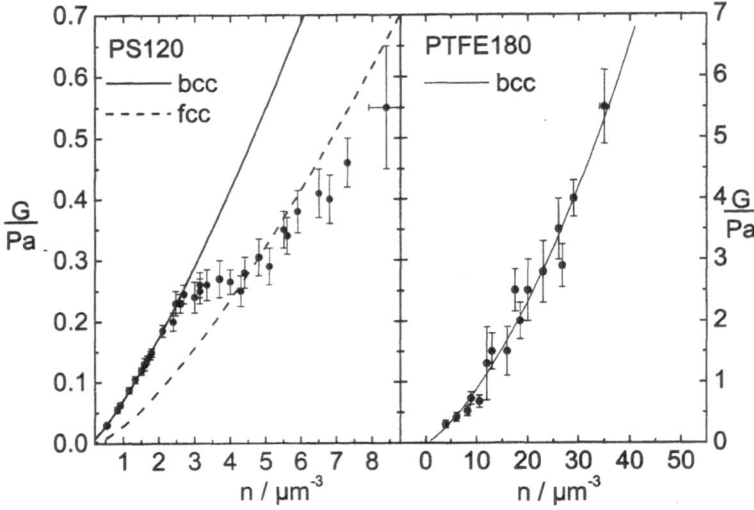

Fig. 2 Shear modulii as a function of *n*. *Left*: PS120. The *solid line* is a fit of the theoretical expression to the body-centred-cubic (*bcc*) data, yeilding $Z^*_G = 474$. The *dashed line* is a prediction for face-centred-cubic (*fcc*) using this value. Note the close agreement irrespective of grain size. A further transition to an as yet unidentified phase is observed at $n \approx 8 \ \mu m^{-3}$ (not shown). *Right*: PTFE180. Note the change of scale. The scatter of the data is somewhat larger, as the lower resonator surface is less well defined with ion-exchange resin beads present at the bottom of the cell. Above $n = 40 \ \mu m^{-3}$ centrifuging the beads leads to long-lived shear-induced structures. The *solid line* is a fit of the theoretical expression for bcc to the data, yielding $Z^*_G = 350$. Note that up to the largest *n* no transition to an fcc local order is observed

the bulk particles as *L* decreases and, simultaneous, a decrease in the interfacial diffusivity as the amorphous state is approached.

Discussion: a new route into the amorphous state

The structural picture emerging is that of a piling of small crystal-like clusters of a few hundred to thousand particles, either adjacent to each other or separated by interfacial regions of frozen long-time dynamics. Although in our case the cluster sizes are somewhat larger, this picture is strongly reminiscent of the Bernal fluid [28] derived from the impossibility of perfect tetrahedral dense packing of HS. Since with long-range repulsions less dense packings are preferred, the bcc local structures observed.

Both this proposed structure and the route taken differ significantly from the results of previous experimental studies [5–7, 13–21] and computer simultaneous [8, 10, 11]. In these, crystallisation was successfully suppressed or at least sufficiently slowed to observe freezing of long-range density fluctuations in an isotropic

Fig. 3 Average linear dimension, *L*, of a cube-shaped crystal as inferred from the peak widths plotted versus *n*. *Left*: PS120: *right*: PTFE180. Note the difference in scale. A PTFE180 bcc (PS120 fcc) grain of $L = 6 \ \mu m$ at $n = 35 \ \mu m^{-3}$ ($n = 6 \ \mu m^{-3}$) contains only $N \approx 7.5 \times 10^3$ (1.3×10^3) particles. The ratio of surface to bulk particles is about 0.5

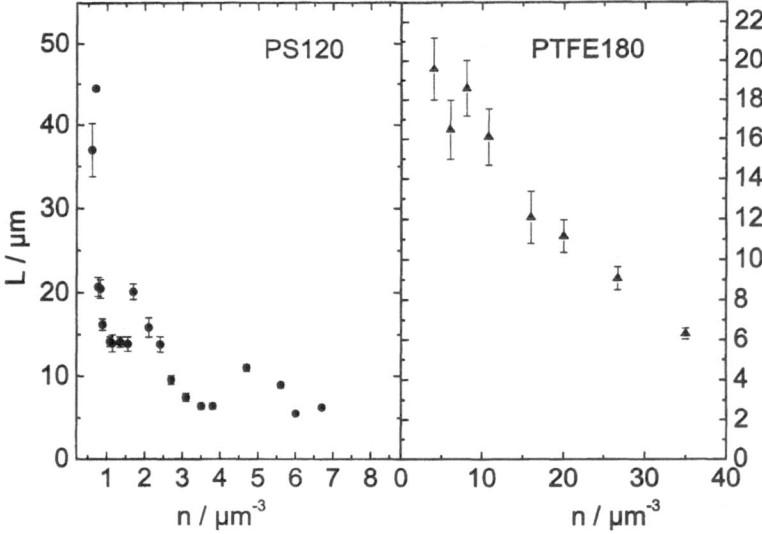

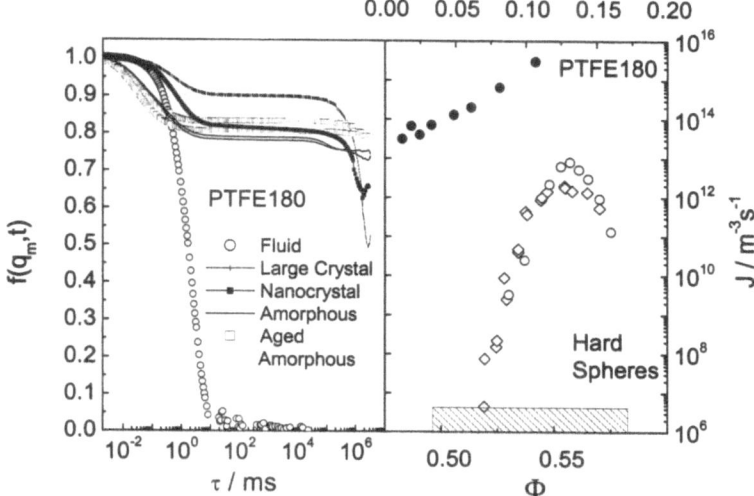

Fig. 4 *Left*: Qualitative differences in the intermediate scattering function $f(q_m, t)$ for differently concentrated PTFE180 measured at the position of the first maximum, q_m, of the static structure factor. The data were ensemble-averaged using the method of Pusey and van Megen [26]. Particle number densities: fluid: $n = 0.01$ μm^{-3}; large crystals: $n = 4$ μm^{-3}; nanocrystals $n = 10$ μm^{-3}; fresh and glass aged 1 year: $n = 35$ μm^{-3}: While multiple scattering effects strongly influence the dynamic data od PS120 (not shown) a qualitatively similar behaviour is observed. *Right*: Semilogarithmic plot of absolute nucleation rate densities, J, for PTFE180 and two hard-sphere (*HS*) samples [4, 7] versus the packing fraction, $\Phi = n(4\pi/3)a^{-3}$. Upper scale: PTFE180: *lower scale*: HS. The *shadowed area* denotes the region where crystallisation is observed in the HS case: $\Phi_F = 0.495 < \Phi < 0.57\Phi_G$. As Φ_G is approached the HS nucleation densities decrease. The J values of the charged sample are much larger than for the two HS systems and show a continuous increase. In contrast to previous studies no observation of the metastable melt is possible

and homogeneous metastable melt. In particular, in several HS cases the nucleation rate density was not only found to be much slower than for charged spheres, but moreover observed to decrease significantly as the GT is approached [4]. In contrast, here the GT predicted [2, 3] by theories of the mode-coupling class is preempted by the formation of crystallites from a still highly diffusive melt. In these, already at low n large-scale density fluctuations are frozen. All that seems left of the GT is that at large n the slow relaxation also cease. Therefore,

the amorphous state is reached on a new route via a nanocrystalline intermediate of coupled structural and dynamic heterogeneity. This is facilitated by a considerable increase in the nucleation rates.

Conclusion

We exploited advanced conditioning techniques and a newly designed multipurpose light scattering instrument to access the GT of charged colloidal spheres. For the first time the competition between crystallisation and vitrification was studied in detail as a function of the increased particle number density. We observed that apart from suppression (for HS, polydisperse systems, etc.) the enhancement of crystal nucleation may also lead to the amorphous state. We hope to have added a new, fascinating and important perspective to the discussion of solidification scenarios of soft matter model systems which also should have implications for an advanced understanding of other fast crystallising but less accessible atomic or molecular materials.

Acknowledgements We thank Clariant, Germany, for the kind gift of PTFE180 and Sonderforschungsbereich SFB 262 and the Materialwissenschaftliches Forschungszentrum, Mainz, for financial support.

References

1. Gutzow I, Schmelzer J (1995) The vitreous state. Springer, Berlin Heidelberg, New York
2. Götze W (1991) In: Hansen J P et al (eds) Liquids, freezing and glass transition. Elsevier, Amsterdam, p 287
3. Cummins H, Li Z, Hwang GYH, Shen GQ, Du WM, Hernendez J, Toa JN (1997) Z Phys B 103:501
4. Palberg T (1999) J Phys Condens Matter 11:R323
5. van Megan W (1995) Transp Theor Stat Phys 24:1017
6. Bartsch E (1995) Transp Theor Stat Phys 24:1125
7. van Megen W, Mortensen TC, Williams SR, Müller J (1998) Phys Rev E 58:6073
8. Lai SK, Ma WJ, van Megen W, Snook IK (1997) Phys Rev E 56:766
9. Bartsch E (1998) Curr Opin Colloid Interface Sci 3:577
10. Löwen H, Hansen JP, Roux JN (1991) Phys Rev A 44:1169
11. Kob W, Barrat JL (1997) Phys Rev Lett 78:4581
12. Ackerson BJ (1983) Physica A 128:221

13. Bartlett P, van Megen W (1994) In: Mehta A (ed) Granular matter Springer, Berlin Heidelberg, New York, p 195
14. Moriguchi I, Kawasaki K, Kawakatsu T (1993) J Phys II 3:1179
15. Bolhuis PG, Kofke DA (1994) Phys Rev E 54:634
16. Meller A, Stavans J (1992) Phys Rev Lett 68:3646
17. Kesavamoorthy R, Sood AK, Tata BVR, Arora AK (1998) J Phys C 21:4737
18. Bonn D, Tanaka H, Wegdam G, Kellay H, Meunier J (1998) Europhys Lett. 45:52
19. Sirota EB, Ou-Yang HD, Sinha SK, Chaikin PM, Axe J, Fujii DY (1989) Phys Rev Lett 62:1524
20. Härtl W, Versmold H, Zhang-Heider X (1995) J Chem Phys 102:6613
21. Beck C, Härtl W, Hempelmann R (1999) J Chem Phys 111:8209
22. Evers M, Garbow N, Hessinger D, Palberg T (1998) Phys Rev E 57:6774
23. Schöpe H-J, Palberg T (2001) J Colloid Interface Sci 234:149–161
24. Schöpe H-J Wette P, Palberg T (1998) J Chem Phys 109:10068 (1998)
25. Aastuen DJW, Clark NA, Swindal JC, Muzny CD (1990) Phase Transitions 21:139
26. Pusey PN, van Megen W (1989) Physica A 157:705
27. Simon R, Palberg T, Leiderer P (1993) J Chem Phys 99:3030
28. (a) Bernal D (1960) Nature 188:908; (b) Bernel D (1960) Nature 185:68

Progr Colloid Polym Sci (2001) 118: 87–90
© Springer-Verlag 2001

G. Odriozola
A. Moncho-Jordá
A. Schmitt
J. Callejas-Fernández
R. Martínez-García
R. Hidalgo-Álvarez

The kinetics of irreversible aggregation processes

G. Odriozola · A. Moncho-Jordá
A. Schmitt (✉) · J. Callejas-Fernández
R. Martínez-García · R. Hidalgo-Álvarez
Grupo de Fluidos y Biocoloides
Departamento de Física Aplicada
Facultad de Ciencias, Universidad de
Granada, Campus de Fuentenueva
18701 Granada, Spain
e-mail: schmitt@ugr.es
Tel.: +34-958-246104
Fax: +34-958-243214

Permanent address: G. Odriozola
Departamento de Química Física y
Matemática, Facultad de Química
Universidad de la República
11800 Montevideo Uruguay
e-mail: odriozol@bilbo.edu.uy

Abstract A sticking probability
model for irreversible aggregation
processes is developed. It allows a
kernel capable of describing not only
the diffusion-limited and reaction-
limited aggregation regimes but also
the whole transition region to be
deduced. According to the definition
given by van Dongen and Ernst, the
kernel establishes $\lambda = 0$ for the entire
range of sticking probabilities. The
model presented gives further insight
into the detailed aggregation
mechanism for slow aggregation
processes.

Key words Aggregation kernel ·
Diffusion-limited cluster
aggregation · Reaction-limited
cluster aggregation · Statistical
physics

The stage of aggregation for a given system may be characterized by the cluster-size distribution, $\vec{N} = (N_1, N_2, N_3 \ldots, N_i, \ldots)$, where N_i denotes the number of i-size clusters. For irreversible aggregation processes in dilute systems, the time evolution of the probability, $P(\vec{N}, t)$, for finding the system in a given state, $\vec{N}$, is given by the nondeterministic mean-field master equation [1, 2]

$$\frac{\mathrm{d}P(\vec{N}, t)}{\mathrm{d}t} = \frac{1}{2V} \sum_{i,j} k_{ij}[(N_i + 1)(N_j + 1 + \delta_{ij})P(\vec{N}_{ij}^*, t)$$
$$- N_i(N_j - \delta_{ij})P(\vec{N}, t)] \ , \tag{1}$$

where V is the volume of the system and N_{ij}^* is defined as

$$\vec{N}_{ij}^* = \begin{cases} (\ldots, N_i + 1, \ldots, N_j + 1, \ldots, N_{i+j} - 1, \ldots) & \text{for } i \neq j \\ (\ldots, N_i + 2, \ldots, N_{2i} - 1, \ldots) & \text{for } i = j \end{cases}.$$

All physical information about the aggregation mechanism is contained in the kernel, k_{ij}, which quantifies the mean rate at which two clusters of size i and j form a $(i + j)$-size cluster.

Initial studies were concerned principally with the diffusion-limited cluster aggregation (DLCA) regime, where the clusters diffuse and form a new bond as soon as they collide. The DLCA kinetics is well described by the Brownian Kernel [2–4]. For this regime, DLCA experiments and computer simulations yield a fractal dimension close to 1.75 and $\lambda = 0$ [3, 5–8]. Here, λ is the homogeneity exponent defined by the relationship $k_{ai,aj} \sim a^\lambda k_{ij}$ for large cluster sizes, where a is a positive constant [9]. Experimental and computer simulated data agree perfectly with the Brownian kernel solutions of the master equation. The analytical expression for this kernel is $k_{ij}^{\mathrm{Br}} = \frac{k_{11}^{\mathrm{Sm}}}{4} (i^{1/d_f} + j^{1/d_f})(i^{-1/d_f} + j^{-1/d_f})$, where $k_{11}^{\mathrm{Sm}} = 8kT/3\eta$ is the Smoluchowski dimer formation rate constant. Here, kT is the thermal energy, η is the solvent viscosity and d_f is the cluster fractal dimension. According to its definition, the Brownian kernel has $\lambda = 0$. It may be written as $k_{ij}^{\mathrm{Br}} \sim (t_{\mathrm{dif}})^{-1}$, where t_{dif} represents the average diffusion time spent by two i- and j-size clusters before they collide.

Reaction-limited cluster aggregation (RLCA) occurs when a large number of cluster–cluster collisions is needed before a bond is formed. From experiments and computer simulations, a fractal dimension of 2.1 is commonly obtained. Nevertheless, a wide range of values for λ is reported in the literature. This values usually lie between 0.5 and 1 [2, 3, 10, 11]. But, is this in agreement with the RLCA nature? According to its definition, λ determines how the rate constants change with the cluster size. The value obtained for the Brownian kernel, $\lambda = 0$, means that $k_{ii} = k_{jj}$ for $i \gg j$ when asymptotic conditions are established. A positive λ implies that the cluster reactivity increases with cluster size, i.e. $k_{ii} > k_{jj}$ for $i \gg j$. Hence, for a sufficiently large cluster size, kernels with positive λ will have larger rate constants than the Brownian kernel. This is, of course, nonphysical since an aggregation regime, which involves not only the diffusion of aggregates but also an additional difficulty for forming new bonds, cannot become faster than that limited only by diffusion [10, 12]. It is no casuality that the RLCA regime is also called the "slow aggregation regime". In conclusion, not only the RLCA regime but also whole transition region from DLCA to RLCA must be characterized by a kernel having $\lambda \leq 0$. Here, we propose a novel aggregation kernel which fulfills these theoretical constraints.

For this purpose, it is convenient to distinguish between cluster–cluster collision and cluster–cluster encounter. In this work, we call "collision" the situation when two clusters touch each other and define "encounter" as a sequence of consecutive collisions between a given pair of clusters. This means that an encounter starts with the first collision and ends when the clusters aggregate or diffuse away. So, a given cluster spends an average time t_{dif} between two consecutive encounters and an average time t_{c} between two consecutive collisions.

The proposed aggregation model is based on the expression $k_{ij} \sim \langle t \rangle^{-1}$, where $\langle t \rangle$ is the average time needed for aggregation by a pair i- and j-size clusters. The clusters are considered to move by Brownian motion and to collide after the average diffusion time t_{dif}. Generally, not every collision leads to aggregation and so a sticking probability, P, must be defined. Consequently, $(1-P)$ is the probability for a collision which does not lead to the formation of a new bond. If the clusters aggregate immediately, they took an average time t_{dif} for the whole process. If they do not, they may collide again or diffuse away in order to collide with a third cluster. Hence, it is convenient to define P_{c} as the probability for the clusters

to collide again. Since the collision cross-section grows with cluster size, one expects that the bigger the clusters are, the bigger P_{c} becomes. The probability for the clusters to diffuse away, i.e. to quit an encounter without forming a new bond, is given by $(1-P_{\text{c}})$. As an example of a more complex aggregation process, the following event may be considered. Two clusters diffuse and collide twice. Then one of them diffuses away and collides three times with a third cluster before forming a new bond. A schematic time and probability diagram of this particular process is shown in Fig. 1. The average time for this event is $t_{\text{ev}} = 2t_{\text{dif}} + 3t_{\text{c}}$ since it consists of two encounters with two and three consecutive collisions, respectively. The probability for this event is given by $P_{\text{ev}} = P(1-P)^4 (1-P_{\text{c}})P_{\text{c}}^3$. In order to determine the average aggregation time, $\langle t \rangle$, it is necessary to consider all possible events weighted by the corresponding probability; hence, $\langle t \rangle$ becomes

$$\langle t \rangle = \sum_{\text{ev}} t_{\text{ev}} P_{\text{ev}} \ . \tag{2}$$

This equation is consistent only if $1 = \sum_{\text{ev}} P_{\text{ev}}$ is verified.

Now, it is necessary to find a general expression for P_{ev} and t_{ev}. For that purpose we construct Table 1, which contains all possible events with fewer than five collisions and the corresponding probabilities. Here, an encounter formed by n collisions is denoted as e(n) and a given event is symbolized by a series of encounters e(n)e(m)e(l)... Hence, the event e(2)e(3) corresponds to the example given in Fig. 1. It should be noticed that some events have the same average time and probability. We call these kinds of events "equivalent events". So, Eq. (2) may be expressed as

$$\langle t \rangle = \sum_{\text{ev}'} q_{\text{ev}'} t_{\text{ev}'} P_{\text{ev}'} \ , \tag{3}$$

where q is the number of equivalent events. Equation 3 is consistent only when $1 = \sum_{\text{ev}'} q_{\text{ev}'} P_{\text{ev}'}$. Using Table 1, the following general expressions may be deduced

$$q_{\text{ev}'}(k, l) = \binom{k-1}{l} \ , \tag{4}$$

$$t_{\text{ev}'}(k, l) = (k-l)t_{\text{dif}} + lt_{\text{c}} \ , \tag{5}$$

$$P_{\text{ev}'}(k, l) = P(1-P)^{k-1} P_{\text{c}}^l (1-P_{\text{c}})^{k-l-1} \ , \tag{6}$$

which lead to

Fig. 1 Schematic time and probability diagram for a particular aggregation event

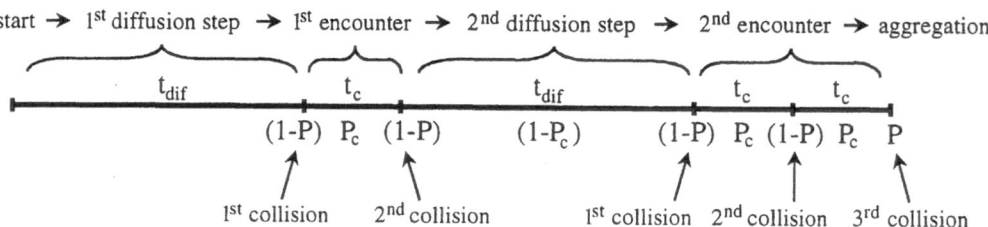

Table 1 Event times and probabilities obtained for events composed by less than five collisions. Notation e(m)e(n) refers to an event composed by two encounters of m and n collisions respectively. q is the number of equivalent events, k is the collisions number and l is the number of collision times, t_c

Event	Time	Probability	q	k	l
e(1)	t_{dif}	P	1	1	0
e(1)e(1)	$2t_{dif}$	$P(1-P)(1-P_c)$	1	2	0
e(2)	$t_{dif}+t_c$	$P(1-P)P_c$	1	2	1
e(1)e(1)e(1)	$3t_{dif}$	$P(1-P)^2(1-P_c)^2$	1	3	0
e(1)e(2), e(2)e(1)	$2t_{dif}+t_c$	$P(1-P)^2P_c(1-P_c)$	2	3	1
e(3)	$t_{dif}+2t_c$	$P(1-P)^2P_c^2$	1	3	2
e(1)e(1)e(1)e(1)	$4t_{dif}$	$P(1-P)^3(1-P_c)^3$	1	4	0
e(1)e(1)e(2), e(1) e(2) e(1), e(2) e(1) e(1)	$3t_{dif}+t_c$	$P(1-P)^3P_c(1-P_c)^2$	3	4	1
e(2) e(2), e(1) e(3), e(3) e(1)	$2t_{dif}+2t_c$	$P(1-P)^3P_c^2(1-P_c)$	3	4	2
e(4)	$t_{dif}+3t_c$	$P(1-P)^3P_c^3$	1	4	3
⋮	⋮	⋮	⋮	⋮	⋮

$$1 = \sum_{ev'} q_{ev'}P_{ev'}$$

$$= \sum_{k=1}^{\infty}\sum_{l=0}^{k-1}\binom{k-1}{l}P(1-P)^{k-1}P_c^l(1-P_c)^{k-l-1} . \quad (7)$$

Here, the limits of the sums are established so that all possible events are taken into account. Considering that $\sum_{x=0}^{y}\binom{y}{x}a^xb^{y-x} = (a+b)^y$, Eq. (7) reduces to

$$1 = \sum_{k=1}^{\infty}P(1-P)^{k-1}[P_c + (1-P_c)]^{k-1}$$

$$= \sum_{k=1}^{\infty}P(1-P)^{k-1} = P\frac{1}{1-(1-P)} = 1 \quad (8)$$

and so is correct.

The average aggregation time may be evaluated by combining Eqs. (3), (4), (5) and (6) which yields

$$\langle t \rangle = \sum_{k=1}^{\infty}\sum_{l=0}^{k-1}\binom{k-1}{l}[(k-l)t_{dif} + lt_c]$$
$$P(1-P)^{k-1}P_c^l(1-P_c)^{k-l-1} . \quad (9)$$

This expression may be transformed into

$$\langle t \rangle = \sum_{k=0}^{\infty}(k+1)t_{dif}P(1-P)^k\sum_{l=0}^{k}\binom{k}{l}P_c^l(1-P_c)^{k-l}$$
$$+ \sum_{k=0}^{\infty}(t_c-t_{dif})P(1-P)^k\sum_{l=0}^{k}l\binom{k}{l}P_c^l(1-P_c)^{k-l} . \quad (10)$$

Taking into account that $\sum_{x=0}^{y}\binom{y}{x}a^xb^{y-x} = (a+b)^y$ and $\sum_{x=0}^{y}x\binom{y}{x}a^xb^{y-x} = ya(a+b)^{y-1}$, one obtains

$$\langle t \rangle = P\left[t_{dif}\sum_{k=0}^{\infty}(k+1)(1-P)^k + (t_c-t_{dif})\sum_{k=0}^{\infty}k(1-P)^kP_c\right] , \quad (11)$$

which finally leads to

$$\langle t \rangle = P\left[\frac{t_{dif}}{P^2} + (t_c-t_{dif})\frac{(1-P)}{P^2}P_c\right] . \quad (12)$$

This equation expresses the average aggregation time as a function of the quantities P, P_c, t_{dif} and t_c. Its reciprocal value is the kernel for irreversible aggregation. Note that t_{dif}, t_c and P_c are, in general, functions of the cluster size. As expected, Eq. (12) leads to t_{dif} for a sticking probability of $P = 1$, i.e. the kernel reduces to the Brownian kernel. When $P \to 0$, the average aggregation time tends to infinity unless $t_c = 0$ and $P_c = 1$, which represents a pathological case. This means that reaction-controlled aggregation processes, characterized by an extremely small sticking probability, would develop on an extremely long time scale, which is absolutely intuitive. Furthermore, Eq. (12) gives average aggregation times which are longer than t_{dif} for all sticking probabilities smaller than unity. This means that the non-DLCA rate constants are always smaller than the DLCA ones. Consequently, Eq. (12) establishes $\lambda \leq 0$ independently of the aggregation regime.

Unfortunately, Eq. (12) is based on quantities which are not directly accessible. In particular, this refers to t_c and P_c and so it becomes necessary to find expressions for both. For this purpose, it is convenient to review the constrains used to deduce the master equation (Eq. 1). This equation, which defines the kernel, is based on a mean-field approximation and is valid only for dilute systems. In such systems, the average diffusion time, t_{dif}, is expected to be much longer than the average time, t_c, spent between consecutive collisions; hence, we simply neglect t_c for further calculations. On the other hand, the probability P_c is related to the mean number of collisions per encounter for a nonaggregating system. This quantity is given by $\mathcal{N} = \sum_{i=1}^{\infty}iP_c^{(i-1)}(1-P_c) = 1/(1-P_c)$. As mentioned earlier, it is reasonable to allow $\mathcal{N}$ to be size-dependent. Since the dependency must be a symmetric function of the cluster sizes i and j, we assume $\mathcal{N}_{ij} = \mathcal{N}_{11}(ij)^b$. Here, $\mathcal{N}_{11}$ is the mean number of monomer collisions per encounter and b is a constant. Substituting P_c in Eq. (12) finally leads to

$$k_{ij} = \frac{k_{11}^{Sm}(i^{1/d_f} + j^{1/d_f})(i^{-1/d_f} + j^{-1/d_f})P\mathcal{N}_{11}(ij)^b}{4(1 + P[\mathcal{N}_{11}(ij)^b - 1])} . \quad (13)$$

For a large cluster sizes, this kernel approaches the Brownian kernel and, hence, has $\lambda = 0$ for all sticking

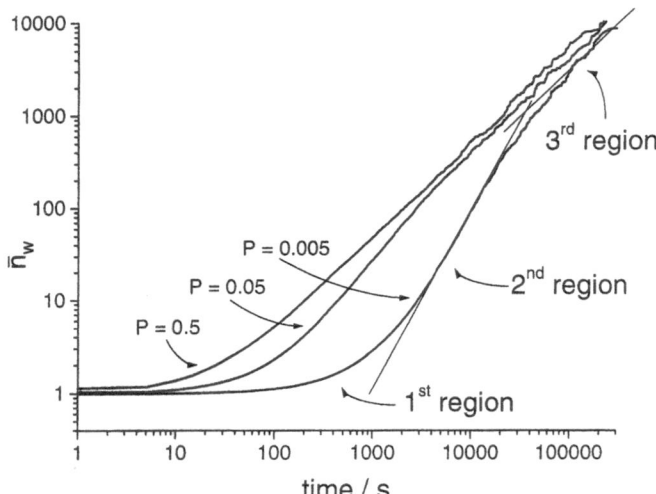

Fig. 2 Time evolution of the weight-average cluster size, $\bar{n}_{\mathrm{w}}$, calculated for the sticking probabilities 0.5, 0.05 and 0.005

probabilities, P. It should be pointed out that for very small P and not too large clusters, i.e. $P[\mathcal{N}_{11}(ij)^b] \ll 1$, an apparent $\lambda_{\mathrm{app}} = 2b$ may be obtained. Here, "apparent" should be understood as not truly asymptotic. This nonasymptotic region extends to longer times the smaller P becomes. This makes it clear that the positive λ values reported for RLCA experiments and computer simulations correspond to λ_{app}.

In order to test the proposed aggregation kernel by Eq. (13), we study the corresponding solutions of the master equation. For this purpose, the stochastic algorithm described in Refs [1, 2] was used to solve Eq. (1) for monomeric initial conditions and a volume fraction of 5×10^{-4}. In order to achieve reliable statistics, the initial number of particles was set to $N_0 = 10^5$. The aggregation kernel is a function of k_{11}^{Sm}, P, d_{f}, $\mathcal{N}_{11}$ and b. Here, k_{11}^{Sm} was calculated from the theoretical expression $k_{11}^{\mathrm{Sm}} = 8kT/3\eta$. This yields $k_{11}^{\mathrm{Sm}} = 11.1 \times 10^{-18}\,\mathrm{m}^3\,\mathrm{s}^{-1}$ for aqueous systems at 293 K. P was set to 0.5, 0.05 and 0.005 and d_{f} to 1.75, 1.9 and 2.1. As we commented earlier, RLCA experiments and simulations have $0.5 < \lambda_{\mathrm{app}} = 2b < 1$ and so $b = 0.35$ was chosen as a

reasonable value. Finally, $\mathcal{N}_{11} = 6.1$ was obtained directly from simulations of non sticking particles.

Time evolutions of the weight–average cluster size, $\sum_{i=1}^{\infty} i^2 N_i / \sum_{i=1}^{\infty} i N_i$, on a logarithmic for sticking probabilities 0.5, 0.05 and 0.005 on algorithmic scale are shown in Fig. 2. For all the curves, three well-defined regions may be distinguished. For short times, $\bar{n}_{\mathrm{w}}$ starts from unity and, as expected, grows faster for larger sticking probabilities. At intermediate times, an apparent asymptotic behaviour appears and remains up to relatively long aggregation times. Here, $\bar{n}_{\mathrm{w}}$ reaches quite large values. For truly asymptotic conditions, λ controls the time evolution of the $\bar{n}_{\mathrm{w}}$ according to the relationship $\bar{n}_{\mathrm{w}} \sim t^{\frac{1}{1-\lambda}}$. Analogously, we may define an apparent λ_{app} for the intermediate region by using the same relationship, i.e. $\bar{n}_{\mathrm{w}} \sim t^{\frac{1}{1-\lambda_{\mathrm{app}}}}$. It can be clearly seen that the slope of the curves, $1/(1 - \lambda_{\mathrm{app}})$, increase for decreasing sticking probabilities. In the limiting case $P \to 0$, a slope of $1/(1 - 2b)$ is expected. This means that λ_{app} grows for decreasing P up to the limiting value of $2b$. The third region corresponds to very long times and extremely large cluster sizes. Here, all curves approach the same real asymptotic behaviour where λ becomes 0. As can be seen, the curves for different P never cross the limiting DLCA curve. So, the proposed kernel predicts an asymptotic long-time behaviour consistent with theoretical considerations. This region is, however, very difficult to achieve in experiments and simulations owing to the extremely large size of the corresponding clusters.

In conclusion, the proposed probabilistic model helps to understand how the RLCA regime works and gives rise to a physically deduced kernel which describes the time evolution of the cluster size distributions for all times, cluster sizes and sticking probabilities. Furthermore, it gives a plausible explanation for the apparent DLCA–RLCA λ contradiction mentioned at the beginning of this work and establishes $\lambda = 0$ for the real asymptotic behaviour independently of the aggregation regime.

Acknowledgements This work was supported by the Comisión Interministerial de Ciencia y Tecnología (project MAT2000-1550-C03-01). G.O. is grateful for a scholarship granted by the European Union (program: αlfa, proposal nº. ALR/B7-3011/94.04–6.017.9).

References

1. Gillespie DT (1977) J Phys Chem 81:2340
2. Thorn M, Seesselberg M (1994) Phys Rev Lett 72:3622
3. Broide ML, Cohen RJ (1990) Phys Rev Lett 64:2026
4. Meakin P (1992) Croat Chem Acta 65:237
5. Broide ML, Cohen RJ (1992) J Colloid Interface Sci 153:493
6. Lin MY, Lindsay HM, Weitz DA, Klein R, Ball RC, Meakin P (1990) J Phys Condens Matter 2:3093
7. Meakin P (1983) Phys Rev Lett 51:1119
8. Odriozola G, Schmitt A, Callejas-Fernández J, Martínez-García R, Hidalgo-Álvarez R (1999) J Chem Phys 111:7657
9. van Dongen PGJ, Ernst MH (1985) Phys Rev Lett 54:1396
10. Lin MY, Lindsay HM, Weitz DA, Ball RC, Klein R, Meakin P (1990) Phys Rev A 41:2005
11. Meakin P, Family F (1987) Phys Rev A 36:5498
12. Ball RC, Weitz DA, Written TA, Leyvraz F (1987) Phys Rev Lett 58:274

Progr Colloid Polym Sci (2001) 118:91–95
© Springer-Verlag 2001

Jianing Liu
Andreas Stipp
Thomas Palberg

Crystal growth kinetics in deionised two-component colloidal suspensions

J. Liu · A. Stipp · T. Palberg (✉)
Institut für Physik der Universität
Mainz Staudinger Weg 7
55099 Mainz, Germany
e-mail: thomas.palberg@uni-mainz.de

Abstract We report first measurements of the crystal growth velocity in deionised two-component aqueous suspensions of charged polystyrene latex spheres. The size ratio was 1:1.3 and mixing ratios up to 18% of the larger particle were investigated. For the pure components limiting growth velocities of ($v_{\infty,120} = 4.8$ μm s^{-1} and $v_{\infty,156} = 2.9$ μm s^{-1}) were observed. In the mixture v drops with increasing mixing ratio even below ($v_{\infty,156}$) of the large minority component. Careful monitoring of the deionisation procedure excludes the explanation of enhanced ionic contamination. Alternatives based on the kinetics of particle attachment and the position of the phase boundary are discussed.

Key words Effective mobility ·
Bragg microscopy · Crystal growth
velocity · Diffusion coefficient

Introduction

The colloid specific length and time scales are much longer than for thermal fluctuations in molecular substances. Owing to this, particles of spherical interaction potential provide a useful model to test mean-field approaches to solidification kinetics. For single-component systems of various repulsive interactions in particular, the validity of the Becker–Döring theory of nucleation [1] and the Wilson–Frenkel (WF) theory of growth [2] have already been tested successfully. For mixed systems phase diagrams show a rich variety, ranging from alloy formation to precipitation glass formation or aggregation [3, 4]. Studies of crystal growth kinetics in such systems are still in demand.

We present here the first measurement on a mixture of size ratio 1:1.3 under conditions of fixed particle concentration. We replaced small by big particles stepwise up to minority fractions of $X_{156} \approx 0.18$. The experiments were conducted under deionised conditions and at a particle concentration where all the samples, including the pure suspensions, were completely solidified at equilibrium. The samples are shear-molten and after the shear ended readily solidified via heterogeneous nucleation at the cell wall with subsequent quasiepitaxial growth. The formerly applied shear orients the nuclei and oriented although twinned, crystals result. Their (1 1 0) plane is parallel to the cell wall with the ⟨1 1 1⟩ direction parallel to the formerly applied flow direction. Growth proceeds inward in the ⟨1 1 0⟩ direction. For the one-component system the velocity was found to obey a WF law:

$$v = v_{\infty}[1 - \exp(-\Delta\mu/k_{\mathrm{B}}T)] \ , \tag{1}$$

where $\Delta\mu$ is the difference in the chemical potential between the melt and the solid and $k_{\mathrm{B}}T$ is the thermal energy. The limiting growth velocity, v_{∞}, was found to be of the order of 2–20 μms^{-1} [5].

The crystals were observed by Bragg microscopy. Details of the version of the technique used here have been given recently [6]. This and other versions of the technique have been employed before to study growth in single-component suspensions, as well as nucleation and ripening or nonequilibrium phase distributions under shear flow [6–11]. It was used here for the first time to study solidification of mixed systems. Examples from the samples discussed here are shown in Fig. 1b–d. Note that in cells of rectangular cross-section four crystals result as sketched in Fig. 1a.

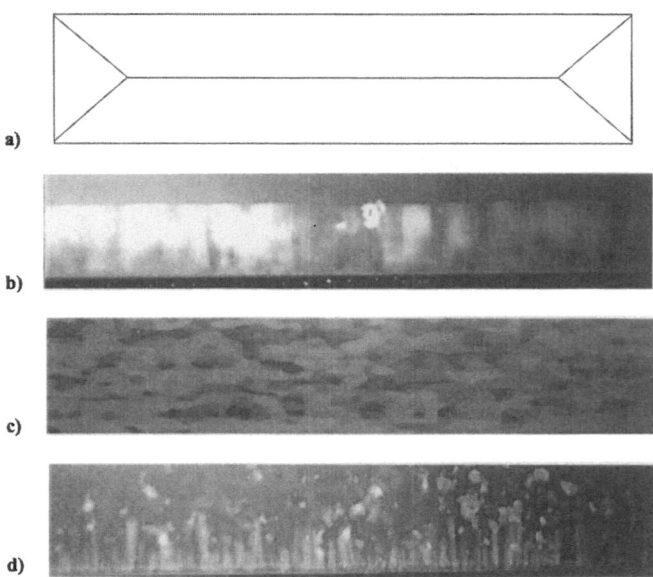

Fig. 1 a Cross section through the sample cell. Note that four crystallites are growing inwards. The cell is observed from the top, thus shifting the cell either yields a side view of crystals growing horizontally or a top view of crystals growing vertically. **b, c** Examples of wall-nucleated crystallites as observed by Bragg microscopy. **b** Side view: high electrolyte content; planar growth front. **c** Top view: high electrolyte content; cloudlike twin pattern. **d** Side view: thoroughly deionised; toothlike growth front

In what follows we first give an outline of the sample conditioning and optical techniques, then present the results and finally discuss the observed dependencies of the measured growth velocities on the deionisation time and the mixing ratio.

Sample conditioning

Two commercially available species of charged polystyrene spheres in aqueous suspension were investigated (henceforth termed PS120 and PS156). Before preparing mixtures of these, they were carefully characterised by various experiments. The most important results are compiled in Table 1.

Two boundary conditions were to be met in that choice. First n should be large enough to be very close to the limiting growth velocity for the pure systems. On the

other hand, it should be lower than the body-centred-cubic (bcc)/face-centred-cubic (fcc) transition for PS120 in order to keep the experiment conceptionally simple. For $n = 0.47 \ \mu m^{-3}$ both pure systems are of bcc structure and v_{120} is only slightly lower than $v_{\infty,120}$. Experiments were then conducted at this fixed n, i.e. PS120 particles were replaced stepwise by PS156 without altering the lattice constant. They were further intended to be conducted under thoroughly deionised conditions.

For deionising the samples we used an advanced, continuous procedure. Details have recently been given elsewhere [13]. In short, the suspension is peristaltically cycled through a closed tubing system connecting various components. It includes an ion-exchange cell filled with a mixed-bed ion-exchange resin (Amberlite, Rohm & Haas, France). This can be bypassed, if desired. A reservoir under an inert gas atmosphere allows the removal or the addition of the suspension to adjust the particle concentration and the sample composition. Further, several measuring cells may be incorporated: a cell for microscopy, a cell for static light scattering, etc.

To control the deionisation process we monitored the conductivity (Bridge WTW535, electrode LTA01, WTW, Weilheim, Germany). The conductivity shows a sharp drop and a constant low value is reached after some 30–60 min. In studies reported previously and in this volume deionisation was usually continued for at least 4 times that period. Sometimes, and in particular at low n, even a shallow minimum is visible [13–15]. Achievement of minimum conductivity, however, is not necessarily identical to reaching the point of complete deionisation. To make this point explicit we show measured growth velocities as a function of deionisation time in Fig. 2.

The initially observed growth velocity drops with increasing X_{156}. For $X_{156} = 0.18$ no growth was observed immediately after reaching minimum conductivity. In general, however, v increases with continued deionisation time and saturates at a constant value after some hours. For the measurements presented later this value was taken as the growth velocity under thoroughly deionised conditions. We note that once this state is reached, further contamination proceeds mainly via CO_2 leaking in through fittings, etc. Interestingly that impurity can be removed on a much faster time scale. Even more importantly, the times to reach minimum conductivities and maximum growth velocities coincide.

Table 1 Pure component properties. $2a_{nom}$: nominal diameter; $2a_h$: hydrodynamic diameter as measured by dynamic light scattering; Z_σ^*: effectively transported charge from conductivity; Z_G^*: effective charge from shear modulus measurements [12]; v_∞: limiting growth velocity at infinite "undercooling"; structure, shear modulus, G, and growth velocity v under the conditions used for the mixture

Sample	$2a_{nom}$/nm	$2a_h$/nm	Z_σ^*	Z_G^*	$v_\infty/\mu m/s$	Structure ($n = 0.47 \ \mu m^{-3}$)	G at $n = 0.47 \ \mu m^{-3}$/Pa	$v/\mu m/s$ ($n = 0.47 \ \mu m^{-3}$)
PS120	120	128	718	474	4.8	Body-centred cubic	0.026	4.1
PS156	156	~	841	615	2.9	Body-centred cubic	0.049	2.9

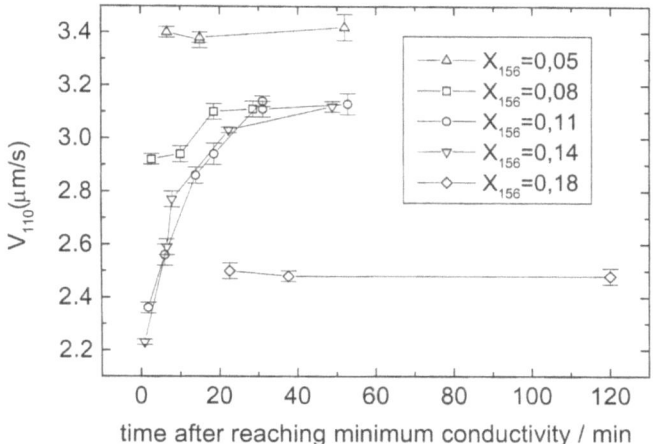

Fig. 2 Growth velocities in different mixtures as a function of deionisation time

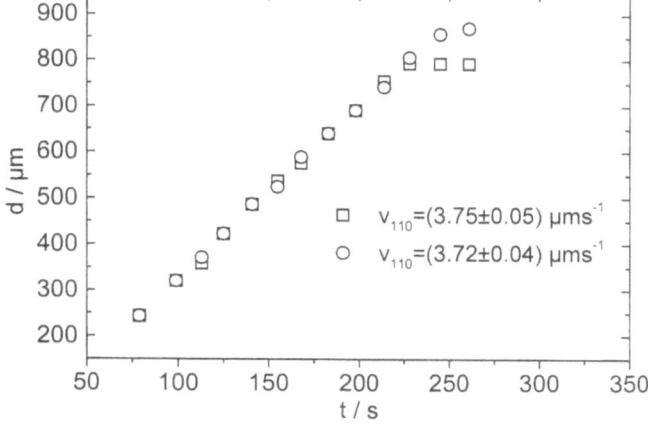

Fig. 3 Extension of different wall crystallites as a function of time elapsed after cessation of shear. While the growth velocities coincide within experimental error, the growth duration is different

Growth measurements

Measurements of the growth velocity and the characterisation of the sample morphology were performed using Bragg microscopy. The optical cell of rectangular cross-section (2×10 mm^2) was mounted on the stage of an optical microscope (Laborlux 12, Leitz, Wetzlar, Germany) equipped with a video camera. The cell was illuminated from below using a cold white light source. The latter was adjusted under angles Θ and φ to induce Bragg reflections of the crystals in the direction of the microscope objective, i.e. normal to the cell wall. In principle, this optical path represents an inversion of the one used in static light scattering [5].

As shown in Fig. 1b–d, properly oriented crystals are visible as bright, colourful regions. The images in Fig. 1a and b were taken just after reaching minimum conductivity. The crystal grows as a compact entity with a rather flat upper surface. The image was taken close to a side wall of the cell. Observation of the central part of the cell reveals a cloudlike pattern. It corresponds to a top view of the wall crystal. In this case the illumination was adjusted such that the Bragg condition was fulfilled for only one of the possible (1 1 0) bcc twin orientations [6]. We thus observe the twin domain pattern of the wall crystal. The lower image was taken as a side view, now after long deionisation. The morphology appears to be column- or toothlike with different extensions in the z-direction. The extension was measured as a function of time and is plotted for two different columns in Fig. 3.

The difference in the final extension is caused by different times of growth abortion, while the growth velocities are nearly identical. Different abortion times are thought to result from some interaction between crystals growing in perpendicular directions, as sketched in Fig. 1a. In contrast, we did not observe a strong

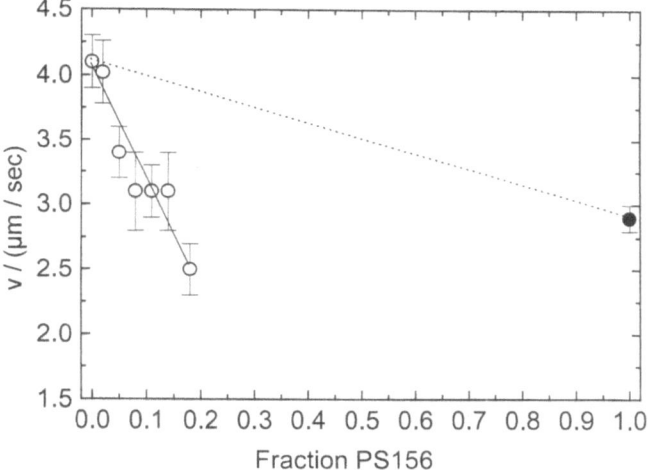

Fig. 4 Maximum growth velocities obtained after thorough deionisation for the pure components and mixtures of different X_{156}. We observe a pronounced drop in v with increasing X_{156}. The *solid line* is a guide for the eye. The *dashed line* gives an expectation for the case that the velocity change is caused only by the changes in the average diffusion coefficient

influence of the composition on the morphology. This and measurements of the properties of a similar system [15] seem to indicate that crystalline alloys of random distribution of components are formed. In particular, the feasibility of Bragg microscopy under unchanged Θ and φ for all compositions shows that the structure remains bcc throughout.

The results of growth measurements for seven suspensions with fractions X_{156} of 0, 0.02, 0.05, 0.08, 0.11, 0.14, 0.18 and 1 are shown in Fig. 4. The growth velocity decreases with increasing X_{156}. In fact, it even falls below the value of pure PS156. Preliminary measurements at slightly elevated salt concentration (where wall crystal growth is the dominant solidification mechanism even at considerably larger X_{156}) confirm this trend and show a minimum growth velocity of some $2.5~\mu ms^{-1}$ at $X_{156} \approx 0.18$.

Discussion

Two points need further clarification. The unexpected increase in v during continued deionisation at constant conductivity and the drop in v with increasing X_{156}.

Since the particle number density stays constant any changes in v have to be attributed to changes in the salt concentration [11]. In particular, we have to suspect a continued decrease of salt concentration at constant or slightly increasing conductivity. To further explore this point we carried out deionisation experiments on aqueous electrolytes to find the rate of anion exchange to significantly exceed the rate of cation exchange. Thus, during deionisation from a pH neutral electrolyte (NaCl) a transient excess concentration of Na^+ appears. Deionisation of a suspension therefore corresponds to the backward performance of an acid–base titration.

Not in all cases, however, was a pronounced minimum observed. In particular, for cases of low charge ratio Z_σ^*/Z the conductivity drops monotonously with time. (Here Z_σ^* is the number of counterions visible in the conductivity experiment and Z the number of dissociated surface groups.) To explain this finding we resort to Hessinger's model of conductivity [16]. It divides the electrical double layer into an outer part, where all ionic species are freely migrating with their bulk mobilities, and an inner part of counterions moving with the particle. The total conductivity is then given by the sum over all particles. Further only the number $Z - Z_\sigma^*/Z$ of bound counterions is conserved, but exchange is allowed between the two parts of the double layer. Consequently at low Z_σ^*/Z added Na^+ will show up with an effective mobility close to that of H^+, while at large Z_σ^*/Z the effective mobility is close to the lower value of Na^+. Coions are assumed to stay outside the inner region and always contribute with their bulk mobility.

The transient excess of Na^+ lowers the average counterion mobility and at large charge ratios a minimum is observed. It disappears as a pure protonic counterion cloud is established. At low charge ratio, the effect is less pronounced and is compensated by the continuous decrease in anion concentration. Therefore, in most of the present cases the conductivity was observed to stay constant during a further decrease in the salt concentration. The latter then translated into an increase in v.

The second point is of more fundamental interest. Why does the growth velocity of the mixtures drop below the values of the pure component? Mixtures were prepared at a particle concentration where for the pure components the limiting velocity was nearly reached. For further discussion it is instructive to recall the expression for the limiting growth velocity.

According to Würth it should be given by $v_\infty = Dd/\ell^2$ [17]; this is also discussed in some detail in Ref. [5]. Here D is an appropriate diffusion coefficient and ℓ is the average distance of a particle in the melt to its target place in the crystal. Growth may be accelerated considerably if a finite thickness of the interface is allowed. Then, the time available for a particle to find its target place is given by the interfacial speed, v, times its thickness, d. Putting this in numbers, i.e. taking $D = 0.1$ D_0 [17] and $\ell = d_{NN}$, the nearest-neighbour spacing, an interfacial thickness, d, of a few layers results [11].

Within the WF approach, two effects may possibly explain our findings. First changes in v_∞ are considered. For the mixtures D_{156} is considerably smaller than D_{120} and, at constant n (implying constant ℓ) and d, a drop in v is expected through the decrease in v_∞. This is indicated by the dashed line in Fig. 4. This alone cannot explain our data. The experimentally observed drop may alternatively be attributed to a change in d. A corresponding physical explanation would be a disturbance of the layer structure by the added minority component.

A second possibility concerns variations in the second term of Eq. (1). We assumed that for both pure samples we are sufficiently far from the phase boundary to observe the limiting velocity, v_∞, or values very close to this. Experiments were carried out after thorough deionisation and at a constant particle density. The previous explanation nevertheless relied on the assumption that the condition of a sufficiently large chemical potential difference between the melt and the solid is also retained for arbitrary composition. In fact it should slightly increase as the more strongly metastable component PS156 is added. This may not be the case. Meller and Stavans [3], for instance, have shown several examples of charged mixtures where the position of the phase boundary in an $n–X$ diagram was not a linear function of X. We may thus be closer to freezing than expected and measure an intermediate (smaller) v than the corresponding v_∞.

Conclusions

Bragg microscopy served as a versatile tool to study the morphology and crystal growth kinetics in mixed colloidal suspensions. It further served as an additional control of the state of deionisation. Referring to Hessinger's conductivity model the interesting observation of a changed growth velocity at constant conductivity was explained with a transient excess of cationic impurities.

With this technique we performed the first quantitative measurements on mixed colloidal crystals. An unexpectedly strong decrease in the growth velocity was observed. Further experiments are now being conducted over a broader range of parameters to discriminate between possible origins. At present we favour either a disturbance of the interfacial structure by the added component or a curved phase boundary as explanations.

Acknowledgements We thank P. Wette, H. J. Schöpe and R. Biehl for helpful discussions. Financial support of the Deutsche Forschungsgemeinschaft (grant no. Pa 459/8-1, 2) is gratefully acknowledged.

References

1. Becker B, Döring W (1935) Ann Phys 24:719
2. (a) Wilson HA (1900) Philos Mag 50:238; (b) Frenkel JZ (1933) Sowjetunion 1:498
3. Meller A, Stavans J (1992) Phys Rev Lett 68:3646
4. Barlett P, vanMegen W (1994) In: Mehta A (ed) Granular matter: Springer, Berlin Heidelberg New York, pp 195–257
5. Palberg T (1999) J Phys Condens Matter 11:R323
6. Maroufi MR, Stipp A, Palberg T (1998) Prog Colloid Polym Sci 108:83
7. (a) Aastuen DJW, Clark NA, Kotter LK, Ackerson BJ (1986) Phys Rev Lett 57:1733; (b) Phys Rev Lett 57:2772 (erratum)
8. Okubo T (1994) In: Schmitz KS (ed) Macro-ion characterization: from dilute solution to complex fluids, ACS symposium series 548. pp 364–392
9. He Y, Olivier B, Ackerson BJ (1997) Langmuir 13:1408
10. Preis T, Biehl R, Palberg T (1998) Prog Colloid Polym Sci 108:129
11. Würth M, Schwarz J, Culis F, Leiderer P, Palberg T (1998) Phys Rev E 52:6415
12. Schöpe HJ, Decker T, Palberg T (1999) J Chem Phys 109:10068
13. Palberg T, Härtl W, Wittig U, Versmold H, Würth M, Simnacher E (1992) J Phys Chem 96:8180
14. Palberg T, Härtl W, Deggelmann M, Simnacher E, Weber R (1991) Prog Colloid Polym Sci 84:352
15. Wette P, Schöpe HJ, Palberg T (2001) Prog Colloid Polym Sci PCPS 72
16 Hessinger D, Evers M, Palberg T (2000) Phys Rev E 61:5493
17. Löwen H, Palberg T, Simon R (1993) Phys Rev Lett 70:1557

Progr Colloid Polym Sci (2001) 118:96–99
© Springer-Verlag 2001

Titanium dioxide nanoparticle films made by using poly(ethylene glycol) oligomers as templates

E. Stathatos
P. Lianos
P. Falaras

E. Stathatos · P. Lianos (✉)
Engineering Science Department,
University of Patras, 26500 Patras
Greece
e-mail: lianos@upatras.gr
Tel: +30-61-997587
Fax: +30-61-997803

P. Falaras
Institute of Physical Chemistry
NCSR "Demokritos", 15310 Athens
Greece

Abstract TiO$_2$ mesostructured films have been made on glass slides by a dip-coating process, using ethanolic solutions containing titanium isopropoxide and poly(ethylene glycol) oligomers of various chain lengths. This convenient and easy procedure provides a means of controlling the size of titania nanoparticles, as well as the fractality and the roughness of the film surface by simply choosing the size of the oligomer chain length.

Key words Titania · Films · Poly(ethylene glycol) templates · Nanoparticles

Introduction

Materials that contain titanium dioxide nanoparticles, either in colloidal solutions or in solid films, have become very popular, owing to a multitude of applications, such as solar cells [1–6], lithium batteries [7, 8], air-purification systems and treatment of wastewater [9–14], catalysis [15–17], biocompatible materials [18, 19], metal recovery [20], laser scatter-gain materials [21], etc. Recently increased attention has been paid to thin nanostructured titanium dioxide films [22–29] since the possibility of making solar cells and air-purification arrays based on these systems seems to be about to become reality. The usual points raised by most recent work are related to the size, shape and order of the nanoparticles that make up the films [22, 24], as well as the morphology, roughness, fractality and thickness of the films. The production of open, highly porous structures is important because of their large surface area, which allows extensive contact with the reaction medium. At the same time, continuity should exist between nanoparticles in order to prevent traps and dead spots that would deplete the efficiency of the film [22]. The size of the nanoparticles is an additional important parameter that affects the electronic properties and the absorption onset of the nanocrystallites [30] and should be taken into account [31].

In the present work, we present TiO$_2$ films made by a very simple method, using poly(ethylene glycol) (PEG) oligomers [13] as templates. Composite organic–inorganic materials, where the organic subphase plays the role of a template to produce mesoporous oxides, is a method that is gaining ground in materials science. Even though surfactants are the most popular choice for templates, polymers appear to offer the possibility of very interesting and unusually structured materials. For example, "coral-like" TiO$_2$ structures have been made by using polymer gels as templates [22]. Well-ordered hexagonal mesoporous silica structures have been obtained using triblock copolymer templates [32], while macroporous materials with highly ordered three-dimensional arrays of spheroidal voids have been made using latex particles as templates [33]. PEG oligomers offer extensive choice and control of the type of films made in their presence. Parameters such as the size of the nanoparticles, film thickness, surface roughness and surface fractal dimension can be easily controlled by simply choosing the length of the PEG chain. The scope of the present work is to demonstrate the great capacity of this novel method.

Materials and methods

All the chemicals used in the present work were from Aldrich and were used as received. TiO$_2$ thin films were made by the following procedure [13]. Ethanol (8 g) was flashed with dry N$_2$ to reduce the amount of dissolved water and oxygen. Then, 1 g PEG followed by 1 g titanium isopropoxide were added under stirring. Finally, a glass slide, previously cleaned in sulfochromic solution, was dipped in the solution and was withdrawn at a speed of 42 mm/min. The film obtained was left to dry in air and was subsequently calcinated at 450 °C in air.

Absorption measurements were made with a Cary 1E spectrophotometer and atomic force microscopy (AFM) images were obtained with a Nanoscope III (Digital Instruments) in tapping mode. The film thickness was measured with a Sloan Dektak IIA stylus profilometer.

Results and discussion

PEG oligomers are soluble in ethanol; however, addition of titanium isopropoxide affects solutions in a way that depends on the PEG chain length. The solution was clear up to PEG-1000 but for longer PEG chains the species present in solution were large enough to give a milky mixture. The composite organic/inorganic films obtained immediately after dipping, as well as the thin oxide films obtained after calcination, followed suit. The films made using short PEG chains were transparent, while the films made with longer PEG chains were milky. Other conditions being equal, the chain length also affected the thickness of the oxide films, which was smaller in the case of shorter chains. The measured film thicknesses are given in Table 1, column 2. The absorption spectra of films made in the presence of different PEG chains are shown in Fig. 1. The film that corresponds to PEG-2000 is not shown, since it scatters light to such an extent that it is practically nontransparent. Figure 1 shows that samples made with longer PEG chains have the tendency to give an absorption onset at longer wavelength. This shift in the absorption onset, albeit small, is justified by the change in the particle size and by the ensuing size effects on the electronic properties of the semiconductor [30].

AFM images of some films made using PEG of different chain lengths are shown in Fig. 2. The films consist of random networks of nanoparticles. The average particle diameter increased with the size of the PEG chain. It was 10 nm in the smallest particles and grew to 50 nm in the largest. The size polydispersity was very narrow when the particles were smaller, while for the larger particles made with the longer PEG chains, the

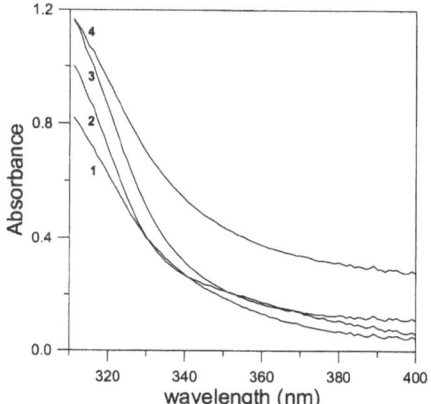

Fig. 1 Absorption spectra of titania films made from ethanolic solutions containing poly(ethylene glycol) (*PEG*): *1* PEG-200; *2* PEG-400; *3* PEG-600; *4* PEG-1000

particle diameter varied between 35 and 50 nm. The average particle diameters are listed in Table 1 (column 3). The packing of the nanoparticles was very dense in the case of smaller sizes but was much less dense in the case of larger particles. The values of the fractal dimension of the surface of each film are listed in column 4 of Table 1. The fractal dimension, f, expresses the surface geometric complexity and, for three-dimensional images, it is expected to vary between 2.0, for a perfectly flat surface, to 3.0, for an infinitely expanded surface. The algorithm employed for fractal analysis divides the surface into a series of triangles, starting with a cell size of 1×1. In the first iteration, the surface is divided into two triangles. The surface area of each triangle is calculated and recorded. In the second iteration, each cell is further divided in half, resulting in eight triangular cells. In the third iteration 32 triangles are defined and so on, until the image has been divided into the maximum number of triangles, each having a cell size of one pixel. As the number of triangles increases, the total surface area of the sample increases. The logarithmic plot of the cell size versus the surface area determines the fractal value of the surface. In fact, the fractal dimension is defined as the slope of the line obtained by plotting the logarithm of the cell size versus the logarithm of the cell surface area. The values of f shown in Table 1 were calculated for images of size $1 \mu m \times 1 \mu m$, as those seen in Fig. 2. It is interesting that the films made using short PEG chains and consisting of small nanoparticles attained relatively high fractal dimensions, i.e. the total surface expansion of the film is very large in that case. As the particles became larger, the fractal dimension decreased. The values of the film roughness, R_{ms}, are given in column 5 of Table 1. R_{ms} is automatically calculated by the instrument and it represents the standard deviation of the z-dimension on the surface of the film. Rough surfaces should have large R_{ms} values. Indeed, as seen in Table 1, R_{ms} was smaller for small and densely packed

Table 1 Various parameters characterizing TiO_2 thin films made by means of poly(ethylene glycol) oligomer templates

Poly(ethylene glycol) chain size	Film thickness (μm)	Average particle diameter (nm)	Fractal dimension	Roughness (nm)
200	0.29	10	2.456	0.84
400	0.43	15	2.380	0.97
600	1.2	30	2.375	1.57
1,000	2.0	35	2.250	3.70
2,000	3.5	35–50	2.177	6.19

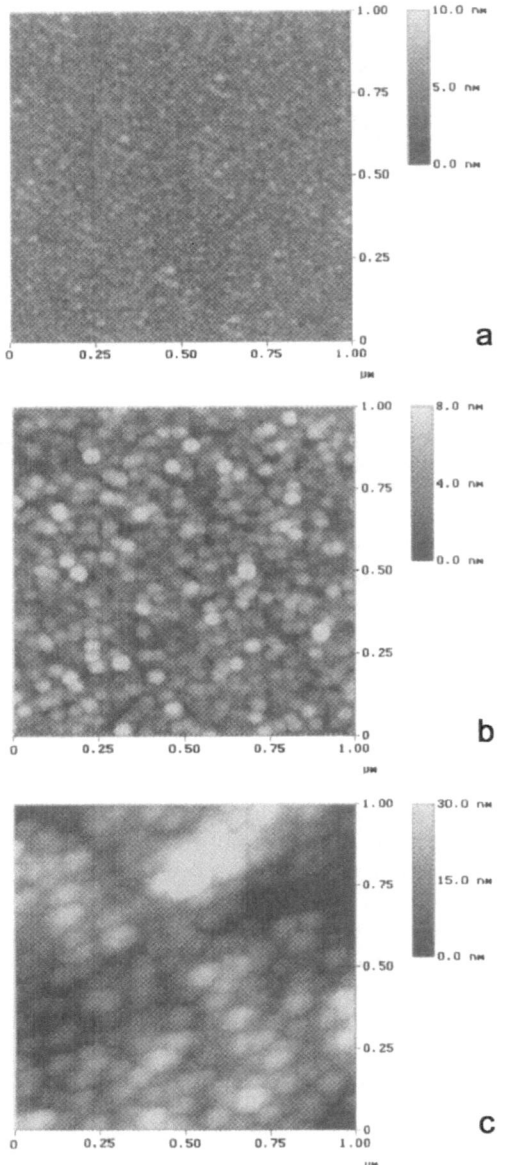

Fig. 2 Atomic force microscopy images of titania films made from ethanolic solutions containing **a** PEG-400, **b** PEG-600 and **c** PEG-2000

particles but rapidly became large in the case of larger and less densely packed particles. f and R_{ms} always varied in opposite directions.

Hydrolysis of titanium isopropoxide takes place both in solution and in air, as soon as the material making the film is withdrawn from the solution. Hydrolysis in ethanolic solution is limited, since the only available water is that solubilized in ethanol, which is further limited by the flow of dry N_2. Thus, the alkoxide hydrolysis and condensation steps are completed in the atmosphere. Partially hydrolyzed titanium isopropoxide is complexed with PEG chains present in the ethanolic solution. This is indicated by the fact that when the PEG chains are long, the complexes grow to a size that causes significant scattering of light. The solutions then become milky and slow precipitation is observed, but gelling was never really obtained in solution. The existence of complexes in solution is also supported by the fact that the final film structure is a random network of almost spherical particles. This means that no organization of the organic and the inorganic phases occurs either in solution or upon deposition and the nanoparticles obtained seem to occur as individual entities. Apparently, exclusively steric reasons dictate the size of the nanoparticles, as the result of the particular combinations with a given chain length. Complex formation is, however, not the only process occurring in solution. Complexation is, apparently, accompanied by clustering of individual complexes. Clustering can occur by physical forces ensuing from interaction between PEG chains but also from, limited, Ti–O–Ti polymerization. Clustering should be responsible for intensive light scattering in the case of relatively long PEG chains and for the formation of thicker films in that case. Cluster formation is also responsible for the large voids in the random particle network (Fig. 2c), the low fractal dimension and the extensive roughness of the film surface (Table 1), in the case of PEG-2000.

PEG oligomers then provide an easy means of making TiO_2 with a highly controlled nanoparticle structure. The question, of course, is how useful such films might be. One can envisage some different applications. It has been found that the films adsorb substances by demonstrating a specificity for elemental ions and small inorganic molecules, particularly, films made with short polymer chains and consisting of relatively small and densely packed nanoparticles. In contrast, they possess no capacity at all for adsorption of organic solutes. This property can make such films useful for metal recovery [20] or photocatalysis of airborne pollutants [13], etc. Such problems are currently being studied in our laboratories.

Acknowledgements We acknowledge financial aid from the program Κ.ΚΑΡΑΘΕΟΔΩΡΗΣ of the University of Patras.

References

1. O'Regan B, Gratzel M (1991) Nature 353:737
2. Zaban A, Micic OI, Gregg BA, Nozik AJ (1998) Langmuir 14:3153
3. Fessender RW, Kamat PV (1995) J Phys Chem 99:12902
4. Argazzi R, Bignozzi CA, Heimer TA, Castellano FN, Meyer GJ (1997) J Phys Chem 101:2591
5. Cao F, Oskam G, Searson PC (1995) J Phys Chem 99:17071
6. Falaras P, Xagas AP, Hugot-Le-Goff A (1998) New J Chem 557
7. Kavan L, Gratzel M, Ruthousky J, Zu kal A (1996) J Electrochem Soc 143:2
8. Krtil P, Kavan L, Fattakhova D (1997) J Solid State Electrochem 1:83
9. Vinodgopal K, Hotchandani S, Kamat PV (1993) J Phys Chem 97:9040
10. Hidaka H, Asai Y, Zhao J, Nohara K, Pelizzetti E, Serpone N (1995) J Phys Chem 99:8244
11. Anderson C, Bard AJ (1997) J Phys Chem 101:2611
12. Vinodgopal K, Wynkoop DE, Kamat PV (1996) Environ Sci Technol 30:1660
13. Negishi N, Iyoda T, Hashimoto K, Fujishima A (1995) Chem Lett 841
14. Tada H, Honda H (1995) J Electrochem Soc 142:3438
15. Ioannides T, Verykios X (1996) J Catal 161:560
16. Ziolkowski L, Vinodgopal K, Kamat PV (1997) Langmuir 13:3124
17. Minero C, Pelizzetti E, Terzian R, Serpone N (1994) Langmuir 10:692
18. Nonami T, Taoda H, Hue NT, Watanabe E, Seda K, Tazawa M, Fukaya M (1998) Mater Res Bull 33:125
19. Roddick-Lanzilotta AD, Connor PA, McQuillan AJ (1998) Langmuir 14:6479
20. Sahyun MRV, Serpone N (1997) Langmuir 13:5082
21. Lawandy NM, Balachandran RM, Gomes ASL, Souvain E (1994) Nature 368:436
22. Caruso RA, Giersig M, Willig F, Antonietti M (1998) Langmuir 14:6333
23. Matsusita SI, Miwa T, Tryk DA, Fujishima A (1998) Langmuir 14:6441
24. Burnside SD, Shklover V, Barbe C, Comte P, Arendse F, Brooks K, Gratzel M (1998) Chem Mater 10:2419
25. Stathatos E, Lianos P, DelMonte F, Levy D, Tsiourvas D (1997) Langmuir 13:4295
26. Lin H, Kozuka H, Yoko T (1998) Thin Solid Films 315:111
27. Sakai H, Kawahara H, Shimazaki M, Abe M (1998) Langmuir 14:2208
28. Thayer AM (1998) Chem Eng News March 9:10
29. Provata A, Falaras P, Xagas A (1998) Chem Phys Lett 297:484
30. Brus L (1986) J Phys Chem 90:2555
31. Stathatos E Tsiourvas D, Lianos P (1999) Colloids Surf A 149:49
32. Zhao D, Feng J, Huo Q, Melash N, Fredrickson G, Chmelka BF, Stuckn GD (1998) Science 279:548
33. Holland BT, Blanford CF, Stein A (1998) Science 281:538
34. Falaras P (1998) Sol Energy Mater Sol Cells 53:163

Progr Colloid Polym Sci (2001) 118: 100–102
© Springer-Verlag 2001

P. Staszczuk
M. Majdan
T. Danielkiewicz
M. Matyjewicz

Investigations of adsorption and porosity properties of pure and modified zeolites by means of thermal analysis and sorptometry techniques

P. Staszczuk (✉) · M. Majdan
T. Danielkiewicz · M. Matyjewicz
Maria Curie-Sklodowska University
Chemistry Faculty, M. Curie-Sklodowska
Sq. 3, 20-031 Lublin, Poland

Abstract The adsorption of Ni^{2+} on zeolites is presented. Using the techniques of thermal analysis and sorptometry information such as the surface capacity, wetting and the nature of the centres was obtained

and good correlations were obtained.

Key words Adsorption · Heterogeneity · Thermal analysis · Pure and modified zeolities

Introduction

The history of zeolites started in the eighteenth century. In 1756 the amateur Swedish mineralogist, A.F. von Cronstedt observed that some minerals emit much water during the heating process. They looked as if they were in a "boiling" state. He called them zeolites, which in Greek means "boiling" stones. In 1840 the first investigations of zeolite hydratation were carried out and in 1842 the composition of natural foyazite was determined. In 1920 studies of other zeolites (e.g. chabazite, mordenite, gmelinite) were carried out. In 1925 it was observed that chabazite is suitable for the separation of water from methanol and the microporous structure of zeolites was determined. In 1930 it was proved using X-ray data that the absorption of small molecules on the zeolite surface depends on the nature of the compound and the sizes of the "windows" in the zeolite inner structure (therefore, zeolites were later called "molecular sieves"). Next, methods of production of synthetic zeolites were introduced and the surface properties were investigated. The most dynamic development took place in Japan, the USA and Europe in 1960–70 when zeolites began to be used as additives to fodder. In the USA methods to remove ammonium ions from water using zeolites were worked out. Since that time industrial applications of natural zeolites have been developed. Their output increased from 13,000 tons in 1975 to 34,000 tons in 1992 in the USA and at present the world zeolite consumption is several hundred thousand tons per year.

Till now over 40 types of natural zeolites have been discovered and synthetic zeolites production exceeds 100 generations, but most of them do not have equivalents in nature. In the early 1990s, a new method for the preparation of porous materials (so-called MCM-41 molecular sieves) with a definite pore shape was introduced. The MCM-41 material, a member of the M41S group, contains a hexagonal array of uniform mesopores but lacks a strict crystallographic order on the atomic level. The MCM-41 molecular sieves have already been used to obtain catalysts for practical application in the process of ethylene and propylene production from methanol. On the other hand, the ion-exchange properties of the zeolites are used in the removal of toxic metals from wasters. Among toxic metals mercury, cadmium, cobalt, nickel, is also worth of mention owing to its toxic properties. The adsorption of metals on zeolites is interesting from a practical and a theoretical point of view, especially if we take into account different models of adsorption: electrostatic and nonelectrostatic.

The quasiisothermal technique of thermal analysis had previously been used to characterize the heterogeneous properties of the materials, e.g. MCM-41 samples synthesized under different conditions [1–4] and pure and modified zeolites.

Methods

The measurements of the programmed thermodesorption of polar (water, *n*-butanol) and nonpolar (benzene, *n*-octane) liquids from pure and modified by adsorption from solution of Ni^{2+} ions CBV 10A zeolite-mordenite ($Na_2O \cdot Al_2O_3 \cdot 10SiO_2 \cdot 6H_2O$) samples (Zeolyst International Co., USA) were made using the simultaneous derivatograph Q-1500D (MOM Hungary) and the method described in Refs. [5, 6]. The Ni^{2+} adsorption value determined by spectrophotometric methods was 3.9×10^{-3} mol/g [7]. In the studies of the liquid thermodesorption process the zeolite samples were wetted with liquid vapors in a vacuum dessicator where the relative vapor pressure, p/p_o, was 1. The Q-thermogravimetric (TG) mass loss and the first derivative of the Q-differential TG (DTG) mass loss curves with respect to temperature and time were recorded. Moreover, the porosity of the samples was estimated by means of nitrogen adsorption at 77.4 K, with the Brunauer–Emmett–Teller and BHJ equations employed using a sorptomat apparatus of the type ASAP 2405 V1.01 (Micrometrics, USA). On the basis of the data obtained the specific surface area, the pore volume and the radius as well as pore size distribution functions were calculated.

Results and discussion

Q-TG mass loss and Q-DTG mass loss curves in relation to temperature obtained during thermodesorption of benzene (Figs. 1, 3) and octane (Figs. 2, 4) from pure (Figs. 1, 2) and modified (Figs. 3, 4) zeolite samples were made.

From these figures it follows that steps and/or inflections of the thermodesorption process, i.e. evaporation of liquids from capillary tubes, pores, "windows" and active centers of pure and modified surface, were obtained. It is worth noting that the Q-DTG curves presented in Figs. 1 and 3 are characterized by high selectivity and resolving power distribution. They can be considered as a certain type of "spectrum" of a

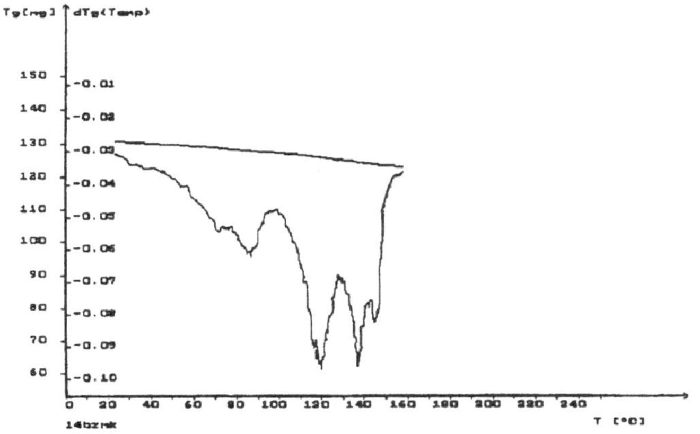

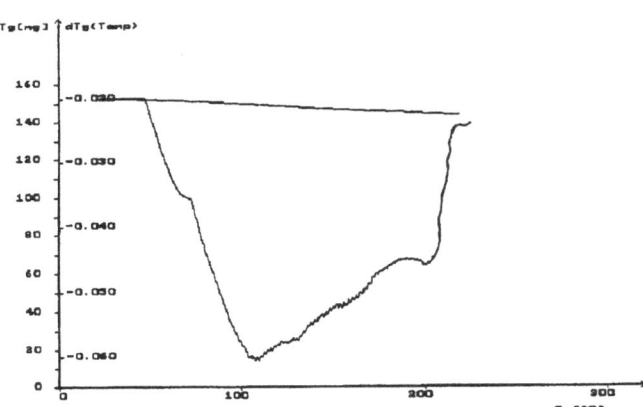

Fig. 1 Q-thermogravimetric (TG) mass loss and Q-differential TG (DTG) mass loss plots versus temperature for thermodesorption of benzene from a pure zeolite sample

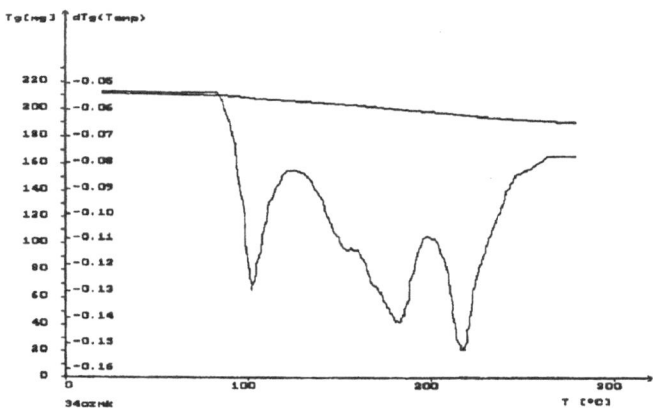

Fig. 3 Q-TG mass loss and Q-DTG mass loss plots versus temperature for thermodesorption of benzene from a modified zeolite sample

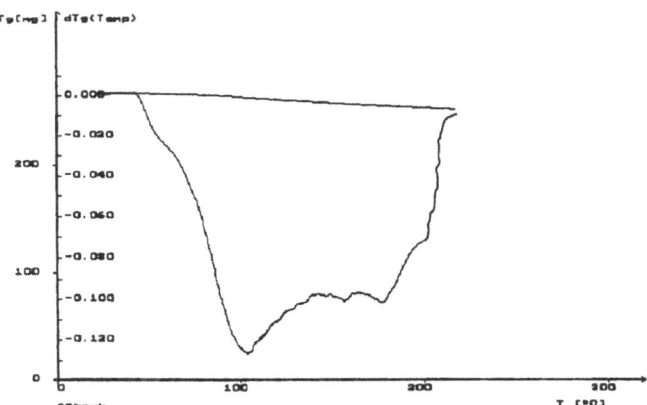

Fig. 2 Q-TG mass loss and Q-DTG mass loss plots versus temperature for thermodesorption of octane from a pure zeolite sample

Fig. 4 Q-TG mass loss and Q-DTG mass loss plots versus temperature for thermodesorption of octane from a modified zeolite sample

Table 1 Adsorption capacity

Sample of zeolites	Sorptometric method: pore radius (Å)	Sorptometric method: maximum of nitrogen adsorption (cm³/g)	Thermal analysis method, liquid desorption (mmol/g)			
			Water	n-Butanol	Benzene	n-Octane
CBV 10A	10.81	5.0	7.3	1.12	0.90	0.55
CBV 10A + Ni^{2+}	18.18	149	18.1	1.27	1.03	0.84

Table 2 Pore volume (cm³/g)

Sample of zeolites	Sorptometric method	Thermal analysis method, liquid desorption			
		Water	n-Butanol	Benzene	n-Octane
CBV 10A	0.21	0.16	0.19	0.13	0.21
CBV 10A + Ni^{2+}	0.14	0.19	0.12	0.1	0.15

thermodesorption process describing an energetic state of liquid molecules on the surface and reflecting the distribution function of the desorption energy of the liquid on the surface of the sample studied [5, 6].

The adsorption capacity, the pore radius and the volume and specific surface area values of pure and modified zeolite samples calculated from thermal analysis and sorptometric data are given in Tables 1 and 2.

The surface properties of the zeolite after modification with nickel has been changed (Tables 1, 2). This fact can be explained as the result of the surface precipitation of the hydroxocompounds of Ni^{2+} of microcrystalline type. At the pH of sorption (the pH after equilibration of the phases was above 6) the precipitation of the hydroxo-chloride of Ni^{2+} or even $Ni(OH)_2$ is possible [7]. The modified zeolite surface has a higher adsorption capacity, a larger pore radius and a smaller pore volume.

capacity (thickness and volume of adsorbed films) and wetting phenomena, the nature of the active centres, the discontinuous change of the adsorption layer properties, the mechanism of surface film destruction, the kinetic thermodesorption of films and its stability. The method presented is very quick and convenient in studies of surface heterogeneity with respect to different adsorbates. It is possible to study the effect of heterogeneity by chemical modification of sorbents on the wettability of solids and the formation of liquid films. From the experimental thermogravimetry results, the adsorption and porosity parameters of the zeolite samples tested were compared with analogous ones obtained by the sorptometry technique and good correlations were obtained.

Acknowledgements Thanks are due to Zeolyst International Co., Valley Forge, Pa., USA for kindly supplying the zeolite samples.

Summary

On the basis of the data presented it was possible to obtain important information concerning the surface

References

1. (a) Korzan K (1996) MSc thesis. UMCS, Chemistry Faculty, Lublin; (b) Danielkiewicz T (1997) MSc thesis. UMCS, Chemistry Faculty, Lublin; (c) Kołodziejczyk J (1998) MSc thesis. UMCS, Chemistry Faculty, Lublin; (d) Kaszak M (1999) MSc thesis. UMCS, Chemistry Faculty, Lublin
2. Staszczuk P, Klinowski J, Danielkiewicz T (1997) In: Staszczuk P (ed) Program and abstracts: 27th International Vacuum Microbalance Techniques Conference. UMCS, Chemistry Faculty, Lublin, p 92
3. (a) Staszczuk P, Danielkiewicz T (1998) Report. Chemistry Faculty, Maria Curie-Sklodowska University, pp 79–80; (b) Staszczuk P, Danielkiewicz T (1999) Report. Chemistry Faculty, Maria Curie-Sklodowska University, pp 65–66
4. Staszczuk P, Danielkiewicz T, Klinowski J (2000) Adsorption Sci Technol 18:307
5. Staszczuk P (1998) J Thermal Anal Cal 53:597
6. Staszczuk P, Głażewski D (1999) J Thermal Anal Cal 55:467
7. Kowalska M, Majdan M (1999) Przem Chem 7:257

Progr Colloid Polym Sci (2001) 118:103–106
© Springer-Verlag 2001

C. A. Bunton
A. Garreffa
R. Germani
G. Onori
A. Santucci
G. Savelli

Relation between the IR spectrum of water and decarboxylation kinetics in microemulsions

C. A. Bunton
Department of Chemistry and
Biochemistry, University of California
Santa Barbara, CA 93106, USA

A. Garreffa R. Germani · G. Savelli
Dipartimento di Chimica
Università di Perugia, Perugia, Italy

G. Onori (✉) · A. Santucci
Istituto per la Fisica della Materia,
Unità di Perugia and Dipartimento
di Fisica, Università di Perugia
Perugia, Italy
e-mail: symbio@pg.infn.it

Abstract The properties of water solubilised by cetyltripropylammonium bromide (CTPABr) and sodium bis(2-ethylhexyl)sulfosuccinate (AOT) in CCl_4 were studied as a function of the number of water molecules per surfactant molecule(W) by the kinetics of decarboxylation of the 6-nitrobenzisoxazole-3-carboxylate ion (6NBIC) and IR spectroscopy in the O–H stretching region in the region of low water ($0 < W < 10$) where the properties depend strongly upon W. Decarboxylation of 6NBIC in CTPABr reverse micelles is strongly inhibited by an increase in W and the first-order constants level off at $W > 1$. The rate effects are ascribed to increases in polarity and hydrogen bonding in the reaction region as water is added and are related to changes in the IR spectrum. For the AOT system, the kinetics is independent of W and this difference from the behaviour of the CTPABr system is ascribed to the difference in the interactions of 6NBIC with surfactant headgroups.

Key words Microemulsions · Cetyltripropylammonium bromide · IR spectrum · Decarboxylation kinetic

Introduction

In apolar organic solvents surfactants often form aggregates which can solubilise large amounts of water and are described as "water-pool" reverse micelles with the number of water molecules per surfactant molecule (W) less than 10–15. Most studies on reverse micelles have focussed on the three-component sodium bis(2-ethylhexyl)sulfosuccinate (AOT)/water/oil system. Little is known about the physicochemical properties of other surfactant systems, in particular as to the state of the solubilised water.

In previous articles we described IR investigations of water/AOT/CCl_4 [1] and water/AOT/n-heptane [2] reverse micelles in the O–H stretching region which indicate that properties of surfactant-trapped water differ from those of bulk water and change strongly with water content if $W \lesssim 6$. At larger W the surfactant-trapped water behaves as in the bulk. The anomalous behaviour of water at low W has been attributed to its

local interactions with Na^+ and sulfosuccinate ions. More recently, properties of water solubilised by cetyltrimethylammonium bromide (CTABr) in CH_2Cl_2 have been studied by kinetics and IR spectroscopy [3]. We investigated the spontaneous decarboxylation of the 6-nitrobenzisoxazole-3-carboxylate ion (6NBIC, **1**) as a kinetic probe (Scheme 1).

This reaction of 6NBIC is very sensitive to changes in the polarity of the medium and hydrogen-bond donation [4], and has been used extensively as a kinetic probe for studying properties of normal and reverse micelles [5].

Results from both the IR and the kinetic approaches [3] emphasise the existence of two types of aqueous domain in reverse micelles which are organised by the surfactant: a bound-water domain with water molecules hydrating the interfacial surfactant headgroups and a bulklike water domain. Decarboxylation of 6NBIC in CTABr reverse micelles is strongly inhibited by an increase in W and the first-order rate constants (k_{obs}) level off at $W > 5$ [3]. The rate effects have been ascribed

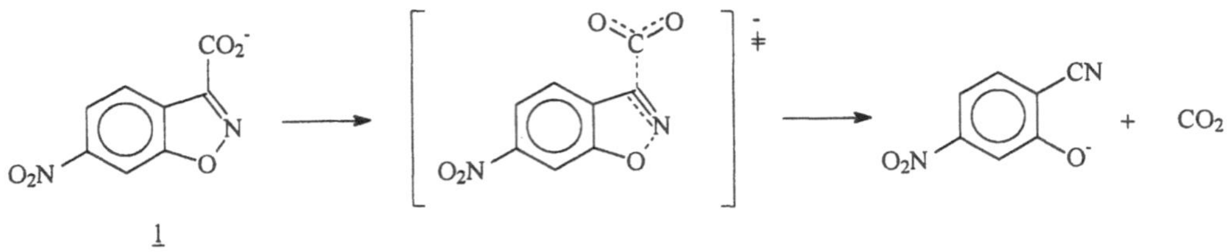

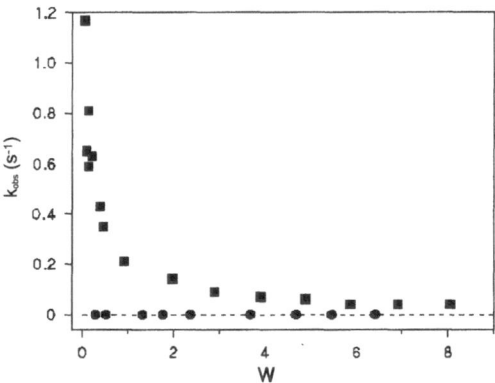

Scheme 1 Decarboxylation of the 6-nitrobenzisoxazole-3-carboxylate ion (**1**)

to an increase in polarity and hydrogen-bond donation in the reaction region as water is added and have been related quantitatively to changes in the IR spectrum.

In the present work the properties of water solubilised by cetyltrypropylammonium bromide (CTPABr) and AOT in CCl_4 were studied kinetically by decarboxylation of 6NBIC and by IR spectroscopy in the O–H stretching region. In CCl_4 almost all the water is solubilised in the reverse micelle and, therefore, the data analysis is more direct than in the earlier work [3]. We note that CTABr is insoluble in CCl_4 and has a low solubility in CH_2Cl_2 which increases with increasing W and measurements in the $CTABr/H_2O/CH_2Cl_2$ system were with $W > 1.5$ [3]. However, CTPABr is soluble in CCl_4 even at $W = 0$, which allowed us to study the region at very low W where the properties of the system depend strongly on W.

Experimental

AOT 99% (Alfa) purified by recrystallization from methanol and dried in a vacuum was stored in vacuum over P_2O_5.

CTPABr prepared by alkylation of tripropylamine by cetyl bromide in CH_3CN was purified by recrystallising it twice from ethyl acetate and was dried in a vacuum at 65 °C.

Bidistilled water and CCl_4 (purity 99.5%) were used without additional purification. The $AOT/H_2O/CCl_4$ and $CTPABr/H_2O/CCl_4$ mixtures were prepared by weight. In order to remove traces of water, solutions of CTPABr in CCl_4 were dried for 3 days over molecular sieves (4 Å).

Decarboxylation of 6NBIC was followed spectrophotometrically at 25.0 °C and 410 nm using HP-8452 diode-array spectrophotometers. The reaction was started by adding substrate as the acid in CCl_4, as described in Refs. [5, 6], with 3×10^{-5} moldm^{-3} 6NBIC and 3×10^{-3} moldm^{-3} $(C_2H_5)_3N$ in the reaction solution. Deprotonation is rapid under these conditions.

IR spectra were recorded at room temperature using a Shimadzu model 470 IR spectrophotometer equipped with a variable path length cell and CaF_2 windows. Typical path lengths were 50–800 μm, although the spectra of water in CCl_4 were taken with path lengths of 2 cm.

The molar extinction coefficient of water is given by $\varepsilon = A/(cd)$, where A is the absorbance, c the water concentration in moles per cubic decimetre and d the cell path length in centimetres.

Results and discussion

Kinetic measurements

The kinetics of the decarboxylation of anionic 6NBIC in water/surfactant mixtures solubilised in CCl_4 was inves-

Fig. 1 Rate constants (k_{obs}) of decarboxylation of the 6-nitrobenzisoxazole-3-carboxylate ion in H_2O/cetyltripropylammonium bromide (*CTPABr*)/CCl_4 (*squares*) and in H_2O/sodium bis(2-ethylhexyl)sulfosuccinate (*AOT*)/CCl_4 (*circles*) systems as a function of W. [CTPABr] = [AOT] = 0.1 moldm^{-3}. The *dashed line* shows the values of k_{obs} in pure water

tigated for two surfactants differing in the headgroup charge: CTPABr (cationic) and AOT (anionic), as a function of water content ($0 < W < 10$). With CTPABr decarboxylation is much faster than in water, and its addition slows the reaction (Fig. 1) in agreement with earlier results [6]. The anionic substrate in the interior of a CTPABr reverse micelle interacts with the CTPA$^+$ headgroup, which assists charge delocalisation in the transition state, and reactions at $W \sim 0$ are faster than those in water ($k_{obs} = 3 \times 10^{-6}$ s^{-1}) by a factor of approximately 10^6.

These interactions weaken with increasing water content and, in the limit of high W, 6NBIC is extensively hydrated. This initial-state stabilisation strongly inhibits decarboxylation. There is a marked decrease in k_{obs} up to $W \cong 1$ and it becomes approximately constant at higher W, indicating that the decrease in k_{obs} is essentially due to association between CTPABr and the first added water molecule. This result is similar to that obtained with CTABr [3].

With AOT, decarboxylation of 6NBIC has a rate similar to that in water (Fig. 1) and there is no catalysis, with $0 < W < 10$ indicating that decarboxylation takes place in the water pool and does not involve the anionic surfactant, consistent with repulsive interactions between anionic 6NBIC and the AOT headgroups.

Fig. 2 O–H stretching band of water solubilised in **a** CCl$_4$ and in H$_2$O/CTPABr/CCl$_4$ at selected values of W. **b** $W = 0.023$, **c** $W = 0.103$, **d** $W = 0.498$

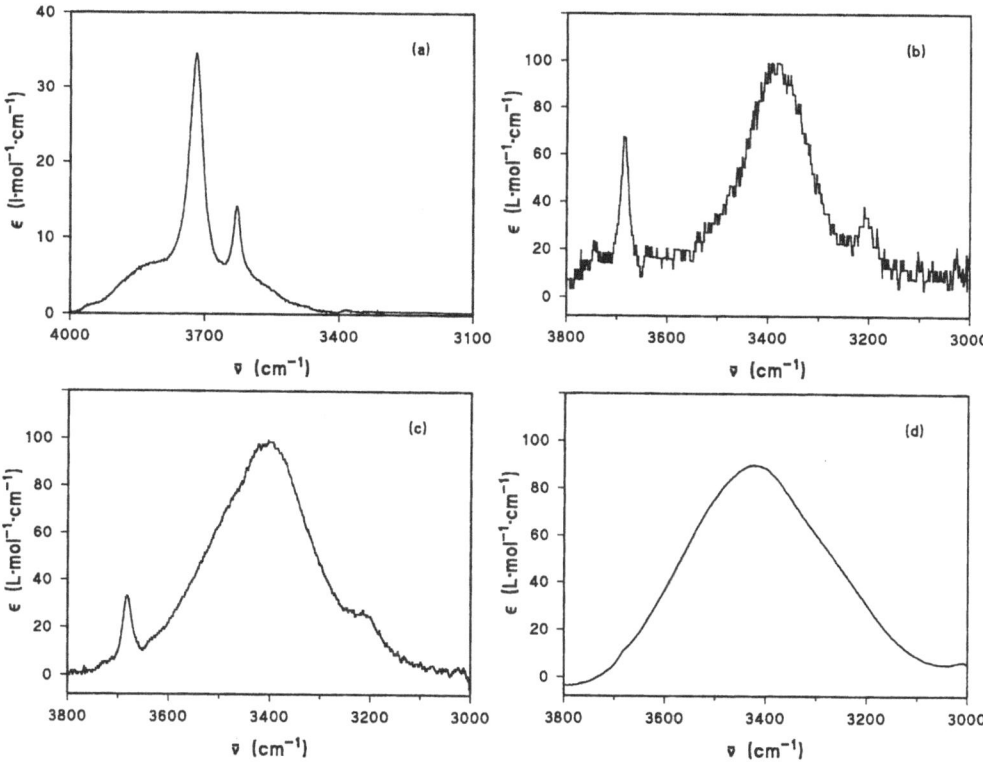

IR measurements

The O–H stretching band for water solubilised in CCl$_4$ and in H$_2$O/CTPABr/CCl$_4$ at selected values of W are shown in Fig. 2.

The molar extinction coefficient of water in CCl$_4$ (Fig. 2a) shows absorptions at 3,720 and 3,625 cm^{-1} assigned to antisymmetric and symmetric O–H stretching vibrations, respectively. At very low W a species is formed as a result of water/CTPABr association and is characterised by three absorption bands at 3,685, 3,390 and 3,206 cm^{-1} (Fig. 2b). The higher-frequency band is narrow and lies slightly shifted from the position of the antisymmetric stretching mode of free water (Fig. 2a). The middle frequency band that characterises this association species is broad and is considerably displaced to frequencies lower than those in the range of free O–H stretching modes. These spectral patterns are similar to those observed in water/alkylammonium halide systems in CCl$_4$ [7] and adequately reveal the nature of the species formed in these systems. The two bands that appear on low water addition (Fig. 2b) show characteristics expected for a 1:1 water/Br$^-$ species of the type

$$
\begin{array}{l}
\mathrm{O-H \cdots Br^-} \\
/ \\
\mathrm{H}
\end{array}
$$

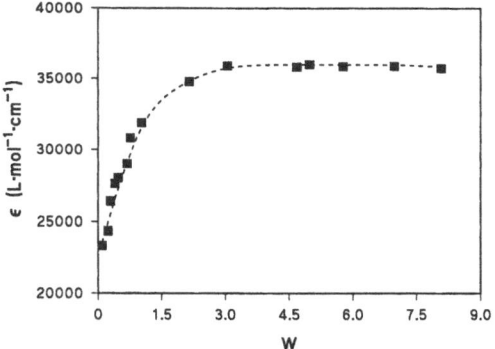

Fig. 3 Integrated molar extinction coefficients, ε_{int}, as a function of W. The *line* is to guide the eye

The sharp, high-frequency band, showing the Lorentzian profile typical of noninteracting species, can be attributed to the stretching of a free OH, while the broad band at 3,390 cm^{-1} is associated with hydrogen-bonded OH. The third band, at 3,206 cm^{-1}, can be attributed to the overtone of the bending vibration [7].

On increasing W the intensity of the high-frequency band, due to free OH of the bound water, decreases (Fig. 2c); finally, at high water concentration (Fig. 2d), water displays more marked "bulk" properties and the absorption band approaches that of bulk water.

The changes in the absorption bands on varying W can be described by means of several spectral parameters. The integrated molar extinction coefficient, $\varepsilon_{int}(W) = \int \varepsilon(\bar{\nu}, W)d\bar{\nu}$, is shown versus W in Fig. 3.

On increasing W, ε_{int} changes significantly up to $W \sim 2$ and then, at higher water content, gradually approaches the value of bulk water. There is a strong correlation between the spectral and kinetic data (Figs. 1, 3) with a close connection between the relation of k_{obs} to W and the properties of water in reverse CTPABr micelles.

We propose that the carboxylate moiety in 6NBIC competes with water for the cationic surfactant headgroups and their counterions. Hydrogen bonding inhibits decarboxylation, which is fastest in the hypothetical situation in which 6NBIC is in contact with CTPA$^+$ in the absence of water. On increasing the amount of solubilised water, there is less interaction of 6NBIC with CTPA$^+$ owing to hydration of CTPA$^+$ and Br$^-$, and k_{obs} gradually decreases until hydration around the interfacial headgroups is almost complete. Our results indicate that the decrease in k_{obs} is due largely to initial hydration of CTPABr according to previous results with CTABr [3].

References

1. Onori G, Santucci A (1993) J Phys Chem 97:5430
2. D'Angelo M, Onori G, Santucci A (1994) Nuovo Cimento D 16:1601
3. Di Profio P, Germani R, Onori G, Santucci A, Savelli G, Bunton CA (1998) Langmuir 14:768
4. Walrafen GE (1972) In: Franks F (ed) Water, a comprehensive treatise, vol 1. Plenum, New York, pp 151–214
5. Kemp DS, Paul KG (1975) J Am Chem Soc 97:7305
6. Germani R, Savelli G, Cerichelli G, Mancini G, Luchetti L, Ponti PP, Spreti N, Bunton CA (1991) J Colloid Interface Sci 147:152
7. Mohr SC, Wilk WD, Barrow GM (1965) J Am Chem Soc 87:3048

Progr Colloid Polym Sci (2001) 118:107–109
© Springer-Verlag 2001

Faouzia Hamdoune
Chaouki El Moujahid
Marie-José Stébé
Christine Gérardin
Piotr Tekely
Claude Selve
Ludwig Rodehüser

Interactions of colloidal grafted silica with transition-metal cations

F. Hamdoune · C. El Moujahid
Faculté des Sciences et Techniques
Département de Chimie, BP 416
Tangier, Morocco

M.-J. Stébé · C. Gérardin · C. Selve
L. Rodehüser (✉)
Laboratoire de Chimie Physique
Organique et Colloïdale, UMR
7565 C.N.R.S.-UHP Nancy I
Faculté des Sciences, BP 239
54506 Vandoeuvre-lès-Nancy cédex, France
e-mail: ludwig.rodehuser@lesoc.uhp-nancy.fr
Tel.: +33-383-912361; Fax: +33-383-912532

P. Tekely
Méthodologie RMN, UPRES A 7042
C.N.R.S.-UHP Nancy I
Faculté des Sciences, BP 239
54506 Vandoeuvre-lès-Nancy cédex, France

Abstract Synthetic silica-bearing organic groups derived from amino acids at their surface have been prepared via the sol–gel technique. The particle size of the colloidal silica obtained in this way has been assessed by different methods. The complexation of divalent metal cations, such as Cu^{2+}, Cd^{2+}, and Pb^{2+}, has been studied in aqueous media by using ion-selective electrodes.

Key words Grafted silica · Colloids · Amino acids · Complexation · Metal cations

Introduction

The production of grafted silica has largely expanded during the past 15 years owing to a multitude of potential applications for this kind of material: chemistry, biology, nonlinear optics, the environment [1–4].

In this work, the coordinating properties of a series of synthetic silica compounds were studied in the presence of different transition-metal cations. The colloid-size silica particles were prepared via the sol–gel technique [5] starting from chemically modified siloxanes (**1**) (Scheme 1). The latter contain amino acids or oligopeptides (aspartic acid, carnosine) as prosthetic groups, and these are known to have a strong affinity for transition-metal cations such as copper(II) [6].

Experimental

Ion-selective electrode (ISE) measurements were made by means of a Radiometer PHM 240 ion meter to which the corresponding ISE and an Ag/AgCl reference electrode had been connected. The

precision was ± 0.1 mV. The measuring vessel was kept at 25 ± 0.1 °C throughout the experiments.

Light scattering measurements were carried out using a home-built computer-controlled instrument equipped with a Malvern correlator.

$$Y\text{—}NH(CH_2)_3Si(OEt)_3$$

1

"sol-gel process" | $MeOH/H_2O/NH_4OH$ + $Si(OEt)_4$ (**TEOS**)

Grafted polymer with prosthetic groups **Y**

Solid-state ^{29}Si and ^{13}C NMR spectra were recorded using a Bruker AM 300 instrument.

Electron micrographs were obtained by transmission electron microscopy at a magnification of typically 20,000.

Results and discussion

After controlled polymerization of the monomers under well-defined conditions the colloidal silica particles bear on their surface the coordinating groups (Table 1); their density per surface unit and the size of the particles depend on the conditions of synthesis [7]. They were characterized by solid-state NMR spectroscopy. The complexing ability of these colloids for the metal cations Cu^{2+}, Cd^{2+}, and Pb^{2+} has been evaluated from ISE measurements and correlated with the particle size determined by light scattering and/or electron microscopy.

ISE measurements

The complexing ability of these colloids for the metal cations Cu^{2+}, Cd^{2+}, and Pb^{2+} was evaluated from ISE measurements. An aqueous solution of known metal concentration (10^{-4} moll^{-1}) was added to a suspension of

Table 1 Prosthetic groups Y in the grafted silica structure

Substituent Y	Formula
Di-boc-Carnosine (1)	
Carnosine (2)	
Aspartic acid β-OBz (3)	
Aspartic acid α-OBz (4)	
Aspartic acid (5)	
Aspartic acid (α-OBz) β-Ala (6)	

Table 2 Particle size (mean values) for synthetic silica grafted with different substituents Y

Group Y	1	2	3	4	5	6
Size (nm)	170	130	530	260	270	550

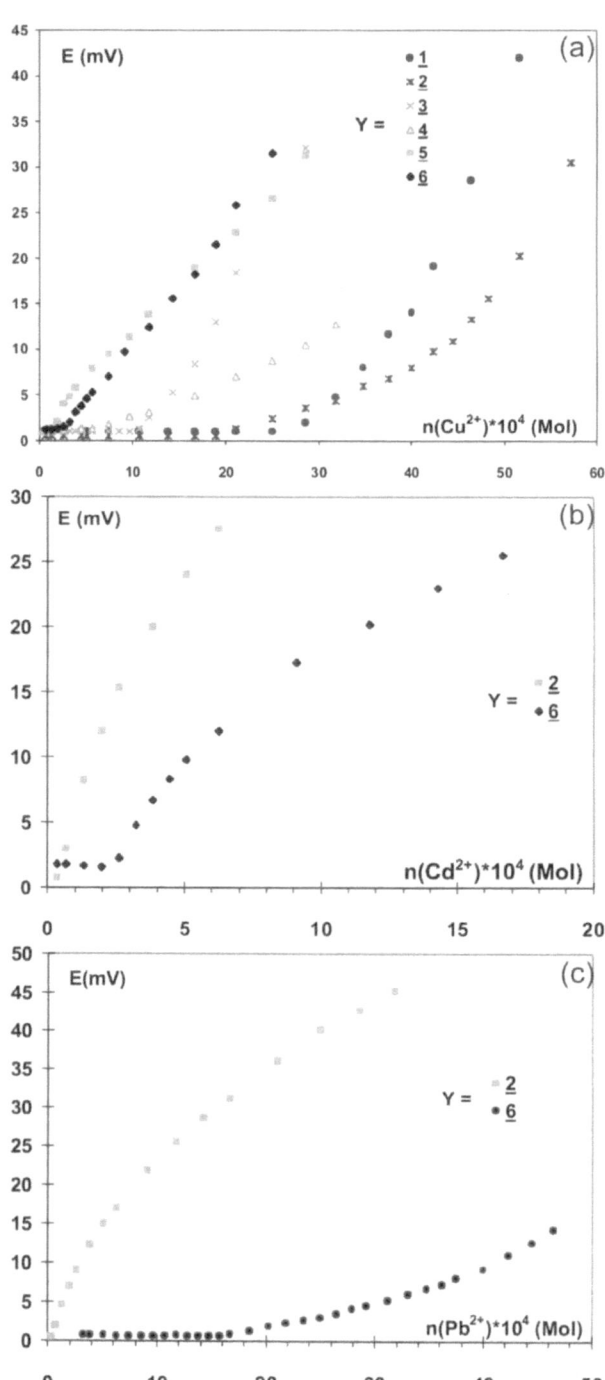

Fig. 1 Ion-selective electrode potential versus metal ion concentration for grafted silica in the presence of **a** copper (II), **b** cadmium(II), and **c** lead(II)

the grafted silica in small increments and the potential of the electrode was monitored at constant ionic strength (10^{-2} M KNO$_3$). The results are shown in Fig. 1.

Since the ISE indicates the activity of the free metal ions, the curves show that copper(II) is coordinated

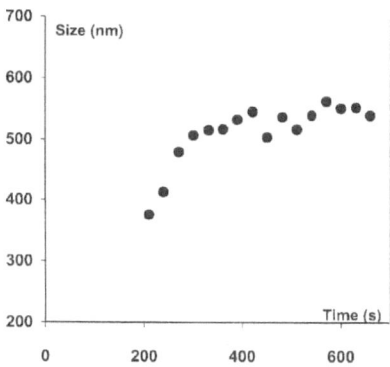

Fig. 2 Variation with time of the particle size of silica grafted with **6**, as assessed by light scattering measurements

most efficiently by silica grafted with carnosine (**1**, **2** in Fig. 1a; free Cu^{2+} appears only at important quantities of total added copper), whereas cadmium and lead are selectively bound by aspartic acid when this group is attached to the silica structure via a β-alanine spacer. Among the latter two, the amount of lead coordinated to **6** is significantly larger than that of cadmium under the same conditions.

Particle size

The particle size of the synthetic silica was assessed by light scattering measurements and, in some cases, by electron microscopy. The results of both methods are coherent. The mean values of the particle sizes found for the different products are shown in Table 2.

The initial size of the polymerized silica varies with time and reaches an upper limit as shown in Fig. 2 for a representative sample of **6**.

An image of aggregated silica particles obtained by electron microscopy is shown in Fig. 3. It can easily be seen that the size of the small spheres in this representative sample is of the same order as that resulting from light scattering experiments.

Conclusions

Among the different colloidal grafted silica studied the compounds containing carnosine derivatives have a

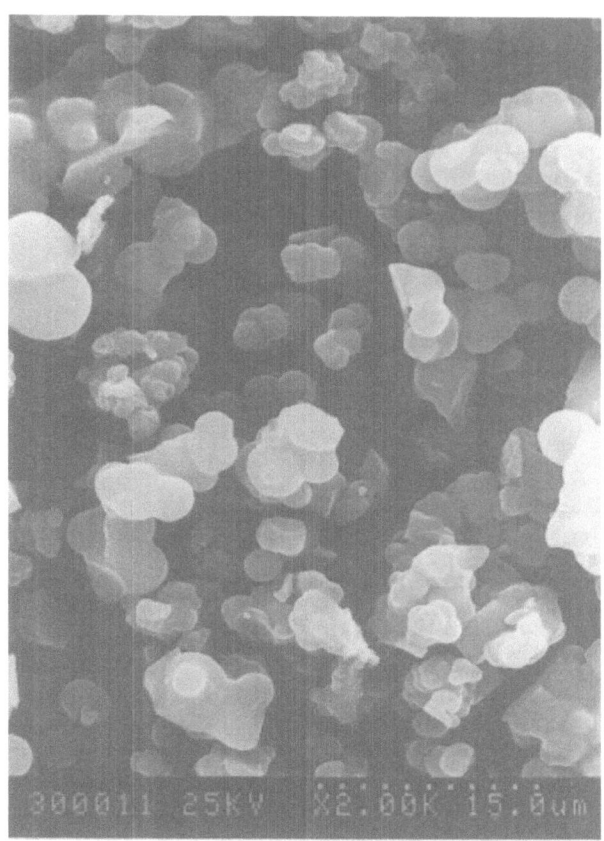

Fig. 3 Electron micrograph of a representative sample of grafted silica after aggregation

pronounced affinity for Cu^{2+} ions, whereas the presence of flexible aspartic acid–β-alanine prosthetic groups on the silica surface favors the complexation of the ions Cd^{2+} and Pb^{2+}. The particle size of the grafted silica does not appear to be related in a significant manner to their complexing properties. The Pb^{2+} ion is bound more efficiently than the smaller Cd^{2+} ion. This observation may be explained by the interaction of lead with more than one coordinating group on the silica surface owing to its larger size favoring chelation of the metal ion, as well as by its lower hydration in aqueous media. These hypotheses have to be confirmed by complementary studies.

Acknowledgements The authors would like to thank the C.N.R.S. for financial support and Alain Kohler for technical assistance with the electron microscopy measurements.

References

1. Badley RD, Ford WT, McEnroe FJ, Assink RA (1990) Langmuir 6:792
2. Richer R, Mercier L (1998) Chem Commun 1775
3. Fowler CE, Lebeau B, Mann S (1998) Chem Commun 1825
4. Elings JA, Ait-Meddour R, Clark JH, Macquarie DJ (1998) Chem Commun 2707
5. Hamdoune F, Gérardin-Charbonnier C, Rodehüser L, Selve C (1999) Amino Acids 17:112
6. Gajda T, Henry B, Delpuech JJ (1992) J Chem Soc Dalton Trans 2313
7. Hamdoune F, El Moujahid C, Rodehüser L, Gérardin C, Henry B, Stébé M-J, Amos J, Marraha M, Asskali A, Selve C (2000) New J Chem 24:1037

Progr Colloid Polym Sci (2001) 118: 110–114
© Springer-Verlag 2001

Y. Chevalier
M.-C. Dubois-Clochard
J.-P. Durand
B. Delfort
P. Gateau
L. Barré
D. Frot
Y. Briolant
I. Blanchard
R. Gallo

Adsorption of poly(isobutenylsuccinimide) dispersants at a solid–hydrocarbon interface

Y. Chevalier (✉)
Laboratoire des Matériaux Organiques à
Propriétés Spécifiques
UMR 5041 CNRS-Université de Savoie
BP 24, 69390 Vernaison, France
e-mail: yves.chevalier@lmops.cnrs.fr
Tel +33-4-78022271
Fax: +33-4-78027187

M.-C. Dubois-Clochard · J.-P. Durand
B. Delfort P. Gateau · L. Barré · D. Frot
Y. Briolant · I. Blanchard
Institut Français du Pétrole
BP 311, 1–4 av de Bois-Préau
92506 Rueil-Malmaison cédex, France

R. Gallo
ENSSPICAM, Faculté Saint-Jérôme
13397 Marseille cédex 20, France

Abstract The adsorption at the solid–xylene interface of poly(isobutenyl-succinimides) (PIBSI) has been studied on carbon black by means of adsorption isotherms and small-angle neutron scattering. Simple diblock PIBSI having various chemical structures and poly(PIBSI) with a comblike structure were compared. The adsorption is due to the hydrophilic polyamine part. It was related to the chemical structure of the dispersants (length of the polyamine part, simple diblock structure versus comblike). The adsorption phenomenon was irreversible at low concentrations; the adsorbed macromolecules are fully stretched and form a mono-layer of 30-Å thickness. The consequences for the colloidal stability of carbon black dispersions in xylene were analyzed by means of quasi-elastic light scattering and rheology measurements.

Key words Adsorption · Solid–liquid interface · Amphiphilic copolymers · Poly(isobutenylsuccinimide) · Carbon black

Introduction

Polymer surfactants such as poly(isobutenylsuccinimide) (PIBSI) derivatives are extensively used as fuel detergents and ashless dispersant additives for lubricants [1]. Their role in fuels is to keep the engine wall clean by preventing the deposition of badly burnt carbonaceous materials [2]. These polymers prevent the aggregation of soot in lubricants in order to keep the lubricant oil fluid for long utilization times [3]. In both applications, the polyamine block of the polymers ensures a strong adsorption onto the deposits or metallic surfaces and the polyisobutenyl block forms a steric barrier preventing adhesion and/or aggregation [4].

Poly(isobutenylsuccinimides)

The PIBSI studied are amphiphilic block copolymers with a polyamine polar part and a nonpolar part made of polyisobutene [5]. Two macromolecular architectures were compared: a diblock and a comblike structure (Fig. 1). The usual diblock structure consists of a polar polyamine block, which ensures the adsorption, and a nonpolar polyisobutene block, which ensures the solubility in hydrocarbon solvents and forms a steric barrier at the surfaces. This diblock copolymer, called PIBSI, is widely used as a fuel or lubricant additive. In the comblike architecture, the diblock polymers were attached together by means of a central backbone with degrees of polymerization, dp, of 9, 13.5 or 18. In both types of polymer, the number-average degree of polymerization of the polyisobutene blocks was 20 and different polyamine blocks derived from diethylenetriamine (DETA), triethylenetetramine (TETA), tetraethylenepentamine (TEPA) and aminoethylpiperazine (AEP) were studied (Fig. 1).

Adsorption isotherms

The adsorption takes place by means of acid–base interactions of the polyamine blocks with acidic sites at the surface of the carbon black. The surface density of the

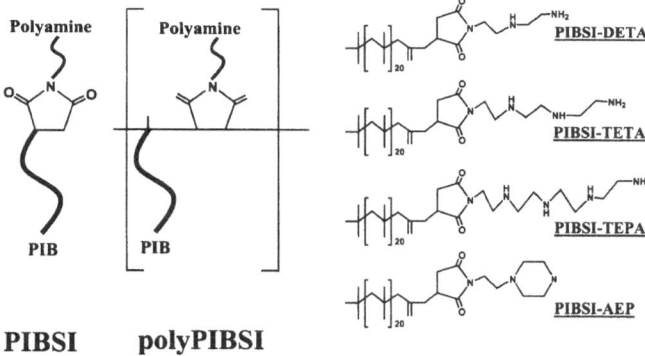

PIBSI   **polyPIBSI**

Fig. 1 Macromolecular architecture of the polymers (*left*) and chemical formulae of poly(isobutenylsuccinimide) (*PIBSI*) (right):diethylenetriamine (*DETA*), triethylenetetramine (*TETA*), tetraethylenepentamine (*TEPA*) and aminoethylpiperazine (*AEP*)

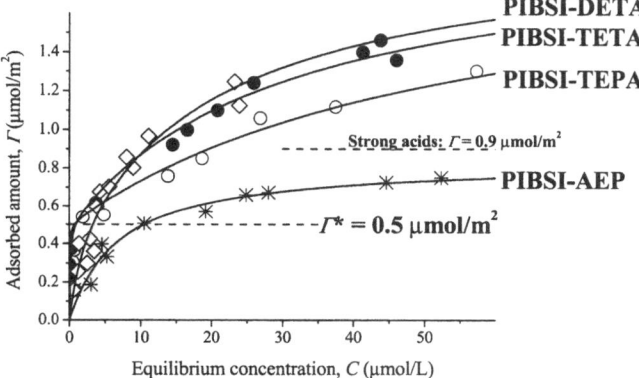

Fig. 2 Adsorption isotherms of the diblock PIBSI at the xylene–carbon black interface at 20 °C: PIBSI–AEP (*stars*); PIBSI–DETA (*diamonds*); PIBSI–TETA (*filled circles*); (○) PIBSI–TEPA (*open circles*). The *continuous lines* are the best fit of the models described in Ref. [9]

acidic sites as estimated by titration was 2.0 μmmol/m^2, a fraction of them amounting to 0.9 μmmol/m^2 were strong acids (carboxylic acids).

Diblock PIBSI

The adsorption isotherms clearly show two regimes (Fig. 2): a steep increase in the amount adsorbed (Γ) at low concentrations and low coverages ($\Gamma < 0.5$ μmmol/m^2), followed by a smoother increase in Γ. The high slope in the dilute regime indicates a strong affinity of the polyamine blocks for the carbon black surface. The second regime of weaker adsorption is reached for a coverage identical for all the diblock polymers and is lower than the surface density of strong acidic sites. This weakening of the adsorption strength was then ascribed to lateral interactions between the polyisobutene tails, in

accordance with the Alexander–de Gennes picture of adsorbed end-functional polymers [6]: In the dilute regime, called the "mushroom regime", lateral interactions are negligible and free adsorption is identical to that of small molecules. At higher coverages, the "brush regime" is entered, where lateral interactions impede the adsorption and the polymer tails stretch because of the lateral interactions. The transition for the mushroom to the brush regime is reached when unperturbed coils are able to cover the whole surface: $\Gamma^* = 1/\left(\pi R_G^2 N_A\right) = 0.5$ mmol/m^2 ($R_G = 10$ Å [7]), right at the coverage where a break in the slope is observed in the adsorption isotherms. The two-step adsorption behavior which was strongly reminiscent of the Alexander–de Gennes picture was already encountered in aqueous systems where the affinity for the surface was very high [8]. In the present case, the AEP derivative, PIBSI–AEP, was quite different from the other PIBSI: it could not enter the brush regime at high polymer concentrations because the affinity of the polyamine group for the surface was too weak. The low adsorption strength was attributed to the poor accessibility of the basic amino groups of the polar moiety containing a secondary amino group and a tertiary one which was buried inside the molecule.

The affinity of the polar part for the surface was measured in the dilute regime where free adsorption takes place. The standard free energy of adsorption, $\Delta_{\text{ads}}G^0$, was estimated by means of a model of the adsorption behavior in Ref. [9]. For the present discussion, the order of increasing affinity for the carbon black surface can be seen on the adsorption isotherms: PIBSI–AEP < PIBSI–DETA < PIBSI–TETA < PIBSI–TEPA. As expected, a longer polyamine yields a stronger adsorption because of multiple binding. The stronger adsorption of polymers with respect to their monomeric analogue is a well-known phenomenon. The slope of the adsorption isotherms in the brush regime was lower because of the lateral interactions between the PIB tails; however, the hindrance effect was more intense for the longer polyamines, although the PIB part was the same. It was quite surprising that the order of affinity in the brush regime was the reverse of that in the dilute regime. A rationale for this paradoxical effect can be found in the polymeric nature of the adsorbing moiety. Because of the multiple binding occurring in the dilute regime, the surface density of free sites available for further adsorption in the brush regime is less for longer polyamines. Further adsorption takes place at less acidic surface sites; however, once a new macromolecules is adsorbed, even on a poorly acidic site, the remaining free amino groups can bind to the surface by means of an exchange with already adsorbed groups. Since an exchange is involved, the energy balance is very low. A dynamic rearrangement inside the adsorbed layer explains the strong binding with a low free-energy balance.

The adsorption was irreversible at coverages $\Gamma < 1$ μmmol/m^2. This was shown in experiments where the solution was diluted with pure solvent: the adsorbed amount remained at a constant level, while it should have decreased, following the adsorption isotherm, in the case of a reversible adsorption. The irreversible adsorption is a well-known consequence of multiple binding; its manifestation was observed at a macroscopic level. However, a dynamic exchange of adsorbing groups inside the adsorbed layer was put forward; the adsorption was irreversible on a macroscopic scale, but reversible on a microscopic (molecular) level (Fig. 3).

Comblike poly(PIBSI)

On going from the diblock PIBSI to the comblike poly(PIBSI), the affinity for the surface observed at low coverages increased dramatically. This was the expected behavior since the number of amino groups, which was moderate in the diblock structure, reached the level of true macromolecules. The two-step feature of the adsorption isotherms was even more pronounced than for the diblock (Fig. 4). Secondly, the mushroom-to-brush transition coverage was higher than for the diblock although the PIB tails were the same. Indeed, the PIB were already close to each other in the comblike polymer, so most of the geometrical constraints that hinder the adsorption in the brush regime were released. $\Gamma^* = 0.8$–0.9 μmmol/m^2 was found, whatever the degree of polymerization. Lastly, the adsorption in the brush regime was definitely less than the diblock in spite of a higher affinity for the surface. It even decreased as a

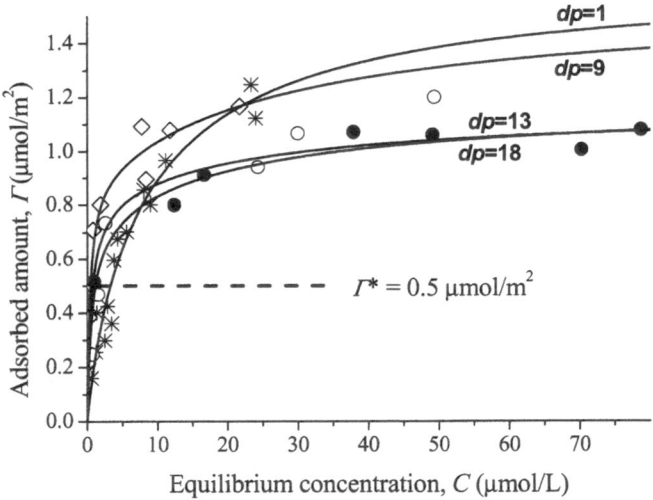

Fig. 4 Adsorption isotherms for diblock PIBSI(DETA) (dp = 1) and comblike poly(PIBSI–DETA)

function of the degree of polymerization. It was speculated that the dynamic rearrangement process which allowed the adsorption of the diblock in the brush regime was no longer operative. The amino groups of high linear density along the polymer backbone have frozen in the structure of the adsorbed layer. This is an example where an increase in the adsorption strength led to less adsorption; the coverage remains moderate, but the brush regime is entered. Different behavior would be observed if the adsorption strength was increased by substituting the polar groups for more reactive ones [8] instead of increasing their number.

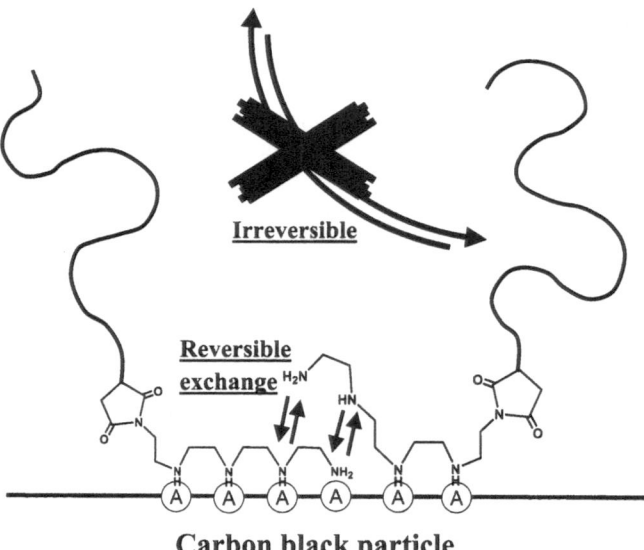

Fig. 3 Sketch of the dynamic phenomena at the interface showing the reversible and irreversible processes

Structure of the adsorbed layer

The structure of the adsorbed layer in the brush regime was measured by small-angle neutron scattering. Thus, at the contrast matching conditions where the scattering length density of the carbon black particles and the solvent are identical, the scattering comes from the polymer only [10]. The residual polymer in solution could be removed by replacing the supernatant by pure solvent without altering the adsorbed layer because of the irreversible nature of the adsorption. The scattered intensity from the adsorbed layer essentially decayed as q^{-2}, showing that the structure was a thin layer. Assuming a homogeneous internal structure of the layer, the scattered intensity is $I(q) \propto q^{-2}\exp(-q^2h^2/12)$ [11]. The measured thickness was $h = 30$ Å, significantly larger than the thickness in the mushroom regime ($2R_G = 20$ Å) and less than a fully extended PIB chain (40 Å). The polymer tails were stretched in the brush regime, as inferred by Alexander and de Gennes.

Colloidal stability of carbon black suspensions

The adsorption of PIBSI derivatives bring about a steric stabilization of the carbon black suspension because of the PIB tails protruding towards the solvent. The colloidal stability was evaluated by measurements of aggregation kinetics. This well-established method, which involves time-resolved optical measurements (turbidity, particle counting, quasielastic light scattering) [12], was adapted to the present case where the carbon black particles strongly absorb light. Thus, conventional measurements by transmission are not possible unless the suspension is extremely dilute. Measurements could be performed at high dilution (10^{-5} weight fraction) but the kinetics were so slow that the experiments lasted 1,000 h [13]. A quasielastic back-scattering device (LID DL 135-45) [14] was used in order to make kinetic measurements for a few hours with fairly concentrated suspensions (1–5 wt%). The result of an experiment was a time-resolved hydrodynamic radius of the particles (Fig. 5).

The aggregation of particles led to the formation of loose superstructures of low density which usually have a fractal internal structure characterized by the fractal dimension, d_f. The faster the aggregation, the lower the density of the particles inside the aggregates, the lower the value of d_f. In the fast coagulation regime, where there is no barrier against the aggregation, diffusion-limited colloidal aggregation (DLCA) occurs and d_f is of the order of 1.78. The steric stabilization by an efficient layer of adsorbed polymer decreases the sticking efficiency of each interparticle collision. In the case of very low sticking probability, denser aggregates are formed by reaction-limited colloidal aggregation and d_f is expected

to be of the order of 2.1. The kinetics of DLCA obeys some universal relationships [15]. In particular, the hydrodynamic radius increases as $R_H = A\, t^{1/d_f}$ for long times.

The carbon black particles were dispersed at $t = 0$ in xylene containing the polymer by means of sonication. The hydrodynamic radius at high polymer contents was 90 nm, far larger than the radius of the elementary particles of carbon black. The initial sonication did not fully disrupte the aggregates of particles, even when large amounts of polymer were present. In the aggregation kinetics, the "primary particles" were small clusters of elementary particles which were irreversibly stuck together.

The domain of polymer concentration where kinetics measurements could be performed was quite narrow because, on one hand, the aggregation at low polymer contents was very fast and large aggregates settled to the bottom of the measurement cell, prohibiting further measurements; on the other hand, the particle size did not vary at high polymer contents. The presence of the polymer has a beneficial effect with respect to the colloidal stability since it was able to fully prevent the aggregation. In the intermediate regime, where the size variation could be measured, the aggregation kinetics followed the DLCA law. The value of $d_f\approx2.1$ was larger than predicted for a DLCA process, suggesting a restructuring of the aggregates faster than the measurement time [16]. Such a phenomenon has already been observed with latex particles stabilized by a polymer [17]. However, the aggregation was obviously not diffusion-limited since it was in the slow coagulation regime.

Lastly, the minimum polymer content required for an efficient stabilization (no size variation) corresponds to an adsorbed amount of the order of 0.8–0.9 μmol/m^2, whatever the polymer (except for PIBSI–AEP). This is in the brush regime.

Notice that PIBSI–AEP was again different since 10 times higher concentrations of polymer were required. PIBSI–AEP does not form a brush because of its low adsorption; however, this failure can be retrieved by using the comblike poly(PIBSI–AEP).

The colloidal stabilization by the polymer brushes persists in concentrated dispersions of carbon black. This can be monitored by rheology measurements on 40 wt% carbon black suspensions. The suspensions containing low levels of polymer were viscoelastic. The flow behavior as measured in a cone–plate rheometer was pseudoplastic. As increasing amounts of polymer were added, both the yield stress and the plastic viscosity decreased progressively. Finally, a Newtonian flow was observed at high polymer contents, when the adsorbed amount was above 0.8–0.9 μmol/m^2. The viscosity was low (5–10 mPas) and did not decrease further with increasing amounts of polymer. Notice that the role of the PIBSI dispersants in lubricants is precisely to

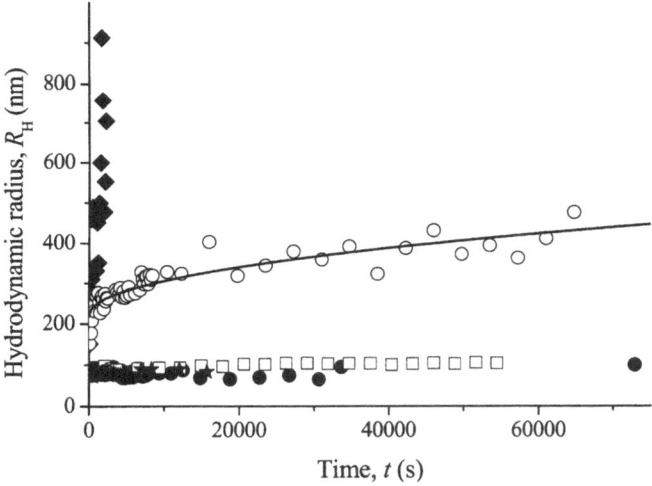

Fig. 5 Aggregation kinetics of carbon black suspensions stabilized with various amounts of PIBSI–TEPA relative to carbon black: 0.0625 (*diamonds*); 0.0714 (*open circles*); 0.125 (*filled circles*); 0.25 (*squares*); 1 (*stars*). The solid line is for $d_f = 2.1$

maintain a low viscosity of the lubricant oil by preventing the aggregation of the soot particles.

Conclusions

The different PIBSI-type polymers adsorbed by their hydrophilic polyamine part on the acidic functions of carbon black particles. The adsorption took place in two steps and was reminiscent of the two regimes of adsorption described by Alexander and de Gennes as the "mushroom" and "brush" regimes [6]. At low coverage, the unperturbed macromolecules adsorbed on independent strong acidic sites. A second regime was reached at higher coverages, where the steric hindrance by the macromolecular tails contributes to the thermodynamics of adsorption. A dynamic reorganization of the bound amino groups inside the adsorbed layer was postulated. At low concentrations, it was shown that the adsorption phenomenon was irreversible, owing to the multifunctional nature of the polar part.

The affinity for the solid surface increased with the number of amino groups in the polar part. Similarly, changing the diblock for a comblike structure led to a much stronger adsorption because of the large density of amino groups along the polymer backbone and since the steric hindrance of adsorption by lateral interactions was released in the comblike structure.

The macromolecules adsorbed as a monolayer. The layer thickness of 30 Å measured in the brush regime showed the stretching of the PIB tails.

Some compounds having amine groups behave quite differently, however. PIBSI–AEP bound rather weakly. It was unable to bind to any available acidic site and could not enter the brush regime.

The adsorption of polymers prevented the aggregation of carbon black particles in colloidal suspensions. Thus, the presence of adsorbed polymer switched the aggregation kinetics into the slow coagulation regime. Aggregation was completely suppressed when the adsorbed amount was above 0.8 μmmol/m^2, whatever the type of PIBSI derivative.

The control of the aggregation had important consequences for the rheology of concentrated carbon black dispersions in xylene. As PIBSI was added to the dispersions, the rheological behavior progressively turned from that of a plastic body with a yield stress to Newtonian behavior with a low viscosity.

References

1. Guibet J-C (1997) Carburants et moteurs. Publications de l'Institut Français du Pétrole, Technip, Paris
2. Owen K, Landells RGM (1989) In:Owen K (ed) Gasoline and diesel fuel additives. Wiley, London, p 23
3. Forbes ES, Neustadter EL (1972) Tribology 5:72
4. (a) Pugh RJ, Matsunaga T, Fowkes FM (1983) Colloids Surf 7:183; (b) Pugh RJ, Fowkes FM (1984) Colloids Surf 9:33
5. Dubois-Clochard M-C (1998) PhD thesis. University Paris VI and Publications de l'Institut Français du Pétrole, Technip, Paris
6. (a) Alexander S (1977) J Phys 38:983; (b) de Gennes PG (1980) Macromolecules 13:1069; (c) de Gennes PG (1987) Adv Colloid Interface Sci 27:189
7. Fox TG Jr, Flory PJ (1951) J Am Chem Soc 73:1909

8. (a) Chevalier Y, Brunel S, Le Perchec P, Mosquet M, Guicquero J-P (1997) Prog Colloid Polym Sci 105:6; (b) Mosquet M, Chevalier Y, Brunel S, Guicquero J-P, Le Perchec P (1997) J Appl Polym Sci 65:2545
9. Dubois-Clochard M-C, Durand J-P, Delfort B, Gateau P, Barré L, Frot D, Briolant Y, Blanchard I, Chevalier Y, Gallo R (2001) Langmuir (in press)
10. Williams CE (1991) In:Lindner P, Zemb T (eds) Neutron, X-ray and light scattering. North-Holland, Amsterdam, p 101
11. Auvray L, Auroy P (1991) In:Lindner P, Zemb T (eds) Neutron, X-ray and light scattering. North-Holland, Amsterdam, p 199
12. Overbeek JTG (1952) In:Kruyt HR (ed) Colloid science, vol 1. Elsevier, Amsterdam, p 278

13. (a) Bezot P, Hesse-Bezot C, Rousset B, Diraison C (1995) Colloids Surf A 97:53; (b) Bezot P, Hesse-Bezot C, Diraison C (1997) Carbon 35:53
14. (a) Pinier F, Woodley B, Patin P, Frot D (1996) US Patent 5,572,321; (b) http://www.l-i-d.com/
15. (a) Weitz DA, Huang JS, Lin MY, Sung J (1985) Phys Rev Lett 54:1416; (b) Lin MY, Lindsay HM, Weitz DA, Klein R, Ball RC, Meakin P (1990) J Phys Condens Matter 2:3093
16. Aubert C, Cannel DS (1986) Phys Rev Lett 56:738
17. Tirado-Miranda M, Schmitt A, Callejas-Fernández J, Fernández-Barbero A (1997) Prog Colloid Polym Sci 104:138

Progr Colloid Polym Sci (2001) 118:115–118
© Springer-Verlag 2001

STRUCTURE AND DYNAMICS AT INTERFACES

A. Schmitt
A. Fernández-Barbero
M. Á. Cabrerizo-Vílchez
R. Hidalgo-Álvarez

A single-cluster light scattering study of fast-aggregating protein-coated polymer colloids

A. Schmitt (✉) · M. Á. Cabrerizo-
Vílchez · R. Hidalgo-Álvarez
Biocolloid and Fluid Physics Group
Department of Applied Physics
University of Granada
Campus de Fuentenueva
18071 Granada, Spain
e-mail: schmitt@ugr.es
Tel.: +34-958-246104
Fax: +34-958-243214

A. Fernández-Barbero
Complex Fluid Physics Group
Department of Applied Physics
University of Almería
Cañada de San Urbano s/n
04120 Almería, Spain

Abstract We show how single-cluster light scattering may be used to study the aggregation kinetics of surface-modified colloidal particles. The surface characteristics of the particles were modified by adsorbing different amounts of bovine serum albumin (BSA). All aggregation measurements were performed at high salt concentration and at the protein's isoelectric point. Single-cluster light scattering was employed to determine how different amounts of BSA adhered on the particle surface affect the aggregation mechanism. The results obtained show that the cluster size distributions scale independently of the degree of surface coverage. It was found that the BSA molecules drastically change the time evolution of the number-average cluster size. Nevertheless, the homogeneity parameter remains practically zero at all degrees of surface coverage and, hence, the aggregation regime is diffusion-like.

Key words Single-cluster light scattering · Colloidal aggregation · Dynamic scaling · Bridging flocculation · Protein adsorption

Introduction

Single-cluster light scattering (SCLS) is a powerful tool for studying colloidal aggregation processes. Its principle of operation is based on the unambiguous relation between the cluster size and the light intensity scattered by individual aggregates at low angle. SCLS allows the cluster size distribution and its temporal evolution to be measured directly and, hence, may be employed to study kinetic aspects of colloidal aggregation processes in detail.

In this work, single cluster light scattering was used to study the influence of the particle surface characteristics on the aggregation behaviour of colloidal suspensions. Aqueous suspension of spherical polystyrene beads were chosen as model systems. The particle surface characteristics were modified by adsorbing different amounts of bovine serum albumin (BSA).

Depending on the electrochemical properties of the aqueous phase, different aggregation phenomena, such as diffusion-limited cluster aggregation (DCLA) or reaction-limited cluster aggregation (RLCA) may be observed for uncovered particles. At intermediate and high surface coverage, however, interparticle bridging and steric effects are expected to alter the aggregation behaviour. This means that the adsorbed macromolecules will change the aggregation regime quite drastically.

The aim of this study was to determine to what extent the different amounts of macromolecules adhered on the surface of colloidal particles affect the aggregation mechanism for samples which initially aggregate in the DLCA regime.

Theoretical background

The time evolution of the cluster size distribution arising during aggregation of dilute colloidal system is generally described by Smoluchowski's equation [1, 2]

$$\frac{dc_n}{dt} = \frac{1}{2}\sum_{i+j=n} k_{ij}c_i c_j - c_n\sum_{i=1}^{\infty} k_{ni}c_i , \tag{1}$$

where $c_n(t)$ denotes the number concentration of aggregates of size n. The aggregation kernel, k_{ij}, quantifies the rate at which i-mers react with j-mers. Smoluchowski's equation considers only the rate at which clusters of size n are formed when two smaller clusters of size i and j collide as well as the rate at which clusters of size n disappear when they react to form larger aggregates.

It should be pointed out that the reaction kernel, i.e. the set of rate constants k_{ij}, contains all the physical information on the aggregation process. In order to obtain further insight into the aggregation process, it is therefore essential to extract information on the aggregation kernel. From an experimental point of view, it is, however, impossible to fit the complete aggregation kernel, i.e. an infinite set of rate constants, to a finite set of data. So, only some general characteristics of the aggregation kernel may be obtained. Therefore, van Dongen and Ernst [3, 4, 5] developed a classification scheme for homogeneous kernels which characterizes the aggregation kernel in terms of two scaling exponents, λ and μ

$$k_{(ai)(aj)} \approx a^\lambda k_{ij} \quad \lambda \leq 2 \; , \tag{2}$$

$$k_{i \ll j} \approx k_0 i^\mu j^v \quad v = (\lambda - \mu) \leq 1 \; , \tag{3}$$

where a is a large positive constant and k_0 a scale factor. The homogeneity exponent, λ, relates the aggregation rate of smaller clusters to the rate at which two larger clusters react; therefore, it controls the overall time evolution of the aggregation process. The scaling exponent, μ, compares the aggregation between large and small clusters with the rate at which two larger aggregates react. For DLCA, $\lambda = 0$ and $\mu < 0$ was found [6–8]. For RLCA, however, values of $\lambda \approx 0.5$ and $\lambda \approx 1$ are reported in the literature [7]. In this case, it is only clear that μ should be positive.

For systems which do not give rise to gel formation, i.e. have $\lambda \leq 1$, the cluster size distribution can be factorized for large aggregates and long times as [3]

$$c_n(t) \sim s^{-2} \Phi(n/s) \; . \tag{4}$$

In this equation, $\Phi(x)$ and $s(t)$ are two independent scaling functions which characterize the aggregation process. The time-independent function, $\Phi(x)$, may be understood as a scaled cluster size distribution which depends only on the normalized cluster size, n/s. The scaling function, $s(t)$, should be related to some average mean cluster size.

The exact analytical solution for $s(t)$ is given by [7]

$$s(t) = \begin{cases} [C_1 + (1 - \lambda)C_2 t]^{1/(1-\lambda)} & \lambda < 1 \\ C_1 \exp(C_2 t) & \lambda = 1 \end{cases} , \tag{5}$$

where C_1 and C_2 are constants. For $\lambda < 1$, $s(t)$ grows as a power law at long aggregation times,

$$\lim_{t \to \infty} s(t) \sim t^{1/(1-\lambda)} \; . \tag{6}$$

The analytical solution for the scaling distribution, $\Phi(x)$, is known only for large and small arguments [7].

For nongelling kernels, $s(t)$ can be expressed as a power of the number-average mean cluster size, $\langle n_n \rangle$, [9]:

$$s(t) \sim \langle n \rangle^\alpha \; , \tag{7}$$

where $\alpha \geq 1$. α may be used as a fitting parameter and varied until $\Phi(x)$ is calculated correctly according to Eq. (4). For further details on this fitting procedure see Ref. [9].

λ may be obtained by combining Eqs. (6) and (7). This yields.

$$\langle n_n \rangle \sim t^{\frac{1}{\alpha(1-\lambda)}} \; , \tag{8}$$

which implies that the number-average mean cluster size grows as a power of time:

$$\langle n_n \rangle \sim t^w \; , \tag{9}$$

where $w = 1/[\alpha(1 - \lambda)]$. Solving for λ, finally gives

$$\lambda = 1 - \frac{1}{\alpha w} \; . \tag{10}$$

Thus, λ may be calculated from the long-time behaviour of $\langle n_n \rangle$, once the fitting parameter α is known.

Materials and methods

Aqueous suspensions of spherical polystyrene particles were used for the aggregation experiments. The particle diameter was 580 ± 27 nm as determined by transmission electron microscopy (TEM). The polydispersity index was 1.005. The untreated suspensions maintained stable owing to charged sulphate groups on the particle surface. The corresponding surface charge density of $-2.4 \pm 0.1 \, \mu C/cm^2$ was measured by conductometric titration.

The particle surface characteristics were modified by adsorbing different amounts of BSA. Samples with 0, 25, 50, 75 and 100% of the surface covered by protein were prepared. Here, a surface coverage of 0% means that the sample were treated exactly the same way as all other samples, with the only difference being that no protein was added.

Aggregation was induced by mixing equal volumes of electrolyte solution and stable sample by means of a Y-shaped mixing device. The final electrolyte concentration was 1 M KCl. An acetate buffer of low ionic strength was employed for fixing the pH of the aggregating samples at pH 5. The initial particle concentration was always 1×10^8 cm^{-3}.

SCLS was employed to monitor the cluster size distribution during aggregation. The SCLS instrument used for this study separates the aggregates by means of hydrodynamic focusing and forces them to flow one by one across a focused laser beam. As the particles pass, they scatter a pulse of light, which is detected at low angle. The instrument counts the light pulses and classifies them according to the scattered light intensity. This makes it possible to monitor the cluster size distribution up to heptamers and to detect the overall number of aggregates. The number-average mean cluster size is calculated by dividing the initial particle concentration, c_0, by the overall particle concentration at a given time, i.e. $\langle n_n \rangle(t) = c_0 / \sum c_n$. For further details on the experimental setup see Refs [8, 10].

Results and discussion

The aim of this study was to determine how different amounts of BSA adhered on the surface of colloidal particles affect the aggregation mechanism. Five samples with different degrees of surface coverage were aggregated at high salt concentration (1 M KCl). Under these conditions, the long-range electrostatic repulsion forces are completely screened and only some residual short-range interactions between the particles may be present.

All aggregation experiment were performed at pH 5. The reason for that lies in the nature of the adsorbed BSA molecules. At a pH higher than pH 5, the BSA molecules are negatively charged and at a pH smaller than pH 5, a positive net charge is observed. Only at the isoelectric point of pH 5 are the BSA molecules neutral overall. This means that at pH 5, the presence if adsorbed BSA molecules does not alter the particle surface charge density and hence, the residual particle interaction.

All the aggregation experiments were performed and evaluated in a similar way. First, the cluster size distribution was monitored by means of SCLS. Then, the dynamic scaling formalism was used to evaluate the data obtained. In the following section the data evaluation for a sample with 75% of its surface covered by proteins is discussed as an example for all different aggregation measurements.

The cluster size distribution as obtained by SCLS is shown in Fig. 1. As one can see, the monomer concentration decreases monotonously owing to the fact that monomers can only disappear as they react with other aggregates. The curves for dimers and larger aggregates go through a maximum since they must first be formed

before they can react with the clusters. The smaller plot in the upper right hand corner of Fig. 1 shows the time evolution of the number-average mean cluster-size on a logarithmic scale. At long times, a power-law dependence according to Eq. (9) can be observed. The best fit for the exponent w leads to 0.7 ± 0.1. The dynamic scaling distribution was calculated from the cluster size distribution according to Eq. (4). Therefore, Eq. (7) was used to correlate the scaled cluster size, $s(t)$, with the number-average cluster-size. The fitting exponent α was then varied until the curves for the different cluster sizes superimposed perfectly and defined a single master curve. This curve was then identified as the dynamic scaling distribution, $\Phi(x)$. The best fit for α gives 1.5 ± 0.2. The results obtained for the exponents α and w can now be used to calculate the scaling exponent, λ, according to Eq. (10). The result obtained is $\lambda = 0.05 \pm 0.19$. This value is in good agreement with the value of $\lambda = 0$ which is generally accepted for DLCA. Nevertheless λ is not sufficient to identify DLCA unequivocally and, therefore, we prefer to denominate the aggregation regime as "diffusion-like".

The experimental results for the whole series of aggregation experiments are summarized in Fig. 2. For all the samples, the data were evaluated in the same way as described in the previous section. The small plot in the upper left-hand corner of Fig. 2 shows the exponent w as a function of the particle surface coverage. w decreases almost linearly from approximately 1 to 0.5 for increasing degree of surface coverage. This means that the number-average cluster size increases linearly in time for the uncovered particles and square-root-like for the completely covered particles. This finding suggests – at

Fig. 1 Time evolution of the number concentration of monomers (□), dimers (○) and trimers (△). The data correspond to a sample with 75% of its surface covered by bovine serum albumin. The sample was aggregated at pH 5 and 1 M KCl. The *continuous lines* are drawn as a guide for the eye. The plot in the *upper right-hand corner* shows the time evolution of the number-average mean cluster size, $\langle n_n \rangle$

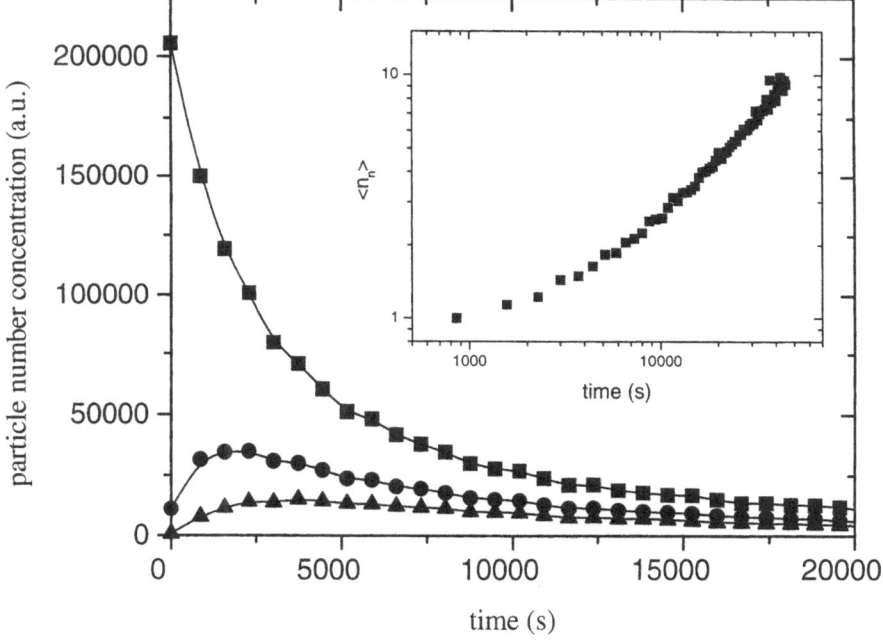

Fig. 2 Homogeneity exponent, λ, as a function of the degree of surface coverage. The plots in the *upper left-hand* and *upper right-hand corners* show the same dependency for the scaling exponent, w, and the fitting parameter, α, respectively

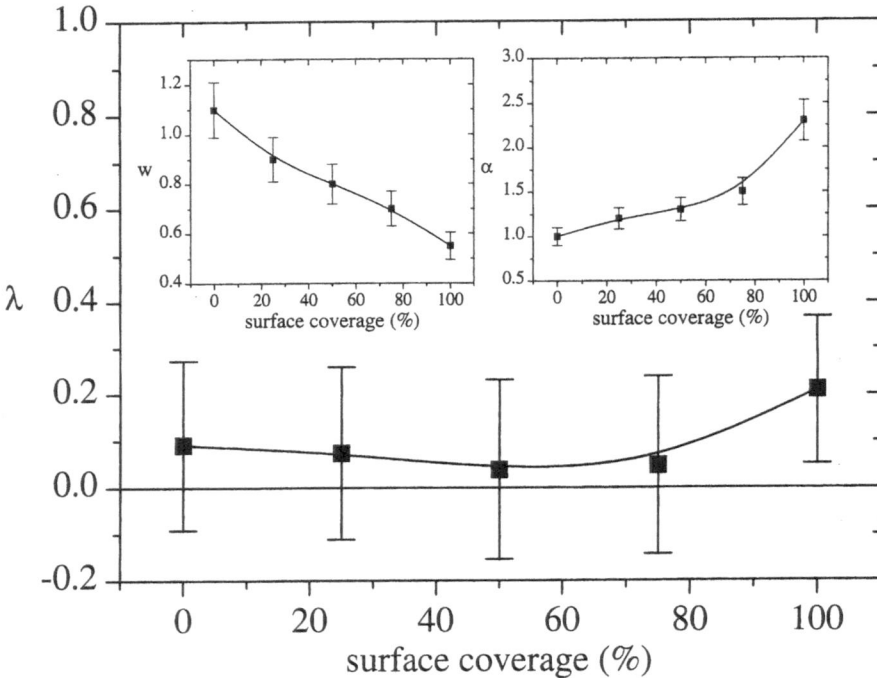

first view – a radical change in the aggregation mechanism. Similar behaviour is found for the fitting parameter α, which is plotted in the upper right-hand corner of Fig. 2. As can be seen, α increases monotonously from 1 to approximately 2 for increasing degree of surface coverage. This once again suggests a drastic change in the aggregation mechanism.

Nevertheless, neither w nor α is related directly to the aggregation kernel and hence, both have to be evaluated with care. Only the homogeneity parameter, λ, has a clearly defined physical meaning and may be interpreted directly. The main plot of Fig. 2 shows λ as a function of the particle surface coverage. As can be seen, λ remains zero for all degrees of surface coverage. Only for the completely covered particles is a slight increase for λ observed. This makes it clear that all the samples are still aggregating in a diffusion-like aggregation regime which is characterized by $\lambda = 0$. Consequently the direct interpretation of the exponents w or α would have led to erroneous results.

Conclusions

The aggregation of protein-covered colloidal particles was studied at high salt concentration by means of SCLS. All aggregation measurements were performed at the protein's isoelectric point in order to keep constant the surface charge density of the particles. The results obtained show that the cluster size distributions scale independently of the degree of surface coverage. It was found that the BSA molecules drastically change the time evolution of the number-average cluster size. Nevertheless, the homogeneity parameter remains practically zero at all degrees of surface coverage and, hence, the aggregation regime is diffusion-like.

Acknowledgements This work was supported by the Comisión Interministerial de Ciencia y Tecnología (CICYT, project MAT2000-1550-CO3-01 and MAT2000-1550-CO3-02). A. S. is grateful for financial support from the Gottlieb-Daimler- and Karl-Benz-Stiftung.

References

1. von Smoluchowski M (1916) Phys Z 17:557
2. von Smoluchowski M (1917) Z Phys Chem 92:129
3. van Dongen PGJ, Ernst MH (1985) Phys Rev Lett 54:1396
4. Meakin P (1998) Fractals, scaling and growth far from equilibrium. Cambridge University Press, Cambridge
5. van Dongen PGJ, Ernst MH (1988) J Stat Phys 50:295
6. Ball RC, Weitz TA, Witten TA, Leyvraz F (1987) Phys Rev E 58:274
7. Broide ML, Cohen RJ (1992) J Colloid Interface Sci 153:493
8. Fernández-Barbero A, Cabrerizo-Vílchez MA, Martínez-García R, Hidalgo-Álvarez R (1996) Phys Rev E 53:4981
9. Schmitt A, Fernández-Barbero A, Cabrerizo-Vílchez MA, Hidalgo-Álvarez R (2000) J Phys Condens Matter A 12:281
10. Fernández-Barbero A, Schmitt A, Cabrerizo-Vílchez MA, Martínez-García R (1996) Physica A 230:53

Progr Colloid Polym Sci (2001) 118: 119–122
© Springer-Verlag 2001

N. Stubičar

Kinetics of growth of submicrometer crystalline lanthanum fluoride particles using the pF-stat method

N. Stubičar
Laboratory of Physical Chemistry
Chemistry Department, Faculty of Science
University of Zagreb, Marulićev trg 19/II
P.O. Box 163, 10 001 Zagreb, Croatia
e-mail: stubica@sirius.phy.hr
          stubicar@phy.hr

Abstract The constant composition method (pF-stat method) was employed to prepare submicrometer-sized (few tens of nanometers to several hundred nanometers), very stable, positively charged LaF_3 particles of narrow size distribution in the acidic region from pH 4 to 4.5. Polarizing microscope pictures (anisotropic clusters) and dynamic light scattering (DLS) data demonstrate that the submicrometer structure is self-similar and a fractal with the dimension going from $D_f = 1.4$ to 2.1 (± 0.05), i.e. the aggregation process proceeds from diffusion-limited to reaction-limited with increasing supersaturation of the solution. The pF-stat rate in acidic solution obeys first-order kinetics ($p = 1.09 \pm 0.08$). Also the rate of growth from the DLS data, i.e. the slope of the change in the size (normalized median hydrodynamic diameter) with mass increment (which is proportional to time), is about 1; more precisely it increases from 0.85 to 1.05 with increasing supersaturation.

Key words Constant composition method of crystal growth · Dynamic light scattering · Hydrodynamic diameter · Lanthanum fluoride · Particle aggregation

Introduction

The constant composition method (CCM), particularly the pF-stat method, is a convenient steady-state technique for preparing large amounts of and/or large size crystalline metal fluorides far from thermodynamic equilibrium, even at very low supersaturation avoiding the nonreproducible effects of nucleation (no secondary nucleation). It has attracted theoretical and experimental attention owing to its potential technological importance [1–4]. It is well known that the working supersaturated solution used for crystal growth must be stable and have a constant composition of all the ionic species until well-defined seeds are introduced so that the growth process starts on their surface. However, the working solution itself has hardly ever been investigated dynamically and structurally in mesoscopic space and on the corresponding time scale. This is the aim of this contribution. Earlier, phenomenological research [4] showed that large micron-sized LaF_3 crystals can be prepared by the CCM method only if the pH of the equimolar supersaturated solution is in the range 5.3 ± 0.6. If it is either lower or higher than the mentioned value, which was attained with a large excess of lanthanum or fluoride ions, the solid phase formation is governed by a different mechanism, leading to systems of different structure. Here systems in the acidic region are characterized.

Experimental methods and procedure

KF and $La(NO_3)_3$ (pro analysi) were from Merck (Darmstadt, Germany). Both potentiometric pF and pH measurements were performed in the pF-stat kinetic experiments using a F^- selective electrode (Metrohm) or a glass electrode with a saturated calomel electrode as a reference. The pH was checked occasionally during the run and at the end of the run. The addition of the titrants (in the concentration ratio 1:3) is proportional to the rate of growth and was governed by the decrease of $a(F^-)$ in the working solution, so the difference against the set potential was 4 mV or a maximum of 6 mV (about 7.5×10^{-5} mol/dm^3 F$^-$). The average of six $\Delta V^{tit}/\Delta t$ (change in the added volume with time) recorded at each double mass of the initial mass of seeds, m_0, (i.e. at each $m/m_0 = 1, 2, ...6$) was used in the calculation of the rate (error about 3%). The experimental setup and the procedure have been described previously [2, 3]. The growth of the submicrometer particles was also followed by measuring the dynamic light scattering (DLS) (d_H, hydrodynamic diameter determination) using a DLS-700S spectrophotometer (Otsuka, Japan) with an Ar$^+$ laser of optimum power 10 mW [3]. A 10-cm^3 aliquot of the original (not diluted) samples, withdrawn at definite times as mentioned, was added to a 2.1-cm diameter cylindrical cell. Furthermore, the systems were examined using a polarizing microscope with a λ-plate (Leitz, Wetzlar, Germany) microscope with an automatic camera [4].

Calculation of supersaturation and growth rate

Kinetic crystal growth experiments were performed at constant supersaturation. The activity of the ions was calculated using the computer program Mineql+, [5]. The constants in Table 1, mainly taken from the Ref. [6], were taken into account.

The relative supersaturation, σ, for this particular (1:3) salt was calculated using the equation

$$\sigma = [c(La^{3+})c^3(F^-)y_\pm^{12}/K_{so}^\phi]^{1/4} - 1 \ , \tag{1}$$

where the average activity coefficient, $y_\pm$, of positive and negative univalent ions was calculated from the simple Debye–Hückel equation.

The growth rate normalized to the mass of seeds and to the specific surface area of seeds (56.13 m^2 g^{-1}) is proportional to the effective order of growth process, p, [2]:

$$J/(\text{mol} - \text{s}^{-1}\text{m}^{-2}) = k(K_{so}^\phi)^{p/4}\sigma^p \ ; \tag{2}$$

hence, in the logarithmic form of Eq. (2) the slope is equal to p.

The kinetics of growth of the submicrometer particles was calculated from the change in the median weight-average (WA) hydrodynamic diameter (median d_H) with time of the withdrawn samples using DLS measurements, as described earlier [3]; (Δd_H median)$_i$/Δd_H median)$_1$. The fractal dimension, D_f, of these particles was determined from the static light scattering data (total Rayleigh ratio as a function of the scattering vector q) using the known relation

$$R(q) \propto q^{-D_f} \ . \tag{3}$$

Table 1 $\log K$ for the complexes and the dissolved solids [6]

	$\log K$ (at 25 °C, $I{\rightarrow}0$)
Complexes:	
LaNO$_3^{2+}$	1.50
OH$^-$	−14.00
LaOH^{2+}	−9.00
H$_2$F$_2$ (aq)	6.77
HF$_2^-$	3.75
HF (aq)	3.17
LaF^{2+}	3.60
LaF$_2^+$	5.56
Dissolved solids:	
LaF$_3$ (s)	−17.95
La(OH)$_3$ (s)	−21.20

Results and discussion

Kinetic pF-stat experiments of crystal growth and characterization of the precipitate in the acidic pH region

The particles of the LaF$_3$ systems in the acidic pH region (pH between 3.95 and 4.5) with excess lanthanum nitrate over potassium fluoride are well characterized by DLS measurements. They are small, submicrometer-size particles with high colloid stability (blue sol) owing to the electrostatic repulsion between positively charged particles, as determined by the electrophoretic mobility measurements (La^{3+} ions in this pH region). The seeds used for pF-stat growth experiments had d_H(w.a.) $= 360 \pm 16$ nm, $dw/dn = 1.01$, the electrophoretic mobility of 3.14×10^{-4} cm^2/Vs and the zeta potential of 38.95 mV. The hydrodynamic diameters were the same or decreased in the initial period of time of the pF-stat experiments (compared to the size of seeds added) until the steady-state was attained, not only in the sense of constant potential (activity of F$^-$ ions) but also in the sense of the particle size and the particle size distribution (PSD). The redistribution of sizes occurs without secondary nucleation. After the steady-state had been achieved (that is here certainly at $m/m_0 = 3, ...6$, i.e. 300% of growth and more) the shifts to higher d_H values of the submicrometer particles accompanied with a narrowing of the PSD with time were determined, as is shown in Fig. 1. The linear increase in the median hydrodynamic diameters (w.a. at 50% frequency) with time is presented in Fig. 2, from which the effective order of reaction, p, is determined to be 1 (within the limit of

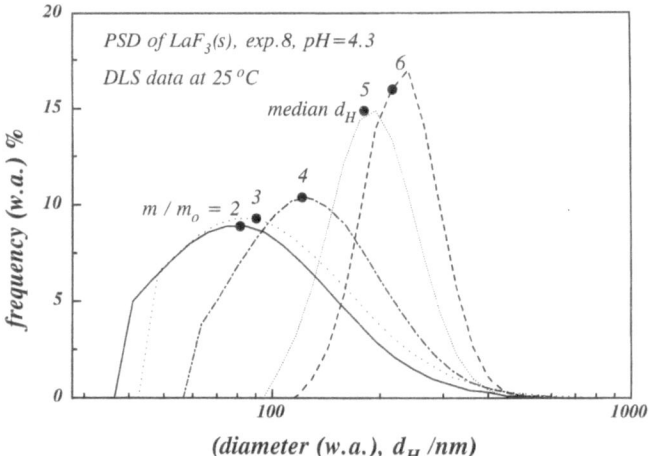

Fig. 1 Particle size distribution (PSD) of the submicrometer LaF$_3$ particles (dynamic light scattering, DLS, data) during the pF-stat growth from the supersaturated solution with equimolar concentrations of lattice ions and pH 4.3, at mass increments $m/m_0 = 2, 3, 4, 5, 6$ (m_0 is the initial mass of the seeds)

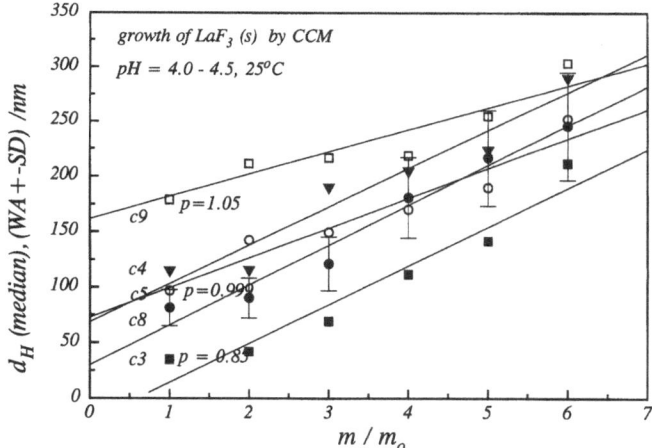

Fig. 2 Kinetics of particle growth by pF-stat experiments (constant composition method, *CCM*) determined as a change in the median hydrodynamic diameter (DLS data) with LaF_3 (s) mass increment. *3, 4, 5, 8* and *9* denote the concentrations of $La(NO_3)_3$ in the supersaturated solution: $3 \times 10^{-4} - 9 \times 10^{-4}$ $moldm^{-3}$ and KF in corresponding equimolar concentrations (1:3)

experimental error), as well as from the pF-stat growth kinetics, according to eq. 2, shown in Fig. 3. The photomicrograph which shows the manner of the aggregation and growth of the submicrometer particles i.e. anisotropic dendrites obtained after water evaporation under a microscope glass from the 8-year stable blue sol at pH~4.3, in region I of the precipitation diagram (where the solubility line coincides with the equimolarity line) was published in Ref. 4 (Figs. 4, 5a). The determination of the fractal dimension of these sols with increasing concentration of lattice ions finally approaching the equimolarity solubility line for a 1:3 salt is

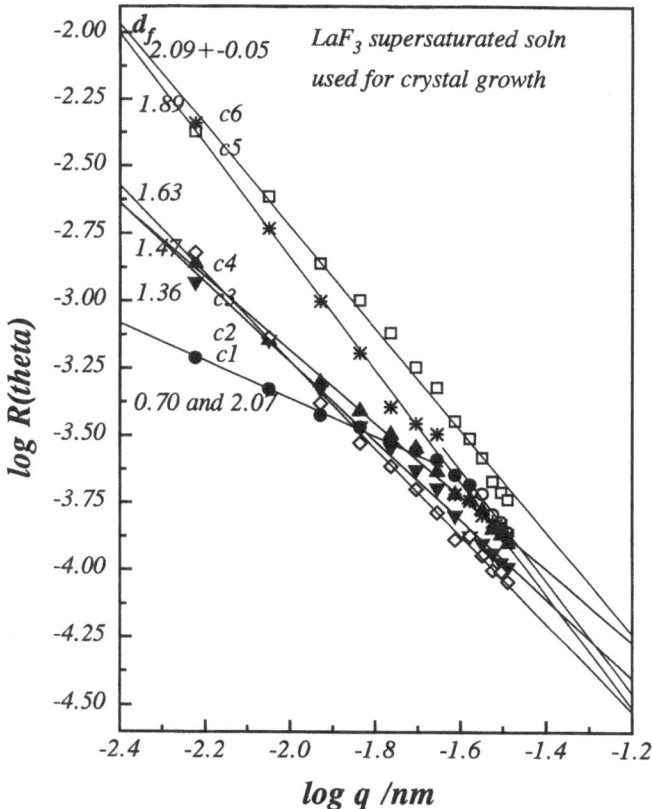

Fig. 4 Fractal dimensions of LaF_3 (s)– particles in supersaturated solutions used for pF-stat growth: total Rayleigh ratio as a function of scattering vector. The concentrations are as for Fig. 2

presented in Fig. 4. The straight lines can be drawn in the double-logarithmic plot of the light scattering intensity presented as the total Rayleigh ratios against the scattering vector, *q*, over a range equivalent to approximately 40–446 nm in real space. The fractal dimension increases with concentration going from the values characteristic for diffusion-limited aggregation finally to those characteristic for reaction-limited aggregation.

Conclusion

The CCM, in this particular case the pF-stat method, is applicable and convenient for the preparation and study of the kinetics of submicrometer particles. The aggregate structures reported here, LaF_3 blue sols in acidic solution over the pH range from 3.95 to 4.5 and over the scale ranging from a few tens of nanometers to over several hundred nanometers, are self-similar and fractal.

Acknowledgements Financial support of the Ministry of Science and Technology of the Republic of Croatia is acknowledged (project no.119 495).

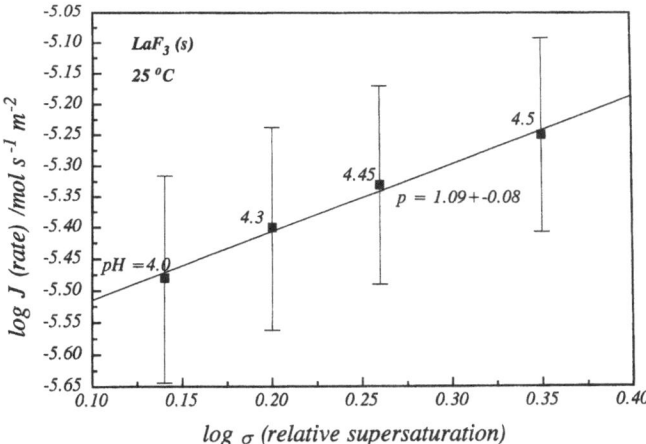

Fig. 3 Kinetics of pF-stat growth rate ($\log J$) as a function of relative supersaturation ($\log \sigma$). *p* is effective order of reaction, as for Fig. 2

References

1. Yoshikawa Y, Nancollas GH, Barone B (1984) J Cryst Growth 69:357–361
2. Stubičar N, Sčrbak M, Stubičar M (1990) J Cryst Growth 100:261
3. Stubičar N (1998) Langmuir 14:4322–4329
4. Stubičar N (1999) Prog Colloid Polym Sci 112:200–205
5. Schecher WD (1989–1994) Mineql+, version 3.01b. Environmental Research Software, Hallowell, Me, USA
6. Högfeldt E (1980) Stability constants of metal-ion complexes, part A: inorganic ligands. IUPAC chemisty data series, no 21. Pergamon, Oxford

Progr Colloid Polym Sci (2001) 118: 123–126
© Springer-Verlag 2001

F. Martínez-López
M. A. Cabrerizo-Vílchez
R. Hidalgo-Álvarez

Colloidal interaction at the air–liquid interface: experimental results

F. Martínez-López
M. A. Cabrerizo-Vílchez
R. Hidalgo-Alvarez (✉)
Dpto. Física Aplicada
Universidad de Granada
Campus Universitario de
Fuentenueva
18071 Granada, Spain
e-mail: rhidalgo@ugr.es
Fax: +34-958-243214

Abstract We present some results obtained in 2D aggregation experiments. Experiments were made using a salt subphase of 2.0 M KBr and pH values ranging from 1.5 to 6.0. We used two latex polystyrene samples with different contact angles (i.e. different hydrophobic properties) and similar particle sizes. Digitised images were obtained at different aggregation times. These images were analysed to obtain the aggregates from the whole image. The aggregation results were compared using the kinetic exponents obtained from the asymptotic behaviour of the total number of aggregates as a function of time. $N \propto t^z$. In contrast to the 3D case, the aggregation at the air–solution interface was faster for the hydrophilic particles than for the hydrophobic ones. These experimental findings were analysed using a model of the interactions between colloidal particles at the air–liquid interface that takes an electrostatic dipolar repulsive term into account between the air-exposed parts of the particles.

Key words Colloidal interaction · Air–liquid interface · Colloidal aggregation · Kinetics of aggregation

Introduction

Colloidal stability is a complex phenomenon that has attracted the interest of the scientific community. The most important theory that explains colloidal stability is the Derjaguin–Landau–Verwey–Overbeek (DLVO) theory [1]. This theory was shown to be able to explain colloidal stability of colloidal particles in the bulk; however, it is not able to explain colloidal stability of colloidal particles at the air–liquid interface. This is so because of the complex nature of the interactions between colloidal particles at the interface. At the interface we found capillary interactions that do not have an analogue in the bulk and we also found a hydrophobic interaction that is not included in the DLVO theory [2].

When comparing the stability behaviour of the same kind of colloidal particles trapped by capillary and electrostatic forces [3, 4] at the air–liquid interface and in the bulk it is usually found that at the air–liquid interface

the colloidal particles are more stable [5]. This experimental finding is not explained if we only use the energetic terms of interaction that are included in the DLVO theory and even if we include hydrophobic and capillary interactions. When we use these interaction terms, the predicted stability for colloidal particles at the interface is lower than for the same particles in the bulk, in contrast to the experimental results. This is mainly due to the increase in the Van der Waals attraction and the decrease in electrostatic repulsion between the double layers of the particles at the interface. This fact is a clear sign of the complexity of the interaction between colloidal particles at the interface.

In this work we report some experimental results corresponding to aggregation experiments that were done at the salt solution–air interface using latex particles. Aggregation was studied as a function of the subphase pH at a high salt concentration, 1 order of magnitude higher than the critical coagulation concentration (ccc) of the same latex particles in the bulk.

Experimental results

Colloidal aggregation experiments at the air–liquid interface were carried out using a small circular cell of Teflon. Salt solution was introduced into the cell and then latex particles were spread at the interface using methanol as a spreading agent. After the evaporation of the methanol a more-or-less uniform monolayer of colloidal particles was obtained. The particles at the interface move randomly and aggregate on contact. After methanol evaporation the cell was cover to prevent air motion affecting the aggregation process. The kinetics of the aggregation was followed using a phase-contact microscope with a charge–coupled–device camera that allows digital images of the aggregation to be captured.

We present results for two latex particles, the characteristics of which are shown in Table 1. Both types of particles have the same size but different charges and contact angles. The last value is especially interesting because of the effect that the contact angle has in the particle exposure to the air phase. So it was expected that both kinds of particle would behave differently.

Colloidal aggregation experiments were made at a salt concentration 2.0 M KBr (1 order of magnitude higher than the ccc of these latex particles, which is around 200 mM for the same salt).

We have studied the effect of the pH of the subphase on the aggregation regime.

A typical image obtained in an aggregation experiment is shown in Fig. 1. Before obtaining information from this picture we had to process the image digitally to distinguish the colloidal particles from the background; this is of fundamental importance for the experimental results.

A lot of parameters can be analysed from the picture obtained in an aggregation experiment. Examples of these are the fractal dimension of the aggregates, their anisotropy, the size density distribution, $n_s(t)$, of the aggregates, i.e. the number of aggregates of size s that are present in the image in a given time, t, etc. We focused our attention on the scaling behaviour of the total number of aggregates included in an image. The total number of aggregates, N, is a function of time that behaves as $N \propto t^z$, where the kinetics exponent z is dependent on the aggregation process followed by the system.

Fig. 1 An example of a digital image obtained in a 2D aggregation experiment at the air–liquid interface. The experiment shown was made with the JL-10 latex at a 2.0 M KBr salt concentration and pH 2.0

The pH dependence of z is shown in Fig. 2 for both latexes. As can be seen a typical characteristic of this dependence is the existence of a maximum. For both latexes the maximum is at about pH 3.0. As can be seen in this figure, the hydrophilic latex, JL10, aggregates under all the experimental conditions faster than the hydrophobic one. This result is in contrast to the experimental findings for bulk aggregation, where hydrophobic latex particles aggregate faster than the hydrophilic ones owing to the water hydration that stabilises the hydrophilic particles.

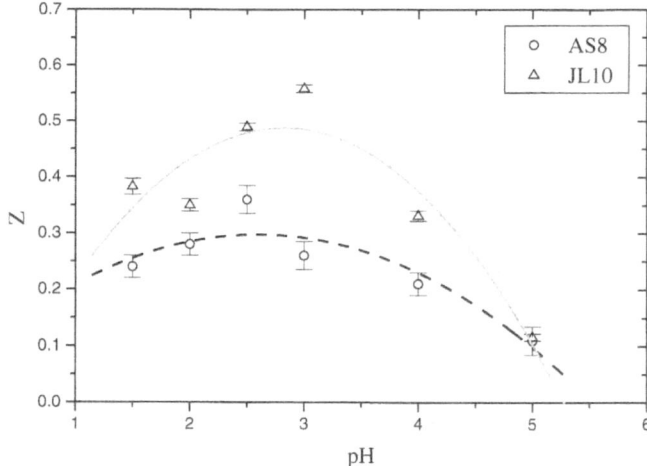

Fig. 2 Kinetics exponent, z, obtained from the scaling behaviour of the total number of aggregates as a function of time. Both kinds of latex particles, hydrophilic and hydrophobic particles, have similar behaviour for the dependence of z as a function of the pH of the subphase. Both cases present a maximum near pH 3.0, but the values are very different. In nearly all the experimental conditions the hydrophilic latex, JL10, aggregates faster than the hydrophobic one, AS8. The curves are only a guide for the eye and they do not correspond to any theoretical approximation

Table 1 Characteristics of the experimental systems used for the aggregation experiments. Two kinds of latex particles were used with different hydrophobic character. Both types of particles have a similar size (around 600 nm) but different charge density (σ) and contact angle (θ)

Latex	Size (nm)	σ ($\mu C/cm^2$)	θ
JL-10	636 ± 16	-10.3 ± 0.4	50
AS-8	580 ± 30	-2.4 ± 0.1	82

So, we have found two main questions relating to the stability of colloidal particles at the air–liquid interface:

1. Why are colloidal particles so stable at the interface?
2. Why in contrast to the bulk case are hydrophilic particles less stable at the interface?

We will try to answer these questions in the next section.

Model of the colloidal particle at the interface

Besides the energetic terms that we referred to in the Introduction we will add to our model of the colloidal particle at the interface a dipolar term of interaction due to the air-exposed parts of the particles [5]. Dipoles originate at least in part from the counterions in the subphase.

So the energetic terms that our model includes are

1. Electrostatic interaction between the immersed part of the particles.
2. van der Waals interaction both between immersed parts and emergent parts, but with a different Hamaker constant.
3. Hydrophobic interaction between immersed parts.
4. Dipolar interaction between emergent parts.

We do not include capillary attraction because for colloidal particles of this size this energetic contribution can be neglected; however, when the aggregation evolves this term might be needed because of their size dependence that scales with the sixth power of the particle radius.

A sketch of the particular model is shown in Fig. 3. We consider a spherical particle that is at the air–liquid interface and which has an air-exposed part that has a dipolar shell.

To compute the terms of the potential of interaction we made different approximations. To obtain the

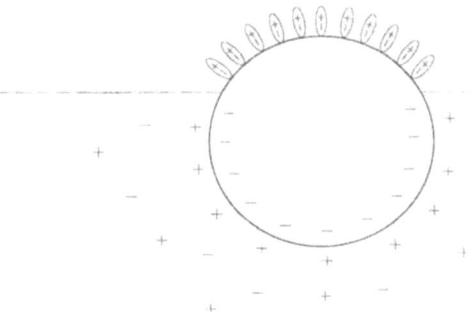

Fig. 3 Model of a colloidal particle at the interface. The spherical particle has a dipolar shell in the air-exposed part. The dipoles originate at least in part from the counterions in the subphase. The interactions also include a double-layer contribution of the particle part in the subphase

electrostatic interaction between the immersed parts of the particles we used the Derjaguin approximation [6] and the expression for the electrostatic interaction between half spaces separated by a distance h as given by Verweey and Overbeek [7]

$$V(h) = \frac{64n^0 k_B T Z^2}{\kappa} \exp(-\kappa h) ,$$

where $Z = \tanh(v e \varphi_0 / 4 k_B T)$, n^0 is the salt concentration, k_B the Boltzmann constant, T the temperature, e the electron charge, κ the inverse of the Debye length, v the ionic valence and φ_0 the surface potential. The computation of the van der Waals interaction was done using the expression obtained from Gregory [8] and Overbeek [9],

$$V(h) = -\frac{A}{12\pi h^2}\left(\frac{1}{1 + bh/\lambda}\right) ,$$

which takes into account the retardation effect. Here A is the Hamaker constant and $b = 5.32$ and $\lambda = 100$ nm. The values of the Hamaker constant used were $A_{air} = 6.6 \times 10^{-20}$ J and $A_{water} = 0.95 \times 10^{-20}$ J. The Christensen and Claesson [10] approximation for the hydrophobic interaction between half spaces was also used,

$$V(h) = W_0 \exp(-h/\lambda_0) ,$$

where W_0 and λ_0 are constants related to the strength and length respectively of the interaction. Typical values for both can be obtained from Ref. [11]. Finally, to obtain the value for the dipolar interaction of the emergent parts of the particles we divided the air-exposed part of each particle into different elements, each of them forming a dipole the length of which is the sum of typical values for the radius of a cation and an anion, and we computed the interaction by using Coulomb's law. This computation needs a 4D integration. With these approximations we can compute the total pair energy of interaction between colloidal particles at the interface. The total pair potential of interaction for both types of particles is shown in Fig. 4. The parameters used in the computation are indicated in the figure legend. This figure shows the potential barrier height is higher for the case of the hydrophobic particles. This indicates that hydrophilic particles will aggregate faster than the hydrophobic ones, as was revealed by the experimental results. The curves shown in Fig. 4 were obtained using the surface potential of these colloidal particles obtained from their mobility at pH 6.0; if we introduced the corresponding values for a lower pH the height of the potential barrier might be lower because of the decrease in the electrostatic repulsion of the immersed parts of the particles. The effect of pH on the potential of interaction within the DLVO approximation is shown in Fig. 5. If we compare Fig. 4 and 5 we can deduce that it is the dipolar interaction that makes the colloidal particles stable at

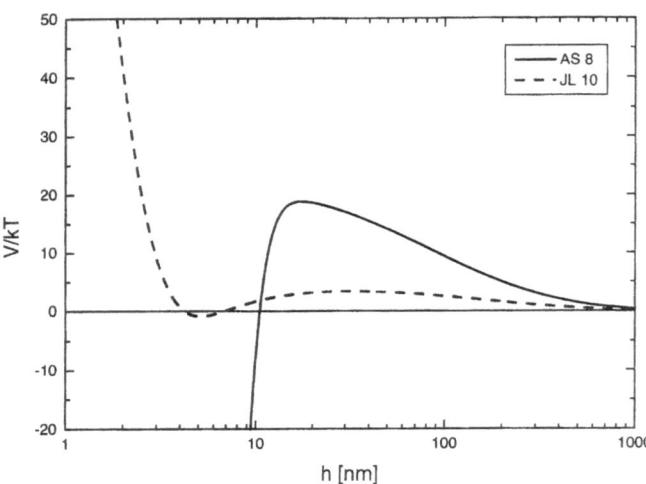

Fig. 4 Total pair potential of interaction for both latexes. The parameters charactering the JL10 latex were contact angle 50°, surface potential −56 mV, surface density of dipoles 0.18 dipoles/nm², $W_0 = 3$ mJ m⁻² and $\lambda_0 = 0.85$ nm. The corresponding values for the AS8 latex were contact angle 82°, surface potential −40 mV, surface density of dipoles 0.18 dipoles/nm², $W_0 = -60$ mJ m⁻² and $\lambda_0 = 1$ nm.

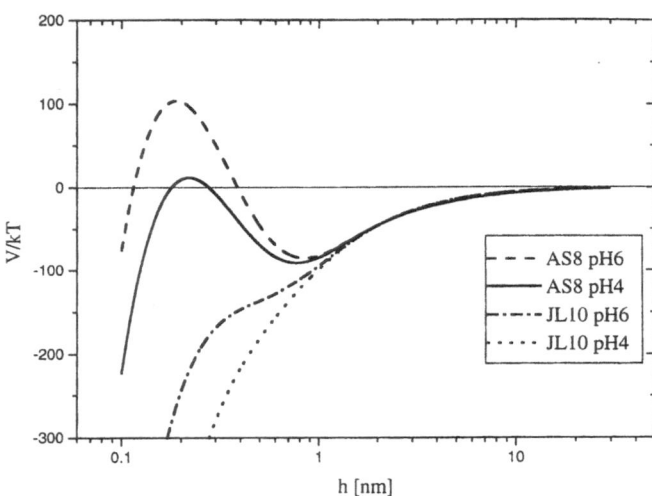

Fig. 5 Derjaguin-Landau-Verwey-Overbeek approximation, i.e. electrostatic interaction between the immersed parts of the particles and van der Waals interaction between immersed parts of the particles and between emergent parts of the particles at the interface. The *curves* correspond to different pH values for both latexes

the air–liquid interface. Both figures correspond to a 2.0 M salt concentration of a 1:1 electrolyte.

Conclusions

Colloidal particles at the air–liquid interface present a stability that is not justified using an analogue with the bulk behaviour. This great stability can be justified if we use a model of the interaction between particles at the air–liquid interface that includes a dipolar repulsion term which originates at least in part from the counterions that

are absorbed from the solution in the air-exposed part of the interfacial particle. This model can justify the colloidal stability of latex particles at the interface and it can also explain the lower stability of hydrophilic particles at the interface when compared with hydrophobic particles. This behaviour, which is different from the bulk behaviour, can be understood if we take into account that hydrophobic particles have a greater exposure to the air phase and thus greater dipolar repulsion.

Acknowledgements Financial support from "Comisión Interministerial de Ciencia y Tecnología (CICYT)", project MAT2000-1550-C03-01, is gratefully acknowledged.

References

1. See for example, Hunter RJ (1987) Foundation of colloid science, vol 1. Clarendon. Oxford
2. Williams DF (1991) PhD thesis. University of Washington
3. Earnshaw JC (1986) J Phys D 19:1863
4. Pleranski P (1980) Phys Rev Lett 45:569
5. Robinson DJ Earnshaw JC (1993) Langmuir 9:1436
6. Derjaguin BV (1934) Kolloidz 69:155
7. Verwey EJW, Overbeek JThG (1948) Theory of stability of lyophobic colloids. Elsevier, Amsterdam
8. Gregory J (1981) J Colloid Interface Sci 83:138
9. Overbeek JTG (1966) Proc Konigs Akad Wetensh B 69:501
10. Christenson HK, Claesson PM (1988) Science 239:390
11. Israelachvili J (1992) Intermolecular and surface forces, 2nd edn. Academic, New York

Progr Colloid Polym Sci (2001) 118: 127–131
© Springer-Verlag 2001

J.-C. Jacquier
P. MacArtain
K. A. Dawson

Hydrophobically modified calcium-induced κ-carrageenan gels

J.-C. Jacquier (✉) · P. MacArtain
K. A. Dawson
Irish Centre for Colloid Science
and Biomaterial
University College Dublin
Belfield Dublin 4 Ireland
e-mail: jean.jacquier@ucd.ie
Tel.: +353-1-7062503
Fax: +353-1-7062127

Abstract κ-Carrageenans are known to form strong but turbid gels in the presence of calcium ions. Nevertheless, the resulting gels show no syneresis and a relatively low melting temperature, which makes them of great interest in food applications. To better understand the mechanism of gelation of these polysaccharides, the effect of chemical modification of the remaining free secondary hydroxyl groups was studied. The formation of hydroxyethyl derivatives increases the hydrophilicity, resulting in less turbid but thermally and physically weaker gels. Alkylation of the κ-carrageenan polymer resulted in increased hydrophobicity, turbidity and brittleness of the gels. Differences in thermal stability and physical strength of both types of modified κ-carrageenan gels were studied by rheology. Turbidity was assessed by UV–vis spectroscopy.

Key words Carrageenan · Polysaccharide · Gel · Hydrophobicity · Thermoresponsive

Introduction

Carrageenans are linear sulphated polysaccharides extracted from marine red algae. They have a basic linear primary structure based on a repeating disaccharide of $\alpha(1-3)$-D-galactose and $\beta(1-4)$-3,6-anhydro-D-galactose. κ-Carrageenan contains one sulphate group per disaccharide unit at carbon 2 of the 1,3-linked galactose unit (Fig. 1) while ι-carrageenan contains an additional sulphate group at carbon 2 of the 1,4-linked galactose unit.

κ- and ι-Carrageenans are two gel-forming carrageenans of enormous importance not only in the food industry [1] but also in the pharmaceutical industry [2]. The thermoreversible gelation of these polyelectrolytes involves a coil-to-helix transition upon cooling, followed by the aggregation of the ordered molecules to form an infinite network [3]. As expected for polyelectrolytes, the counterion plays a major role in the gelation process. For instance, it is traditionally noted that ι-carrageenan is calcium-sensitive, while κ-carrageenan is more potassium-sensitive. Despite these traditional notions, a limited number of studies have dealt with the effect of calcium on the gelation of κ-carrageenan and have led to confusing results [4–6]. More recently, Michel et al. [6] have looked carefully at the phase diagram of κ- and ι-carrageenans in the presence of various cations. In the case of κ-carrageenan in the presence of calcium, they found that turbidity occurred above a low (about 0.02 M) salt concentration for all the carrageenan concentrations studied. Rheology experiments at a high concentration of polysaccharide (about 20 g/l) showed strong but turbid gels with an elastic modulus not influenced by the salt concentration above a calcium concentration threshold. For systems with much lower concentrations of polysaccharide (5 g/l), we recently showed a maximum in the elastic modulus of the calcium-induced κ-carrageenan gels corresponding to a stoecheiometric ratio, reminiscent of competition between gelation and aggregation or phase separation, the latter favoured at high calcium concentration. Very recently Ramakrishnan and Prud'homme [7] showed that hydrogen-bonding interactions control the helix formation, while Coulombic interactions dominate the mechanism of gelation.

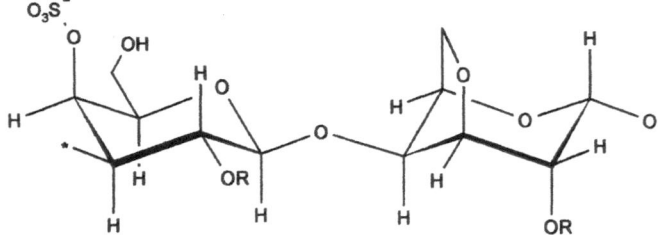

Fig. 1 Repeating disaccharide structure of κ-carrageenan derivatives. Unmodified (R = H), hydroxyethyl carrageenan (eo-cgn) (R = —CH$_2$CH$_2$OH), octyl carrageenan (C$_8$-cgn) (R = —C$_8$H$_{17}$)

Here we report the hydrophobic modification of the κ-carrageenan polymer with the view of assessing the role of hydrophobic interactions in the gelation process. To do so, we proceeded to increase the hydrophobicity of the polysaccharide by etherification of the free secondary hydroxyls groups with 1-chlorooctane or to reduce the hydrophobicity by etherification with 2-chloroethanol. The two modified polysaccharides were studied in terms of their gelling behaviour by rheology and the turbidity of the gel formed was studied by UV–vis spectroscopy.

Experimental

Materials

The κ-carrageenan used was a food-grade product from Quest (Deltagel P378). The molecular weight was determined by viscosimetry as 415,000 [7] ([η]$_{0.1M\ NaCl}$ = 672 ml/g). The samples were purified as previously described according to a procedure derived from Ramzi et al. [8].

Hydroxyethyl carrageenan

The reaction of κ-carrageenan with 2-chloroethanol (99% grade from Aldrich) was adapted from Guiseley [9]. Carrageenan (S$_g$) was added to 100 ml aqueous solution containing 1.2 g NaOH and 0.5 ml sodium borohydride Venpure (12% w/w in 14 M NaOH, from Aldrich). The solution was heated to 80 °C under mechanical stirring for 1 h to ensure complete dissolution of the polymer and 2.64 g 2-chloroethanol in 11 ml water was added dropwise over 15 min. The reaction was maintained at 80 °C for 2 h before being cooled and neutralised with a 3 M acetic acid solution. Purification of the resulting hydroxyethyl carrageenan (eo-cgn) was performed as for the unmodified polymer in order to obtain the sodium form. This treatment did not result in substantial degradation of the polymer chain as determined by viscosity measurements ([η]$_{0.1M\ NaCl}$ = 649 ml/g).

Elemental analysis of the product led to inconclusive estimation of the reactivity yield, but on the basis of published data [9] we estimated a reactivity of 40% per disaccharide unit.

Octyl carrageenan

The reaction procedure of κ-carrageenan with 1-chlorooctane (99% grade from Aldrich) was the same as for the hydroxyethyl derivative. Once again, no degradation of the polymer chain was observed ([η]$_{0.1M\ NaCl}$ = 680 ml/g). Elemental analysis of the

product led to an estimation of the reactivity yield of 7% per disaccharide unit. This somewhat low yield is explained by the poor solubility of the chlorinated reagent in the reaction medium.

Methods

Rheology

A dynamic stress rheometer (Rheometric Scientific) with parallel plates (40 mm diameter, 0.2 mm gap) was used for all the rheology experiments. A vapour barrier was used to prevent solvent evaporation. All the samples were aqueous solutions of carrageenan derivative at 5 g/l with different amounts of calcium chloride (up to 0.04 M).

The melting transitions and gelation of the various carrageenan samples were studied by dynamic oscillatory measurements during temperature ramps between 5 and 60 °C (rate 1 °C/min) using the Peltier heating option. In order to be in the linear viscoelastic domain, these experiments were performed at a low frequency (1 rad/s) and with a low stress (typically 1–5 Pa). The resulting strains on the gels were then in the 1% range.

To estimate more accurately the elastic shear modulus, G', of the gels at 5 °C, dynamic oscillatory measurements were performed at constant frequency (1 rad/s) during stress ramps (0.1 Pa to break point). The values of the elastic modulus obtained were well within the plateau region and typically at strains of 1%.

Turbidity

Turbidity was assessed using a UNICAM Helios α UV–vis spectrometer equipped with a jacketed cell holder maintained at 10 ± 0.1 °C with a circulating water bath. Carrageenan samples were poured in 1 cm fluorimeter polystyrene cuvettes purchased from Sigma (cutoff 340 nm). After zeroing on water, the transmittance was recorded between 350 and 600 nm. As the transmittance typically increased monotonically over the entire spectrum range, the transmittance at 450 nm was taken as indicative. A temperature of 10 °C was chosen rather than 5 °C, as excessive condensation developed at this lower temperature, which made the reading of the transmittance quite irreproducible and sometimes meaningless.

Results and discussion

Thermal stability

The melting transition of the carrageenan samples was studied by measuring the elastic (G') and storage (G'') moduli while imposing a temperature ramp, and the melting temperatures were taken at the moduli crossover.

The evolution of the melting temperatures of the various carrageenan derivatives with increasing concentrations of calcium chloride is shown in Fig. 2. Figure 3 is a linearisation of the curves in Fig. 2 following the Ferry–Eldridge model [10].

We can see clearly that the enthalpy associated with the melting process is constant over the range of temperature investigated and furthermore that all the samples show a similar enthalpy (slopes in Fig. 3). The chemical modification of the carrageenan samples does not seem to alter significantly the mechanism of gelation. The effect of hydrophobic modification on the thermal

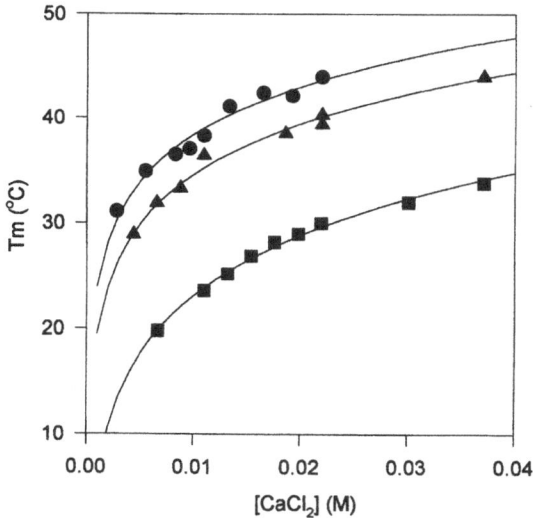

Fig. 2 Evolution of the melting temperature of carrageenan gels with calcium. Unmodified (●), eo-cgn (■), C_8-cgn (▲)

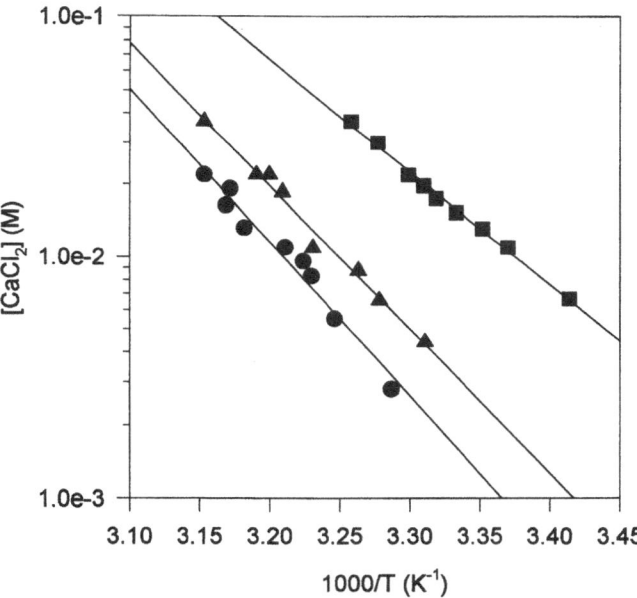

Fig. 3 Ferry–Eldridge plot of the evolution of the melting temperature of carrageenan gels with calcium. Same symbols as in Fig. 2

steric hindrance in helix formation owing to the long alkyl chains. In order to better understand this phenomenon, we compared the evolution of the gel strength for the various derivatives.

Gel strength

The evolution of the elastic modulus of the three carrageenan samples at 5 °C with varying calcium concentration is shown in Fig. 4.

As noted previously, G' of unmodified carrageenan passes through a maximum at an "optimal" calcium concentration of around 0.015 M, which corresponds roughly to one calcium per disaccharide unit. This behaviour is maintained for both carrageenan derivatives except that the maximum is reached at a slightly higher calcium concentration.

The hydrophilic derivative eo-cgn shows a 20-fold decrease in G' when compared to the unmodified carrageenan and is not affected as much by high calcium concentrations. The hydrophobic derivative C_8-cgn has G' values similar to the unmodified carrageenan for low calcium concentrations, but the gel strength is less dramatically affected at high calcium concentrations.

To better understand this phenomenon of "optimal" calcium concentration, we proceeded to measure the evolution of the yield stress of the various samples with calcium concentration (Fig. 5). To do so, we recorded the dynamic stress sweeps of each sample at 5 °C and under a constant frequency. As can be seen from Fig. 5, where the

stability of these carrageenan samples is less clear. As previously described [9], the hydrophilic modification results in a large decrease in thermal stability, the melting temperatures of the eo-cgn samples being on average 15 °C lower than the unmodified samples over the entire range of calcium concentration studied. However, the hydrophobically modified octyl carrageenan (C_8-cgn) sample also showed some slight decrease in thermal stability when compared to the unmodified sample. One explanation for such behaviour might reside in some

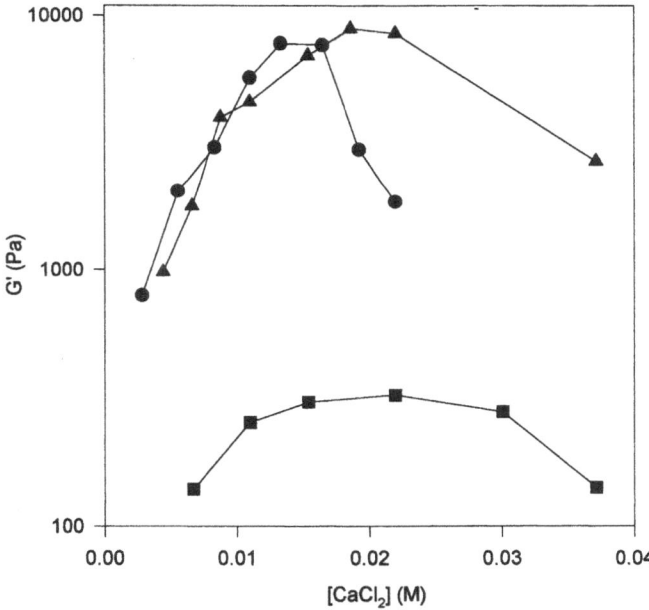

Fig. 4 Evolution of the elastic modulus (G') of the carrageenan gels with calcium at 5 °C. Same symbols as in Fig. 2

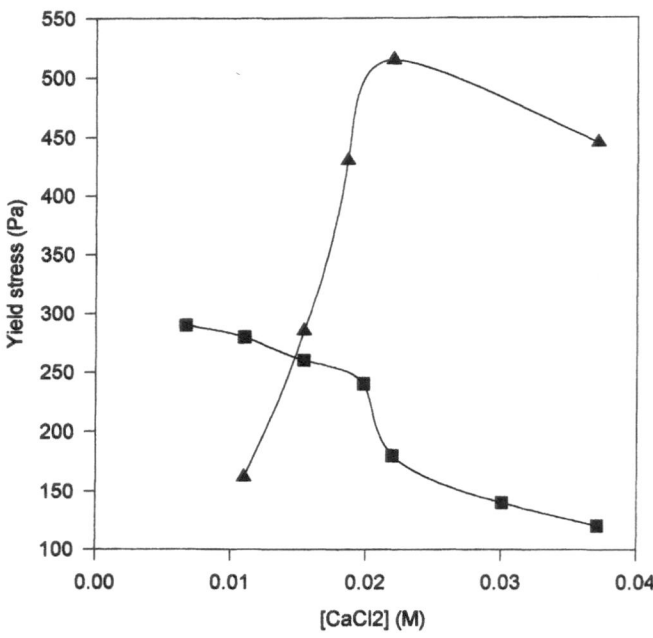

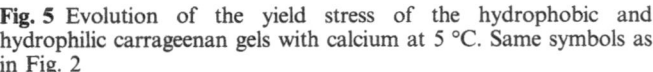

Fig. 5 Evolution of the yield stress of the hydrophobic and hydrophilic carrageenan gels with calcium at 5 °C. Same symbols as in Fig. 2

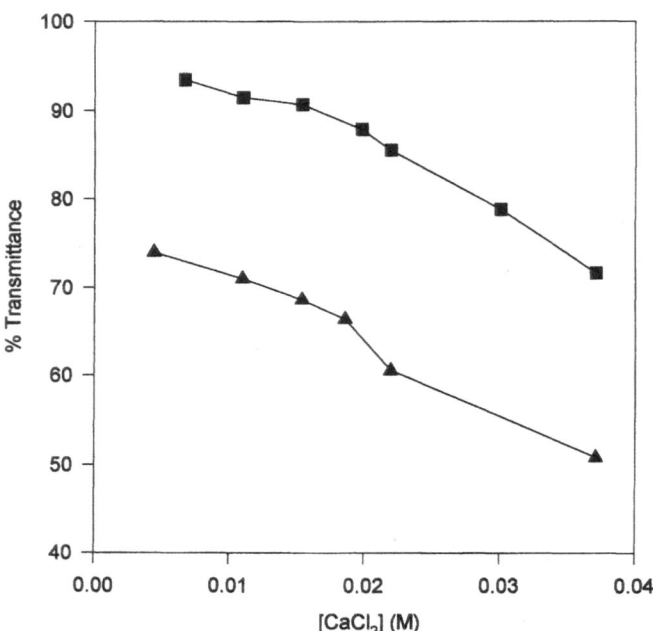

Fig. 6 Evolution of the turbidity at 450 nm of the carrageenan gels with calcium at 10 °C. Same symbols as in Fig. 2

results for eo-cgn and C$_8$-cgn are plotted, the behaviour of the samples is affected by their hydrophobicity. The hydrophilic sample (eo-cgn) shows a decrease in yield stress over the entire range of calcium concentration studied, with some marked decrease at the "optimal" concentration of 20 mM. The relatively low values of yield stress for this derivative hide the high compliancy of the samples as expressed in maximum strains of the order of 100%. The hydrophobic sample shows markedly different behaviour, the yield stress increasing up to the "optimal" calcium concentration and decreasing afterwards. For comparison, the maximum deformation strains of these samples were in the region of 15%, a low value which reflects their brittleness.

On visual inspection, we noted that the gels made of C$_8$-cgn with high calcium concentrations were very turbid, a phenomenon reminiscent of aggregation or precipitation. In order to assess if the rheological findings for the elastic moduli, the yield stress and the maximum strains had any connection to the transparency, we studied the optical behaviour of the different samples at 450 nm in order to quantify their turbidity.

Turbidity measurements

Turbidity was assessed as the percentage transmittance of the samples at 450 nm and 10 °C. The evolution of the turbidity as a function of added calcium is shown in Fig. 6.

Once again, the behaviour of the samples is bimodal. More importantly, the general trend for both modified carrageenan samples points to some relation between hydrophobicity and turbidity: the more hydrophobic the polymer, the more turbid the gel.

In both cases, turbidity increases (the percentage transmittance decreases) slightly up to about 0.020 M calcium and drops afterwards. The early slight decreases can be explained in terms of thermal stability. Indeed, all the samples are not turbid at high temperature, when melted. As the temperature of the experiment is quite close to the melting temperature, the samples with low amounts of calcium show very slight turbidity. As the amount of calcium increases above this value of one calcium ion per disaccharide unit, the turbidity decreases more sharply.

Conclusion

This series of experiments has revealed interesting behaviour in calcium-induced hydrophobically modified carrageenan gels. The gels are strong and thermally stable below a calcium concentration corresponding roughly to one calcium ion per disaccharide unit. In this region, calcium ions induce gelation. Above this concentration however, calcium seems to induce phase separation and precipitation. There seems, therefore, to be a balance between gelation and precipitation in calcium-induced carrageenan gels: the more hydrophobic the polymer the more prone it is to precipitation; the more hydrophilic the polymer, the more prone it is to gelation. This overall behaviour of hydrophobically modified κ-carrageenan gels puts some limitation on their potential use as hydrophobic drug delivery systems, unless one can put the precipitation process to one's advantage, as in the case of associative microgels.

Acknowledgements J.C.J. is grateful to the Higher Education Authority of Ireland for supporting this work. P.M. acknowledges the support of the Irish American Partnership.

References

1. Glicksman M (1979) In: Blanshard JMV, Mitchell JR (eds) Polysaccharides in food. Butterworths, London, pp 185–204
2. Guo J-H, Skinner GW, Harcum WW, Barnum PE (1999) Pharm Sci Technol Today 6:254
3. Guenet JM (1992) Thermoreversible gelation of polymers and biopolymers. Academic, London
4. Morris VJ, Chilvers GR (1983) Carbohydr Polyon 3:129
5. Watase M, Nishinari KJ (1991) J Texture Stud 12:427
6. Michel AS, Mestdagh MM, Axelos MAV (1997) Int J Biol Macromol 21:195
7. Ramakrishnan S, Prud'homme RK (2000) carbohydr Polyon 43:327
8. Ramzi M, Borgstrom J, Piculell I (1999) Macromolecules 32:2250
9. Guiseley KB (1978) Modified kappa-carrageenan. US Patent 4,096,327
10. Eldridge JE, Ferry JD (1954) J Phys Chem 58:992

Progr Colloid Polym Sci (2001) 118: 132–135
© Springer-Verlag 2001

BIOCOLLOIDS

Christos Ritzoulis
Eric Dickinson
Malcolm J. W. Povey
Yongtao Wang

Ultrasonic studies of the development of flocculation in mixed sodium caseinate and Tween 20 emulsions

C. Ritzoulis · E. Dickinson (✉)
M. J. W. Povey · Y. Wang
Procter Department of Food Science
University of Leeds, Leeds LS2 9JT, UK
e-mail: e.dickinson@leeds.ac.uk
Tel.: +44-113-2332956

Abstract Ultrasonic velocity and attenuation measurements were performed for a series of emulsions containing sodium caseinate and Tween 20. The kinetics of creaming were followed in order to establish relationships between depletion flocculant concentration and the onset of creaming. An interesting phenomenon was noticed where attenuation throughout the bulk of the emulsion dropped from very high values to lower ones with time. This was more intense at higher concentrations of Tween 20 (more flocculated emulsions). The lower attenuation values corresponded approximately to the theoretically predicted attenuation spectra for flocculated emulsions. The explanation for the higher attenuation at the beginning of the lifetime of the emulsions seems to be related to the formation of a transient gel spanning throughout the structure, eventually breaking up to produce discretely creaming flocs.

Key words Emulsions ·
Ultrasound · Caseinate · Delayed creaming · Phase transition

Introduction

An emulsion stabilised against coalescence by means of an adequate quantity of surface-active material and against Ostwald ripening by means of a very insoluble dispersed phase is prone to destabilisation through flocculation and creaming. In the case of excess surfactant, the main threat to emulsion stability is depletion flocculation [1]. Mixed excess surfactant systems can exert significant synergism towards depletion flocculation [2].

Destabilisation may be apparent as phase separation or enhanced creaming. It has been reported in previous work [3–6] that creaming in strongly flocculated emulsions will only appear after a certain lag phase. The behaviour is suspected to involve the formation of an intermediate unstable gel network [7]. Sometimes the process of formation of two thermodynamically incompatible phases, which eventually leads to creaming, can be visualised as a fluid–solid transition [8].

Correlations between the strength of flocculation and the concentration of the surface-active materials in mixed sodium caseinate and Tween 20 emulsions have previously been established in our laboratory [2]. It was found that increasing concentrations of Tween 20 in a caseinate-based emulsion can enhance depletion flocculation through a combination of unadsorbed surfactant micelles, desorbed caseinate, and/or mixed caseinate–surfactant aggregates.

Ultrasound measurement techniques have evolved during the last few years as a novel convenient method of obtaining data on complex liquid or liquidlike systems [9]. These techniques offer a very interesting alternative to techniques such as microscopy, electrical conductivity, and light scattering methods, as they are noninvasive, do not require sample dilution [10, 11] and are thus able to detect otherwise undetectable phase transitions [9, 12].

Materials and methods

Spray-dried sodium caseinate (less than 6% moisture, 0.05% calcium) was obtained from de Melkindustrie (Veghel, The Netherlands). Sunflower oil was purchased from the local supermarket and was subsequently purified by passing through a Florisil packed column, as proposed by Gaonkar [13]. Tween 20 (50% lauric acid, balanced by myristic, palmitic and stearic acid), tris(hydroxymethyl)aminomethane hydrochloride (Trizma-HCl) buffer (greater than 99% purity) and Florisil (activated magnesium silicate) were purchased from Sigma (St Louis, Mo.).

The emulsion aqueous phase was prepared by dissolving an appropriate amount of sodium caseinate in 0.025 M Trizma-HCl aqueous solution at pH 6.8. Purified sunflower oil was added to make a 30% oil-in-water premix; this was homogenised by three successive passes through a Shields high-pressure homogeniser in order to prepare a fine, stable 2% caseinate emulsion of the desired average droplet size. A Malvern Mastersizer was used to ensure that all the emulsions prepared were of a uniform average droplet size. Analysis of the scattering data revealed that the size distribution of the droplets was bimodal, with the main peak having a maximum at about 1 μm and a smaller peak with a maximum at about 0.2 μm for all the emulsions. d_{32} was calculated to be 0.35 μm. Tween 20 was then added as required under very gentle stirring. Care was taken throughout to maintain the emulsions at 25 °C.

The velocity and attenuation of ultrasound pulses were measured at 5-mm vertical intervals along the entire length of a calibrated cell (280 mm) for emulsion samples stored for different lengths of time as described elsewhere [9].

For the measurements of the attenuation spectra, the emulsions were placed in the sample cell of the FSuper ultrasound attenuation spectrometer, and attenuation-frequency spectra were collected at regular time intervals until creaming started to appear. Where appropriate, the core–shell model proposed by Hemar et al. [14], and modified by McClements et al. [15] for polydisperse systems, was used to predict the level of flocculation and/or the size of the flocs in the emulsion. Physical properties of the materials involved were taken from Coupland and McClements [16], Povey et al. [17], and the laboratory's Malvern Ultrasizer database. Attenuation measurements were carried out in the creaming rig at different heights, as well as in the specialised attenuation measurement instrument, the FSuper. The latter is made of a sample chamber immersed in a water bath, with a transducer/receiver mounted on the wall of the instrument.

Explicit solutions of the equations correlating attenuation over a range of frequencies with droplet size, flocculation level, floc size, and droplet packing, taking into account the thermal layers of the dispersed particles, were obtained through a Microcal Mathcad routine kindly provided by Julian McClements (University of Massachussetts, Amherst, Mass., USA).

Results and discussion

A comparison of the time needed for creaming to appear in a series of emulsions with increasing flocculation level is displayed in Fig. 1. As has been described before [2], increasing amounts of Tween 20 when added into a sodium caseinate stabilised emulsion lead to stronger flocculation and more enhanced creaming. We define as the onset of creaming the time when the phase-separation boundary reaches one-tenth of the emulsion height. This time for creaming to start occurring is plotted as a function of Tween 20 concentration (increasing flocculation level, see earlier).

As the strength of the depletion interaction increases with the Tween 20 concentration, the population of the creaming droplets increases. However, the time that it

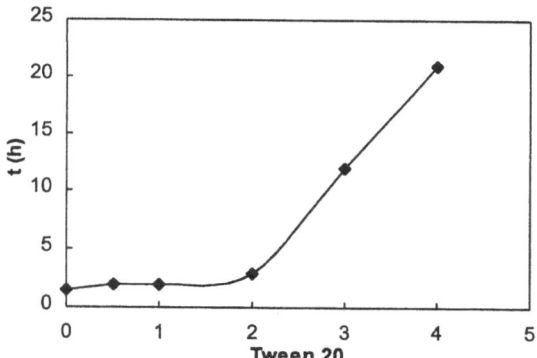

Fig. 1 Effect of surfactant concentration on delayed creaming. The time required for the phase-separation boundary ("serum") to reach one-tenth of the emulsion height in the creaming rig tube is plotted against the Tween 20 concentration

takes for a more flocculated emulsion to cream also increases. Hence, for a nonflocculated emulsion containing only 2% caseinate, a droplet-poor region appears at the bottom of the tube in less than 5 h, while in a far more flocculated 2% caseinate and 4% Tween 20 emulsion, the same region appears only after more than 20 h.

Ultrasound attenuation measurements were then used as a probe to detect structural changes in an emulsion before creaming starts to occur. A precise attenuation spectrometer, the FSuper, was used to continuously measure the attenuation spectra of the emulsions. The time dependence of the attenuation at 3 MHz is displayed in Fig. 2. An examination of the graph reveals an interesting observation. For emulsions with a highly flocculated character (i.e., 3% Tween 20), an initial high attenuation of about 100 Npm^{-1} is apparent, which is eventually lowered to about 20 Npm^{-1}. The latter values, but not the former, are roughly consistent with the theoretical value expected from the thermal field overlap

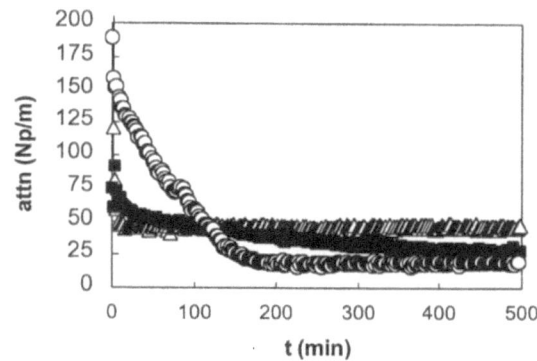

Fig. 2 Time-dependent changes in attenuation at 3 MHz for a series of 2% caseinate emulsions containing different amounts of Tween 20: no surfactant (*triangles*); 1% Tween 20 (*squares*); 3% Tween 20 (*circles*)

theory. This transition from high values to lower, theoretically expected values is less pronounced as the flocculation of the emulsions becomes weaker. Hence, there is practically no time-dependent change in the case of the very weakly flocculated 2% caseinate alone emulsion. These results are in good quantitative agreement with results from an attenuation measurement apparatus mounted in the creaming rig, which was used to obtain the data for Fig. 1.

Since no new components have been added to the system and no gravitational separation has yet occurred, the time-dependent changes in Fig. 2 should reflect structural rearrangements within the flocculated emulsion. If the attenuation changes were related to the formation of flocs, then fitting the experimentally obtained attenuation spectra to a theoretical model for the prediction of attenuation spectra of an emulsion under various states of flocculation should give us valuable insight into precreaming structural changes. We used the model developed by Hemar et al. [14] and McClements et al. [15]. This model assumes that spherical flocs are homogeneously distributed within a continuous phase. Droplet size, polydispersity, floc size, and floc packing are incorporated into this model. Flocs are initially considered as scattering bodies, and effects due to the overlap of the thermal fields on the approach of two droplets (or viscoinertial effects) are taken into consideration.

As can be seen in Fig. 3, a satisfactory fit was obtained for the final attenuation in the flocculated systems, using the model described earlier, assuming idealised spherical flocs of roughly similar (here monomodal) droplet size distribution. The initial stages present an attenuation much too high for us to be able to fit the results to the simple model we are using here. Thus, the high attenuation values imply a continuous phase having substantially different physical properties from that of water

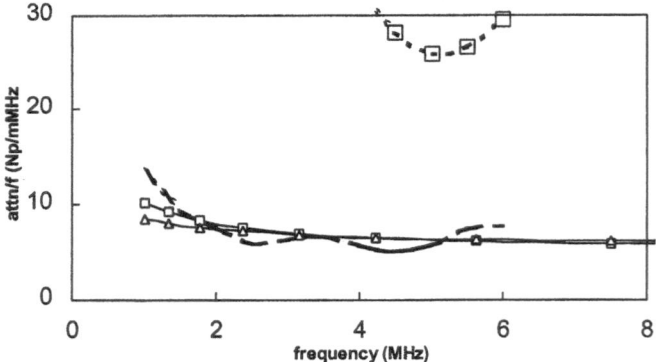

Fig. 3 Theoretical predictions of the attenuation spectra for emulsions: fully flocculated, floc size 16 μm (*triangles*); nonflocculated (*squares*). Experimental data: 2% caseinate and 3% Tween 20, just prepared (*broken line with squares*); 2% caseinate and 3% Tween 20, 300-min old (*broken line*)

containing spherical isolated flocs. Thus, a new model taking into account an entirely different and more complex continuous phase will be required to predict attenuation in the transient gels.

It is known [18, 19] that macroscopic destabilisation processes can involve a transient stage whereby the dispersed particles are connected in a continuous gel network. In a phase-inversion process this network eventually becomes the continuous phase of the colloid [20]. The network collapses because of the energy stored, as the elastic component forces the redistribution of the two phases of the colloid, which move under gravity with different velocities [19]. The slower-moving phase then separates from the faster-moving phase owing to this mobility difference. The idea of a structure-spanning network as an intermediate stage before the appearance of creaming seems to be a plausible explanation for the delay before the onset of creaming in depletion-flocculated emulsions [21]. The acoustical properties of this regime should be very different from the ones assumed by a continuous aqueous phase containing individual spherical entities made of flocculated droplets, which the structure will eventually break up into. The system starts off as a phase-inverted emulsion, with the aqueous phase clustered inside a sticky droplet network, as is expected for viscoelastic phase separation.

On the basis of these observations, we can propose a mechanism to account for the behaviour of a system with a strong depletion interaction before the onset of creaming (Fig. 4). In the initial stage, a nonflocculated emulsion (I) builds, under the influence of the flocculant (II), into a continuous gel-like network. The evolution of gel formation and gel breakup should be similar to the one presented by Tanaka [20] for a polymer solution. The transient particle gel should cover the entire volume of the container (II). Owing to differences in the mobility of the asymmetric mixture of the gel network and the trapped aqueous phase, the gel should gradually collapse on its own to give a continuous aqueous phase with polydisperse spherical-like flocs which then cream fast (III).

We can propose two possible sources for the high attenuation of the initial stages of the life of the flocculated emulsion. Firstly, the sound absorption/scattering from a structure-spanning elastic network of interconnected droplets (Fig. 4, II) should be entirely different from that of a continuous aqueous phase with dispersed flocs, as described by our models. Enhanced attenuation can be attributed to the acoustical properties of the structure itself. As the structure collapses to its component flocs, and creaming starts to manifest itself, the attenuation values drop. These values correspond to a system with flocs dispersed into a continuous aqueous phase (Fig. 4, III). A second possible explanation is that the attenuation can be attributed to air bubbles which are trapped into the gel network and are released with

Fig. 4 Proposed schematic representation of the evolution of the transient gel formation and breakup during the precreaming stages of the life of an emulsion

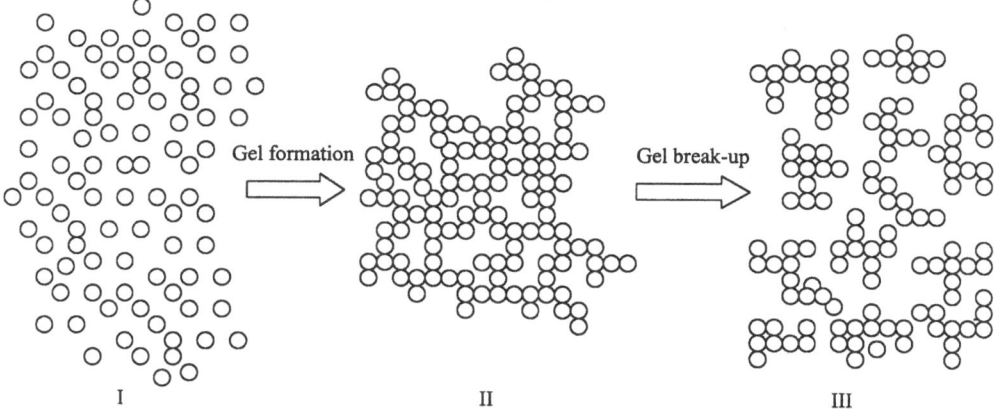

subsequent lowering of attenuation as the structure-spanning network inverts to a flocculated emulsion (Fig. 4, II, III). In either case, we can follow, directly or indirectly, the same evolving viscoelastic phase transition. We are currently working to understand which of the two mechanisms is the most plausible.

Conclusions

Ultrasonic velocity scanning has been used to monitor the delay in creaming of a series of model emulsions with increasing flocculation levels. Emulsions with stronger flocculation were found to exhibit significant delays in creaming. Precreaming changes in an emulsion can be detected as changes in attenuation. In flocculated emulsions, the very high initial attenuation values

eventually change to low values before creaming starts to become manifest. The latter values correspond to a flocculated emulsion, while the initial high values should correspond to a gel network where water is trapped as isolated pockets. We can propose a mechanism to account for the delayed creaming in heavily flocculated emulsions. Initially an elastic network is formed. This structure rearranges itself with time, and the gel eventually breaks up to individual flocs. Further work is still needed to accurately model the attenuation of these particle gels.

Acknowledgements C.R. thanks the Commission of the European Communities for the provision of a Marie Curie Research Training Grant (FAIR-CT98-5013). We thank Julian McClements for the provision of the computer routine to calculate particle size distributions from the attenuation spectra.

References

1. Dickinson E, Golding M (1997) J Colloid Interface Sci 191:166
2. Dickinson E, Ritzoulis C, Povey MJW (1999) J Colloid Interface Sci 212:466
3. Poon WCK, Starrs L, Meeker SP, Moussa A, Evans RML, Pusey PN, Robins MM (1999) Faraday Discuss 112:143
4. Glasrud GG, Navarrete RC, Scriven LE, Macosko CW (1993) AIChE J 39:560
5. Allain C, Cloitre M, Wafra M (1995) Phys Rev Lett 74:1478
6. Parker A, Gunning PA, Ng K, Robins MM (1995) Food Hydrocolloids 9:333
7. Poon WCK (1998) Curr Opin Colloid Interface Sci 3:593
8. Bibette J, Roux D, Nallet F (1990) Phys Rev Lett 65:2470
9. Povey MJW (1997) Ultrasonic techniques for emulsion characterization. Academic, San Diego
10. Dukhin AS, Goetz PJ (1996) Langmuir 12:4336
11. McClements DJ (1996) Langmuir 12:3454
12. Wines TH, Dukhin AS, Somasundaran P (1999) J Colloid Interface Sci 216:303
13. Gaonkar AG (1989) J Am Oil Chem Soc 66:1090
14. Hemar Y, Herrmann N, Lemaréchal P, Hocquart R, Lequeux F (1997) J Phys II 7:637
15. McClements DJ, Hemar Y, Herrmann N (1999) J Acoust Soc Am 105:915
16. Coupland JN, McClements DJ (1997) J Am Oil Chem Soc 74:1559
17. Povey MJW, Golding M, Higgs D, Wang Y (1999) Int Dairy J 9:299
18. Poon WCK, Haw MD (1997) Adv Colloid Interface Sci 73:71
19. Tanaka H (1997) Phys Rev E 56:4451
20. Tanaka H (2000) J Phys Condens Matter 12:R207
21. Manoj P, Fillery-Travis AJ, Watson AD, Hibberd DJ, Robins MM (1998) J Colloid Interface Sci 207:283

Progr Colloid Polym Sci (2001) 118: 136–140
© Springer-Verlag 2001

BIOCOLLOIDS

Anna Stradner
Sara Romer
Claus Urban
Peter Schurtenberger

Aggregation and gel formation in biopolymer solutions

A. Stradner (✉) · S. Romer · C. Urban
P. Schurtenberger
Department of Physics
University of Fribourg
1700 Fribourg, Switzerland
e-mail: anna.stradner@unifr.ch
Tel.: +41-26-3009120
Fax: +41-26-3009747

S. Romer
Polymer Institute, ETH Zürich
8092 Zurich, Switzerland

Abstract We report an investigation of the microscopic structure and dynamics of biopolymer gels and relate them to the macroscopic viscoelastic properties of such systems. Biopolymer solutions and gels represent one of the most interesting class of gelling systems since they are of major industrial and scientific interest. We performed a systematic study using concentrated solutions of casein micelles which we destabilized and investigated during the process of gelation using diffusing wave spectroscopy (DWS) and rheological measurements. An analysis of the light scattering data shows a significant increase in the characteristic decay time of the correlation functions during the sol–gel transition. For the analysis of the DWS data we developed an algorithm which, based on the so-called microrheology approach, determines the viscoelastic properties $G'(\omega)$ and $G''(\omega)$ of the gel. A comparison of the results obtained with DWS and measurements with a rheometer shows excellent agreement of both approaches. We demonstrate that we can clearly link the changes observed in the microscopic dynamics to the formation of a macroscopic gel with drastically modified viscoelastic properties.

Key words Diffusing wave spectroscopy · Colloidal gels · Sol–gel transition · Casein micelles

Introduction

Acidified milk products such as yoghurt are popular and important food products. For such acid-induced gels the process variables (composition of the milk, gelation temperature, heat pretreatment, kind of acidification, etc.) have an enormous impact on the gel process as well as on the physical properties of the final gel. However, we currently lack clear understanding of the relationship between the microscopic structural dynamic properties of the macromolecular constituents and the resulting macroscopic mechanical and rheological properties of the gel. In milk the main part of the proteins, the caseins, are organized into micelles, which are sterically stabilized by a layer or "brush" of κ-casein molecules. The κ-caseins extend their C-terminal part into the solution,

causing the micelles to repel each other on close approach. A number of studies have demonstrated that the casein micelles have properties very similar to those of sterically stabilized hard-sphere suspensions [1–4]. To destabilize these micelles and to induce the gelation process one can either remove the stabilizing hairs enzymatically or acidify the milk. We followed the second route and used glucono-δ-lactone (GDL), where the hydrolysis of GDL to gluconic acid shifts the pH to lower values. This leads to a "collapse" of the stabilizing brush, which now is no longer able to prevent aggregation of the particles.

In this work we focus on the gel formation of acidified fat-free and fat-containing milk which we followed by classical rheological oscillatory experiments as well as with diffusing wave spectroscopy (DWS). DWS is a quite recent development in dynamic light scattering with turbid suspensions and allows in situ, nondisturbing monitoring of the whole process. It works in the limit of very strong multiple scattering, where a diffusion model can be used to describe the propagation of the light through the sample. Using a diffusion approximation it is possible to directly relate the correlation function of the scattered light to the average mean-square displacement (MSQD), $\langle \Delta r^2(\tau) \rangle$, of the single scatterers in turbid suspensions [5–7]. An algorithm based on the optical microrheology approach [8] is furthermore used to calculate the frequency dependence of the storage and loss moduli from DWS data over more than 5 decades. These results are compared to $G'(\omega)$ and $G''(\omega)$ of the final gel obtained from classical rheological experiments.

Materials and methods

Materials and preparation

For the fat-free milk, 10.45 g Nilac skimmed milk powder (NIZO, Ede, The Netherlands) was dispersed in 100 g deionized water at 20 °C and stirred at 45 °C for 1 h using a magnetic stirrer, which leads to a protein solution with a concentration of 3.47% by weight. After adding 0.02 wt% sodium azide the solution was kept for 12 h at 4 °C. The sample was then allowed to equilibrate at 32 °C for 2 h before 1.1 wt% GDL was added in order to induce gelation.

The fat-containing milk was prepared by mixing appropriate amounts of fat-free milk (Nilac milk with 5.87 wt% protein, same preparation procedure as previously) and full-fat milk (pasteurized full-fat milk from the supermarket containing 3.9 wt% fat and 3.2 wt% protein), which results in milk containing 2 wt% fat and 4.5 wt% protein. Prior to GDL addition (1.1 wt%) and subsequent measurements the fat-containing milk was heat-pretreated for 10 min at 90 °C and then quickly cooled to 32 °C.

Methods

All the experiments were performed at a temperature of 32 °C. The rheological measurements for the fat-free milk were performed using a USD 200 rheometer (Paar Physica) with a double-gap Couette system. The time dependence was measured with a frequency of 10 Hz and an amplitude of 10%. The frequency sweep was performed with an amplitude of 1%. The time-dependent moduli for the fat-containing milk were determined with an MCR 300 rheometer from Paar Physica using a double-gap Couette system at 1 Hz and 1% amplitude.

DWS experiments were performed with a solid-state laser (Verdi 2 W, Coherent, $\lambda = 532$ nm) with an expanded beam ($d \approx 7$ mm). After passing a rectangular cuvette the light was collected in transmission geometry with a single-mode fiber after a polarizer, which is perpendicular to the incident beam polarization. This ensures that only multiply scattered light is detected. Because of possible after-pulsing effects of the detector the signal is split and fed into two photomultipliers (Hamamatsu). Finally, a digital correlator (ALV-5000E, ALV) performs a pseudo-cross-correlation measurement. The transport mean free path, l^*, a measure for the sample turbidity, was obtained from a static transmission measurement relative to a sample of known l^*.

The pH measurements for the fat-containing milk were performed with a standard pH meter (PHM210, Radiometer Analytical, France).

Results and discussion

Typical DWS results from measurements on fat-free milk at different stages of the gelation process are shown in Fig. 1. The correlation functions $g(\tau) - 1$ are given in Fig. 1a and the corresponding MSQDs $\langle \Delta r^2(\tau) \rangle$ are given in Fig. 1b. The first measurement immediately after the addition of GDL (open circles) reflects the diffusion of the free, stabilized casein micelles. Initially, we then observe a faster decay of the correlation function, which corresponds to an increase of the diffusion coefficient (open triangles). This is due to a decrease in the particle size owing to the brush collapse and the decreasing voluminosity of the casein micelles caused by the addition of GDL. As the casein micelles are no longer stabilized, they then start to aggregate, which can be seen from the slowing of the particle diffusion owing to the formation of clusters.

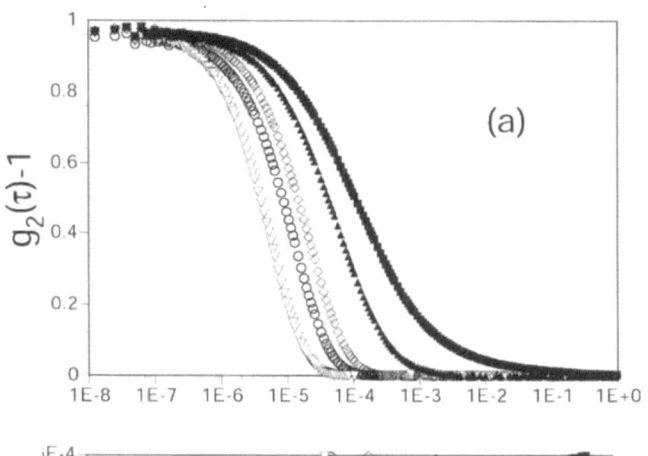

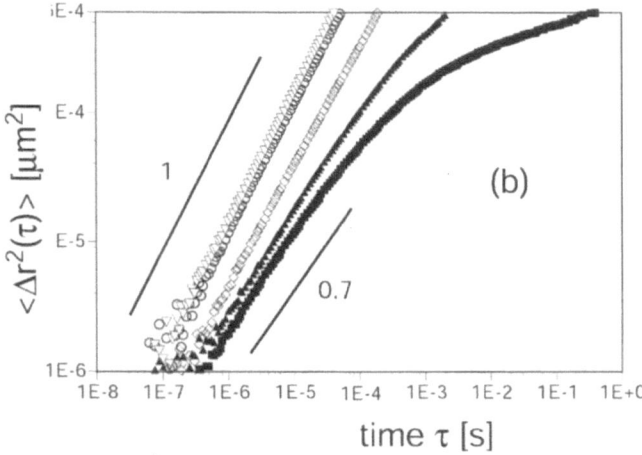

Fig. 1 a Intensity autocorrelation functions $g_2(\tau) - 1$ and **b** the corresponding mean square displacements, $\langle \Delta r^2(\tau) \rangle$, for fat-free milk at different stages of the aggregation and gelation process. The *open symbols* correspond to times prior to the gel point and the *filled symbols* are measurements in the gel (symbols ordered with increasing time: *open circles, open triangles, open diamonds, filled triangles, filled squares*)

The particle dynamics in the stable milk as well as in the aggregating suspensions prior to the gel point (open symbols) exhibit the typical characteristics of free particle diffusion owing to Brownian motion. This is reflected by an almost exponential decay of the correlation functions and leads to a linear dependence of the MSQD on time (indicated by a line with slope 1 in Fig. 1b). However, at the gel point a quite dramatic change in the particle dynamics occurs, and the short-time behavior changes from Brownian to a subdiffusive motion (filled symbols). The MSQD is now well described by a stretched exponential. At short times this results in a power-law behavior $\langle \Delta r^2(\tau) \rangle \sim \tau^p$ with an exponent $p \approx 0.7$ (Fig. 1b). Similar results have already been found in concentrated suspensions of almost monodisperse polystyrene spheres that undergo a sol–gel transition [9].

The same experiments were repeated for the fat-containing milk, and the results are summarized in Fig. 2. The correlation functions and the corresponding MSQDs are plotted at different times during the destabilization process from the stable milk to the final gel. In contrast to the fat-free milk, the initial increase in the casein micelle diffusion coefficient cannot be observed any more owing to the contribution from the additional population of fat droplets. However, we clearly observe the slowing of the average diffusion coefficient owing to cluster formation and the crossover of the particle dynamics at the gel point. This is again visible in the short-time behavior of $\langle \Delta r^2(\tau) \rangle \sim \tau^p$, where we also find a decrease in the exponent p from 1 for the stable milk (open symbols) to about 0.7 for the gel (filled symbols).

In rheology it is common practice to define the gel point as the point where the elastic properties (represented by the storage modulus G') dominate over the viscous properties (represented by the loss modulus G'') [4]. A comparison between the time-resolved rheological measurements and the DWS experiments demonstrates that the qualitative change in microscopic particle dynamics indeed coincides with a dramatic change in the macroscopic viscoelastic properties of the samples at the gel point. This is shown in Figs. 3 and 4, where G' and G'' measured at a single oscillation frequency and p obtained from DWS are plotted as a function of time. In the case of the fat-free milk (Fig. 3) we observe a steep increase in G' approximately 5.8 h after GDL addition, indicating the transition from a sol to a gel (Fig. 3a). Figure 3b shows that at the same time p drops from 1 to about 0.7.

The results obtained for fat-containing milk are plotted in Fig. 4, and they qualitatively show the same feature. However, owing to heat-pretreatment the gel point is located at an earlier time. Moreover, the resulting values of G' are significantly higher when compared to the fat-free milk, and the sol–gel transition appears to be broader. This is visible in the time evolution of G' and G'' (Fig. 4a) as well as of p (Fig. 4b). We currently lack a detailed and quantitative understanding of the effects of particle polydispersity on the sol–gel transition in attractive particle suspensions. Moreover, it is clear that the fat droplets may experience a different interaction potential than the casein micelles and are possibly playing the role of a "filler" particle. Nevertheless our results clearly demonstrate that the sol–gel transition in fat-containing milk exhibits the same similarities in the link between microscopic dynamics and macroscopic elastic properties as the much simpler model systems latex suspensions and fat-free milk.

Recently it has been demonstrated that one can extract the viscoelastic properties of complex fluids from DWS measurements by relating the time evolution of the MSQD to the frequency-dependent elastic and viscous moduli of the medium [8]. By describing the motion of a scattering particle in a viscoelastic medium in terms of a

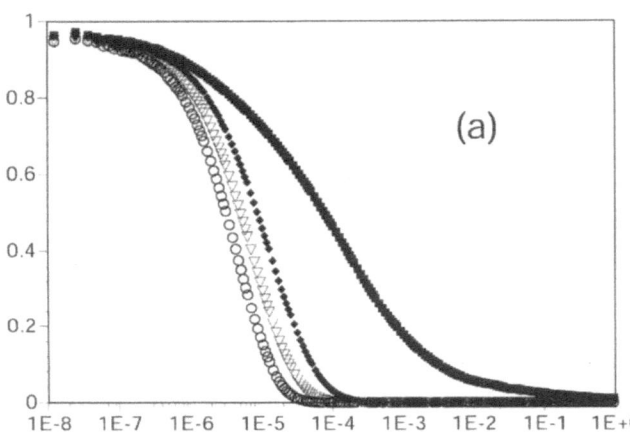

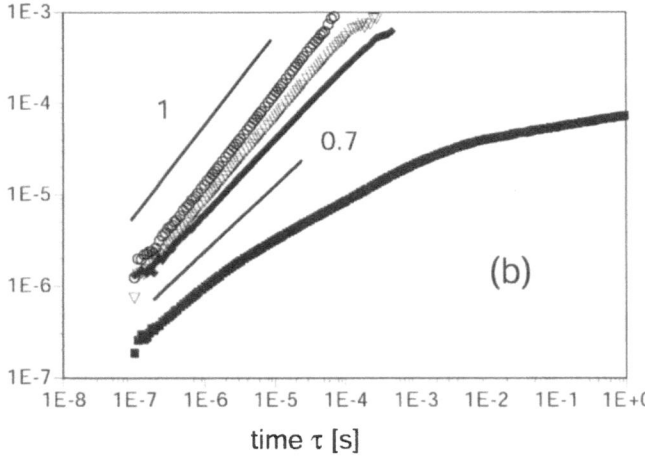

Fig. 2 a $g_2(\tau)-1$ and **b** the corresponding $\langle \Delta r^2(\tau) \rangle$ for fat-containing milk at different stages of the aggregation and gelation process. The *open symbols* correspond to times prior to the gel point and the *filled symbols* are measurements in the gel (symbols ordered with increasing time: *open circles, open triangles, filled triangles, filled squares*)

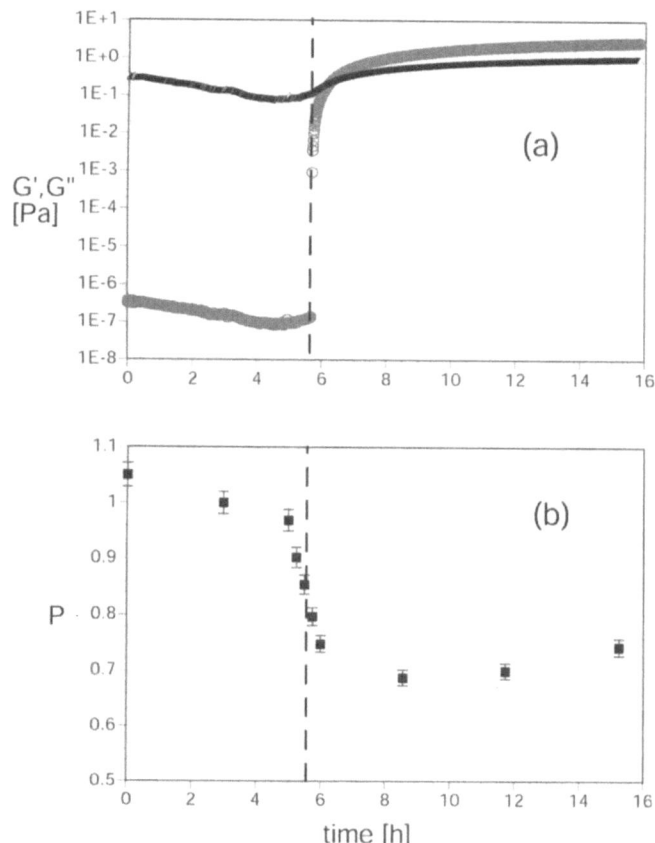

Fig. 3 Comparison of results from classical rheology and diffusing wave spectroscopy (*DWS*) during the yoghurt-making process in fat-free milk. **a** Time evolution of the storage modulus, $G'(t)$, (*big gray circles*) and the loss modulus, $G''(t)$, (*small black triangles*) obtained from an oscillating rheological measurement (10 Hz, 10% amplitude). **b** Time evolution of the exponent p obtained from DWS. The enormous increase in G' and the drop in the exponent coincide (*dashed line*) and indicate a link between microscopic particle dynamics and macroscopic sol–gel transition and viscoelastic properties

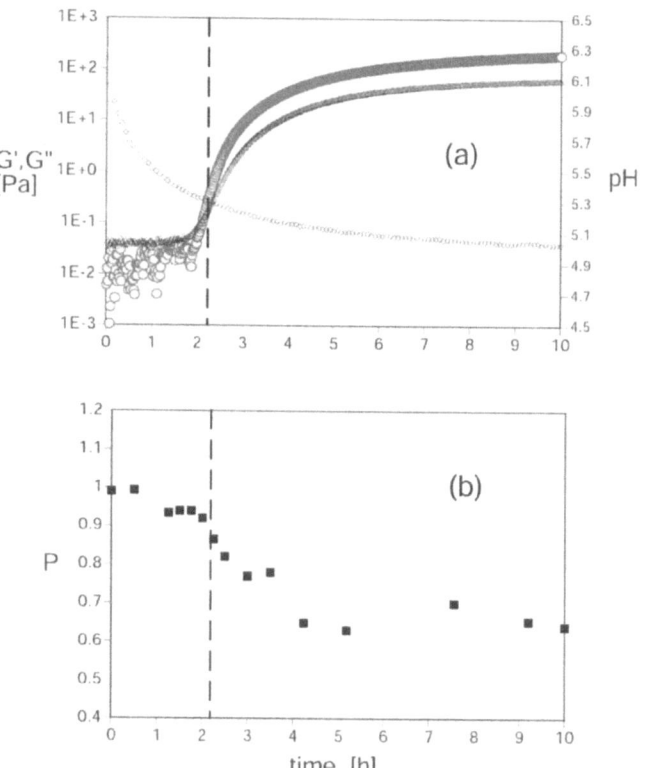

Fig. 4 Comparison of results from classical rheology and DWS during the yoghurt-making process in fat-containing milk. **a** Time evolution of $G'(t)$ (*big gray circles*) and $G''(t)$ (*small black triangles*) obtained from an oscillating rheological measurement (1 Hz, 1% amplitude). **b** Time evolution of the exponent p obtained from DWS. Also shown is the pH of the milk which was followed simultaneously during the entire process (*open diamonds* in **a**)

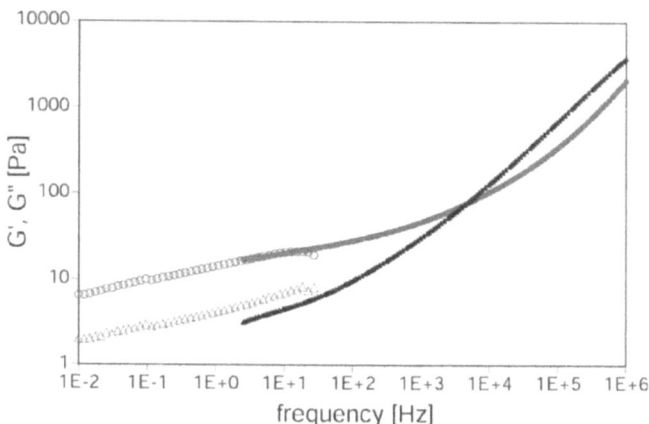

Fig. 5 Comparison of the frequency-dependent shear moduli in the final gel of fat-free milk obtained from classical rheological measurements with a double-gap Couette system (G', *open triangles*; G'', *open circles*) and DWS (G', *filled triangles*; G'', *filled circles*) using the optical microrheology approach

generalized Langevin equation with a memory function the time evolution of the MSQD can be related to the frequency-dependent storage, $G'(\omega)$, and loss, $G''(\omega)$, moduli of the medium. As the "probe particles" used in DWS have colloidal dimensions, the accessible range of frequencies extends to much higher values and can be as large as 10^{-3} rad/s$\leq\omega\leq10^6$ rad/s. Moreover, this technique ensures that the true equilibrium module is measured as the method intrinsically relies on thermal fluctuations only. The results from a corresponding analysis of the time dependence of $\langle\Delta r^2(\tau)\rangle$ of the final gel of fat-free milk are plotted in Fig. 5 together with $G'(\omega)$ and $G''(\omega)$ obtained from a classical rheological frequency sweep. We observe quite good agreement between optical and mechanical measurements. At low frequencies G' dominates, reflecting the solidlike behavior of the system. At high frequencies G'' dominates G'. The crossover ($f_0\approx4{,}500$ Hz) reflects the change in the

excitations of the network. At frequencies lower than f_0 the full network responds, while at higher frequencies the shear moduli reflect the single particle rather than the network response. We are currently performing a systematic investigation on the applicability of the microrheology approach to gelled milk samples at different times, where we compare the frequency dependence of G' and G'' measured with a classical rheometer with the results from optical rheology.

Conclusion

Our results clearly demonstrate the existence of a link between the local short-time particle dynamics measured with DWS and the macroscopic viscoelastic properties of particle gels in milk and colloidal model suspensions. Moreover, we have been able to show that the optical microrheology approach can be extended to complex food systems such as the yoghurt-making process in fat-free milk. The enormous frequency range over which rheological data can be obtained by optical methods is shown in Fig. 5. This is even more remarkable when looking at the relatively short duration of an individual DWS measurement, which can be as short as 120 s. Moreover, it is important to point out that these measurements are completely noninvasive, and thus ideally suited to investigate systems that undergo a sol–gel transition and/or weak gels.

Acknowledgements We gratefully acknowledge financial support from the Swiss National Science Foundation, COST (Action P1) and the Swiss Federal Office for Education and Research. We are indebted to C.G. de Kruif for valuable discussions and for providing the Nilac milk powder used in this study.

References

1. Holt C, Horne DS (1996) Neth Milk Dairy J 50:85
2. de Kruif CG (1999) Int Dairy J 9:183
3. Lucey JA, Singh H (1998) Food Res Int 30:529
4. Horne DS (1999) Int Dairy J 9:261
5. Maret C, Wolf PE (1987) Z Phys B 65:409
6. Pine DJ, Weitz DA, Chaikin PM, Herbolzheimer E (1988) Phys Rev Lett 60:1134
7. Weitz DA, Pine DJ (1993) In: Brown W (ed) Dynamic light scattering. Clarendon, Oxford, pp 652–720
8. Mason TG, Gang H, Weitz DA (1997) J Opt Soc Am A 14:139
9. Romer S, Scheffold F, Schurtenberger P (2001) Phys Rev Lett 85:4980

Progr Colloid Polym Sci (2001) 118: 141–144
© Springer-Verlag 2001

BIOCOLLOIDS

Temperature dependence of salt-induced aggregation in poly(ethylene oxide)-coated latex particle suspensions

A. Di Biasio
F. Bordi
C. Cametti

A. Di Biasio · F. Bordi · C. Cametti
Istituto Nazionale per la Fisica della
Materia, Unita' di Romal, Rome, Italy

A. Di Biasio
Dipartimento di Matematica e Fisica
Universita' di Camerino, Camerino, Italy

F. Bordi
Dipartimento di Medicina Interna
Universita' di Roma "Tor Vergata"
Rome, Italy

C. Cametti (✉)
Dipartimento di Fisica
Universita' di Roma "La Sapienza"
Rome, Italy

Abstract The influence of temperature on the kinetics of aggregation of poly(ethylene oxide) coated latex particles induced by adding appropriate amount of simple electrolyte has been investigated by means of dynamic light scattering measurements. The hydrodynamic radius of the resulting aggregates markedly depends on the molecular weight of the polymer adsorbed on the polystyrene particle surface. We investigated the influence of poly(ethylene oxide) of three different molecular weights (from 8 to 5×10^3 kD) in the temperature interval from 10 to 50 °C. A bridging mechanism has been suggested as responsible, at an intermediate surface coverage, for enhancing the aggregation rate. The influence of the polymer solubility in this process is briefly discussed.

Key words Latex suspensions · Aggregation · Dynamic light scattering · Poly(ethylene oxide) adsorption

Introduction

In our previous work [1], we investigated the effect of the addition of a water-soluble polymer, namely poly(ethylene oxide) (PEO) of different molecular weights, on the stability of a dilute polystyrene suspension, whose aggregation was induced by adding appropriate amounts of uni-univalent electrolyte solution [2–4].

It is well known that when polymers are added to a colloidal suspension, the resulting stability of the whole system can be greatly modified as a consequence of a bridging flocculation effect, i.e., when coadsorption of a polymer chain originally on one particle onto a second particle occurs. This effect, resulting in a reduction in the magnitude of the interparticle electrostatic repulsion, enhances the rate of successful encounters between particles and produces an aggregation process with the formation of clusters of increasing size. In fact, in the salt-induced aggregation of a charge-stabilized polystyrene suspension [5], the addition of appropriate amounts of simple salt reduces the electrical double-layer thickness around each colloidal particle with a decrease in the height of the repulsive energy barrier between two approaching particles, leading to diffusion-induced rapid aggregation (diffusion-limited cluster aggregation, DLCA).

The polymer adsorption on the colloidal particle surface (even at low surface coverage) modifies the structure of the double layer, since latex particles, carrying terminally anchored polymer chains, allow polymer bridging to occur, resulting in considerable enhancement of the flocculation (or coagulation) rate [6–8]. It was also found that, depending on the polymer molecular weight, the polymer adsorption may diminish the coagulation rate [9], yielding a sterically stabilized dispersion when the bridging effect is hindered by complete polymer coverage.

These two concomitant effects produce a very complex phenomenology, being further complicated by varying the temperature below or above the θ temperature, which establishes the polymer solubility in the aqueous solvent.

In this note, we extend our previous investigation on the PEO-coated latex particle aggregation induced by salt addition [1], by varying the temperature of the system from 10 to 50 °C, to cross the temperature which divides the temperature range into poor ($T < \theta$) and good ($T > \theta$) regions, corresponding to poor solvent and good solvent, respectively. For the systems under investigation the θ temperature is approximately 42 °C [10]. The θ temperature affects the attractive forces between colloidal particles, both those due to polymer–polymer (segment–segment) attraction and those due to bridging attraction. As one goes from poor solvent ($T < \theta$) to good solvent ($T > \theta$), the polymer–polymer interaction vanishes and repulsive forces dominate, yielding latex stabilization. Conversely, depending on the percentage of the surface coverage, for temperatures above the θ temperature, the formation of bridges enhances particle–particle attraction, producing a latex destabilization. In this picture, the θ temperature plays a key role in the stability of colloidal dispersions (known as electrosteric stabilization) where electrostatic effects as well as those arising from the nature of the adsorbed polymer layer are present.

In this work, we analyze, by means of a dynamic light scattering technique, the time evolution of the hydrodynamic radius of the clusters formed from the aggregation of PEO-coated polystyrene particles, induced by NaCl electrolyte solution, at different temperatures from 10 to 50 °C. Monodisperse polystyrene latex particles were chosen as a model hydrophobic surface, providing an ideal surface for the study of the change in hydrodynamic radius upon cluster adsorption. The effect of the θ temperature on the stabilization of the latex dispersion is briefly discussed.

Experimental

Materials

Polystyrene latex dispersions were commercially available products from Dow Chemical Co, and were used without further purification. The mean diameter of the latex particles employed, as determined by light scattering methods, was (910 ± 20) Å. Each particle is negatively charged with a surface charge density of about 3 μC/cm^2. Three (PEO) samples with nominal molecular mass 8 kD, 200 kD, and 5 MD were used as received. These were supplied by Sigma Chemical Co. The adsorption experiments were carried out by mixing the latex suspension (with a volume fraction of $\Phi = 2 \times 10^{-5}$, corresponding to a solid content of about 2×5^{-5} g/ml) with the polymer solution at the appropriate polymer concentration. The suspension was allowed to equilibrate for 5 h.

Polymer adsorption

In order to give the PEO concentration with respect to the latex content, we used an index (P/L) defined as the ratio of the mass of the polymer added to the suspension per unit volume to the fractional volume of the latex suspension. This index (with dimension of grams per milliliter when multiplied by the volume of the single latex particle equals the mass of PEO polymer per latex particle. In the present study, we investigated PEO-coated latex suspensions by varying P/L from 0 to 10, corresponding to values of polymer mass from 0 to 4×10^{-15} g/particle. The amount of polymer adsorbed as a function of the equilibrium polymer concentration in the aqueous phase was measured by means of an

isothermal titration microcalorimetry technique. The details are given elsewhere [1]. For example, for polymer of molecular mass $M_w = 200$ kD and a particle of 1920 Å in diameter, the saturation in the surface particle coverage occurs at a polymer concentration of about 0.9 g/l, corresponding to a polymer concentration of about 5.5 mg/m^2 and to a ratio P/L of about 0.2. For the polymer concentration employed (P/L from 0 to 10), the maximum in the particle coverage is reached for all the systems investigated and, in some cases, largely exceeded.

Dynamic light scattering measurements

The hydrodynamic radius of the diffusing clusters was measured by means of a standard dynamic light scattering experiment. The average decay rate, related to the hydrodynamic radius of the diffusing particles, is calculated from the initial slope of the intensity autocorrelation function, $G_2(t)$, expanded in terms of cumulants. Owing to the low polymer concentration, no correction was made for the viscosity of the PEO solution, assumed to be that of water at the temperature of the experiment. The details of the dynamic light scattering measurements are given elsewhere [1].

Results and discussion

In all the systems investigated, the electrostatic aggregation process takes place in the presence of 0.4 M NaCl electrolyte solution, causing the reduction (or the elimination) of the electrostatic repulsion between latex particles. Under these experimental conditions, the adsorption of PEO chains on the particle surface, depending on the polymer molecular weight, causes different aggregation behaviour, which can be summarized as follows:

1. At low molecular mass ($M_w = 8$ kD) (Fig. 1), the increase in the hydrodynamic radius owing to the polymer adsorption is small for all the values of

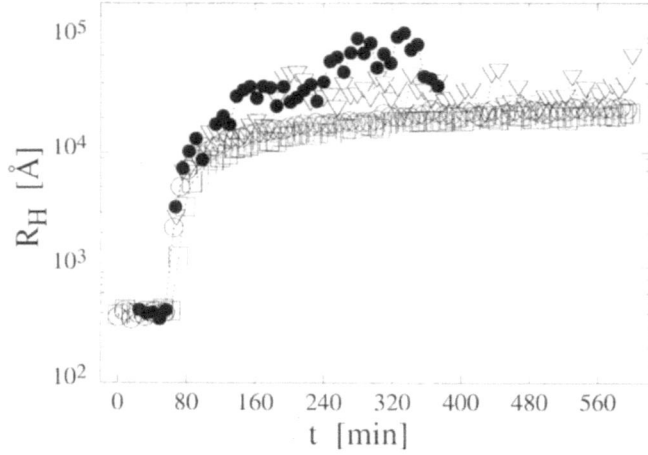

Fig. 1 The time evolution of the hydrodynamic radius of poly(ethylene oxide) (*PEO*)-coated latex particles at different temperatures. *open symbols*: in presence of PEO of molecular weight 8 kD, with $P/L = 10$; (*Open squares*): 10 °C; (*open circles*): 30 °C; (*open triangles*): 50 °C. *Filled circles*: in the absence of PEO, at temperature of 50 °C. The aggregation is induced by 0.4 M NaCl electrolyte solution

the parameter P/L we investigated (up to $P/L = 10$). The salt-induced aggregation proceeds very quickly, reaching a steady-state within times of the order of some tens of minutes. The increase in the temperature from 10 to 50 °C does not affect the dynamics of the aggregation and no marked effect is evidenced by crossing the θ temperature. In this case, in the coverage saturation condition ($P/L = 10$), the bridging mechanism, if it exists, should be negligible and only the reduction of the electrostatic repulsion dominates the aggregation.

2. At intermediate molecular mass ($M_w = 200$ kD) (Figs. 2, 3), the PEO adsorption causes an increase in the

hydrodynamic radius of the order of 15–20% and the salt-induced aggregation strongly depends on the fraction of polymer coverage. For high values of P/L ($P/L = 10$) (Fig. 2) the polymer adsorption produces a marked stabilization of the latex suspension. The hydrodynamic radius of the resulting aggregates maintains a constant value over the whole time interval investigated (up to 1000 min) at low temperature (10 °C) and a somewhat higher value at higher temperature (50 °C). In this case, a steric stabilization occurs and the effect of the temperature increase produces only a small increment in the average size of the aggregates. Very different behavior is evidenced at a lower P/L value ($P/L = 1$) (Fig. 3), where the partial coverage of the particle surface produces a marked temperature-dependent aggregation. There is a progressive enhancement of the cluster formation as the temperature is increased, with a dramatic increase at temperatures higher than the θ temperature.

3. Finally, at higher molecular mass ($M_w = 5000$ kD), after the initial increase owing to the adsorbed polymer, the hydrodynamic radius undergoes a very small increase, roughly temperature-independent. This behavior appears even for low polymer concentration ($P/L = 0.1$) (Fig. 4), although, in this case there is an evident effect (a rate enhancement) due to temperature. With polymer of high molecular mass, the latex suspension becomes stable by the effect of steric hindrance, owing to repulsion of loops and tails of the adsorbed polymers. At high molecular mass, the polymer flattening on the particle surface makes the bridging mechanism inefficient. Following the Flory–Huggins theory for polymers in solution [11], the free energy, ΔG, associated with two

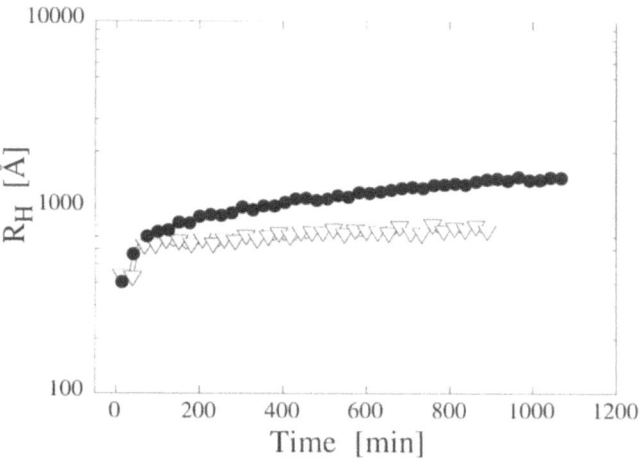

Fig. 2 The time evolution of the hydrodynamic radius of PEO-coated latex particles at different temperatures in the presence of 0.4 M NaCl electrolyte solution. The PEO molecular weight is 200 kD and $P/L = 10$. 10 °C (*triangles*); 50 °C (*circles*)

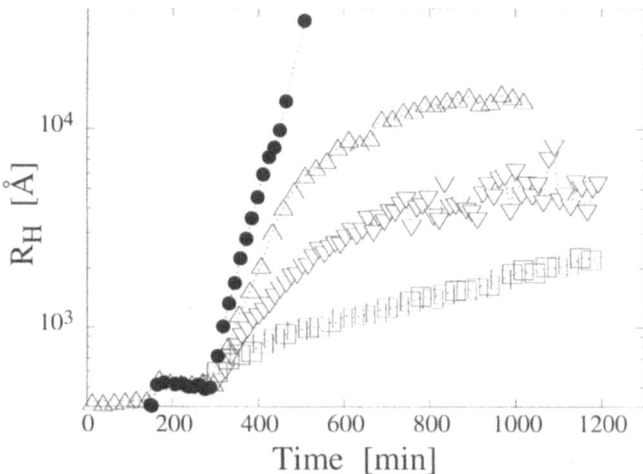

Fig. 3 The time evolution of the hydrodynamic radius of PEO-coated latex particles at different temperatures in the presence of 0.4 M NaCl electrolyte solution. The PEO molecular weight is 200 kD and $P/L = 1$. 10 °C (*squares*); 30 °C (*down triangles*); 40 °C (*up triangles*); 50 °C (*circles*)

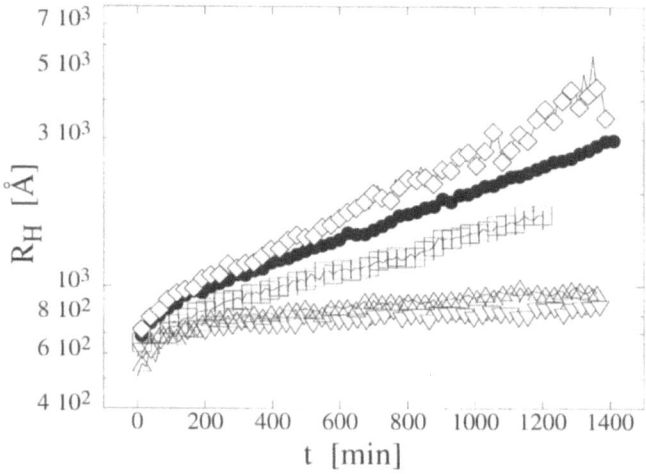

Fig. 4 The time evolution of the hydrodynamic radius of PEO-coated latex particles at different temperatures in the presence of 0.4 M NaCl electrolyte solution. The PEO molecular weight is 5 MD and $P/L = 0.1$. 10 °C (*down triangles*); 20 °C (*up triangles*); 30 °C (*squares*); 40 °C (*circles*); 50 °C (*diamonds*)

approaching particles covered with an adsorbed polymer (in analogy with polymers in the limit of $M_w \to \infty$) can be written as

$$\Delta G = 2k_B T \left(\frac{V_p^2}{V_d V_w} \right) \Psi \left(1 - \frac{\theta}{T} \right) V_l \, , \tag{1}$$

where V_l is the volume of the lens-shaped region formed during the particle overlapping, V_w and V_p are the partial molar volumes of the water and of the polymer in the domain of the polymer-coated particle, whose volume is V_d, while Ψ takes into account the entropy change in the process. θ is the "theta" temperature, i.e., the temperature at which the solution behaves as an ideal solution. Equation (1) defines two different behaviors, below the θ temperature, where $\Delta G < 0$ implies polymer–polymer aggregation, and above the θ temperature, where $\Delta G > 0$ implies that the overlapping is opposed and the system undergoes entropic stabilization. According to this model, we expected that with an increase in temperature, the aggregation kinetics was become slower and the suspensions stabler. This behavior is contradicted by the experimental findings and this means that an additional aggregation process is present. In the case of PEO $M_w = 5$ MD, with a partial coverage ($P/L = 1$) we observe a reaction-limited cluster aggregation (RLCA)

regime from 10 °C, when the system is stable, to a temperature of 50 °C, where the aggregation proceedes with a progressively increasing rate. Owing to the partial polymer coverage, the bridging process must be invoked. Moreover, the presence of an adsorbed layer at the particle interface ensures that no desorption process induced by temperature takes place.

At higher coverage ($P/L = 1$), the bridging effect is reduced and the temperature-dependent effect completely balanced by the steric stabilization.

With PEO of lower molecular mass, ($M_w = 200$ kD) and with partial coverage ($P/L = 1$) the aggregation regime is strongly dependent on temperature, continuously varying from a RLCA regime at 10 °C to a DLCA regime at 50 °C. In particular, the change in the aggregation regime occurs between 40 and 50 °C, close to the θ temperature (42 °C). Also in this case, the aggregation must be imputed to the bridging process.

Finally, when the coverage is greater ($P/L = 10$), a substantial stable system is observed.

With PEO of molecular mass $M_w = 8$ kD, the aggregation kinetics is fast for all the temperatures investigated and does not appreciably deviate from that occurring in the suspension with $P/L = 0$. It must be noted, however, that, within the experimental uncertainty, the thickness of the adsorbed layer is very small.

References

1. Di Biasio A, Bordi F, Cametti C (1999) Colloids Surf A 160:189–198
2. Fleer GJ, Lyklema J (1983) In: Panfitt GD, Rochester CH (eds) Adsorption from solution at the solid/liquid interface. Academic, New York, p 123
3. Tadros T (ed) (1987) Solid-liquid dispersions. Academic, New York
4. Goodwin JW, Buscal R (eds) (1995) Colloidal-polymer particles. Academic, New York
5. Ohshima H, Furusawa K (eds) (1998) Electrical phenomena at interfaces. Dekker, New York
6. Uemura Y, MacDonald PM (1996) Macromolecules 29:63–69
7. Zhao J, Brown W (1995) Langmuir 11:2944–2959
8. Voegtli LP, Zukoski CF (1991) J Colloid Interface Sci 141:92–108
9. Barlet A, Horn D, Geiger W, Kern G (1994) Prog Colloid Polymer Sci 95:161–167
10. Napper DH (1977) In: Kerker M, Rowell RL, Zettlemoyer AC (eds) Colloid and interface science, vol. 1. Academic, New York, p 18
11. Hiemenz PC, Rajagopalan R (1997) Principles of colloid and surface chemistry. Dekker, New York

Progr Colloid Polym Sci (2001) 118: 145–148
© Springer-Verlag 2001

BIOCOLLOIDS

New biosurfactants containing β-lactam and triazole rings

Stéphane Auberger
Christine Gérardin
Ludwig Rodehüser
Chantal Finance
Céline Nicolazzi
Lourdes Pérez
Maria-Rosa Infante
Maria-Angeles Manresa
Claude Selve

S. Auberger · C. Gérardin
L. Rodehüser · C. Selve (✉)
Laboratoire de Chimie-physique
Organique et Colloïdale UMR 7565
Faculté des Sciences
UHP-Nancy I, BP 239
54506 Vandoeuvre-lès-Nancy, France
e-mail: claude.selve@lesoc.uhp-nancy.fr
Tel.: +33-3-83912360
Fax: +33-3-83912532

C. Finance · C. Nicolazzi
Laboratoire de Microbiologie Moléculaire
UMR 7565. Faculté des Sciences
Pharmaceutiques et Biologiques
54000 Nancy, France

L. Pérez · M.-R. Infante
Centro de Investigation y Desarollo
(CSIC), Departamento de tensioactivos
c/Jorge Girona 18–26
08034 Barcelona, Spain

M.-A. Manresa
Laboratori de Microbiologia
Facultat de farmacia
Universitat de Barcelona
08028 Barcelona, Spain

Abstract We report the synthesis and the properties of a new class of monobactams containing a triazole ring, that are both bio- and surface-active materials. A standard procedure for β-lactam synthesis from fatty amines, aminoacid, and 2,2-bis(hydroxymethylpropionic acid), involves an intramolecular reaction. After introduction of an azido group, the reaction with acetylenic derivatives leads to a triazole ring. The critical micelle concentrations of aqueous solutions have been determined; antibiotic and antiviral activities have been evaluated.

Key words β-Lactam · Bioactive surfactant · Triazole · Antibiotic · Antiviral activity

Introduction

The discovery of nocardicine A by Aoki et al. [1] and aztreonam (Scheme 1) showed that monocyclic β-lactams, collectively known as monobactams, can have antibiotic activity. This activity is poor but is compensated by the unique effect they can induce on certain microbial cell membranes [1, 2].

Surfactants have a great importance in numerous biological processes. For example, a correlation between biological properties, surface properties, and molecular structure has been found; however, little is known about this tridimensional relation. Our quest for new nonconventional surfactants for various biomedical applications

led us to synthesize bioactive compounds with structures similar to nocardicins [3–5].

We present here the preparation and the study of original trimodular biosurfactants of type I (Scheme 2).

These compounds present a hydrophobic part introduced by an ester or amide linkage with an aminoacid, a junction modulus which corresponds to β-lactam, and a hydrophilic part which contains a triazole. Effectively, triazole derivatives have been reported to have pharmacological activity as anti-β-lactamase, bactericides, and viricides [6, 7, 8, 9].

In view of this fact and with the hope of developing bioactive compounds with high potency, a series of novel

Scheme 1 Structures of two examples of monobactams

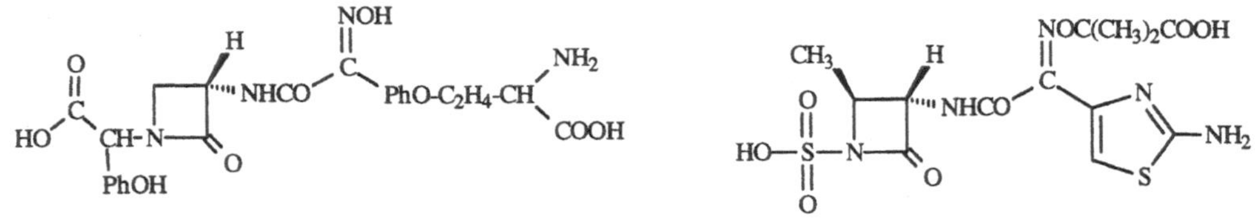

Nocardicine A

Aztreonam

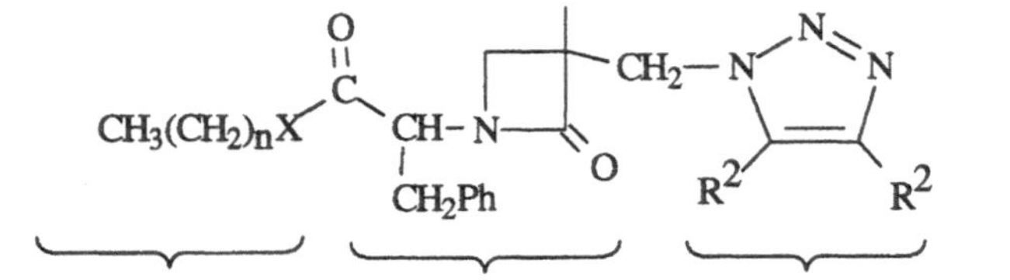

Hydrophobic part   Junction modulus   Hydrophilic part

1

Scheme 2 Target structures

surfactants were synthesized. Triazole might be expected to enhance the biological activity of these compounds.

Synthesis

The compounds were synthesized from 2-hydroxymethyl-2-methylpropionic acid in five steps. Selective activation of one of the primary hydroxyl groups was accomplished by the formation of alkoxy tris(dimethylamino)phosphonium (ATDP) salts (**3**) from the corresponding diol [10].

Treatment of **3** with excess potassium carbonate in refluxing anhydrous acetone yields the monobactams (**4**) (Scheme 3).

Scheme 3 Synthesis of monobactams

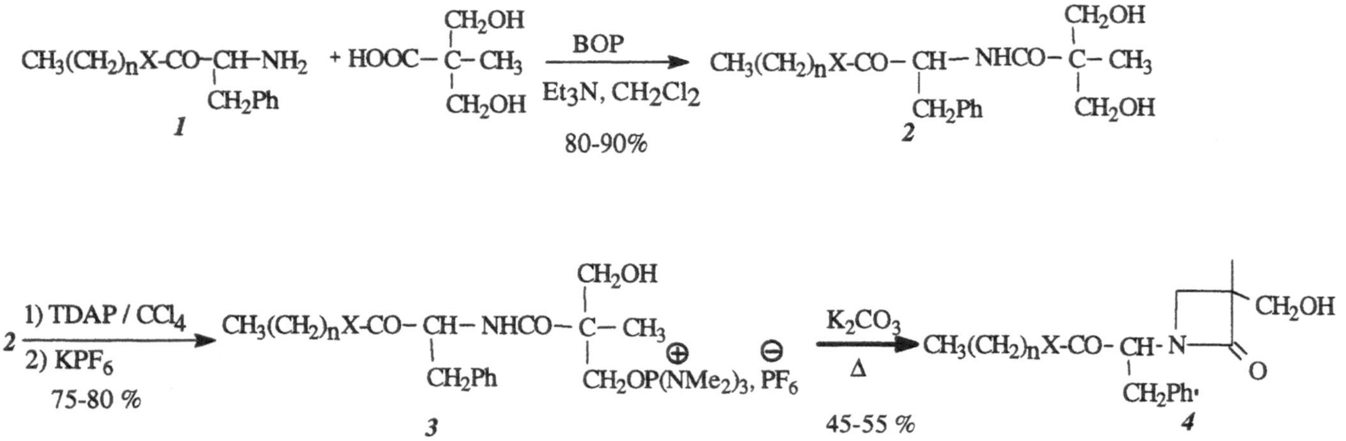

Activation by ATDP salts followed by treatment with sodium azide and refluxing in toluene gives the azido compound **6**. The reaction with acetylenic derivatives allows the surfactants **7** to be obtained [11] (Scheme 4). The results are presented in Table 1.

Physicochemical properties

The surfactant properties of the aqueous solutions were evaluated by surface tension measurement (γ) carried out using the Wilhelmy method (Dognon–Abribat tensiometer).

We studied the behavior of aqueous solutions of compounds **8**, the precursors **7** being insoluble in water (Fig. 1).

Compounds **8a** and **8b** reduce the surface tension (γ_{sat}) of water from 72 to about 30 mNm^{-1}. The solubility in water was sufficient to obtain a γ versus $\log c$ plot that was used to determine their critical micelle concentration (cmc). At room temperature **8a** and **8b** have cmc of $2 \cdot 10^{-2}$ and $5 \cdot 10^{-5}$ moll^{-1}, respectively.

Antiviral activity

The presence of an azido group on these structures incited us to determine the potential antiviral activity. We used the Elisa test on human (fibroblastic) cells (MRC-5) and human cytomegalovirus (CMV) (Table 2).

Scheme 4 Synthesis of biosurfactant compounds

Table 1 Synthesis of azido compounds and the triazole ring

Compound	n	X	R^1	Yield (%)
6a	9	O	–	90
6b	13	O	–	92
6c	9	NH	–	90
7a	9	O	H	92
7b	13	O	H	96
7c	9	NH	H	85
7d	9	O	CH_3	70
7e	13	O	CH_3	75

The determination of the therapeutic index showed that these products are interesting and we have to continue this investigation.

Antibacterial activity

The antibiotic activity on the basis to the minimum inhibitory concentration of some of the compounds was tested on different strains of gram-positive and gram-negative bacteria.

The antibacterial activity was evaluated for compounds which contain monobactams and

- A hydroxyl group [compounds **4** noted $C_xO\beta(OH)$].
- An azido group [compounds **6** noted $C_xO\beta(N_3)$].
- A triazole ring and a carboxylic group [compounds **7** noted $C_xO\beta T(H_2)$].
- A triazole ring and a carboxylate group [compounds **8** noted $C_xO\beta(Na_2)$].

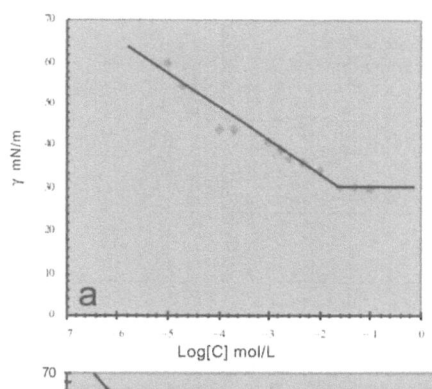
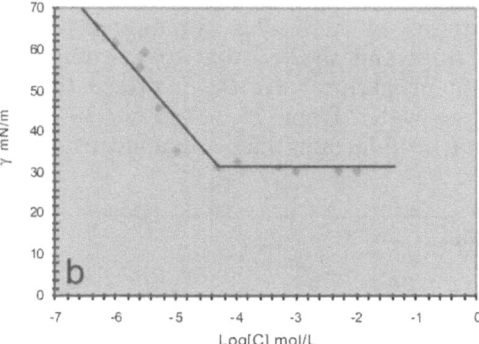

Fig. 1 Plot of γ versus log c of compound 8

The results are summarized in Table 3 with ampicillin as a reference.

All the compounds examined had antibiotic activity. Higher activity is obtained with compounds which contain an azido group or a triazole ring.

148

Table 2 Evaluation of antiviral activity. *CI*: inhibitory concentration; *CL*: cytotoxycity

Compound	Solubility in water (moll^{-1})	Concentrations (for ELISA and MTT tests) (μM)	CI$_{50}$ (μM)	CL$_{50}$ (μM)	Therapeutic index
C$_{16}$Nβ(N3)	$<10{-}8$	0.25–83	0.85	83	0.01
C$_8$Oβ(N3)	$<10{-}6$	1–512	104	>500	0.2
C$_{10}$Oβ(N3)	$<10{-}6$	1–512	4.5	230	0.02
C$_{10}$Nβ(N3)	$<10^{-6}$	0–2,048	180	225	0.8
C$_{18:1}$Oβ(N3)	$<10^{-6}$	0–2,048	>256	1,050	–
C$_{18:1}$Oβ(OH)	$<10^{-6}$	0–2,048	100	313	0.3

Table 3 Evaluation of antibacterial activity (minimum inhibitory concentration in mg/l)

	C$_1$O β(N3)	C$_8$O β(N3)	C$_{12}$O β(N3)	C$_{16}$O β(N3)	C$_{10}$N β(N3)	C$_{10}$N β(T)H2	C$_{10}$N β(T)Na2	C$_1$O β(OH)	C$_8$N β(OH)	C$_8$O β(OH)	Ampicillin
Bacillus cereus ATCC	16	32	–	16	8	64	32	32	128	128	8
Bacillus subtilis	>256	16	>256	>128	8	32	16	32	–	128	16
Bacillus pumillus	–	64	–	>128	32	16	32	16	–	–	–
Staphylococcus aureus	16	4	4	8	64	32	64	64	>128	64	32
Staphylococcus epidermidis	–	–	>256	–	–	–	–	–	>128	64	64
Streptoccocus faecalis	–	128	–	4	–	–	16	–	>128	>128	128
Proteus mirabilis	256	64	256	64	128	64	64	64	–	–	–
Escherichia coli	8	4	4	32	64	64	32	16	>128	128	>128
Pseudomonas aeruginosa	256	64	64	64	64	64	64	64	>128	128	>256
Bordetella bronchiseptica	–	–	–	–	–	–	–	–	–	128	>128
Enterobacter aerogenes	8	8	8	8	–	–	–	–	–	–	128
Salmonella thiphimurium	4	4	8	8	–	–	–	–	–	–	0.25
Candida albicans	–	–	–	–	–	–	–	64	>128	128	64
Candida rugosa	–	–	–	–	–	–	–	–	–	–	–
Candida lipolitica	–	32	–	8	8	32	16	32	–	–	–
Candida tropicalis	–	128	–	4	–	–	16	8	–	–	–
Microccocus luteus	8	8	8	8	–	–	–	–	–	–	0.5

Conclusion

We established a simple and efficient procedure for the synthesis of molecules containing β-lactam and triazole rings and showed that these compounds have surfactant properties since they lowered the interfacial tension of water from 72 to about 30 mNm^{-1}. All the synthetic β-lactams had a significant and selective antibiotic activity; compounds which contain a triazole ring present higher activity probably because of their anti-β-lactamase rule. Moreover, the results obtained with the test of antiviral activity are promising.

Acknowledgements S.A. thanks ANRT and Salveco Society. This work was generously supported by ANRT and the Salveco Society. (contract CIFRE).

References

1. Aoki H, Sakai H, Kohsaka M, Konomi T, Kubochi Y, Iguchi E, Imanaka H (1976) J Antibiot 29:492
2. Kawamoto I, Miyauchi M (1992) In: Japanese technology reviews (section E: biotechnology), vol 2. Gordon and Breach, pp 121–124
3. Molina L, Pérani A, Infante M-R, Manresa M-A, Maugras M, Achilefu S, Stébé M-J, Selve C (1995) J Chem Soc Chem Commun 1279
4. Molina L, Gérardin-Charbonnier C, Selve C, Stébé M-J, Maugras M, Infante M-R, Torres R-L, Manresa M-A, Vinardell P (1997) New J Chem 21:1027
5. Gérardin-Charbonnier C, Auberger S, Molina L, Achilefu S, Manresa M-A, Vinardell P, Infante M-R, Selve C (1999) Prep Biochem Biotechnol 29: 257
6. Revankar GR, Solan VC, Robino RK (1981) Nucleic Acids Res Symp Ser 9:65
7. Makabe O, Suzuki H, Unezawa S (1977) Bull Soc Chem Soc 50:2689
8. Wigerinck P, Aershot P, Claes P, Balzarini J, Declerq E, Herdewyn P (1989) J Heterocycl Chem 26:1635
9. Adam Y, Huet J (1992) Traité de Chimie Thérapeutique "Médicaments antibiotiques" Medicales Internationales (ed) Tec et Doc Lavoisier 2:207
10. Molina L, Gérardin-Charbonnier C, Cartier A, Infante M-R, Selve C (1997) J Chim Phys 94:1159
11. Chrétien F, Gross B (1982) J Heterocycl Chem 19:263

Progr Colloid Polym Sci (2001) 118: 149–152
© Springer-Verlag 2001

Hydrophilic matrices as carriers in felodipine solid dispersion systems

E. Karavas
E. Georgarakis
D. Bikiaris
T. Thomas
V. Katsos
A. Xenakis

E. Georgarakis
Section of Pharmaceutics and Drug Control
Department of Pharmacy
Aristotle University of Thessalonike
54006 Thessalonike, Greece

E. Karavas · A. Xenakis (✉)
Institute of Biological Research
and Biotechnology
National Hellenic Research Foundation
48 Vas. Constantinou Ave.
11635 Athens, Greece
e-mail: arisx@eie.gr
Tel.: +30-1-7273762
Fax: +30-1-7273758

T. Thomas · V. Katsos
Pharmathen Pharmaceuticals Ltd.
6, Dervenakion Str.
15351 Pallini Attikis, Greece

D. Bikiaris
Department of Chemistry
Aristotle University of Thessalonike
54006 Thessalonike, Greece

Abstract Hydrophilic matrices or hydrocolloids are polymers which swell on contact with aqueous solutions and dissolve slowly from the surface forming a gel mass. Several studies have been carried out in the past few years on the use of hydrocolloids in controlled release formulations. The present study used three modified celluloses, carboxymethyl cellulose sodium, hydroxyethyl cellulose (HEC), and hydroxypropylmethyl cellulose (HPMC) in systems using the dihydropyridine felodipine, which is slightly soluble in water, as the active ingredient. This study was concerned with solid dispersions, which were prepared following the dissolution method using a common solvent. The drug–polymer interactions were studied using differential scanning calorimetry and IR techniques, as well as high-performance liquid chromatographic purity after storage in strength conditions. Neither significant interactions nor degradation of the active ingredient was observed after storage at 40 °C for 3 months. In addition, felodipine release from the solid dispersion systems was studied and the factors influencing release, such as the drug–polymer ratio, interactions, and polymer properties were investigated. HPMC was observed to promote a more significant retard and a more linear release of the active ingredient than HEC. Finally, the natural mixtures presented a larger variation and high relative standard deviation values.

Key words Hydrocolloids · Hydrophilic matrices · Felodipine · Solid dispersion systems · Controlled release

Introduction

Generally, dihydropyridines are insoluble in water, and for this reason several studies [1] have been carried out in the past few years aimed at the optimization of their bioavailability via different systems and carriers. Felodipine [2] is a dihydropyridine whose pharmacokinetic parameters [3–5] advocate the development of an extended release formulation. It is well established that the most critical parameter of bioavailability in an insoluble compound is its dissolution rate.

In this study, three hydrophilic matrices, carboxymethyl cellulose sodium (CMCS), hydroxyethyl cellulose (HEC), and hydroxypropylmethyl cellulose (HPMC), in

the form of solid dispersions with felodipine (FL) as the active ingredient were studied. Interactions, compatibilities, and incompatibilities as well as the release profile of the active ingredient in these systems were studied.

Solid dispersions were prepared by applying the dissolution method [6, 7] using absolute ethanol and purified water as the solvents, which can legally be used in the manufacture of pharmaceuticals.

The selection of the three carriers was based on the characteristic property of hydrocolloids to expand in aqueous solutions, forming a gel mass on the external surface, which dissolves gradually [8, 9]; therefore, the development of a homogenic system of the active ingredient with the hydrocolloid is expected to result in a gradual dissolution of the active ingredient, following the dissolution rate of the polymer. Simultaneously, the sparse distribution of the active ingredient in the polymeric mass may promote the dissolution rate independently crystalline form and particle size of the active ingredient.

The aims of this study were, firstly, to present the release profile of FL from solid dispersion systems containing the three hydrocolloids and, secondly, to investigate the interactions, compatibilities, and incompatibilities of these systems.

Experimental

Materials

FL with 99.9% of active ingredient was supplied by Medichem. HPMC with a viscosity of 4,000 cps was supplied by Colorcon. HEC was supplied by Hercules and had a viscosity of 1%, 2,000 cps, whilst CMCS was supplied by Akzo. All other reagents were of pro analysi grade, whilst all the experiments were carried out under low light so as to prevent light degradation of FL.

Preparation of the systems

Solid dispersion systems of 50:50, 75:25, and 90:10 FL–polymer ratios were prepared as follows. The polymer was distributed in equal quantities of water until a colloidal dispersion was formed, and an equal quantity of absolute ethanol was added to each dispersion. The relative quantities of FL were added after having been diluted in absolute ethanol. The dispersions were mixed, exposed to ultrasonication, and then heated to 40 °C for 48 h until the solvent was fully evaporated. After drying, the solid dispersions were pulverized. It must be noted that CMCS precipitates with ethanol, and for this reason only the 50:50 solid dispersion system was prepared. In addition, for comparison purposes, 50:50 natural mixtures were prepared by simple mixing of the active ingredient and the polymers. All the systems were assayed before use, using the reverse-phase high-performance liquid chromatography (RP-HPLC) method at 237 nm.

Interactions and compatibilities

Differential scanning calorimetry (DSC) and Fourier transform (FT) IR methods were employed for the study of interactions using a Shimadzu DSC-50Q fast quenching differential scanning calorimeter, with a heating rate of 20 °C/min and a Biorad FTS-45A FT IR spectrometer. The compatibility between the polymer and the active ingredient was investigated by comparing the chromatographic purity of the systems with those of the active ingredient using RP-HPLC at 237 nm after storage at 40 °C for 3 months.

Release profiles

A modified dissolution apparatus II USP using a stationary disc (paddle over disc) at 100 rpm, 500 ml 0.1 M phosphate buffer pH 6.5, and containing 2% Tween was used [10, 11]. Sampling was carried out at every hour, whilst the percentage of active ingredient release was measured using the UV spectrum at 237 nm.

Results and discussion

Interactions, compatibilities, and incompatibilities

The polymers used in this study are cellulose compounds which are of particular interest for the manufacture of oral pharmaceutical dosage forms. It has been shown several times that dissolution or, in general, the behaviour of certain active ingredients may be altered by interactions with excipients [12, 13]. Cellulose compounds used as excipients contain hydroxyl groups, which may possibly form hydrogen bonds with the carbonyl groups of FL as well as with aromatic ring chlorides. In order to confirm such bonding, the solid dispersions prepared were investigated using DSC, used specifically for the investigation of this phenomenon [14]. An indicative thermogram is presented in Fig. 1.

FL was shown to possess a melting point of 150 °C, whilst the cellulose excipients presented a broad endothermic peak, the maximum of which was unstable. The maximum of HEC was approximately 123 °C, that of HPMC was 105 °C, whilst that of CMCS was 132 °C. This broad peak does not characterize the melting point of the compounds, but loss of adsorbed water, since, due to their strong hydrophilic nature, large quantities of moisture are absorbed, which are difficult to remove. This observation was corroborated by the study of the pure compounds using thermogravimetric analysis, which confirmed a loss of mass in this region. This peak was also noted, albeit distinctively weaker, in the compound mixture, owing to the loss of moisture during drying of the samples, and presented a maximum at slightly lower temperatures. In addition, the FL peak was seen at a maximum of 147 °C. The difference was too small to confirm interactions with the excipients. For this

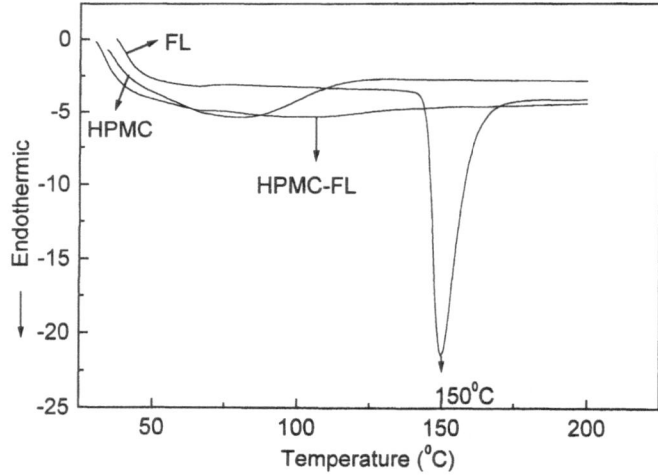

Fig. 1 Differential scanning calorimetry thermogram of pure felodipine (*FL*), pure hydroxypropylmethyl cellulose (*HPMC*), and 50:50 HPMC–FL solid dispersion

reason, the mixtures were additionally studied using spectroscopy. FL produced intense peaks, characteristically those at 3,373 cm^{-1} of the secondary amino groups (>NH) and at 1,689 and 1,698 cm^{-1} of the esteric carbonyls, which, owing to resonance, were at shorter wavelengths than expected. For the cellulose compounds characteristic absorbencies of the >C–O group in the finger print region (1,000–1,200 cm^{-1}) appeared as a triple peak. No significant shift in the peak absorbencies of the mixtures or in the relevant carbonyl groups of FL was observed. It was therefore concluded that no significant interactions between excipients and the active ingredient occur during the manufacture of the tablets.

The chromatographic purity of all the samples remained unchanged after storage at 40 °C for 3 months. No increase in the peaks that could be attributed to impurities or degradation products was observed.

These results indicate that the three polymers do not present chemical incompatibilities or strong interactions with the active ingredient, and as a result are considered to be suitable as carriers of solid dispersion systems with FL.

Release profiles

Attention was focused on the HEC and HPMC polymer systems. For the investigation of the effect of the drug–polymer ratio on the rate of release of the active ingredient the mean dissolution time (MDT) values were studied. The MDT is defined as the time required for 50% of the active ingredient to be released from the dosage form and indicates the reduction in the release rate of the active ingredient, as caused by the carriers. A comparison of drug release presented by the carriers HEC and HPMC was carried out with a direct comparison of the release diagrams.

All the solid dispersion systems showed homogeneity in release between the different units, with relative standard deviation (RSD) values smaller than 5%. This did not occur with the natural mixtures, probably because of the different placement of powders in the capsule and the diphasic property of the systems. It is therefore concluded that the natural mixtures are of interest only when tabletting by compression is carried out.

As the 50:50 systems promote rapid drug release (MDT 0.7 h for HPMC, 1.5 h for HEC), which does not satisfy the pharmacokinetic requirements of FL, interest was thus focused on the systems containing HPMC and HEC with polymer–drug ratios 75:25 and 90:10. The FL release from the 75:25 solid dispersions for a period of 4 h is presented in Fig. 2. Approximately 70% of the active ingredient was released from both systems. The HEC solid dispersion system gave a rapider drug release at initial times, with a MDT value of 1.8 as opposed to 2.2 for the HPMC system.

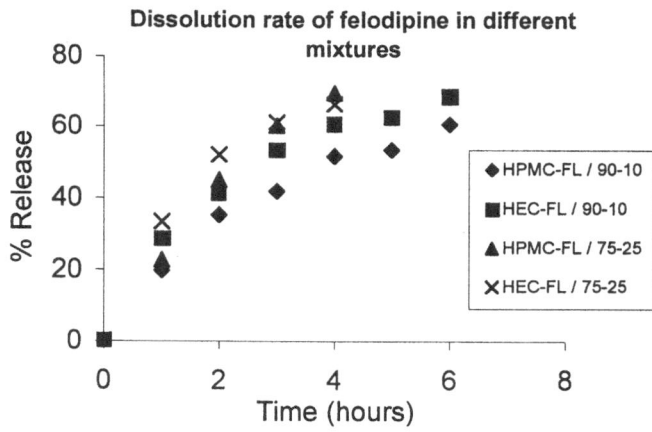

Fig. 2 Release profile of FL in solid dispersions of MPMC–FL and hydroxyethyl cellulose (*HEC*)–FL

This behaviour results in the HPMC solid dispersion system having a more linear drug release (with a correlation coefficient $R^2 = 0.97$), nearing zero-order release kinetics. Similarly, the 75:25 HEC system released the active ingredient with less linearity ($R^2 = 0.88$).

The corresponding FL release from the 90:10 systems (polymer–drug ratio) is presented in Fig. 2 for a 6-h time period. The 90:10 HPMC–FL system released approximately 60% of the active ingredient over 6 h, whilst the 90:10 HEC–FL system released approximately 70%. In the 4-h time period, where the 75:25 solid dispersion systems were studied, drug release was found to be approximately 50% for the 90:10 HPMC–FL system and 60% for the 90:10 HEC–FL system. The 90:10 HEC–FL system showed a rapider and greater drug release than the 90:10 HPMC–FL system, with a MDT value of 2.8 h as opposed to the MDT value of 4 h for the latter system. Additionally, in the 90:10 polymer–drug ratio systems, the HPMC system presented a more linear release ($R^2 = 0.92$), whilst the HEC system gave a correlation coefficient of 0.88.

The HPMC system therefore gave a slower drug release than the HEC system in both polymer–drug ratio systems, which may be attributed to the higher viscosity of the former cellulose (4,000 cps) as opposed to the latter (2,000 cps) in identical 1% w/v aqueous solutions. In this respect, it can be stated that the two systems do not present significant differences in interactions according to the results of this study. In addition, the HPMC systems presented a more linear drug release than those of HEC, which may be attributed to the morphology and physicochemical traits of each polymer.

In order to investigate the effect of the polymer–drug ratio on the release of FL from the systems, the MDT values were obtained for each system in relation to the percentage of each hydrophilic matrix in the systems and are recorded in Fig. 3.

152

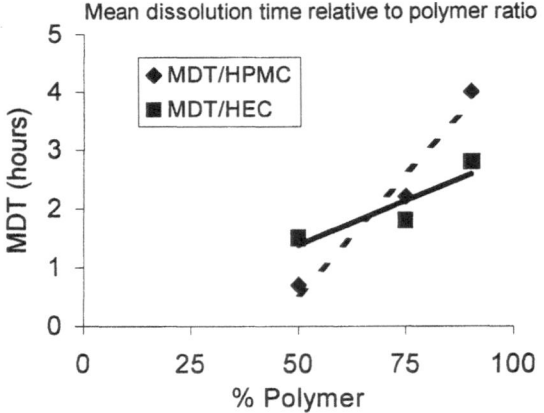

Fig. 3 Mean dissolution time (*MDT*) values relative to the polymer percentage for HPMC–FL and HEC–FL solid dispersion systems

In both systems it was observed that an increase in the percentage of the polymer in the system corresponded to a reduction in the release rate of the active ingredient with increasing MDT values. The HPMC systems produced slower drug release, with a corresponding increase in the percentage of polymer, than the HEC systems. The increase in the MDT values appeared to be almost linear for the HPMC system ($R^2 = 0.96$), which did not occur in the HEC systems ($R^2 = 0.81$).

The 0.0802 and 0.0304 slopes for the HPMC–FL and HEC–FL systems, respectively, confirm that the polymer percentage effect is more significant in the former system than in the latter.

Conclusions

The three modified celluloses, HPMC, HEC, and CMCS, as observed from the studies on the interactions and compatibilities–incompatibilities, are suitable as excipients for the active ingredient FL. HPMC and HEC are compatible with FL at all ratios in solid dispersion systems. The systems with a polymer amount of over 75% present particular interest. HPMC can be characterized as a more suitable polymer than HEC, as it promotes a more linear release of the active ingredient, whilst the percentage in which it is added to solid dispersion systems presents a more critical effect on the drug release profiles. Therefore in order to promote retarded drug release from the HPMC system, a much smaller quantity of HPMC would be required as opposed to the HEC system. Finally, the natural mixtures present large variations from unit to unit (RSD > 15%), most probably owing to placement in the capsule, the variation in the contact surface with the dissolution medium, as well as the variation in total porosity. The fact that the natural mixtures are diphasic systems (excipient–active ingredient) must be taken into account. Natural mixtures would be of interest only in those cases of tablet manufacture where these factors would be controlled.

Acknowledgements The study was funded by Pharmathen Pharmaceuticals Ltd., Pallini, Attikis.

References

1. Suzuki H, Sunada H (1997) Chem Pharm Bull 45:1688
2. (1997) European Pharmacopoeia, 3rd edn. European Department Quality of Medicines. Council of Europe, Strasbourg, p 847
3. Blychert E (1992) Blood Press Suppl 2:1
4. Videback LM, Jacobsen IA (1997) Int J Clin Pharmacol Ther Nov:35
5. Edgar B, Lundborg P, Regardly CG (1987) Drugs 34:16
6. Chiw WL, Riegelman S (1987) J Pharm Sci 60:1281
7. Ford JL (1986) Pharm Acta Helv 61:69
8. Conte U, Maggi L, Colombo P, La Manna A (1993) J Controlled Release 26:39
9. Peppas N, Sahlin J (1989) Int J Pharm 57:169
10. Wingstrand K, Abrahamson B, Edgar B (1990) Int J Pharm 60:151
11. Abrahamson B, Johansson D, Torstensson A, Wingstrand K (1994) Pharm Res 11:8
12. Ling GN, Ochaenfeld MM, Walton C, Bersinger TJ (1980) Physiol Chem Phys 12:3
13. Ozeki T, Yuasa H, Kanaya Y (1997) Int J Pharm 155:209
14. Okhamafe AO, York P (1988) J Pharm Sci 77:438

Progr Colloid Polym Sci (2001) 118: 153–156
© Springer-Verlag 2001

Y. Rochev
T. Golubeva
A. Gorelov
L. Allen
W. M. Gallagher
I. Selezneva
B. Gavrilyuk
K. Dawson

Surface modification for controlled cell growth on copolymers of *N*-isopropylacrylamide

Y. Rochev (✉) · T. Golubeva
A. Gorelov · L. Allen · K. Dawson
Irish Centre for Colloid Science
and Biomaterials,
Department of Chemistry
University College Dublin
Belfield, Dublin 4, Ireland
e-mail: rotchev@yahoo.com
Tel.: +353-1-762408
Fax: +353-1-762415

Y. Rochev · A. Gorelov
I. Selezneva · B. Gavrilyuk
Institute of Theoretical and Experimental
Biophysics of Russian Academy of Science
Pushchino, Russia

L. Allen · W. M. Gallagher
Conway Institute Biomolecular and
Biomedical Research
Department of Pharmacology
University College Dublin
Belfield, Dublin 4, Ireland

Abstract We investigated the behaviour of cell cultures on the surface of thermoresponsive polymers. We synthesised a series of copolymers of *N*-isopropylacrylamide and *N-tert*-butylacrylamide (NtBA) with ratios of 85:15, 65:35 and 50:50. Increasing the amount of NtBA results in a reduction in the lower critical solution temperature as determined by microcalorimetric methods and an increase in surface hydrophobicity. Experiments determined that human epithelial cell growth was almost identical on surfaces with a higher degree of hydrophobicity. Preconditioning of surfaces with cell culture medium containing serum promoted cell adhesion on more hydrophilic polymers.

Key words Cell adhesion · Thermoresponsive polymer · Protein adsorbtion

Introduction

Advances in molecular and cell biology, combined with polymer science and engineering, are pushing the field of biomaterials into new applications. A new class of responsive polymers which respond to environmental stimuli has been developed. One such class of responsive polymers are thermoresponsive polymers, polymers that respond to changes in temperature. A subclass of these are thermally reversible polymers based on *N*-isopropylacrylamide (NIPAM).

Poly(NIPAM) (PNIPAM) is a well-known temperature-sensitive polymer, exhibiting a lower critical solution temperature (LCST) at 32 °C in water. PNIPAM is soluble below this temperature and becomes insoluble, owing to the destruction of hydration around the polymer chains and increasing intra- and intermolecular hydrophobic interactions, at temperatures higher than the LCST [1]. This unique thermoresponsive property of PNIPAM and its copolymers makes it particularly relevant as a novel method for drug delivery, as well as peptide and protein delivery [2, 3].

When surfaces are modified with thermoresponsive PNIPAM, the surfaces show temperature-responsive wettability changes in aqueous solutions. This phenomenon creates a substrate, which can interchange between being hydrophilic at temperatures below the LCST and hydrophobic at temperatures above the LCST. Modification of surfaces by addition of polymers that are thermoresponsive at physiologically relevant temperatures creates a novel cell culture substrate and cell recovery system. Anchorage-dependent cells adhere and grow on PNIPAM in its hydrophobic state at 37 °C and these cells detach from such surfaces upon lowering of the temperature below the LCST [4, 5]. By utilising this property, one can efficiently recover cells without the use of destructive proteolytic agents such as trypsin or other chemical agents. Removing the need for such agents is especially important when preforming primary cell culture and experimental protocols where initial cell attachment and spreading are used to assess candidate biomaterials.

Although thermoresponsive polymers have been used as novel cell culture substrates, the ability of adherent cells to grow on such substrates is poor in comparison to the more conventional cell culture grade polystyrene [6]. So, the problem of improving thermoresponsive substrates for mammalian cell cultivation is a current topic. The purpose of our work was to study cell adhesion and growth on a series of thermoresponsive copolymers with different ratios between PNIPAM and *N-tert*-butyl-acrylamide (NtBA).

Materials and methods

Copolymer preparation

Materials

NIPAM (99%, Acros Organics, New Jersey, USA) and NtBA (purum, Fluka Chemie, Switzerland) were recrystallised from hexane and dried at room temperature in vacuum. *N,N′*-Azobis(isobutyronitrile) (AIBN), (Phase Separation, Queensferry, Clwyd, UK) was recrystallised from methanol. Benzene was dried under sodium and was distilled before use. All the other solvents were reagent grade and were purified by conventional methods.

Polymerisation procedure

A series of copolymers was prepared by radical polymerisation of NIPAM and NtBA using AIBN (0.01 mol AIBN/mol monomer) as initiator in benzene (10 w/w %) under argon. After polymerisation

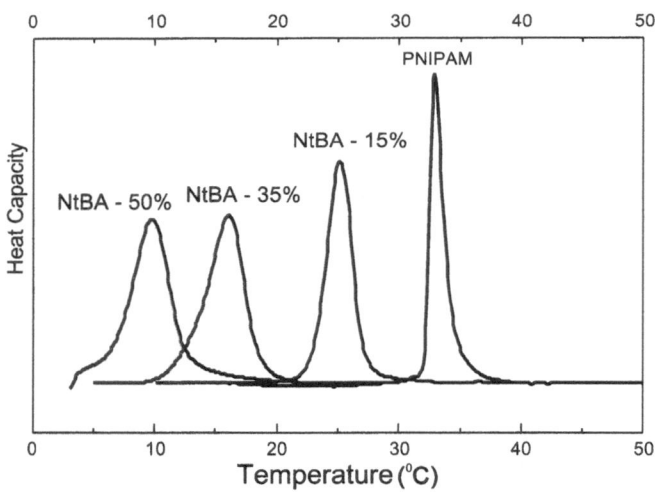

Fig. 1 Thermograms of the copolymers with different *N*-isopropyl-acrylamide (*NIPAM*)/*N-tert*-butylacrylamide (*NtBA*) ratios. The heating rate is 1 K/min

at 60 °C for 24 h, the mixture was precipitated in diethyl ether. Precipitation was repeated three times using acetone as a solvent and hexane as a nonsolvent, and the product was dried at room temperature in a vacuum.

Microcalorimetric measurements

To prepare the polymer samples for calorimetric measurements, a weighed sample of polymer was dissolved in distilled water by careful mixing at 5 °C over a 12 h period. Each sample was prepared in triplicate. Calorimetric measurements were carried out using a DASM-4 differential adiabatic scanning microcalorimeter (NPO BIOPRIBOR, Pushchino, Russia) with a cell volume of 0.47 ml. Measurements were taken in the temperature range 5–50 °C at a constant heating rate of 1 K/min and at a constant excess pressure of 2.0 atm. The calorimetric traces were digitised. For each experiment, two heating scans were normally carried out. The first scan was done with both calorimetric cells filled with water. This gave an instrumental base line. In the second scan, one of the cells contained the solvent and the other cell contained a polymer solution in the same solvent. During this scan, the difference in the heat capacities of the sample and the solvent was measured as a function of temperature. The transition temperature was taken as the temperature maximum of the thermogram. Thermograms of all the copolymers are presented in Fig. 1. The observed cooperative transitions were characterised by the transition temperature and half-width of transition shown in Table 1.

Film preparation

PNIPAM–NtBA films were cast in (20-mm diameter) wells of 12-well polystyrene tissue culture grade dishes from a 5% (w/w) solution of polymer in dry ethanol (54.5 μl/well). The ethanol was allowed to evaporate for 24 h in a laminar flow hood creating polymer films 5 μm thick.

Cell culture

Human epithelial (HeLa) cells were maintained in Dulbecco's modified eagles medium (DMEM) containing 10% (v/v) foetal bovine serum (FBS), 50 mg/l penicillin, 50 mg/l streptomycin and 4 mM L-glutamine. The cells were cultured at 37 °C in a humidified atmosphere containing 95% air and 5% CO_2 and were left undisturbed until the cells had adhered and spread. The medium was changed every second day until the cell monolayer had reached 70% confluence, at which point the cells were harvested for experimentation or reseeding.

Cell adhesional growth on copolymer films

Two millilitres HeLa cell suspension (50,000/ml) in DMEM containing 10% (v/v) FBS, 50 mg/l penicillin, 50 mg/l streptomycin and 4 mM L-glutamine was incubated at 37 °C in a humidified

Table 1 Transition temperature and half-width transition of *N*-isopropylacrylamide (*NIPAM*)/*N-tert*-butylacrylamide (*NtBA*) copolymers

Copolymers	Lower critical solution temperature (°C)	$T_{1/2}$ (°C)
PNIPAM	32.9	1.2
NIPAM/NtBA 85:15	25.1	2.3
NIPAM/NtBA 65:35	16.1	3.5
NIPAM/NtBA 50:50	9.8	3.8

atmosphere containing 95% air and 5% CO_2 and was left undisturbed with the polymer surface for 24 h. After 24 or 72 h, the culture medium was removed and the dishes were washed three times with prewarmed (37 °C) sterile phosphate buffered saline (PBS). Cells on the surface of the polymer films were harvested after 24 and 72 h by trypsinisation (0.25% trypsin for 5 min). The cells were examined microscopically. Once all the cells had detached, 150 μl medium was added to deactivate the trypsin. The number of viable cells was counted using a haemocytometer counting chamber (improved Neabauer model) under a Nikon TMS phase-contrast microscope. Cell viability was determined by the addition of a Trypan Blue solution [Trypan Blue (0.4%), Sigma Chemicals].

Protein preincubation

The polymer surfaces were incubated with cell culture medium with or without FBS at 37 °C in a humidified atmosphere containing 95% air and 5% CO_2 and were left undisturbed with the polymer surface for 72 h. After 72 h, the cell culture medium was removed and the surface was washed three times with warmed (37 °C) PBS, removing any trace of cell culture medium and serum. Two millilitres HeLa cell suspension (50,000/ml) in cell culture medium without serum was incubated at 37 °C in a humidified atmosphere containing 95% air and 5% CO_2 and was left undisturbed with the polymer surface for 24 h. Only serum proteins which have absorbed to the surface of polymer films should have an affect on the ability of HeLa cells to adhere. The cells were harvested after 24 h and the number of cells was determined as described earlier.

Results

The relationship between surface physical chemistry and cell adhesion and growth is still not completely clear. Surface characteristics such as hydrophobicity, surface energy, texture, surface charge and chemical composition are all known to play roles in cell adhesion. The influence of the degree of surface hydrophobicity has been greatly studied [7] and there appears to be an optimum range of surface energy (as determined by contact-angle measurements), which promotes mammalian cell adhesion. To date, the study of hydrophobicity has involved the use of different materials, such as glass, polystyrene, etc., with different degrees of surface hydrophobicity; however, such materials vary greatly in surface chemistry, making the study of the influence of hydrophobicity alone difficult. We synthesised a series of copolymers on the basis of NIPAM and NtBA which should display similar surface chemistry with different degrees of hydrophobicity at 37 °C.

Increasing the amount of the hydrophobic monomer NtBA in the copolymer lowers the LCST as demonstrated in Fig. 1. From this thermogram, we can also conclude that increasing the amount of NtBA increases the half-width of the transition, the numerical values for which are represented in Table 1. From this, we determined that the polymer films cast from these copolymers display increasing hydrophobicity with

increasing amount of NtBA. On the basis of this premise, we investigated the ability of HeLa cells to adhere on copolymer films (absence of serum) and on films with preadsorbed proteins (presence of serum). The numbers of adherent cells after 24 h are given in Table 2.

Table 2 Adherence of human epithelial cells from the medium to untreated polymers surfaces (−serum) and those preconditioned with serum (+ serum) after 24 h. The initial number was 10^5 cells per dish

Copolymer	+ Serum	− Serum
Control	110,000 ± 20,000	65,000 ± 4,000
NIPAM/NtBA 85:15	85,000 ± 43,000	40,000 ± 7,000
NIPAM/NtBA 65:35	20,000 ± 7,000	55,000 ± 5,000
NIPAM/NtBA 50:50	26,000 ± 7,000	57,000 ± 7,000
	18,000 ± 15,000	5,000 ± 1,000

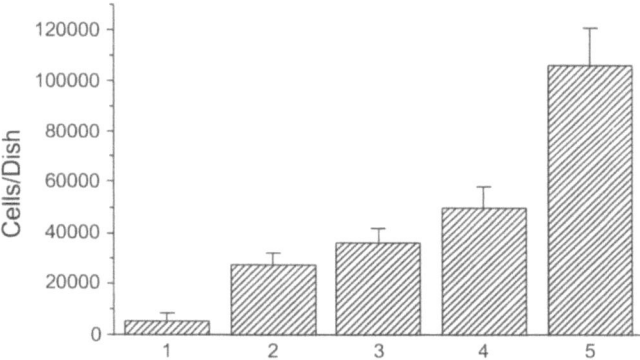

Fig. 2 Adhesion of human epithelial (*HeLa*) cells on copolymers 24 h after seeding. The control is tissue culture grade polystyrene. Poly(N-isopropylacrylamide) (*PNIPAM*) (*1*), NIPAM/NtBA 85:15 (*2*), NIPAM/NtBA 65:35 (*3*), NIPAM/NtBA 50:50 (*4*), control (*5*)

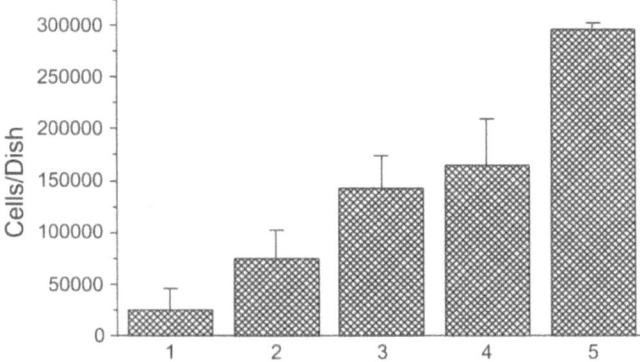

Fig. 3 Growth of HeLa cells on copolymers 72 h after seeding. The control is tissue culture grade polystyrene. PNIPAM (*1*), NIPAM/NtBA 85:15 (*2*), NIPAM/NtBA 65:35 (*3*), NIPAM/NtBA 50:50 (*4*), control (*5*)

The results demonstrate similar cell adhesion on the polymers and the control, with the exception of PNIPAM, in the case where surfaces were not perincubated with serum. So, it may be concluded that in this case the surface chemistry is not crucial for cell adhesion. Protein preadsorbed from serum does promote cell adhesion to less hydrophobic substrates, like PNIPAM, NIPAM/NtBA with ratio 85:15, and to the control, but decreases cell adhesion to NIPAM/NtBA copolymers with ratios 50:50 and 65:35. This means that cell adhesion to the polymer substrate is determined by preadsorbed proteins. The poor cell adhesion on PNIPAM can be explained by the proximity of the transition temperature of PNIPAM to the temperature of cell cultivation. It was observed that PNIPAM films were unstable, which can be explained by the high content of water in the film. The results of cell growth on copolymer films after 24 and 72 h are shown in Figs. 2 and 3, respectively.

In these experiments, cells were seeded on nontreated polymer surfaces in cell culture medium with 10% serum.

The results after 24 h are similar to the results with cell adhesion on non-pre-incubated surface without serum. A more hydrophobic surface and control demonstrate some advantage and PNIPAM again shows poor cell adhesion. We see an increasing number of cells after 72 h on each copolymer. The best cell growth is demonstrated on tissue culture grade polystyrene. Cell growth on the copolymers with monomer ratios 85:15, 65:35 and 50:50 are similar. These results demonstrate that in the absence of serum all the copolymers show the same adhesive properties except PNIPAM. Hydrophobicity can be responsible for cell adhesion, but the main factor in cell–copolymer interaction is protein adsorption.

Acknowledgement The authors wish to thank the EU INCO–Copernicus Program (grant no. NIC15-CT96-0756) for partial support of this project.

References

1. Heskins M, Guillet JE, James EJ (1968) Macromol Sci Chem A 2:1441
2. Kohori M, Sakai K, Aoyagi T, Yokoyama M, Yamato M, Sakurai Y, Okano T (1999) Colloids Surf B 16:195
3. Bromberg LE, Ron ES (1998) Adv Drug Deliv Rev 31:197
4. Rollason G, Davies J, Sefton M (1993) Commun Biomater 14:153
5. Kikuchi A, Okuhara M, Karkikusa F, Sakurai Y, Okano TJ (1998) Biomater Sci Polym 9:1331
6. Piskareva OA, Rochev YA, Gavrilyuk BK, et al (1999) Biofizika 44:281
7. Valk P, Pelt A, Busscher H, Jong H, Wildevuur C (1983) J Biomed Mater Res 17:807

Progr Colloid Polym Sci (2001) 118: 157–162
© Springer-Verlag 2001

I. Lynch
K. A. Dawson

Elastically ineffective chain formation in networks at high initiator concentration

I. Lynch · K. A. Dawson (✉)
Irish Centre for Colloid Science and
Biomaterials, Department of Chemistry
University College Dublin, Belfield
Dublin 4, Ireland
e-mail: kenneth@fiachra.ucd.ie
Tel: +353-1-706 2300
Fax: +353-1-706 2415

Abstract The shrinking kinetics of terpolymer gels composed of 10:20:70 N-$tert$-butylacrylamide:N,N'-dimethylacrylamide: N-isopropylacrylamide were found to be sensitive to the initiator concentration in the pregel solution. The gels were synthesised by redox polymerisation using ammonium peroxydisulphate and N,N,N',N'-tetraethylenediamine, resulting in SO_4^- groups at the free end of the forming chains. Increasing the initiator concentration resulted in significantly faster shrinking times. The transition temperature and the continuity of the transition were not affected by the initiator concentration, and the degree of swelling at low temperatures was only slightly increased. Thus, there was no evidence of increased osmotic pressure owing to increased charge density. However, the shrinking process was dramatically different for the gel at higher initiator concentration, being 2 orders of magnitude faster, and preventing the formation of a surface skin layer at high temperatures. It is postulated that this resulted from an increased number of growing chains being formed at higher initiator concentration, resulting in an increased number of elastically ineffective chains. It is further postulated that these free chains have increased mobility compared to the network chains, resulting in faster shrinking of gels formed at higher initiator concentration.

Key words Gels · Initiator concentration · Shrinking · Elastically ineffective chains

Introduction

Introduction of charged comonomers into a gel is well known to affect the degree of swelling of the gel in a good solvent, the temperature at which the transition occurs and the discontinuity of the temperature-induced volume phase transition [1–6]. The increase in the degree of swelling is a result of the osmotic pressure owing to dissociation of ions. The increased transition temperature is a result of the polymer–water interactions remaining favourable to a higher temperature, as the charged groups form hydrogen bonds with water. The increased discontinuity of the shrinking transition also results from the contribution of osmotic pressure owing to counterions to the equilibrium condition of the gel.

In all the experiments conducted on ionic gels, the charged groups have been introduced as comonomers, such as acrylic acid [6] or sodium acrylate [1]. The nature of the polymerisation in such cases is that the charged groups end up randomly distributed throughout the gel, and thus their effect is evenly felt throughout the gel. Recent experiments by Kokufuta et al. [7] have shown that the charge distribution has a significant effect on the swelling behaviour of N-isopropylacrylamide (NIPA)–acrylic acid gels. They prepared gels with three distinct patterns of charge distribution. The first gel had a typical random distribution of charges. The second and third gels had the charges localised along the polyacrylic acid chains, which were then incorporated into the gels. The gel types are represented schematically in Fig. 1. The

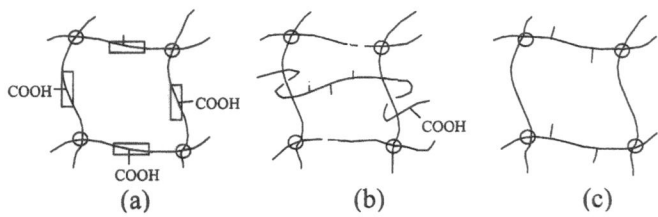

Fig. 1a–c Schematic illustration of the three polyelectrolyte gels consisting of *N*-isopropylacrylamide (NIPA) and acrylic acid residues. **a** Random distribution, **b** polyacrylic acid chains and NIPA monomers in the pregel solution, **c** polyacrylic acid chains and poly(NIPA) chains in the pregel solution

charge distribution was found to have a significant effect on the swelling behaviour and the volume phase transition of the gels [7].

The use of radical initiators has been shown to result in the introduction of charged groups into a gel [8, 9]. The process of initiation using a redox couple, such as ammonium peroxydisulfate (APS) and *N*,*N*,*N'*,*N'*-tetramethylethylenediamine (TEMED), results in the APS molecule breaking at the –O–O– bond into two radical species (SO_4^-). Each of the radicals attack the –C=C– bond of a monomer, leaving a radical on the second carbon of the monomer unit, and so the chain propagates, and leaving a charged group at the "free-end" of the chain. Thus, the gel ends up with a significant number of charged groups located at the free ends of the polymer chains within the gel network.

The effect of increasing the number of charged groups by increasing the initiator concentration on the shrinking kinetics of a 10:20:70 *N-tert*-butylacrylamide (BAM): *N*,*N'*-dimethylacrylamide (DAM):NIPA gel was investigated. This gel composition was selected as it has a continuous transition at 34.5 °C, and its shrinking process has been described in detail [10]. The gels had identical monomer, cross-link and promoter concentrations, with only the initiator concentration varied. Initiator concentrations of 1 and 3 wt% of the initial monomer concentration were used.

Experimental

Materials

NIPA monomer (purity above 99%) from Phase Separations (Clwyd, UK) was recrystallised twice from hexane. BAM, DAM and *N*,*N'*-methylenebisacrylamide from Fluka (Dorset, UK) were used as supplied. APS from Aldrich (Dorset, UK) was used as supplied. TEMED from Sigma (Dorset, UK) was used as supplied. Paraffin oil from BDH (Dublin, Ireland) was washed with deionised water before use. Fluorescein isothiocyanate (FITC) dextrans (molecular weight 9,500, 19,500 and 42,000) from Sigma (Dorset, UK) were used as supplied. All the water used was of Milli-Q (Millipore) quality and was degassed before use.

Synthesis of spherical gel beads

Submillimetre spherical gel beads were prepared by inverse polymerisation, according to the method of Matsuo and Tanaka [1] as described previously [10]. All the gels had 0.8 wt% cross-linking, and the pregel concentration was 700 mM monomer (490 mM NIPA, 70 mM BAM, 140 mM DAM) in 3 mg water. The pregel solution was degassed under vacuum and 15 μl TEMED was added. The initiator concentration was 40 mg (or 120 mg) in 2 mg water, of which 70 μl was added.

Temperature jumps

An aliquot of the gel beads in water (containing a range of gel sizes) was used. The gels were inserted onto a "thermoslide" made of conductive glass and heated by applying a current (described previously [10]). The temperature was increased by a temperature jump from a starting temperature of 30 °C to final temperatures in the range 34–40 °C. The change in diameter (either the diameter at time t divided by the initial diameter (D_t/D_0) or the renormalized diameter [$(D_t/D_0 - D_f/D_0)/(1 - D_f/D_0)$]) was plotted as a function of time, from which the relaxation time (time taken for the gel to reach its equilibrium size at the new temperature) was established.

Rheology

The viscoelastic properties of the gels of various initiator concentrations were determined using a Rheometric Scientific SR2000 dynamic stress strain-controlled rheometer. Parallel plates of 40-mm diameter were used. Frequency sweeps (0.01–10 Hz, stress 0.5 Pa) and stress sweeps (0.5–100 Pa, frequency 0.16 Hz) were performed.

Instron (penetration) tests

The gel strength was determined using an Instron. The gels were compressed to 50% of their initial thickness in order to determine their deformation and breaking point. The crosshead speed was 0.5 mm/s, the load cell was 100 N, the plunger diameter was 35 mm and the load range was 10 N.

Pore size experiments

The pore size distributions of the gels were characterised by the solute-exclusion technique using FITC–dextran fractions as molecular probes, as described by Wu et al. [11]. Gel disks of diameter 20 mm and thickness 3 mm were prepared for initiator concentrations of 1, 2 and 3 wt% of the initial monomer concentration and were placed into sodium azide solutions containing different FITC–dextran fractions (200 μg/ml). The FITC–dextran concentration was monitored daily using a PerkinElmer LS50B luminescence spectrometer (485 nm excitation, 515 nm emission) until the FITC–dextran concentration stopped changing. The gel disks were transferred to 100 ml sodium azide solution for exhaustive extraction, and the concentrations were determined. The volumes of the pores inaccessible and accessible to each FITC–dextran fraction were determined as described previously [12].

Results and discussion

The equilibrium shrinking curves of the slightly charged (1 wt% initiator) and highly charged (3 wt% initiator) 10:20:70 BAM:DAM:NIPA gels are shown in Fig. 2. It is

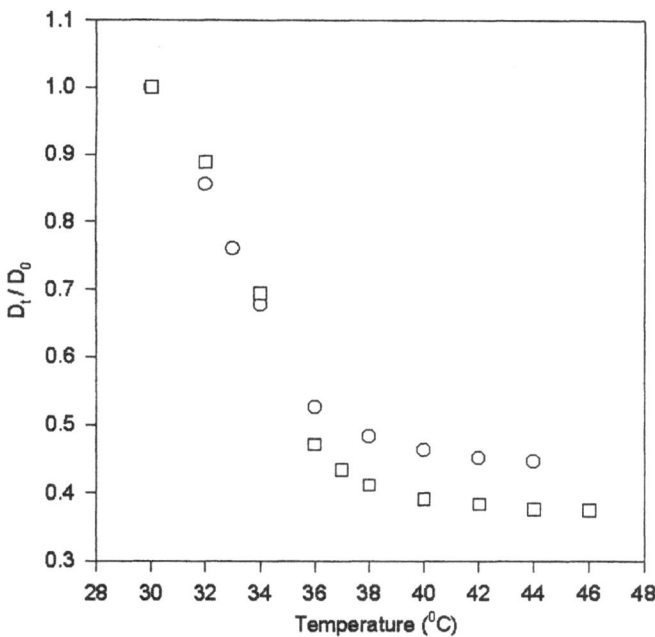

Fig. 2 Comparison of equilibrium shrinking curves for slightly charged (*circles*) and highly charged (*squares*) 10:20:70 *N-tert*-butylacrylamide:*N,N'*-dimethylacrylamide:NIPA 200-μm gel beads

clearly seen that the transition temperature is not affected by the increased charge. Also, there is no increase in the discontinuity of the transition. This is in contrast to the evidence of other researchers, where the presence of as little as 1 mol% charged comonomers was sufficient to alter the transition temperature and discontinuity [13]. However, the presence of additional charged groups is indicated by the increased degree of swelling in the highly charged gel, as a result of the increased osmotic pressure owing to dissociation of the charged groups (as the curves are plotted as a function of final diameter, it is seen as an increase in the degree of shrinking).

The shrinking kinetics of the 10:20:70 BAM:DAM: NIPA gels were determined following temperature jumps to the range of temperatures 30–40 °C. The shrinking curves obtained were extremely dependent on the gel composition. In the case of the weakly charged gel, the shrinking process was exponential up to T^*, the temperature at which the gel was fully collapsed, and became two-staged at temperatures above T^*. The onset of two-stage shrinking behaviour was due to the surface of the gel shrinking faster than the bulk gel and forming a skin layer which prevented the bulk water from being released. A detailed description of the shrinking process of the weakly charged gel was given previously [10].

In the case of the highly charged gel there were several important differences in the shrinking behaviour. Firstly, the shrinking times were significantly faster in the highly charged gel than in the slightly charged gel (2 orders of magnitude faster). Secondly, there was no skin layer

formation in the highly charged gel, even at temperatures well above the transition temperature, and thus no appearance of bubbling. This meant that the shrinking process was exponential at all temperatures, with no evidence of a two-stage shrinking pattern. Finally, there was an interesting pattern formation on the gel surface during the initial few seconds of shrinking in the highly ionised gels.

The rapid shrinking times of the highly charged gels compared to the slightly charged gels are shown in Table 1. The series of photographs in Fig. 3 compare the shrinking processes of both the weakly and the highly charged 10:20:70 BAM:DAM:NIPA gels. The lack of a skin layer in the highly charged gel is clearly seen as compared to the weakly charged gel. There is no appearance of opacity and no bubbling at all.

According to Flory [14] any real network must contain terminal chains bound at one end to the network via a cross-linkage and terminated at the other end (the "free end") by a primary molecule. Such an arrangement is shown in Fig. 4. Since polymerisation is initiated by the free-radical initiator, it is reasonable to assume that the "free end" is in actual fact terminated by an SO_4^- group. Thus, it is postulated that the effect of increasing the initiator concentration is to increase the number of elastically ineffective "free-end" terminal chains. These free chains presumably extend into the solvent as "Flory" type coils at low temperature and shrink with increasing temperature. It is postulated that the introduction of additional terminal chains into the network is responsible for the increased speed of shrinking of the highly charged 10:20:70 BAM:DAM:NIPA gels. In analogy to the graft-chain gel prepared by Kaneko et al. [15] it is suggested that the free-end chains respond rapidly to changes in their surrounding temperature and as they shrink to minimise the polymer–water interactions they pull the bulk network with them, speeding up the collapse transition.

An increase in the number of elastically ineffective chains should result in a weaker (less elastic) gel. The elasticity of a gel network can be determined easily by

Table 1 Relaxation times, τ (seconds), for different charged gels calculated from the experimental data using curve fitting

Gel description	Temperature jump temperature (°C)					
	34	35	36	37	38	40
Slightly ionic						
130 μm	20	18	17	17	17	6
180 μm	27	28	29	28	29	16
212 μm	38	43	41	48	40	21
Highly ionic						
150 μm	20	17	12	10	9	6
200 μm	28	26	17	14	12	8
250 μm	36	28	20	13	11	9

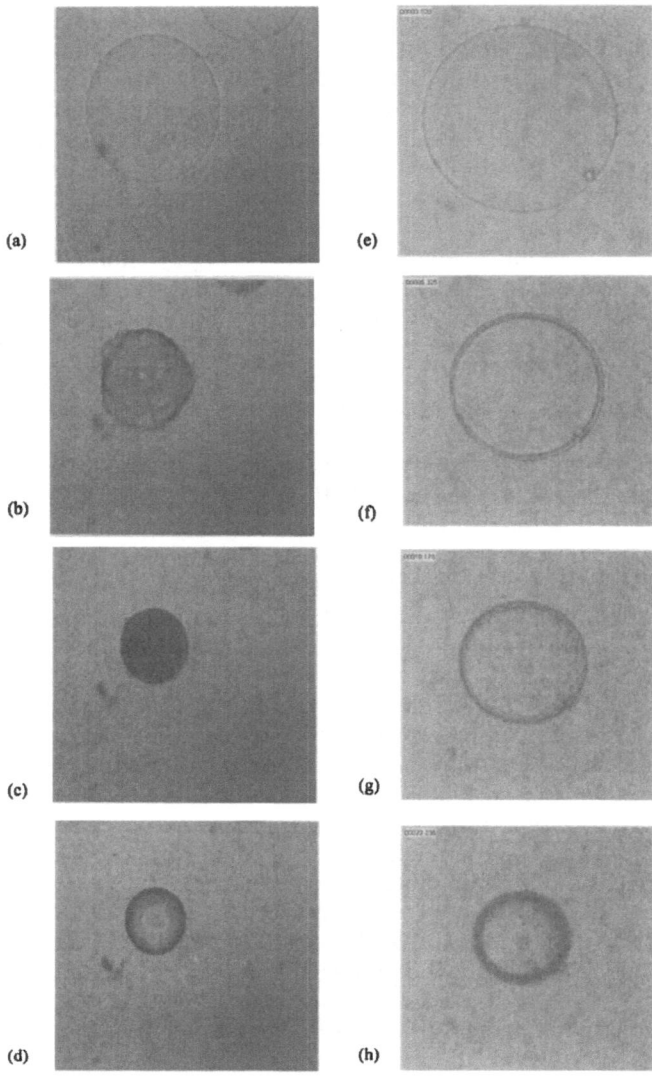

(a)

(b)

(c)

(d)

(e)

(f)

(g)

(h)

Fig. 3a–h Comparison of the shrinking process following a temperature jump from 30 to 40 °C on 200-µm gel beads. 1 wt% (slightly ionic) gel **a** 0 s, **b** 30 s, **c** 70 s and **d** 2,000 s; 3wt % (highly ionic) gel **e** 0 s, **f** 3 s, **g** 12 s and **h** 20 s

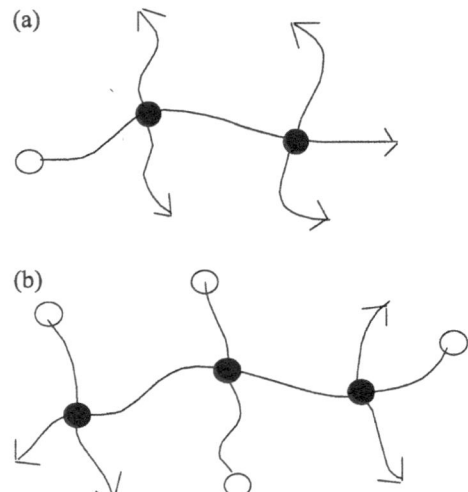

(a)

(b)

Fig. 4a, b Schematic representation of effect of increased initiator concentration on the gel network. The *filled circles* indicate cross-linkages, the *open circles* represent a terminus or "free end" consisting of a charged initiator residue and the *arrows* signify continuation of the network structure. **a** Low initiator concentration, **b** high initiator concentration. (Redrawn from Ref. [14].)

rheology, where the sample is deformed by shear. The viscoelastic properties of the 1 and 3 wt% initiator gels were determined using a stress sweep and a frequency sweep. The 3 wt% initiator gel was expected to have a lower elastic modulus than the 1 wt% initiator gel, as a result of having a large number of elastically ineffective chains; however, as shown in Fig. 5, the elastic modulus (G') is virtually identical for both gels, suggesting that either the initiator concentration has no effect on the network structure or that the increased number of free chains is compensated for by some other change in the network structure. Since it has been shown that there are dramatic differences in the shrinking behaviour of the gels, there is clearly some major difference in the network structures. Thus we postulate that the increased number of free chains results in there being fewer monomer units available for incorporation into the network chains. This would result in a shorter chain length between cross-link units and thus a smaller overall pore size.

An interesting way of determining the number of elastically ineffective chain segments contained in a network was described by Dubrovskii and Rakova [16]. They measured the elasticity of nonionic and weakly ionic acrylamide-based gels as a function of initial total monomer concentration, c_0, by a penetration method. They found that networks prepared with low c_0 contain a large number of elastically ineffective segments, while at high c_0 "effective" perfect networks were formed. The effect of initiator concentration on elasticity was determined by penetration (compression) in order to confirm the hypothesis that the increased initiator concentration resulted in an increased number of terminal or "free-end" chains. As shown in Fig. 6, there is a significant difference in the elasticity of the two gels. The 3 wt% gel breaks at a lower force and at a lower degree of displacement than the 1 wt% gel, suggesting that it has a larger number of elastically ineffective chains and is thus an elastically weaker gel. This confirms the original hypothesis that the increased initiator concentration results in a network with a lot of elastically ineffective chains.

The pore size distribution of the gels was determined by solute exclusion according to the method of Wu et al. [11]. FITC dextrans of molecular weight 9,500, 19,500 and 42,000 were used as probes. It can be clearly seen in

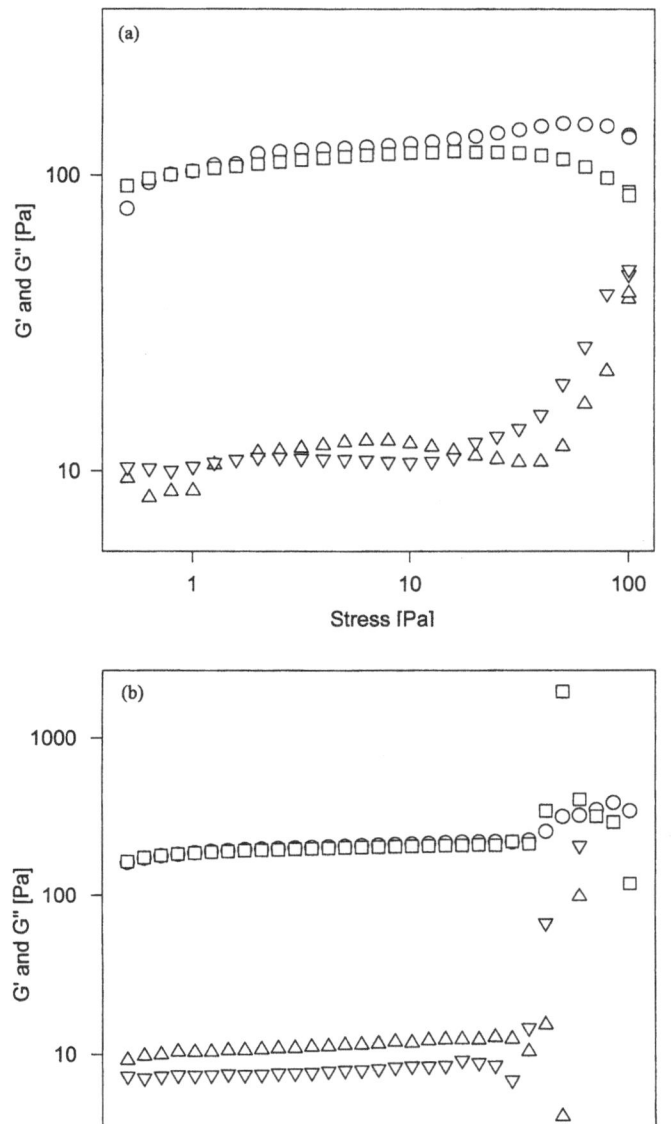

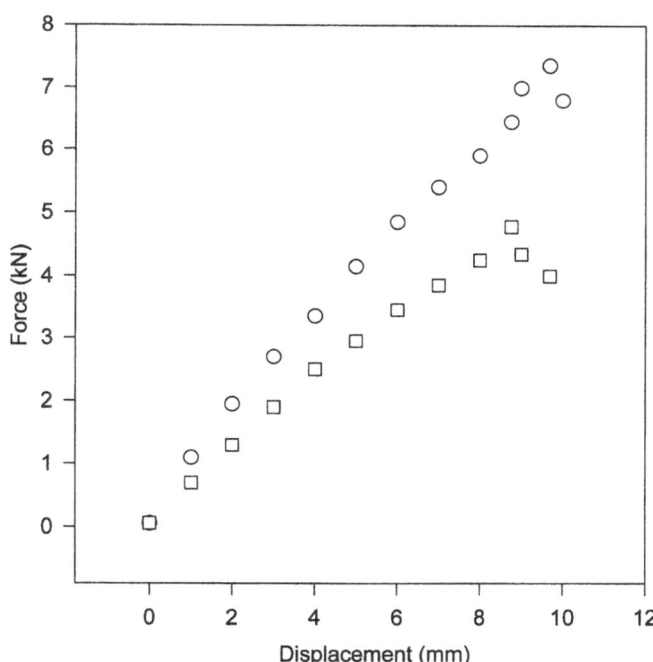

Fig. 6 Determination of elastically effective chain formation by penetration (Instron). 1 wt% initiator gel (*circles*), 3 wt% initiator gel (*squares*)

Fig. 5a, b Elasticity (*G'*) of gels measured by rheology. 1 wt% initiator gel (*squares*), 3 wt% initiator gel (*triangles*). **a** Stress sweep, **b** frequency sweep

presence of dangling chains reduces the number of monomer units available for network chains, resulting in shorter chain lengths between cross-links, and, thus, smaller pores.

Conclusion

The shrinking kinetics of 10:20:70 BAM:DAM:NIPA gels prepared at two different initiator concentrations were compared and it was found that increasing the initiator concentration dramatically altered the shrinking process by preventing the formation of a surface skin layer at high temperatures. The shrinking times of the gels with high initiator concentration were significantly shorter than those of gels with lower initiator concentration.

It was proposed that a higher initiator concentration caused increased numbers of elastically ineffective chains to form, and that these chains are freely mobile and thus can respond to environmental changes faster than the confined network chains. This would account for the faster shrinking process and the lack of a skin layer forming at higher temperatures, as the mobile chains would act as nuclei for the shrinking process.

The presence of increased numbers of elastically ineffective chains at a high initiator concentration was confirmed by penetration tests, where the degree of displacement of the highly ionic gel was significantly less

Fig. 7 that increasing the initiator concentration resulted in a decreased pore size in the network. In the case of the probe of molecular weight 9,500, 80% of the pores of the 1 wt% initiator gel are accessible, while fewer than 40% are accessible when the 3 wt% initiator is used. Thus, the pore size can be clearly seen to decrease with increasing initiator concentration, confirming that the

162

Fig. 7 Pore size distribution of gels with increasing initiator concentration

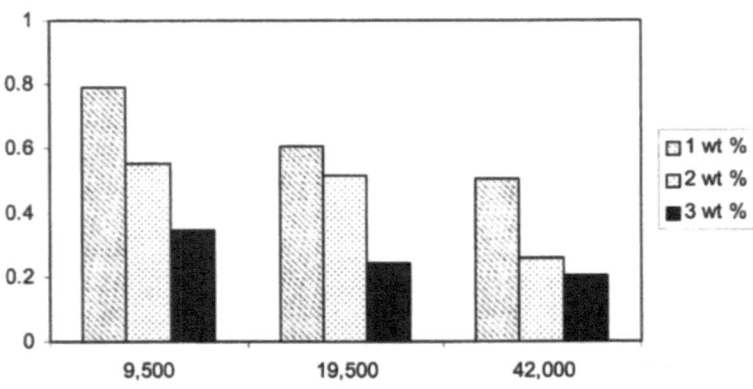

that that of the weakly ionic gel. However, the overall gel strength or elasticity as measured by rheology was not significantly different between the gels. This indicated that the increased number of elastically ineffective chains was somehow compensated for by the network structure. It was proposed that the highly ionic gel had a smaller pore size, owing to there being shorter chains between cross-links. This was confirmed by solute exclusion, and thus confirms the dangling chain structure proposed here for gels formed at high initiator concentration.

Acknowledgements This work was funded by INCO–Copernicus grant no. IC15CT96-0756. The authors thank Mick Cooney (Department of Food Science, University College Dublin) for use of the Instron and Alexander Gorelov and Elaine Duggan. Interesting discussions with B. Lindman, P Tartaglia and F. Sciortino which took place in the scope of the Cost P1 action are acknowledged.

References

1. Matsuo ES, Tanaka T (1988) J Chem Phys 89:1695–1703
2. Hirotsu S (1988) J Chem Phys 88:427–431
3. Ilavsky M, Hrouz J, Ulbrich K (1982) Polym Bull 7:107–113
4. Ilavsky M (1981) Polymer 22:1687–1691
5. Tsuchida E (1974) Makromol Chem 175:603
6. Hirose H, Shibayama M (1998) Macromolecules 31:5336–5342
7. Kokufuta E, Wang B, Yoshid R, Khokhlov AR, Hirata M (1998) Macromolecules 31:6842–6854
8. Bokiaas G, Durand A, Hourdet D (1998) Macromol Chem Phys 199:1387–1392
9. Saunders BR, Vincent B (1999) Adv Colloid Interface Sci 80:1–25
10. Lynch I, Gorelov A, Dawson KA (1999) Phys Chem Chem Phys 1:2103–2108
11. Wu XS, Hoffman AS, Yager P (1992) J Polym Sci A Polym Chem 30:2121–2129
12. Lynch I, Dawson KA submitted to Langmuir
13. Inomata H, Goto S, Saito S (1990) Macromolecules 23:4887–4888
14. Flory PJ (1979) Principles of polymer chemistry, Cornell University Press, Ithaca
15. Kaneko Y, Sakai K, Kikuchi A, Yoshida R, Sakarui Y, Okano T (1995) Macromolecules 28:7717–7723
16. Dubrovskii SA, Rakova GV (1997) Macromolecules 30:7478–7486

Progr Colloid Polym Sci (2001) 118: 163–167
© Springer-Verlag 2001

R. S. Dias
B. Lindman
M. G. Miguel

Interactions between DNA and surfactants

R. S. Dias · B. Lindman
M. G. Miguel (✉)
Chemistry Department, Coimbra
University, 3004-535 Coimbra, Portugal
e-mail: mgmiguel@ci.uc.pt

R. S. Dias · B. Lindman
Physical Chemistry 1, Center for Chemistry
and Chemical Engineering
P.O. Box 124, 221 00 Lund, Sweden

Invited lecture, 14th Conference of the
European Colloid and Interface Science
Patras, Greece, 2000

Abstract The interaction between
DNA and alkyltrimethylammonium
bromides of different chain lengths is
studied, both on a macroscopic and
on a single-molecule level. The phase
maps for aqueous DNA–surfactant
systems as well as some interesting
salt effects are presented. Some
preliminary results on the structure
of DNA–surfactant complexes are
also given. Studies involving DNA
and surfactant mixtures are also
conducted.

Key words DNA–surfactant
systems · Catanionic mixtures ·
DNA–amphiphiles complexe ·
Phase behavior

Introduction

Mixtures of water-soluble polymers and surfactants are
present in a large number of systems in nature and
industrial applications, such as in foods, pharmaceutical
formulations, cosmetics, detergents, paints, etc., and
these systems have been the subject of a large number of
studies [1, 2]. The interactions between amphiphilic
molecules and biologically active polyelectrolytes have
received particular attention not only from the physical
chemistry viewpoint, but also were specially investigated
in biomedical studies. Within this group we consider
DNA–cationic surfactants systems. These systems have a
number of applications, such as the purification of DNA
by condensation and precipitation [3], DNA renatur-
ation and ligation [4], and selective separation of DNA in
the presence of RNA [5]. Surfactants can also be used for
positive charging of neutral liposomes for gene delivery
purposes [6]. Owing to the growing interest in this field
and numerous applications of these systems several
studies have been presented in the literature concerning
DNA interaction with cationic amphiphiles [7–13]. In
spite of this, a number of facts are still not quite clear in
these systems. The mechanism of surfactant binding and

the structure of the complex are some examples. In our
work an attempt is made to understand these and other
points. The surfactants used are cetyltrimethylammonium bromide (CTAB), tetradecyltrimethylammonium bromide (TTAB), and dodecyltrimethylammonium
bromide (DTAB).

DNA association with cationic surfactants

Some applications of the DNA–cationic surfactant
systems are based in the fact that DNA phase-separates
associatively with the amphiphiles. The formation of a
precipitate, or phase concentrated in both components,
has been reported for several systems of polyelectrolytes
and oppositely charged surfactants [14–16]. The electrostatic interactions between them are obviously strong
and lead to strong association. Surfactant aggregates
induced by the polymer will act as counterions, thereby
reducing the charge of the complex and the entropic
driving force for mixing and the interpolymer repulsions
[2]. However, in contrast to other polyelectrolyte–
surfactant systems reported [15, 16] the precipitate does
not redissolve with an excess of surfactant. Other

164

information obtained is that the precipitate is formed at very low amounts of DNA and minor surfactant concentrations, far below the surfactant critical micelle concentration (cmc). Polyelectrolyte–oppositely charged surfactant systems are known to have a critical aggregation concentration (cac) lower than the cmc of free surfactants, often by orders of magnitude. The fact that the cationic surfactant binding occurs preferentially to anionic polyelectrolytes of high charge density further enhances this behavior. These results are presented schematically in Fig. 1. These phase maps, studied by turbidimetry, are presented in a simplified two-dimensional representation. Since the amount of water in these systems is extremely high, this type of representation provides a better visualization.

The salt effect

The same systems were studied with the addition of salt [17]. The results are presented in Fig. 2. It is a commonly accepted viewpoint that the cac of the polyelectrolyte–oppositely charged surfactant systems increases on addition of salt [18]. This is due to a weakened interaction between the polymer and the surfactant induced by the stabilization of free micelles and a screening of the electrostatic interactions. With this we would expect a decrease in the two-phase region and not the observed broadening of it. By looking at the DNA–TTAB phase map, for example, we observe that the lines of precipitation of the system with and without salt cross instead of being dislocated one in relation to the other (Fig. 3); this is unexpected behavior. We believe that the knowledge of the DNA–surfactant complex structure can help us in understanding this trend.

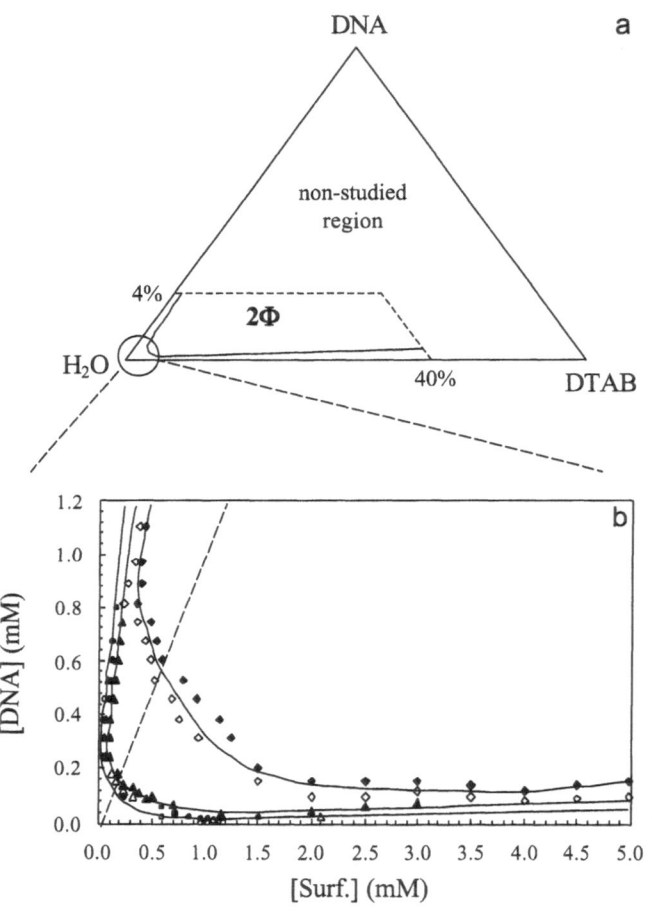

Fig. 1 a Schematic representation of the isothermal pseudoternary phase diagram for the DNA–dodecyltrimethylammonium bromide (*DTAB*)–water system. There is a phase separation into two phases in almost the entire region considered. **b** Expanded view of the water-rich corner of the system (*diamonds*), including the DNA–tetradecyltrimethylammonium bromide (*TTAB*)–water (*triangles*), and DNA–cetyltrimethylammonium bromide (*CTAB*)–water (*circles*) systems for comparison. The *open symbols* refer to the clear one-phase solutions and the *filled symbols* to two-phase samples. The *dashed line* indicates the charge neutralization. $T = 25$ °C. From Ref. [17]

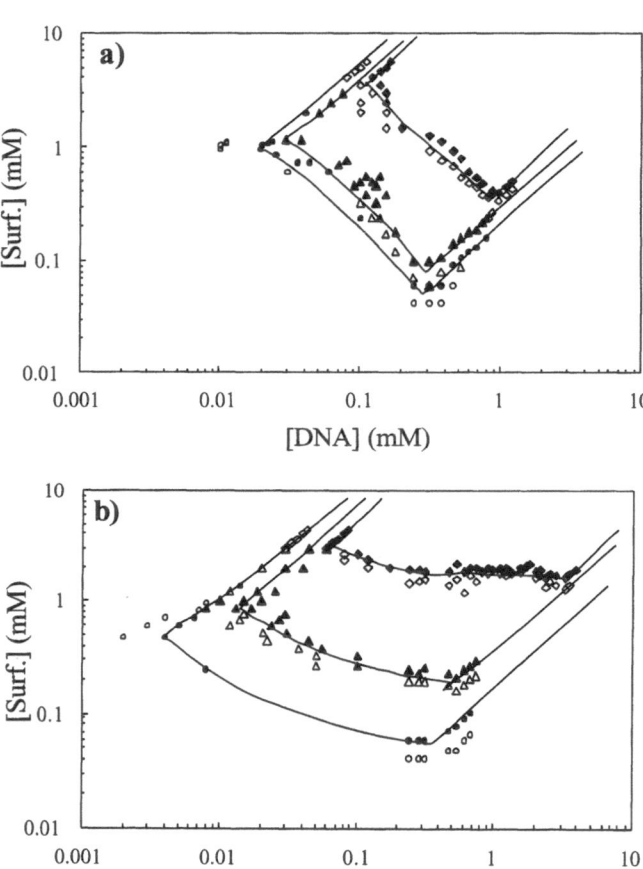

Fig. 2 a Same as Fig. 1b, but the phase map is represented on a logarithmic scale for better visualization of the effect of surfactant chain length on the phase transition. **b** Phase map for the three systems in the presence of 0.1 M NaBr. The *circles* represent samples without salt and the *triangles* samples in the presence of salt. The *open symbols* refer to the clear one-phase solution and the *filled symbols* to two-phase samples. $T = 25$ °C. From Ref. [17]

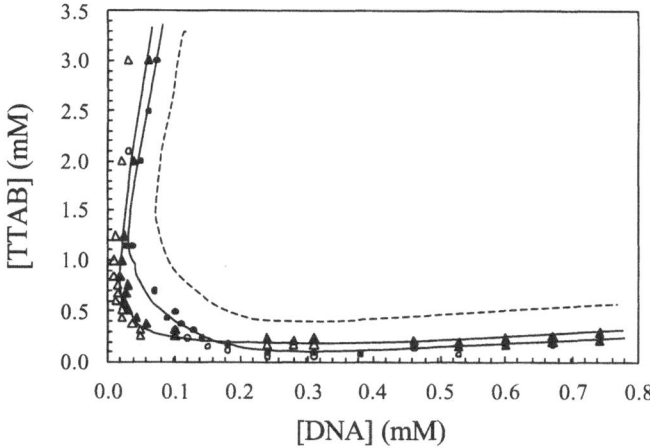

Fig. 3 Effect of the addition of NaBr on the phase behavior of the DNA–TTAB aqueous system. The *circles* represent samples without salt and the *triangles* samples in the presence of salt. As before, the *open symbols* refer to the clear one-phase solution and the *filled symbols* to two-phase samples. The *dashed line* represents the expected direction of change for the system with salt. $T = 25$ °C. Redrawn from Ref. [17]

DNA–surfactant complexes

In spite of all studies made with these systems [8–13], the structure of the complex is not fully known. It is agreed [9, 19] that the complexes formed have a loosely packed hexagonal structure in which the DNA induces and stabilizes the formation of rodlike micelles. In fact, our preliminary results point in that way (Fig. 4a). We can also see that the structure is more ordered for the longer-chain surfactant. The same experiments were performed with the addition of salt and the results are very similar (Fig. 4b). The only difference appears to be that with salt the hexagonal lattice becomes less loose. This could be due to the salt osmotic pressure.

Studies were also conducted within the two-phase region to investigate the dependence of the amount of the precipitate and water in the supernatant on the variation of the DTAB/DNA mixing concentration ratio [17]. We obtained some interesting results. The amount of precipitate reaches the maximum amount for a mixing ratio of approximately 1.0 and remains constant with further addition of surfactant, at least for the region studied. This indicates that the excess of surfactant added to the samples remains in the supernatant, probably as free micelles, and can give an explanation for the nonredissolution of the DNA–surfactant complexes. Since there is no binding of the surfactant to the complex after its neutralization, an inversion of the complex charge is not observed, as happens with many similar systems [15, 16]. The same studies were made with the addition of salt, but again there were no major differences between the results. The precipitation starts for higher amounts of surfactant,

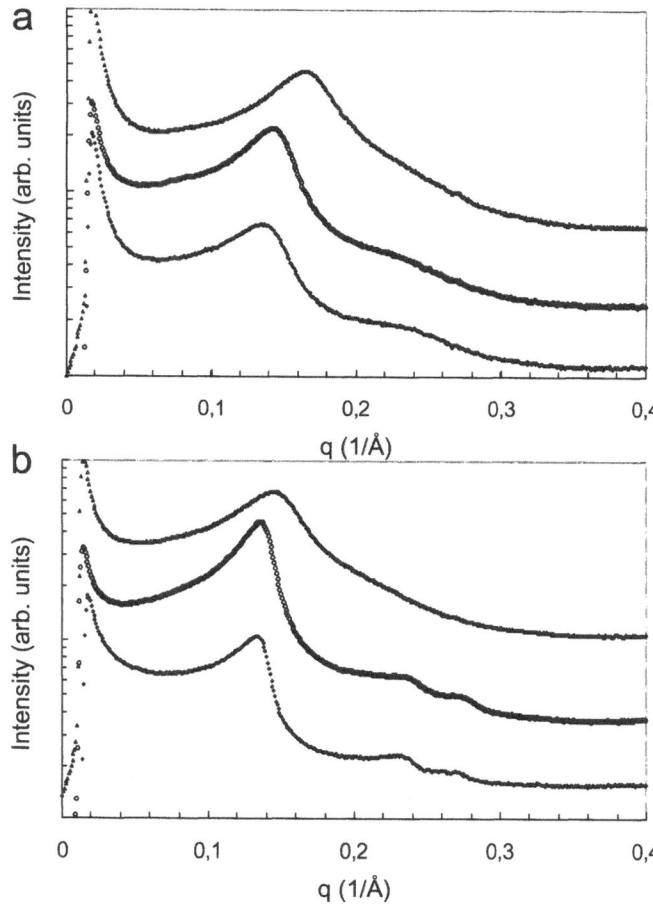

Fig. 4 Small-angle X-ray scattering diffractograms of DNA–surfactant complexes **a** in the absence and **b** in the presence of NaBr. These preliminary results show a loosely packed hexagonal structure. The *filled diamonds* correspond to the DNA–CTAB complexes, the *open circles* to DNA–TTAB, and the *filled triangles* to the DNA–DTAB system. $T = 25$ °C

as expected, and there is a slight decrease in the amount of water in the samples, which corresponds to the salt contribution to the weight of the sample and means, probably, that the major fraction of salt is dissolved in the supernatant. Further studies are being conducted on this.

DNA compaction in the presence of cationic surfactants

Fluorescence microscopy is a technique that has recently been used in the study of DNA conformational behavior in the presence of several cosolutes, having as its main advantage the visualization of single molecules in solution. DNA molecules in aqueous solution present an extended conformation, flowing in the solution and exhibiting a relatively slow wormlike motion, the designated unfolded coil conformation. With the addition of cationic surfactants, for example, the DNA molecules undergo compaction into a globular state. The interesting point in these systems is that the DNA molecules undergo a discrete, or a first-order, phase transition [20, 21]. Associative phase behavior is well documented for mixed aqueous systems of a polyelectrolyte and oppositely charged surfactants [14, 15, 18] and so is the effect of surfactant on the polyion conformation [15, 16]. However the findings for flexible polyions of low charge density differ qualitatively from what we find for DNA. In these cases it can be inferred that while surfactant binding is cooperative, in that it results in surfactant self-assembly aggregates and is characterized by a well-defined cac, there is an essentially uniform distribution of surfactant aggregates among the different polyions. Phase separation takes place when the net charge of the polyion–surfactant aggregates attains a low value and the surfactant distribution explains why rather high amounts of surfactant are needed for phase separation in these cases. For DNA, the very low values of surfactant concentration at which phase separation starts demonstrate a different binding situation and either complete or no binding: as binding to DNA starts, further binding is facilitated and one DNA double helix molecule is saturated before binding starts at another, i.e. there is double cooperativity. This is directly confirmed by the fluorescence microscopy results, demonstrating the coexistence of original extended coils and globules compacted to the final state. Our results for the DNA–CTAB, DNA–TTAB, and DNA–DTAB systems show that this coexistence interval is narrower for DNA-CTAB aqueous mixtures and becomes wider for the shorter-chain surfactant. It is shown in Fig. 5 that a larger amount of DTAB is needed to induce the compaction of the DNA molecules, which is in agreement with our previous results.

Forthcoming work

Since many details in these systems are not well understood more experiments are being conducted in our groups. One of them is the study of the DNA interactions with alkyltrimethylammonium surfactants of even shorter chain length, such as decyltrimethylammonium bromide and octyltrimethylammonium bromide. Some preliminary phase diagrams are presented in Fig. 6. Hexyltrimethylammonium bromide was found not to precipitate DNA. As expected the area of the two-phase region decreases with the decrease in the surfactant hydrophobicity. For the short-chain surfactant the association is no longer observed.

One problem with these cationic systems is the toxicity of the surfactants, which reduces substantially their applications. An effort is being made to use more biocompatible amphiphiles. The use of surfactant mixtures is another way of dealing with this problem. By reducing the amount of cationic amphiphile we decrease the toxicity of the systems. Cationic liposomes, for example, made of neutral and cationic lipids, are well known as gene delivery vehicles. Owing to the undeniable importance of these systems, the interaction between DNA and surfactant mixtures is being carefully studied.

Interactions between DNA and surfactant mixtures

We studied the DNA conformational behavior in the presence of nonstoichiometric mixtures of two oppositely charged surfactants, CTAB and sodium octyl sulfate [22]. The aqueous mixtures of these surfactants exhibit a rich

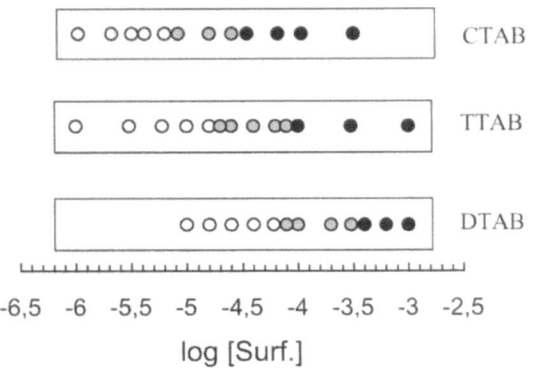

Fig. 5 DNA conformational behavior in the presence of cationic surfactants CTAB, TTAB, and DTAB. The DNA charge concentration was maintained at 0.5 μM. The *open circles* correspond to the coil conformational state of DNA and the *filled circles* to the presence of globular molecules. The *shaded circles* represent the coexistence between elongated coils and compacted globules. $T = 25$ °C. From Ref. [17]

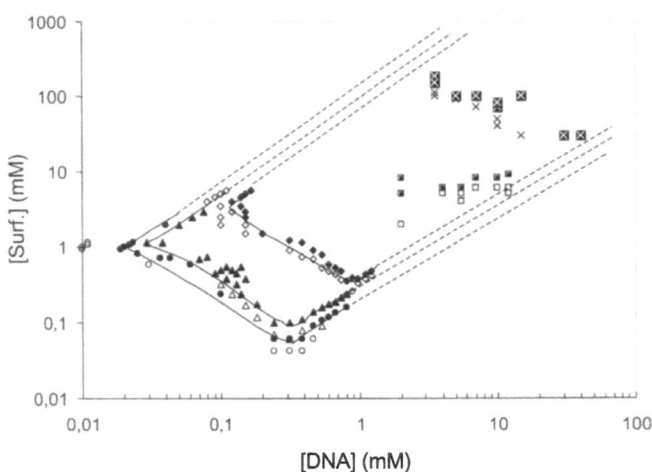

Fig. 6 Same as Fig. 2a with some preliminary results on the DNA–decyltrimethylammonium bromide–water (*squares*) and DNA–octyltrimethylammonium bromide–water (*crosses*) systems. As before, the *open symbols* correspond to one-phase solution and the *filled symbols* to two-phase samples. $T = 25$ °C

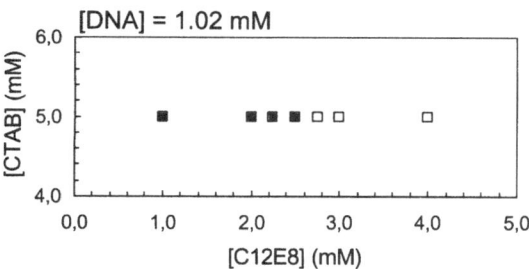

Fig. 7 The addition of a nonionic surfactant prevents the precipitation of DNA–cationic surfactant. A simplified two-dimensional representation of the preliminary results is given for the aqueous system of DNA–octaethyleneglycol mono-*n*-dodecyl ether–CTAB. The *filled symbols* refer to two-phase samples and the *open symbols* to clear one-phase solution. $T = 25\ °C$

phase behavior, with the formation of surfactant aggregates with different charges and geometries [23, 24]. We were mainly interested in the phases of the vesicles. We observed that the presence of anionic-rich vesicles in aqueous DNA solutions does not lead to any associative interaction and does not affect the DNA conformational behavior; however, in the cationic-rich phase, DNA is in the globular state and it adsorbs onto the surface of the vesicles [22]. Moreover, we observed that the change in the molar ratio between cationic and anionic surfactants in solution leads to DNA unfolding. Owing to the interesting results and promising use in several applications other studies are being conducted on this system, such as phase diagrams and the complex structure determinations.

Another interesting part in the study of the DNA–cationic surfactant interactions is the addition of a nonionic surfactant. We observed that the addition of octaethyleneglycol mono-*n*-dodecyl ether induces a weakening in the interaction between the polyion and the oppositely charged surfactant and is capable of preventing the precipitation of the systems at high enough concentrations. Some preliminary results are presented in Fig. 7. During this project other amphiphiles will be used.

Acknowledgements This work was supported by grants from JNICT and Praxis XXI (PRAXIS/BD/21227/99), the Caloust Gulbenkian Foundation, the Swedish Research Council for Engineering Sciences, and the Center for Amphiphilic Polymers in Lund. Sergey Mel'nikov is thanked for valuable participation in the initiation of this project and for useful discussions. We are grateful to Filipe Antunes and Stina Lindman for technical assistance.

References

1. Goddard E, Ananthapadmanabhan K (eds) (1993) Interactions of surfactants with polymers and proteins. CRC, Boca Raton
2. Jönsson B, Lindman B, Holmberg K, Kronberg B (1998) Surfactants and polymers in aqueous solution. Wiley, New York
3. Trewavas A (1967) Anal Biochem 21:324–329
4. Pontius BW, Berg P (1991) Proc Natl Acad Sci USA 88:8237–8241
5. Morimoto H, Ferchmin P, Bennett E (1974) Anal Biochem 62:437
6. Lasic DD (1997) Liposomes in gene delivery. CRC, Boca Raton
7. Hayakawa K, Santerre JP, Kwak JCT (1983) Biophys Chem 17:175–181
8. Gorelov AV, Kudryashov ED, Jacquier J-C, McLoughlin DM, Dawson KA (1998) Physica A 249:216–225
9. Ghirlando R, Wachtel EJ, Arad T, Minsky A (1992) Biochemistry 31:7110–7119
10. Spink CH, Chaires JB (1997) J Am Chem Soc 119:10920–10928
11. Jacquier J-C, Gorelov AV, McLoughlin DM, Dawson KA (1998) J Chromatogr A 817:263–271
12. Buckin V, Kudryashow E, Morrissey S, Kapustina T, Dawson K (1998) Prog Colloid Polym Sci. 110:214–219
13. Mel'nikov SM, Sergeyev VG, Yoshikawa K (1995) J Am Chem Soc 117:2401–2408
14. Chu D, Thomas JK (1986) J Am Chem Soc 108:6270–6276
15. Thalberg K, Lindman B (1989) J Phys Chem 93:1478–1483
16. Ilekti P, Piculell L, Tournilhac F, Cabane B (1998) J Phys Chem 102:344–351
17. Dias R, Mel'nikov S, Lindman B, Miguel M (2000) Langmuir 16:9577–9583
18. Lindman B, Thalberg K (1993) In: Goddard E, Ananthapadmanabhan K (eds) Interactions of surfactants with polymers and proteins. CRC, Boca Raton, p 203
19. Mel'nikov SM, Sergeyev VG, Yoshikawa K (1997) In: Pandalai SG (ed) Recent research developments in chemical sciences. Trivandrum, India, p 69–113
20. Mel'nikov S, Sergeyev V, Mel'nikova Y, Yoshikawa K (1997) J Chem Soc Faraday Trans 107:6917–6923
21. Mel'nikov S, Sergeyev V, Yoshikawa K, Takahashi H, Hatta I (1997) J Chem Phys 107:6917–6923
22. Mel'nikov SM, Dias R, Mel'nikova YS, Marques EF, Miguel MG, Lindman B (1999) FEBS Lett 453:113–118
23. Brasher LL, Herrington KL, Kaler EW (1995) Langmuir 11:4267–4277

Progr Colloid Polym Sci (2001) 118: 168–171
© Springer-Verlag 2001

F. Boffi
A. Bonincontro
E. Bultrini
F. Cherubini
A. Congiu-Castellano
G. Mossa
G. Onori
A. Santucci

Liposome–DNA complex investigated by dielectric spectroscopy and circular dichroism

F. Boffi · A. Bonincontro (✉)
E. Bultrini · A. Congiu-Castellano
INFM-Dipartimento di Fisica Università di
Roma "La Sapienza", P. le A. Moro 2
00185 Rome, Italy
e-mail: adalberto.bonincontro@uniroma1.it

F. Cherubini · G. Onori · A. Santucci
INFM-Dipartimento di Fisica, Università
di Perugia, Perugia, Italy

G. Mossa
CNR-Istituto di Medicina Sperimentale
Rome, Italy

Abstract Cationic liposomes are efficient carriers of genetic material and because of this they have been studied by different biophysical and biochemical approaches. In the present article dielectric spectroscopy (DS) and circular dichroism measurements on cationic liposome–DNA mixtures are reported. These techniques are suitable to characterise the formation of a DNA–liposome complex, which is the first step in the transfection process. In particular, the estimation of a significant structure parameter of DNA, the so-called "persistence length", is possible by DS. Our results confirm that the link between cationic liposomes and DNA is direct and of an electrostatic nature. The maximum complexation is at electroneutrality as reported in the literature. Finally, an interesting structural effect on DNA was evidenced by our measurements. When the biopolymer is partially bonded to liposomes, its free stretches exposed to solvent appear rigider with an increase in the persistence length.

Key words Liposome–DNA · Dielectric spectroscopy · Circular dichroism

Introduction

Structural and dynamic studies of cationic liposome–DNA complexes are of particular interest in gene-therapy research [1, 2]. Cationic liposomes can be used as vehicles for genetic material transport into cell nuclei and represent a safer alternative to viral gene transfection. Most studies regarding the transfection process and the structure of the complexes have focused on evaluating the efficiency of several liposomes as a function of different physical–chemical parameters [3–6]. A systematic study using different cationic liposomes was started by Colosimo et al. [7]. In the present article dielectric spectroscopy (DS) and circular dicroism (CD) measurements on cationic liposome–DNA mixtures are reported. These techniques are suitable to characterise the formation of a liposome–DNA complex, which is the first step in the transfection process. In particular, the estimation of a significant structure parameter of DNA, the so-called "persistence length", is possible by DS [8]. This work represents a first attempt to correlate the formation of the complex with structural modifications induced on DNA.

Materials and methods

Calf thymus NaDNA from Sigma was used. It was diluted in deionised water and then sonicated with a Vibra Cell sonifier from Sonics and Materials (0.5 g/l concentration, 1-min sonication time) to obtain DNA fragments. The molecular-weight distribution of DNA was determined by agarose gel electrophoresis and was widely spread from 0.5 to 2 kilobases.

(1,2-Bis(oleoyloxy)-3-(trimethylammonium)propane) (DOTAP) lipids from Sigma were diluted in deionised water and then sonicated for 20 min in order to obtain monolamellar liposomes. The dimensions of these liposomes were checked by light scattering measurements and were shown to have a diameter distribution centred at about 250 nm.

In dielectric experiments the permittivity, ε', and the dielectric loss, ε'', were measured by a computer-controlled HP 4194A impedance analyser in the frequency range 10^5–10^8 Hz. The measuring cell was a section of a cylindrical waveguide, which could be partially filled with about 1 ml sample solution. The system behaves as a waveguide excited far beyond its cutoff frequency mode, and therefore only the fading wave that is generated at the interface between the coaxial line and the solution is used in the measurement. Cell constants were determined by measurements with electrolyte solutions of known conductivity, following well-defined procedures reported in the literature [9].

CD spectra were obtained with a Jasco 715 spectropolarimeter, using a 1-mm-width quartz cell for the solutions in the ultraviolet range 190–260 nm.

All the experiments were performed at the controlled temperature of 25 °C.

Results and discussion

We measured the complex dielectric constant of cationic liposome DOTAP–DNA mixtures in aqueous solution at different concentrations. The concentration of DNA was fixed to a value of 0.25 g/l, corresponding to a nucleotide concentration of 0.76 mM, and the content of liposome was varied form 0 to 1.5 g/l, i.e. up to 2.1 mM DOTAP. In terms of the relative concentration of charges, this means that we varied the ratio of molarities of liposome and nucleotide DNA from 0 to 2.8.

In the range of frequencies used in our experiments, the DNA suspensions show a typical dielectric relaxation (Fig. 1a) due to counterion oscillations along definite stretches of the polyelectrolyte, the persistence length [8, 10–12]. A valid model interpreting this phenomenon is the Mandel model [13]. In this model the persistence length, b, is related to both the relaxation frequency, f^*, and the dielectric increment, $\Delta\varepsilon$, according to the following equations:

$$f^* = \frac{\pi KTu}{2b^2} \; , \tag{1}$$

$$\Delta\varepsilon = \frac{e^2 gNc\gamma b^2}{36 KT\varepsilon_0} \; , \tag{2}$$

where K is the Boltzmann constant, u is the mobility of the counterions, g is the average fraction of bound counterions, N is the total number of monovalent counterions of charge e per polyelectrolyte molecule, c is the macromolecule concentration per cubic metre and γ, which may be assumed to be close to unity, is the ratio of the effective electric field acting on the polyion and the average Maxwell field in the solution.

In the same frequency interval the liposome solution also exhibits a dielectric relaxation, similar to that of DNA and probably caused by an analogous mechanism at the interface with solvent.

If we observe all the curves of Fig. 1, we can note that the relaxation of DNA reduces as the presence of DOTAP is increased. At a molar ratio close to unity, i.e. electroneutrality between liposome and DNA, the dispersion is near to vanishing. For a higher content of DOTAP, a relaxation appears again (Fig. 1d). Our interpretation is that for DNA molarity prevailing on the liposome, a part of the biopolymer binds DOTAP, saturating its charges. The residual DNA exposed to the solvent produces a relaxation with a reduced dielectric increment. When the negative charges of the phosphate groups of DNA are completely neutralised by an equal number of positive charges of the cationic liposomes, no dielectric dispersion is evident. For such a condition, the

Fig. 1 Permittivity, ε' (*left*), and dielectric loss ε'' (*right*), of DNA–liposome complexes in aqueous solution as a function of frequency, at the liposome–DNA (L/D) molarity ratios of **a** 0, **b** 0.6, **c** 0.9 and **d** 2.8

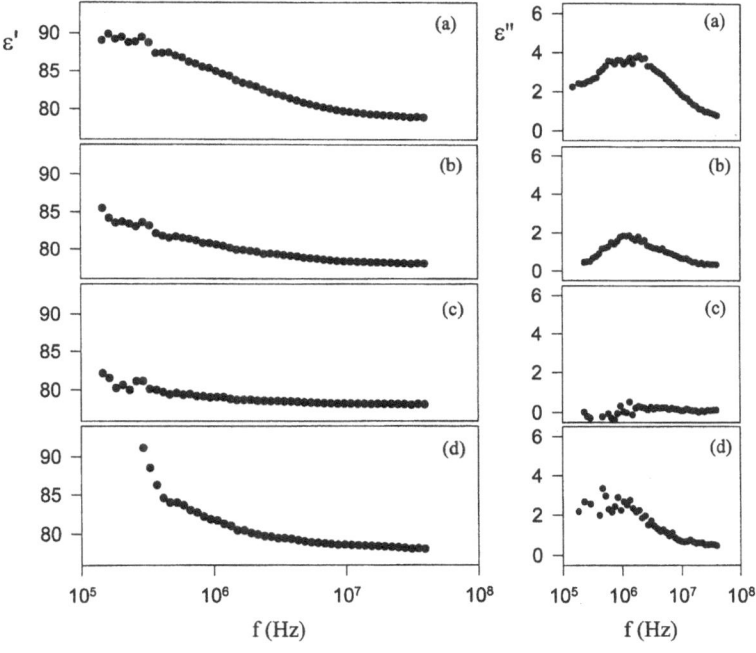

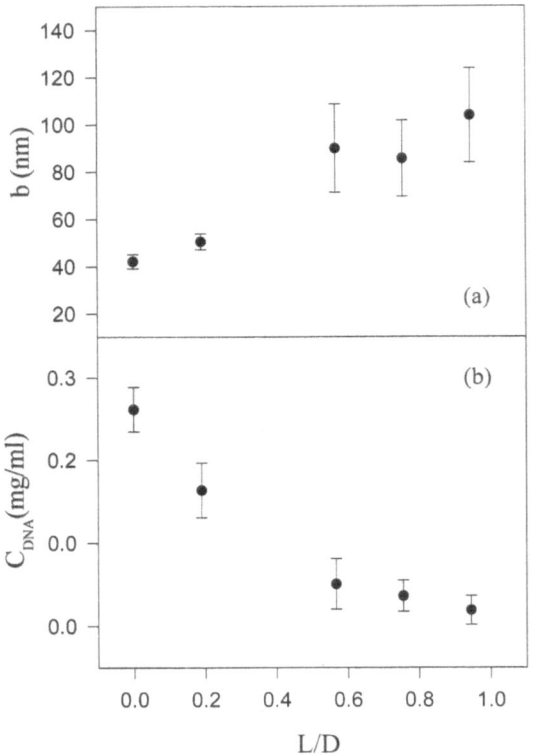

Fig. 2 a Persistance length and **b** concentration of free DNA as a function of L/D molarity ratio

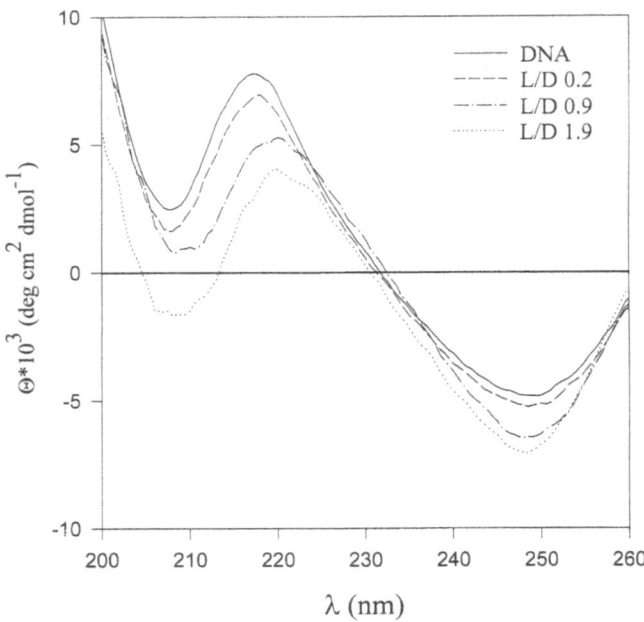

Fig. 3 Circular dichroism spectra of DNA at different L/D molarity ratios. The lipid background was subtracted for each spectrum

maximum complexation between DOTAP and DNA is realised. If the liposome content is increased a dispersion owing to the free liposomes is measured.

Liposomes have an equal number of charges on both external and internal surface, so the complete neutralisation by the DNA charges at a liposome–DNA molarity ratio close to unity implies the breaking of the liposome monolamellar structure during the interaction, and probably the subsequent formation of multilamellar structures. This interpretation seems to be in agreement with the model proposed by Radler et al. [14].

Using the Mandel model we analysed the dispersion of DNA exhibited by samples up to a liposome–DNA molarity ratio of about 0.9. The persistence length, b, obtained from the relaxation frequency (Eq. 1 of the model) shows an increasing trend (Fig. 2a). This implies a structural effect on free DNA induced by the part bonded to the liposomes. This increase in b denotes a greater rigidity of the stretches of DNA exposed to the solvent, when a part of the polyelectrolyte is linked with the liposome.

In Eq. 2 of Mandel model the DNA concentration assumes, in our case, the significance of free DNA concentration. We calculated this free DNA concentration by the experimental dielectric increments, using the

values of the persistence length, b, obtained from the relaxation frequency. The results are shown in Fig. 2b. As can be seen, the fraction of free DNA goes to zero as the ratio of the DOTAP–DNA molarities goes towards unity.

From CD measurements (Fig. 3) in the UV region a change in the spectra of the complexed DNA was recorded. This confirms that the interaction with liposomes induces an effect on the structure of DNA. Note that in DS the signal is due to charged groups of free DNA, while the CD spectra give information on the global DNA structure.

Conclusions

Our results confirm that the link between cationic liposomes and DNA is direct and of an electrostatic nature. The maximum complexation is at electroneutrality as reported in the literature. Finally an interesting structural effect on DNA was evidenced by our measurements. When the biopolymer is partially bonded to liposomes, its free stretches exposed to solvent appear rigider with an increase in the persistence length.

Acknowledgements This work was supported by a grant from the Ministero della Università e della Ricerca Scientifica e Tecnologica. We are grateful to C. Cametti and G. Risuleo for useful discussions and technical help (light scattering measurements, C.C., agarose gel electrophoresis measurements, G.R.).

References

1. Schreier H, Sawyer MS (1996) Adv Drug Deliv Rev 19:73
2. Ferrari S, Moro E, Pettenazzo A, Behr JP, Zacchelo F, Scarpa M (1997) Gene Ther 4:1100
3. Zabner J, Cheng SH, Meeker D, Launspach J, Balfour R, Perricone MA, Morris JE, Marshall J, Fasbender A, Smith AE, Welsh MJ (1997) J Clin Invest 6:1529
4. Caplen NJ, Kinrade E, Sorgi F, Gao X, Gruenert D, Geddes D, Coutelle C, Huang L, Alton EWFW, Williamson R (1995) Gene Ther 2:603
5. Felgner JH, Kumar R, Sridhar CN, Wheeler CJ, Tsai YJ, Border R, Ramsey P, Martin M, Felgner PL (1994) J Biol Chem 269:2550
6. Wheeler CJ, Felgner PL, Tasi YJ, Marshall J, Sukhu L, Doh SG, Hartikka J, Nietupski J, Manthorpe M, Nichols M, Plewe M, Liang X, Norman J, Smith A, Cheng SH (1996) Proc Natl Acad Sci USA 93:11454
7. Colosimo A, Serafino A, Sangiuolo E, Di Sario S, Bruscia E, Amicucci P, Novelli G, Dallapiccola B, Mossa G (1999) Biochim Biophys Acta 1419:186
8. Baker-Jarvis J, Jones CA, Riddle B (1998) Electrical properties and dielectric relaxation of DNA in solution. NIST technical note 1509, USA
9. Athey TW, Stuckley MA, Stuckley SS (1982) IEEE Trans Microwave Theor Tech 30:82
10. Vrengdenhil T, Van der Touw, Mandel M (1979) Biophys Chem 10:67
11. Bonincontro A, Cametti C, Di Biasio A, Pedone F (1984) Biophys J 45:495
12. Pedone F, Bonincontro A (1991) Biochim Biophys Acta 1073:580
13. Mandel M (1977) Ann NY Acad Sci 303:74
14. Radler JO, Koltover I, Salditt T, Safinya CR (1997) Science 275:810

Progr Colloid Polym Sci (2001) 118:172–176
© Springer-Verlag 2001

Irena Maliszewska
Kazimiera A. Wilk
Bogdan Burczyk
Ludwik Syper

Antimicrobial activity and biodegradability of N-alkylaldonamides

I. Maliszewska · L. Syper
Institute of Organic Chemistry
Biochemistry and Biotechnology
Wrocław University of Technology
Wybrzeże Wyspiażskiego 27
50-370 Wrocław, Poland

K. A. Wilk · B. Burczyk (✉)
Institute of Organic and Polymer
Technology, Wrocław University of
Technology, Wybrzeże Wyspiażskiego 27
50-370 Wrocław, Poland
e-mail: burczyk@itots.ch.pwr.wroc.pl
Tel.: +48-71-3203423
Fax: +48-71-3203678

Abstract A number of N-alkylaldo-namides, i.e., nonionic saccharide-based surfactants such as N-alkyl-N-methylgluconamides, N-alkyl-N-methyllactobionamides and N-alkyl-N,N'-bis[(3-D-aldonamidopro-pyl]amines were evaluated for environmental responses. All the N-alkylaldonamides were screened for antimicrobial activity against selected gram-positive (*Staphylococcus aureus* PCM 1944, *Sarcina lutea* PCM 1947, *Bacillus subtilis* PCM 1949, gram-negative (*Escherichia coli* PCM 2057, *Serratia marcescens* PCM 549, *Pseudomonas putida* PCM 2124) bacterial strains and some fungi (*Saccharomyces cerevisiae*, *Penicillum citrinum*, *Aspergillus niger*). It was generally recognized that the compounds tested were completely inactive towards fungi. N-Alkyl-N-methylgluconamides and N-alkyl-N-methyllactobionamides were practically inactive towards the bacteria studied. On the other hand, dicephalic saccharide-based surfactants were active against gram-positive cocci, much more strongly than the N-alkyl-N-methylaldonamides. Using the closed-bottle test (OECD guideline 301D), the biodegradability of these surfactants was determined. N-Alkyl-N-methylaldonamides may be considered to be readily biodegradable by activated sludge. The dicephalic surfactants studied showed slower biodegradation rates and they may be considered to be inherently biodegradable. Members of this class of saccharide surfactants exibit biological properties needed for environmental acceptance.

Key words Nonionic surfactants · N-Alkyl-N-methylaldonamides · N-Alkyl-N,N-bis[(3-D-aldonamido) propyl]amines · Antimicrobial activity · Biodegradability

Introduction

Nonionic surfactants with a carbohydrate moiety such as a hydrophilic group (so-called saccharide- or sugar-derived surfactants) have attracted much attention for both biological and environmental reasons as they are highly biodegradable, nontoxic and biocompatible surface-active agents [1–4]. They show interesting performance properties and have found many applications in detergents, dishwashing agents and personal care products [5, 6]. Among many saccharide-based surfac-tant structures those containing an amide group in their molecules have recently been investigated, i.e., N-alka-noyl-N-methylglucamines [1, 7, 8] and N-alkyl-N-meth-ylaldonamides [9–11]. N-Alkanoyl-N-methylglucamines have already been commercialized [8] and their environmental properties have been determined [12].There is a lack, however, of antimicrobial and biodegradability data of N-alkyl-N-methylaldonamides.

Recently, we synthesized three series of N-alkylaldo-namides: N-alkyl-N-methylgluconamides (C_n-MGA), N-alkyl-N-methyllactobionamides (C_n-MLA) (Fig. 1)

Fig. 1 Structures of *N*-alkyl-*N*-methylaldonamides

RN(CH₃)X		
R	X	Acronym
C_nH_{2n+1} n =10, 12, 14, 16, 18, oleyl	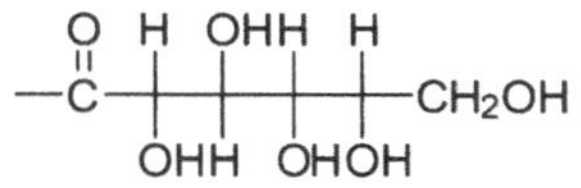	C_n–MGA
C_nH_{2n+1} n =10, 12, 14, 16, 18, oleyl	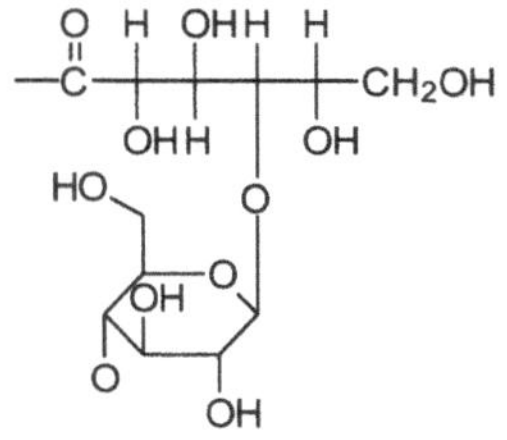	C_n–MLA

[13, 14] and *N*-alkyl-N,*N*-bis[(3-D-aldonamido)pro-pyl]amines of dicephalic type (C₁₂-DGA, -DGHA, -DLA, Cₙ-DLA) (Fig. 2) [15, 16]. In this contribution we report the antimicrobial and biodegradability prop-erties of these compounds.

Experimental

Materials

All the saccharide-based surfactants tested were synthesized as described previously [13–16].The bacterial strains used for deter-mination of the antimicrobial activity were obtained from the Polish Collection of Microorganisms (PCM).The activated sludge biomass used in the tests was from a municipal sewage plant located near Wrocław.

Measurements and procedures

Antimicrobial activity

The antimicrobial activities of the *N*-alkylaldonamides and the nonionic conventional oligooxyethylenated alcohols were evaluat-ed by the agar dilution method [17]. Nutrient agar and mycological agar were used for bacteria and fungi, respectively.

The gram-positive bacterial species were: *Staphylococcus aureus* PCM 1944, *Sarcina lutea* PCM 1947 and *Bacillus subtilis* PCM 1949. The gram-negative bacterial species were:*Escherichia coli* PCM 2057, *Serratia marcescens* PCM 549 and *Pseudomonas putida* PCM 2124. The fungal strains were *Saccharomyces cerevisiae*, *Penicillum citrinum* and *Aspergillus niger*. The incubation was carried out for 48 h at 37 °C for gram-positive bacteria or 72 h at 28 °C for the other species.

Biodegradability

The biodegradability of the *N*-alkylaldonamides as well as of the conventional oxyethylenated alcohols was examined using the closed-bottle test according to OECD test guideline 301D using acivated sludge [18]. The theoretical oxygen demand (TOD) was calculated according to International Standard ISO 9408 (annex A) [19]. The biodegradation was calculated by the following formula:

$$Biodegradation(\%) = (BOD/TOD) \times 100 ,$$

where BOD is the biochemical oxygen demand.

Results and discussion

N-Alkyl-*N*-methylgluconamides (C_n-MGA), *N*-alkyl-*N*-methyllactobionamides (C_n-MLA), and *N*-alkyl-N,*N*-bis[(3-D-aldonamido)propyl]amines (C₁₂-DGA, -DGHA, -DLA, C₁₆-DLA, C₁₈-DLA) constitute a group of nonionic surfactants which can be easily synthesized from renewable and low-cost raw materials [13–16]. They show, like oxyethylenated aliphatic alcohols, high sur-face activity as measured by their critical micelle concentration values [13–15], but in contrast to the latter their 0.1 wt% aqueous solutions do not show a cloud point up to the boiling temperature of the solutions [10, 14, 15]. Their solubility is, however, limited as shown by their Krafft points. As expected the lactobionamides are better water-soluble compounds than the respective gluconamides [14, 15].

Antimicrobial activity

Six *N*-alkyl-*N*-methylagluconamides and six *N*-alkyl-*N*-methyllactobionamides, differing in the alkyl chain length as well as five *N*-alkyl-N,*N*-bis[(3-D-aldonami-do)propyl]amines differing both in the alkyl chain length (C₁₂, C₁₆, C₁₈) and the sugar moiety (gluconamide DGA, glucoheptonamide DGHA, lactobionamide DLA) were screened for antimicrobial activity against various bac-terial and fungal strains. The antimicrobial activity of these amides was determined by means of their minimum inhibitory concentrations, which are the lowest concen-trations at which the tested microorganisms do not show

174

Fig. 2 Structures of *N*-alkyl-N,*N*-bis[(3-D-aldonamido) propyl]amines

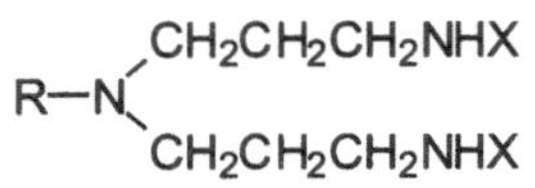

$$R-N\begin{matrix} CH_2CH_2CH_2NHX \\ CH_2CH_2CH_2NHX \end{matrix}$$

R	X	Acronym
C_nH_{2n+1} N =12	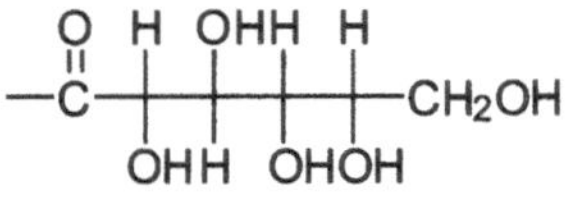 (aldonamide structure)	C_{12}-DGA
C_nH_{2n+1} N=12	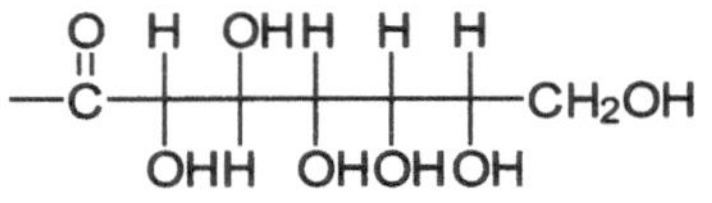 (aldonamide structure)	C_{12}-DGHA
C_nH_{2n+1} n =12, 16, 18	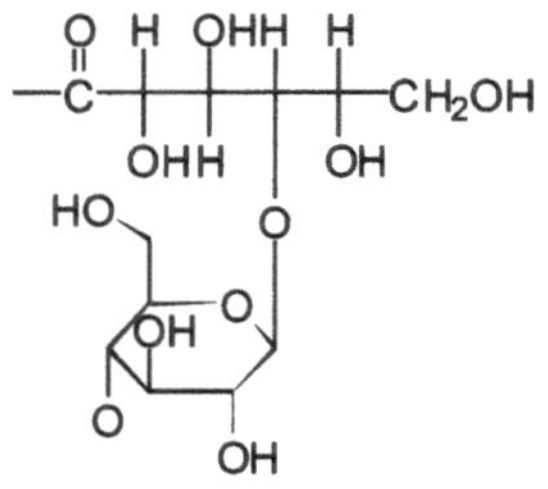 (lactobionamide structure)	C_{12}-DLA C_{16}-DLA C_{18}-DLA

visible growth. All the surfactants tested were completely inactive towards the fungi investigated: *S. cerevisiae*, *P. citrinum* and *A. niger* up to a concentration of 512 mg/l. Moreover, it appeared that the *N*-alkyl-*N*-methylgluconamides and the *N*-alkyl-*N*-methyllactobionamides were practically inactive towards the bacteria studied as well, which follows from the data collected in Table 1. The only exception is the weak activity observed for C_{10}-MGA against gram-negative bacteria and for C_{10}-MLA against gram-positive bacteria. This is in accord

Table 1 Minimum inhibitory concentration (mg/l) of *N*-alkylaldonamides

Surfactant	Gram-positive bacteria			Gram-negative bacteria		
	S. aureus	*B. subtilis*	*S. lutea*	*E. coli*	*S. marcescens*	*P. putida*
C_{10}-MGA	>512	>512	>512	256	512	512
C_{12}-MGA	>512	>512	>512	>512	>512	>512
C_{14}-MGA	>512	>512	>512	>512	>512	>512
C_{16}-MGA	>512	>512	>512	>512	>512	>512
C_{18}-MGA	>512	>512	>512	>512	>512	>512
Oleyl-MGA	>512	>512	>512	>512	>512	>512
C_{10}-MLA	512	512	256	>512	>512	>512
C_{12}-MLA	>512	>512	>512	>512	>512	>512
C_{14}-MLA	>512	>512	>512	>512	>512	>512
C_{16}-MLA	>512	>512	>512	>512	>512	>512
C_{18}-MLA	>512	>512	>512	>512	>512	>512
Oleyl-MLA	>512	>512	>512	>512	>512	>512
C_{12}-DGA	4	512	8	>512	512	>512
C_{12}-DGHA	8	512	256	>512	512	256
C_{12}-DLA	64	512	256	>512	512	512
C_{16}-DLA	64	512	32	512	256	>512
C_{18}-DLA	32	512	32	512	256	>512
$C_{10}E_4$	64	64	64	>512	>512	>512
$C_{12}E_5$	>512	>512	512	>512	>512	32

with other observations which showed that the solubility of the surfactant is better the greater its bactericidal activity [20].

Dicephalic-type surfactants (C_n-DGA, -DGHA, -DLA) showed, on the other hand, a broad spectrum of antimicrobial activity towards gram-positive cocci: *S. aureus* and *S. lutea*, much more strongly than the *N*-alkyl-*N*-methylaldonamides and the reference oxyethylenated aliphatic alcohols did (Table 1). This may be explained by the fact that the dicephalic-type surfactants, irrespective of the two amide groups present in their molecules, are tertiary amines which behave in aqueous solutions like cationic surfactants [21]. It is worth mentioning that the antimicrobial activity of C_{12}-DGA and C_{12}-DGHA against *S. aureus* is similar to that of cetyltrimethylammonium bromide [20].

Biodegradability

It has been reported that both alkyl polyglycosides [4] and alkyl glucamides [12] are completely biodegradable and show favorable ecotoxicological properties. For these reasons they do not constitute an environmental risk [4]. In the case of *N*-alkylaldonamides there is a lack, to our knowledge, of biodegradability and aquatic toxicity data. In this study, screening tests on ultimate biodegradation were performed according to OECD guideline 301D [18]. This is a closed-bottle test in which the BOD and the TOD are determined. The results obtained are shown in Table 2. From the data collected it follows that *N*-alkyl-*N*-methylgluconamides and *N*-alkyl-*N*-methyllactobionamides may be classified as readily biodegradable surfactants. This is in accord with the data of van Ginkel et al. [22], who reported that the oxyethylenated fatty amines and amides pass the level of 60% ultimate biodegradation. The biodegradation rates of the *N*-alkyl-*N*-methylaldonamides studied are, however, a little lower in comparison with those of oxyethylenated aliphatic alcohols.

The data obtained for dicephalic-type surfactants differ substantially from those of *N*-alkyl-*N*-methylaldonamides (i.e., C_n-MGA and C_n-MLA) (Table 2). Their biodegradation rates are markedly slower than those of *N*-alkyl-*N*-methylaldonamides. This may be understandable when taking into account that these surfactants are tertiary amines whose biodegradation takes place at slower rates [23, 24]. Nevertheless, they may be classified as inherently biodegradable surfactants.

Table 2 Biodegradation of *N*-alkylaldonamides

Surfactant	Theoretical oxygen demand (gO/g)	Biochemical oxygen demand (gO/g)	Biodegradation (%)
C_{10}-MGA	2.02	1.3	64.4
C_{12}-MGA	2.12	1.6	75.5
C_{14}-MGA	2.21	1.6	72.4
C_{16}-MGA	2.29	1.5	65.5
C_{18}-MGA	2.36	1.5	63.5
Oleyl-MGA	2.33	1.5	64.3
C_{10}-MLA	1.75	1.0	57.1
C_{12}-MLA	1.84	1.2	65.2
C_{14}-MLA	1.91	1.2	62.8
C_{16}-MLA	1.99	1.2	60.3
C_{18}-MLA	2.05	1.2	58.5
Oleyl-MLA	2.03	1.2	59.1
C_{12}-DGA	1.80	0.7	38.8
C_{12}-DGHA	1.74	0.4	22.9
C_{12}-DLA	1.60	0.9	56.2
C_{16}-DLA	1.74	0.6	34.4
C_{18}-DLA	1.70	0.5	29.4
$C_{10}E_4$	2.33	2.07	88.8
$C_{12}E5$	2.48	2.13	85.8

Conclusions

1. *N*-Alkyl-*N*-methylgluconamides (C_n-MGA) and *N*-alkyl-*N*-methyllactobionamides (C_n-MLA) were practically inactive towards the bacteria studied. The dicephalic-type saccharide-derived surfactants showed a broad spectrum of antimicrobial activity only towards gram-positive cocci, much more strongly than the *N*-alkyl-*N*-methylaldonamides did.
2. All the surfactants tested were completely inactive towards the fungi studied, *S. cerevisiae*, *P. citrinum* and *A. niger*, up to a concentration 512 mg/l.
3. *N*-Alkyl-*N*-methylgluconamides (C_n-MGA) and *N*-alkyl-*N*-methyllactobionamides (C_n-MLA) may be considered to be readily biodegradable. Their biodegradation rates are, however, a little lower than those of oxyethylenated aliphatic alcohols. The dicephalic-type surfactants studied showed slower biodegradation rates than *N*-alkyl-*N*-methylaldonamides.
4. The surfactants studied seem to fulfill the requirements of green chemistry to be environmentally safe.

Acknowledgements Support of this work by the Polish State Committee for Scientific Research, grant no. 3T09B 057 13, is gratefully acknowledged.

References

1. Hildreth JEK (1982) Biochem J 207:363
2. Ricco-Lattes I, Lattes A (1997) Colloids Surf A 123–124:37
3. Matsumura S, Imai K, Yoshikawa S, Kawada K, Uchibori T (1990) J Am Oil Chem Soc 67:996
4. Steber J, Guhl W, Stelter N, Schröder FR (1995) Tenside Surf Deterg 32:515
5. Andree H, Middelhauve B (1991) Tenside Surf Deterg 28:413

6. Busch P, Hensen H, Tesmann H (1993) Tenside Surf Deterg 30:116
7. Kalkenberd H, (1988) Tenside Surf Deterg 25:8
8. S&D News (1997) Inform 8:40
9. Pfannemueller B, Welte W (1985) Chem Phys Lipids 37:227
10. Zhang T, Marchant RE (1996) J Colloid Interface Sci 177:419
11. Arai T, Takasugi K, Esumi K (1996) Colloids Surf A 119:81
12. Stalmans M, Matthijas E, Weeg E, Morris S (1993) SÖFW J 119:794
13. Syper L, Wilk KA, Sokołowski A, Burczyk B (1998) Prog Polym Sci 110:199
14. Burczyk B, Wilk KA, Sokołowski A, Syper L (2000) J Colloid Interface Sci DOI 10.1006/jcis.2001.7704
15. Wilk KA, Syper L, Burczyk B, Sokołowski A, Domagalska BW (2000) J Surf Deterg 3:185
16. (1999) Polish Appl Patent P-331294
17. Bristline RG, Maurer EW, Smith FD, Linfield WM (1980) J Am Oil Chem Soc 57:98
18. (1981) OECD guidelines for testing chemicals 301D. Closed-bottle test. Organization for Economic Cooperation and Development
19. (1991) International standard ISO9408. Water quality evaluation in an aqueous medium of the "ultimate" aerobic biodegradability of organic compounds method by determining the oxygen demand in the Closer respirometer
20. Infante MR, Perez L, Pinazo A (1998) In: Holmberg K (ed) Novel surfactants. Dekker, New York, p 103–107
21. Van Ginkel CG (1995) In: Karsa DR, Porter MR (eds) Biodegradability of surfactants. Blackie, London, p 187
22. Van Ginkel CG, Stroo CA, Kroon AG (1993) Tenside Surf Deterg 30:213
23. Yoshimura K, Machida S, Mosuda F (1980) J Am Oil Chem Soc 57:238
24. Van Ginkel CG, Pomper MA, Stroo CA, Kroon AG (1995) Tenside Surf Deterg 32:355

Progr Colloid Polym Sci (2001) 118: 177–179
© Springer-Verlag 2001

Marjan Bele
Miran Gaberscek
Jernej Drofenik
Robert Dominko
Stane Pejovnik

Gelatin-modified surfaces in selected electronic components

M. Bele (✉) · M. Gaberscek · J. Drofenik
R. Dominko · S. Pejovnik
National Institute of Chemistry
Hajdrihova 19, 1000 Ljubljana
Slovenia

S. Pejovnik
Faculty of Chemistry
and Chemical Technology
University of Ljubljana
Askerceva 5, 1000 Ljubljana
Slovenia

Abstract The gelatin content in graphite–gelatin composites was determined from the difference in the mass of graphite particles before and after the treatment with gelatin and by thermogravimetric analysis. It was found that the content of gelatine has a maximum at those pH values of gelatin solution which correspond to the isoelectric point of gelatin. The result is correlated with previous findings concerning gelatin adsorption on various substrates as a function of pH. The "hairy" structure of gelatin allows deposition of different types of small particles (e.g. carbon black) and, under appropriate conditions, may lead to the formation of thick (0.3 μm) and dense layers. On the basis of the good adsorption of gelatin on graphite surfaces, we prepared anodes for lithium accumulators. In graphitic anodes, gelatin serves both as a binder between particles and as a surface modifier, which leads to lower irreversible losses of charge owing to anode passivation.

Key words Gelatin · Adsorption · Particle deposition · Lithium-ion batteries

Introduction

Gelatin is a product that has been used for hundreds of years. In recent decades it has been extensively used in the food, photographic, and pharmaceutical industries. [1]. Gelatin is produced by the acid or alkali denaturation of collagen which originally forms a triple helix, consisting of three associated left-handed helices [2]. It is well known that gelatin is an amphoteric polyelectrolyte which contains both acidic and basic amino groups which, in solution, transform into the corresponding negatively and positively charged groups. In water, gelatin can form structures such as random coils or random coils with a collagen-folded structure. The structure depends mainly on the amino acid composition, the concentration of the gelatin, and on temperature and pH of the solution [1–4]. In many instances gelatin is used as a stabilizer in both colloidal dispersions and emulsions. In the present contribution we show some examples of gelatin use in electronic components.

Materials and methods

TIMREX SFG44 special graphite obtained from TIMCAL (Sins, Switzerland) was used for the preparation of anode composite materials. The Brunauer–Emmett–Teller (BET) surface of this material is 4.2 m^2/g and the average particle size is 44 μm (data supplied by TIMCAL). Gelatin was used as the surface modifier and the binder of the carbon particles in the anode preparation. Medium gel power gelatin, 180 g Bloom, type A, derived from pigskin was obtained from Fluka no. 48722 (Fluka Chemie, Buchs, Switzerland). Two types of surfactants were used; cationic surfactant cetyltrimethylammonium bromide (CTAB), Aldrich no. 85,582-0 (Aldrich, Milwaukee, Wiss., USA) and anionic surfactant sodium di(isooctyl)succin-L-sulfonate (AOT, Cytec Industries, West Paterson, N.J., USA). Metallic lithium foil (0.75-mm thick) was obtained from Alfa Aesar (Johnson Matthey, Karlsruhe, Germany). For comparison, classical graphitic anodes with 5 wt% Teflon as a binder were prepared according to a procedure described elsewhere [5].

The charge density was determined by charge-compensating polyelectrolyte titration using a KOLB (Chem. Fabrik, Dr. W. Kolb, Hedingen, Germany) charge detector and a 665 Dosimat (Metrohm, Herisau, Switzerland) titrator. The method is based on a stoichiometric reaction between a polycation and a polyanion

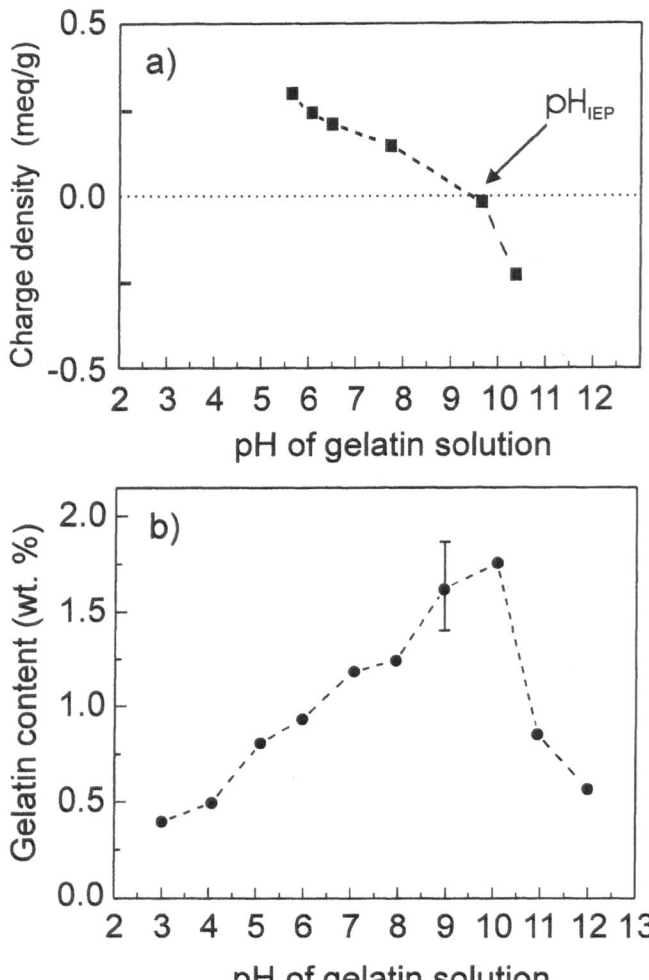

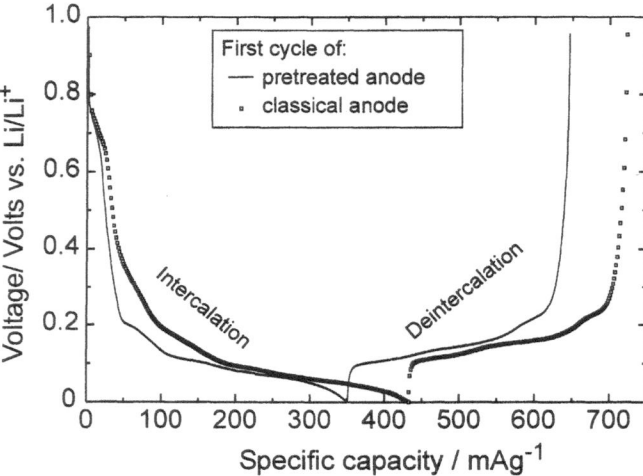

Fig. 2 First charge of the anode prepared from the gelatin-pretreated graphite particles and of the conventional graphite anode prepared with Teflon as a binder. The current density was 50 μAcm^{-2} (corresponding to about C/7). The irreversible loss is roughly related to the difference between the charge needed for intercalation and deintercalation of lithium into and out of the graphite anode

Fig. 1 a Charge density as a function of pH for gelatin type A. **b** Gelatin content in graphitic anode material as function of pH. The *curve* represents average values of a series of thermogravimetric and gravimetric determinations. The maximum *error bar* is indicated at pH 9

[6, 7]. AOT (0.002 M) and CTAB (0.002 M) were used as the standard anionic and cationic polyelectrolyte solutions, respectively. The charge density data were normalized to 100% solids in the solution and were expressed in milliequivalents of positive or negative charge per mass unit of gelatin.

The gelatin content in the graphite–gelatin composites was determined from the difference in the mass of the graphite particles before and after the treatment with gelatin and by thermogravimetric analysis. Thermogravimetric analysis was performed in the temperature range from 30 to 700 °C in an air flow of 18 l/min and at a heating rate of 10 °C/min using a Mettler TA 3000 apparatus (Mettler, Switzerland).

The electrochemical tests were performed using a laboratory-made three-electrode testing cell as described elsewhere [5]. The working electrode was a graphite electrode prepared as described previously, while the counterelectrode and the reference electrode were made of metallic lithium. The electrolyte used was a 1 M solution of LiPF$_6$ in EC:DMC (1:1 ratio) as received from Merck (Darmstadt, Germany). The electrochemical measurements were

carried out using a Solartron 1286 electrochemical interface (Solartron Mobrey, UK). The constant current during cell cycling was 50 μA (corresponding to C/7), while the geometrical surface area of the working electrode was always 0.5 cm^2.

Results and discussion

The shape of gelatin has a significant influence on gelatin adsorption. An important parameter which controls the shape of gelatin macromolecules is the pH value of the gelatin solution during adsorption. It has been found (Fig. 1) that the adsorption of gelatin on graphite has a maximum at those pH values of gelatin solution at which gelatin has low charge density, viz., near the isoelectric point (IEP). The maximum content of gelatin in graphite–gelatin composites was found to be 1.7 wt%, which corresponds to 4 mg gelatin per 1 m^2 graphite if the BET surface is used for the calculation. This value is within the range 1–6 mg/m^2, i.e., the range found in studies of adsorption of different kinds of gelatin onto different substrates [8]. As shown by Kamiyama and Israelachvili [4], near the IEP gelatin is not expanded owing to the electrostatic attraction between the positively and negatively charged groups of adjacent segments.

The fact that the maximum adsorption of gelatin occurs near the IEP was already demonstrated in our previous work [9], in which adsorption of gelatin on printed circuit boards used in the electronics industry was studied. It was assumed that after gelatin adsorption, the loops and the tails of the gelatin molecules were sticking out into the solution. Later on, this was confirmed in a set of experiments [10] in which atomic force microscopy was used to clarify the nature of gelatin adsorption. Presumably, the "hairy" structure of gelatin serves as a steric stabilizer of a colloidal dispersion. Furthermore, the gelatin-modified surface of a substrate can attract many kinds of particles because gelatin contains many kinds of active groups: cationic, anionic and nonionic. However, the properties of gelatin are not sufficient if we want to form a thick carbon black layer on a substrate – one also has to to optimize the properties of the carbon black dispersion using surfactants and salts [11]. The surfactant is not only needed to keep the particle suspension stable, but it also swells the polyelectrolyte layer to a higher distance from the surface. By contrast, the salt controls the electrostatic properties of the particles in the dispersion which screen the electric field of the surfaces [10]. Under such conditions, the deposition (sometimes called the "substrate-induced deposition") of thick and rinse-proof coatings can only occur on modified substrate surfaces, but not on other surfaces, and, of course, coagulation cannot occur in the bulk dispersion. At the end of deposition, the deposited layers of carbon black may reach a thickness of the order of 0.3 μm. Such coatings can be used as a starting point for subsquent electroplating of copper [12, 13].

Gelatin-modified graphite particles can also be used for preparation of electrodes in lithium rechargeable batteries. One obvious role of gelatin is that it can act as a binder between carbon particles. Gelatin is a more efficient binder than polymers, such as Teflon or poly(vinylidene difluoride), so only 1 wt% of gelatin is needed to bind carbon particles into the anode; if Teflon is used, one needs at least 5 wt%. We have discovered a second, even more important role of gelatin in carbon anodes. It is well known that strong passivation of the carbon anode in the first cycle is one of the drawbacks of lithium accumulators. The strong passivation means large irreversible losses of lithium because the lithium is used for chemical reaction instead of for storage of charge. The first charge–discharge cycle of a classical carbon anode is compared with the cycle obtained using a gelatin-pretreated carbon anode in Fig. 2. If carbon black is pretreated, the irreversible loss drops to more than 40% of the theoretical capacity. It seems that gelatin molecules induce the formation of a more uniform and, hence, thinner passive film during battery cycling. A thinner film means that less lithium is consumed for the chemical reaction of passivation.

Conclusions

The use of gelatin in surface pretreatment can be recommended because it has good adsorption properties. The "hairy" structure serves as an adhesion agent for depositing different types of small particles and, under appropriate conditions, may lead to the formation of thick (0.3 μm) carbon black layers. In lithium accumulators, gelatin can serve as an effective binder and, beyond that, as an initiator for nucleation in passive film growth.

Acknowledgements This research was sponsored by NATO's Scientific Affairs Division in the framework of the Science for Peace Programme. Financial support from the Ministry of Science and Technology of Slovenia is also fully acknowledged.

References

1. Alleavitch J, Turner WA, Finch CA (1989) In: Elvers B, Hawkins S, Ravensroft M, Rounsaville JF, Shulz G (eds) Ullmann's encyclopaedia of industrial chemistry, gelatin. VCH, Weinheim, p 307
2. Kramer RZ, Bella J, Mayville P, Brodsky B, Berman HM (1999) Nature Struct Biol 6:454
3. Engel J (1985) In: Pearson AM, Dutson TR, Bailey AJ (eds) Advances in meat research folding and unfolding of collagen triple helices. AVI, New York, p 152
4. Kamiyama Y, Israelachvili J (1992) Macromolecules 25:5081
5. Gaberscek M, Bele M, Drofenik J, Dominko R Pejovnik S (2000) Electrochem Solid-State Lett 3:171
6. Wassmer KH, Schroeder U, Horn D (1991) Makromol Chem 192:553
7. Kam S, Gregory J (1999) Colloids Surf A 159:165
8. Vaynberg KA, Wagner JN, Sharama R, Martic P (1998) J Colloid Interface Sci 205:131
9. Bele M, Pejovnik S, Besenhard JO, Ribitsch V (1998) Colloids Surf A 143:17
10. Bele M, Kocevar K, Pejovnik S, Besenhard JO, Muševisc I (2000) Langmuir 16:(in press)
11. Bele M, Kocevar K, Muševisc I, Besenhard JO, Pejovnik S (2000) Colloids Surf A 168:231
12. Besenhard JO, Meyer H, Gausmann HP (1991) German Patent DE 41 13 407
13. Bele M, Besenhard JO, Pejovnik S, Meyer H (1999) German Patent, DE 197 31 184 C2

Progr Colloid Polym Sci (2001) 118: 180–183
© Springer-Verlag 2001

Electrophoretic mobility and swelling behaviour of 2-acrylamido-2-methylpropane sulphonic acid/poly(*N*-isopropylacrylamide) microgel particles

M. J. García-Salinas
M. S. Romero-Cano
F. J. de las Nieves

M. J. García-Salinas · M. S. Romero-Cano
F. J. de las Nieves (✉)
Complex Fluids Physics Group
Department of Applied Physics
University of Almería
04120 Almería, Spain
e-mail: fjnieves@ual.es
Tel.: +34-50-015434
Fax: +34-50-015434

Abstract A microgel system has been characterized by measuring the size (by photon correlation spectroscopy) and the electrophoretic mobility in different conditions. The microgel particles are poly(*N*-isopropylacrylamide) cross-linked with *N,N'*-methylene bisacrylamide, and the anionic charge of this system is increased by copolymerization with 2-acrylamido-2-methylpropane sulphonic acid. Both the electrophoretic mobility and the hydrodynamic diameter of the particles were measured as a function of pH and electrolyte concentration. Although a weak charge was detected by titration, no trend was observed for the size or mobility against pH variations. In order to apply Ohshima's electrophoretic theories for soft particles to fit the mobility data, the swelling behaviour of the particles was taken into account.

Key words Soft core–shell particles · Swelling · Electrophoretic mobility

Introduction

Colloidal microgels are considered as a model for understanding soft particles (i.e. hard spheres covered with a layer of polyelectrolytes), and are also of great interest and importance in many industrial applications [1, 2].

In this work, a basic characterization and size and electrophoretic mobility measurements have been carried out for core–shell microgel particles. First, titration yielded both strong and weak charge; however, the influence of the latter could not be detected later: no trend was observed for the size or the mobility against pH variations. Next, the size and the mobility were measured as a function of electrolyte (NaCl) concentration. Finally, Ohshima's electrophoretic theory [3] for soft particles was applied to fit the experimental data. Very good agreement was obtained when the charge density was calculated considering the measured variable shell size and when the drag coefficient, λ, was considered to be discontinuous.

Materials and methods

The colloidal system used in this work is a poly(*N*-isopropylacrylamide) (polyNIPAM) microgel, which was synthesized using 2-acrylamido-2-methylpropane sulphonic acid (AMPS) as comonomer, *N,N'*-methylene bisacrylamide (BA) as cross-linker and amonium persulphate (APS) as initiator [4]. The monomers (3.974 g NIPAM, 0.383 g AMPS, 0.21 g BA) were dissolved in 300 ml distilled water and dry N_2 was bubbled for 15 min; 0.21 g APS was added and the reaction was allowed to proceed for 5 h at 70 °C and 350 rpm. After the synthesis the microgel was filtered through glass wool and then cleaned by dialysis over several days [5].

As the particles swell in a good solvent, the particle diameter (transmission electron microscopy) was measured in different conditions, obtaining values from 313 ± 33 nm (standard conditions) to 105 ± 9 nm (when the temperature of the sample was 60 °C and the NaCl electrolyte concentration was 0.5 M). The particle size was also obtained by photon-correlation spectroscopy (PCS) as a function of pH and electrolyte concentration with a Malvern Zetamaster S device. This device is also used for the electrophoretic mobility measurements, which were also carried out as a function of pH and electrolyte concentration. The sizes and the mobilities were taken as the average of at least ten measurements, considering their standard deviation as the experimental error.

Results

The particle charge is assumed to be uniformly distributed in the shell. Figure 1 shows the forward and back conductimetric titration, which gave a strong-acid density charge of 129 ± 15 μEq/g and a weak-acid density charge of 240 ± 80 μEq/g. For comparison, the surface charge density for a polystyrene latex of 20 μC/cm^2 and 280-nm diameter gives a lower value, around 40 μEq/g. The origin of the weak-acid charge is not clear. First, all the chemicals used in the synthesis were of analytical grade and the cross-linker was also recrystalized, so no residuals of acrylic acid were expected. On the other hand, carboxyl groups might originate from sulphate groups (from the initiator) in two ways: when a Kolthoff reaction takes place [6], hydroxyl groups appear that can be finally oxidized into carboxyl groups; and directly from the oxidation of sulphate groups for high temperatures and long reaction times [7]. For our experimental conditions the generation of these groups is not very probable [8]. Finally, some authors have reported the appearance of a weak charge as a consequence of the cleaning process when dialysis is used [9].

To test the possible influence of this weak charge on the electrokinetic and swelling behaviour, size and mobility measurements as a function of pH were carried out; no clear trend was found in either case. These results are shown in Figs. 2 and 3. The buffer solutions were adjusted to keep the ionic strength constant in both cases (the electrical conductivity of the suspension was 200 μS/cm). The particle size (PCS) is plotted against pH for three temperatures in Fig. 2. No trend is observed with pH variation, but the influence of temperature and electrolyte concentration is clearly seen. Higher temperatures and/or higher electrolyte concentrations mean lower sizes. The dotted lines are the average sizes for each set. The influence of electrolyte concentration is shown for the highest temperature, 45 °C. When no buffer is used, the pH values required are obtained directly, and no electrolyte is added to equalize the ionic strengths, so the electric conductivity of the samples is lower (100 μS/cm) and thus the measured size is bigger. The same happens for all temperatures.

The mobility data versus pH obtained for 25 °C are shown in Fig. 3. The dotted line is the average mobility value, and for all cases deviations from it are all small and within the experimental error. We can conclude that changes in pH, and thus in the total particle charge, do not affect considerably the electrophoretic mobility or the particle size.

The experimental and theoretical electrophoretic mobility data versus electrolyte (NaCl) concentration for 25 °C are shown in Fig. 4. In order to explain these results, Ohshima's theory for soft particles was used [3]. From the general mobility expression, one of the limiting cases leads to the following equation [10]:

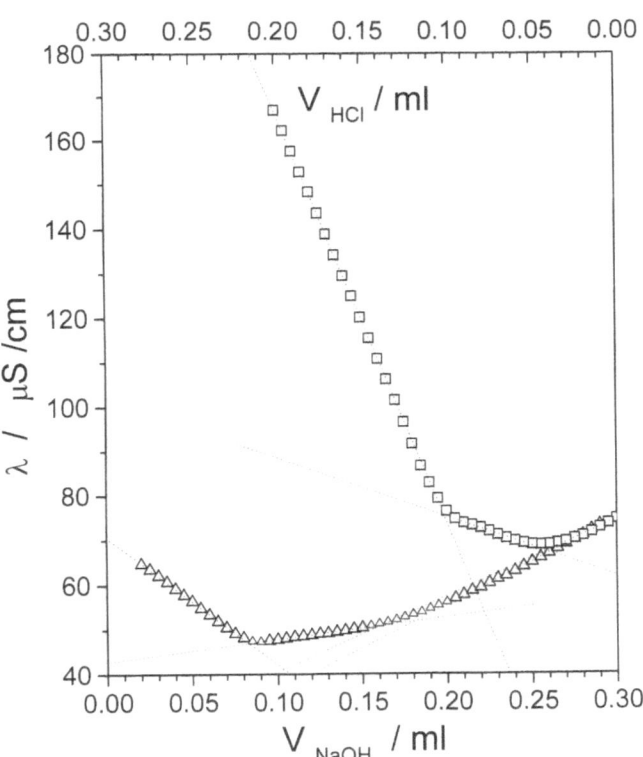

Fig. 1 Forward (*triangles*) and back (*squares*) titration curves of the microgel, with 20.8 mM Nacl and 44.2 mM HCl solutions, respectively

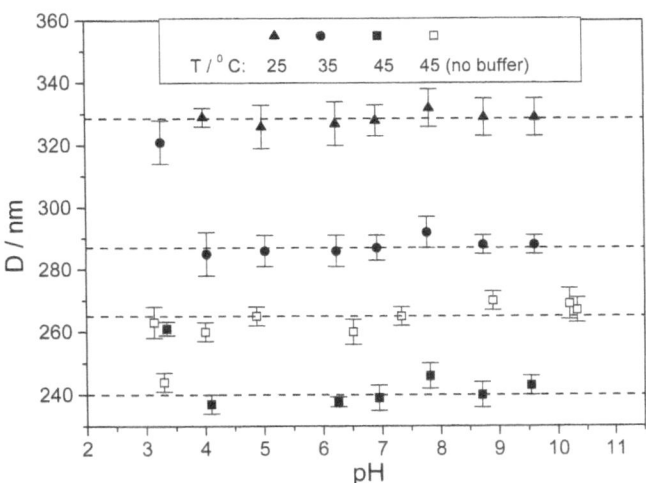

Fig. 2 Particle diameter (photon-correlation spectroscopy) against pH for 25 °C (*triangles*), 35 °C (*circles*) and 45 °C (*squares*) for the same ionic strength (200 μS/cm). The *open squares* are particle diameters for 45 °C and a lower ionic strength (100 μS/cm). The *dotted lines* stand for the average diameter of each set

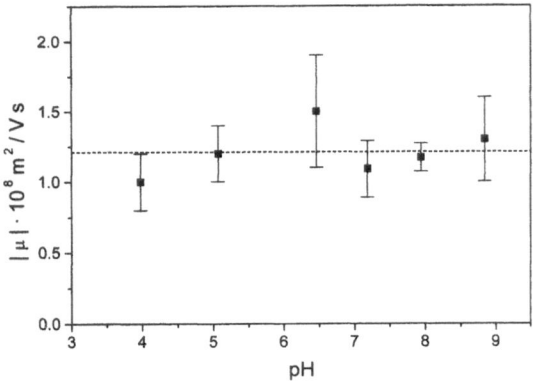

Fig. 3 Electrophoretic mobility against pH for 25 °C. The *dotted line* is the average value

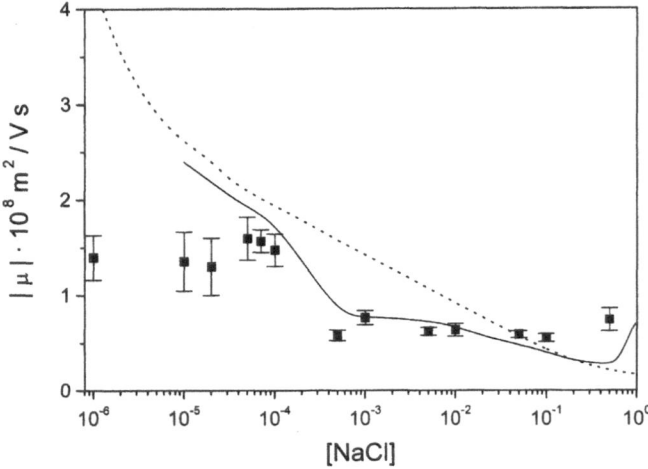

Fig. 4 Electrophoretic mobility against NaCl concentration for 25 °C. The lines are theoretical fits (see text)

Fig. 5 Influence of particle size on theoretical electrophoretic mobility. The *lines* are calculations with Eq. (1), considering a constant charge density (*dotted line*) or a variable charge density depending on size (*solid line*). *Inset*: particle diameter against NaCl concentration for 25 °C

$$\mu = \frac{\varepsilon_r \varepsilon_0}{\eta} \frac{\psi_0/\kappa_m + \psi_{DON}/\lambda}{1/\kappa_m + 1/\lambda} f\left(\frac{d}{a}\right) + \frac{\rho_{fix}}{\eta\lambda^2} \ , \quad (1)$$

where η is the viscosity, ε_0 is the vacuum dielectric permittivity and ε_r that relative to the medium, ρ_{fix} is the uniform volume charge density in the polyelectrolyte shell and the λ is the drag coefficient ($1/\lambda$ is referred to as the softness parameter) related to the frictional coefficient of the polyelectrolyte layer γ:

$$\lambda = \left(\frac{\gamma}{\eta}\right)^{\frac{1}{2}} \ . \quad (2)$$

The function $f(d/a)$ varies from 2/3 to 1 as the thickness of the shell, d, increases with respect to the core radius, a:

$$f\left(\frac{d}{a}\right) = \frac{2}{3}\left(1 + \frac{a^3}{2b^3}\right) = \frac{2}{3}\left(1 + \frac{1}{2(1+d/a)^3}\right) \ . \quad (3)$$

κ_m can be interpreted as the Debye–Hückel parameter of the shell, κ being the Debye–Hückel parameter:

$$\kappa_m = \kappa\left[1 + \left(\frac{\rho_{fix}}{2zen^\infty}\right)^2\right]^{1/4} \ . \quad (4)$$

Finally, ψ_{DON} is the Donnan potential in the shell and ψ_0 is the "surface" potential (i.e. the potential in the limit between the particle and the solution) [11].

Equation (1) holds when $\kappa a \gg 1$, $\lambda a \gg 1$, $\kappa d \gg 1$ and $\lambda d \gg 1$.

In most experimental contributions where Eq. (1) is used to fit mobility data, the fit parameters ρ_{fix} and λ are considered to be constant for a fixed temperature over the whole range of electrolyte concentrations [12, 13]. We show in Fig. 5 how the theoretical prediction from Eq. (1) can change when a variable charge density is

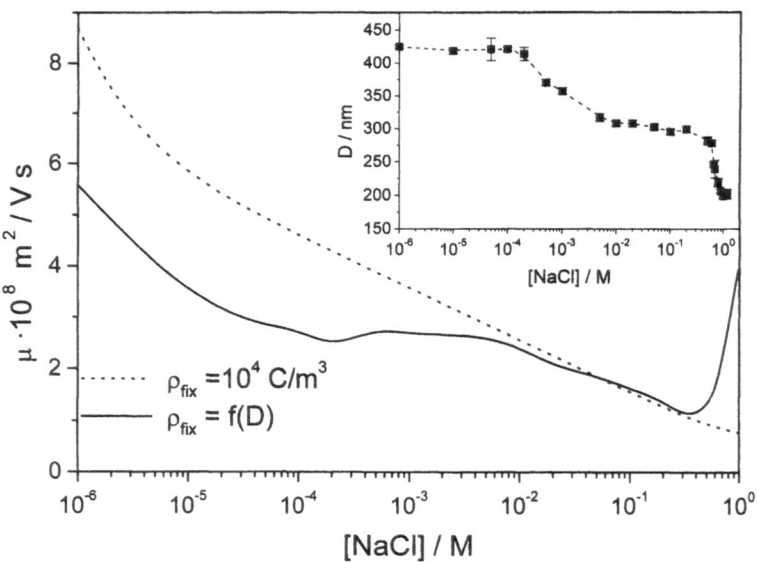

Table 1 Charge density, ρ_{fix}, and drag parameter, λ, for theoretical data in Fig. 4

	ρ_{fix}	λ_1	λ_2
Dotted line	44×10^6 C/m^3	8×10^9 m^{-1}	8×10^9 m^{-1}
Solid line	Depending on size	8×10^9 m^{-1}	20×10^9 m^{-1}

considered. The two mobility curves in Fig. 5 are obtained in the following way. If the charge density ρ_{fix} is taken to be constant, the mobility theoretically obtained is the dotted line ($\rho_{\text{fix}} \cong 10^4$ C/m^3, $\lambda = 5 \times 10^9$ m^{-1}); however, it has been shown here that the particle size changes with electrolyte concentration, so the shell charge density must decrease if the particles swell and vice versa. The inset shows specifically the particle size as a function of electrolyte concentration, and with this information the charge density can be calculated from the total titrated value, q_{T}, and shell size: $\rho_{\text{fix}} = q_{\text{T}}/V_{\text{shell}}$. When this quantity is used in Eq. (1), the mobility obtained is the solid line. There is still one more consideration to fit the experimental mobility data. The drag parameter, λ, depends on the volume fraction of polymer in the shell, so some changes are expected as a consequence of size variations. The behaviour predicted for λ in our conditions, according to Brinkman and Cohen-Stuart models [14, 15], is basically a step function, the discontinuity being a steep increase around 1 mM, so two λ values should be considered.

With all these facts considered, we can go back to Fig. 4, where the theoretical curves are obtained as follows. The dotted line is obtained with constant charge density (obtained from titration and medium shell size) and constant λ value, which is left as a fit parameter. The solid line is obtained as explained before, with a variable ρ_{fix} and two λ values, λ_1 and λ_2, as parameters. All of them are listed in Table 1.

Within the range of validity of Eq. (1), the agreement with experimental data has been improved with the consideration of the influence of particle size changes in the shell charge density and drag parameter.

Conclusions

A core–shell microgel system has been characterized by measuring the size and the mobility as a function of pH and electrolyte concentration. The weak-acid charge detected seems to have no influence on the size or the mobility, because no trends were detected with pH variation. The experimental mobility data were fitted using Ohshima's theory, and a good prediction was obtained when the particle size was taken into account for the charge density calculation.

Acknowledgements This work was supported by the Comisión Interminiaterial de Ciencia y Tecnología under project MAT 2001-2767 and by Acción Integrada with the University of Bristol, HB 1998-0225. The authors thank Alberto Fernández-Nieves for the synthesis of the microgel at the Univesity of Bristol.

References

1. Murray MJ, Snowden MJ (1995) Adv Colloid Interface Sci 54:73
2. Saunders BR, Vincent B (1999) Adv Colloid Interface Sci 80:1
3. Ohshima H (1995) Adv Colloid Interface Sci 62:189
4. Crowther HM (1998) PhD dissertation. University of Bristol
5. Fernández-Nieves A, Fernández-Barbero A, Vincent B, de las Nieves FJ (2000) Macromolecules 33:2114
6. Kolthoff IM, Miller IK (1951) J Am Chem Soc 73:3055
7. Bastos-González D, Hidalgo-Álvarez R, de las Nieves FJ (1996) J Colloid Interface Sci 177:372
8. Goodall AR, Hearn J, Wilkinson HC (1977) Polym Sci Chem Ed 15:2193
9. Wilkinson MC, et al (1999) Adv Colloid Interface Sci 81:77
10. Ohshima H, Kondo T (1989) J Colloid Interface Sci 130:281
11. Ohshima H (1994) J Colloid Interface Sci 163:474
12. Ohshima H, Makino K, Kato T, Fujimoto K, Kondo T, Kawaguchi H (1993) J Colloid Interface Sci 159:512
13. Makino K, Yamamoto S, Fujimot K, Kawaguchi H, Ohshima H (1994) J Colloid Interface Sci 166:251
14. Brinkman HC (1949) Research 2:190
15. Cohen Stuart MA, Waajen FHWH, Cosgrove T, Vincent B, Crowley TL (1984) Macromolecules 17:1825

Progr Colloid Polym Sci (2001) 118:184–189
© Springer-Verlag 2001

A. Forgiarini
J. Esquena
C. González
C. Solans

Formation and stability of nano-emulsions in mixed nonionic surfactant systems

A. Forgiarini · J. Esquena · C. Solans (✉)
Dept. Tecnologia de Tensioactius
Instituto d'Investigaciones Químiques
i Ambientals de Barcelona, CSIC, Jordi
Girona 18-26, 08034 Barcelona, Spain
e-mail: csmqci@cid.csic.es
Tel.: + 34-93-4006159
Fax: + 34-93-2045904

C. González
Dept. Enginyeria Química i Metal.lúrgica
Universitat de Barcelona, Martí
i Franquès 1, 08028 Barcelona, Spain

A. Forgiarini
Facultad de Ingeniería, Universidad
de Los Andes, Mérida 5101, Venezuela

Abstract The formation of nano-emulsions has been studied in water/mixed nonionic surfactant/oil systems using two emulsification methods. In one method, the composition was kept constant and the temperature was changed (phase-inversion temperature, PIT, method), while in the other method, water was added dropwise to a solution of the mixed surfactants in oil at constant temperature (method B). The droplet size and stability were determined as a function of surfactant mixing ratio, W_1, at 25 °C. The droplet size of nano-emulsions obtained by the PIT method is practically independent of W_1 and falls in the range 60–80 nm. In contrast, the droplet size of nano-emulsions prepared by method B, is highly dependent on W_1 and varies between 60 and 300 nm. At W_1 values where the PIT or the hydrophile–lipophile balance temperature (T_{HLB}) of the system is close to 25 °C, the droplet sizes of the nano-emulsions are similar for both emulsification methods. There are three equilibrium phases of the latter compositions: an aqueous micellar solution or oil-in-water microemulsion (W_m), a lamellar liquid-crystalline phase and an oil phase (O) in addition, these nano-emulsions showed higher kinetic stability than those with lower W_1 values (higher T_{HLB}) and consisting of two liquid phases (W_m + O).

Key words Nano-emulsions · Emulsification · Phase-inversion temperature · Hydrophile–lipophile balance temperature · Mixed nonionic surfactant

Introduction

Emulsions are thermodynamically unstable liquid/liquid dispersions formed, generally, by water, oil and surfactant mixtures [1, 2]. The type of emulsion that forms in a water/poly(oxyethylene glycol) alkyl ether surfactant/oil system is highly dependent on temperature. At low temperatures, the surfactant is mainly soluble in water giving rise to emulsions of the oil-in-water (O/W) type. These emulsions separate into two liquid phases: an aqueous micellar solution or oil-in-water microemulsion (W_m) and an excess oil phase (O) (Winsor I system). At high temperatures, the surfactant is preferentially soluble in oil and the emulsions are of the water-in-oil (W/O) type. Two phases are also present at equilibrium: an aqueous phase (W) and a reverse micellar solution or W/O microemulsion phase (O_m) (Winsor II system). At an intermediate temperature, the so-called hydrophile–lipophile balance (HLB) temperature (T_{HLB}) or the phase-inversion temperature (PIT), the hydrophilic–lipophilic properties of the nonionic surfactant are balanced [3]. At the HLB temperature, a transition from O/W to W/O emulsions takes place. The emulsions consist of three liquid phases in single-surfactant systems at this temperature: excess water and oil phases and a bicontinuous microemulsion phase (D) (Winsor III system) [4–6]. It is well know that emulsions showing Winsor III phase equilibria. (e.g. near or at the T_{HLB}) are very unstable owing to the extremely low values of the interfacial tension between the different phases [7–11]. In a mixed nonionic surfactant system, the hydrophilic–lipophilic properties are dependent on the surfactant mixing ratio at constant temperature. In the dilute region of these systems, the structure of the third phase may be a bicontinuous microemulsion [4, 6] or a liquid-crystalline phase [12, 13].

Nano-emulsions are a class of emulsions with a droplet size in the range 50–500 nm and have attracted a great deal of attention in recent years because of their wide range of practical applications [14–16]. Nano-emulsions are, generally, stable against sedimentation or creaming, owing to their small droplet size. The main ageing process in nano-emulsion destabilization is usually Ostwald ripening [17, 18]. In Ostwald ripening the larger droplets grow at the expense of smaller droplets owing to the different Laplace pressure in droplets of different sizes. The Ostwald ripening rate, according to the Lifshitz, Slezov and Wagner theory [19–21] is given by $\omega = dr^3/dt = 8C_\infty \gamma V_m D/9\rho RT$, where r is the average droplet radius, t is time, γ is the interfacial tension, D and C_∞ are the diffusion coefficient and the solubility of the droplet phase material in the bulk phase, respectively, V_m its molar volume, ρ is the density and T the temperature. This equation predicts a linear relationship between r^3 and t.

We have reported, recently, nano-emulsion formation in a water/technical grade nonionic surfactant/oil system, at certain W/O ratios, when water is added stepwise to an oil/surfactant mixture [22]. However, when oil is added to water/surfactant mixtures, nano-emulsions are not obtained. These results could not be explained in terms of the equilibrium properties since the final composition of the emulsions was the same. It was suggested that the phase transitions during emulsification could play a key role in nano-emulsion formation. In order to gain a better understanding of nano-emulsion formation, it was considered of interest to determine the relationship between the emulsification method, the nano-emulsion droplet size and stability in systems with mixtures of pure nonionic surfactant.

Experimental

Materials

Homogeneous nonionics surfactants, tetraethylene glycol dodecyl ether ($C_{12}E_4$), tetraethylene glycol tetradecyl ether ($C_{14}E_4$), tetraethylene glycol hexadecyl ether ($C_{16}E_4$) and hexaethylene glycol dodecyl ether ($C_{12}E_6$) were supplied by Nikko Chemicals Co (Japan). n-Decane (purity above 99%) and NaCl (purity above 99.5%) were obtained from Merck. Water was deionized by Milli-Q filtration.

Methods

HLB temperature

T_{HLB} was determined by conductivity with a Crison model 525 conductimeter, with a Pt/platinized electrode. The samples were prepared with an electrolyte solution (NaCl 10^{-2} M) instead of pure water. The conductivity of samples with different surfactant mixing ratios was measured as a function of temperature.

Phase behavior

Samples with different surfactant mixing ratios and constant total surfactant concentration of 5 wt% and a W/O ratio of 80:20 were prepared and sealed in vials. The samples were stirred and kept in a thermostatted bath at 25 °C until equilibrium was reached. The presence of liquid-crystalline phases was detected by using crossed polarizers.

Emulsion formation

The emulsions were prepared by two low-energy emulsification methods (Fig. 1). In the PIT method the sample is kept at T_{HLB} and the O/W emulsion is obtained by quickly lowering the temperature to 25 °C. In method B, water is added little by little with agitation to a solution of surfactant plus oil at a constant temperature of 25 °C.

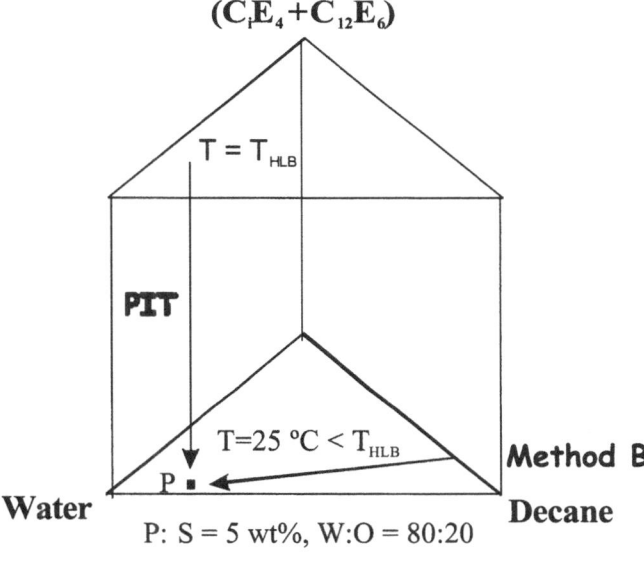

Fig. 1 Schematic representation of two emulsification methods: phase-inversion temperature (*PIT*) and method B

Droplet size

The emulsion droplet size was determined by dynamic laser light scattering (Malvern 4700).

Stability

The emulsion stability was assessed by measuring the emulsion droplet size as a function of time at 25 °C.

Results and discussion

T_{HLB} and emulsion type

Conductivity plots as a function of temperature are shown in Fig. 2 for one of the systems studied, water/($C_{16}E_4$ + $C_{12}E_6$)/decane, at three values of the surfactant mixing ratio, W_1 (defined as the weight fraction of the most lipophilic surfactant over the total surfactant).

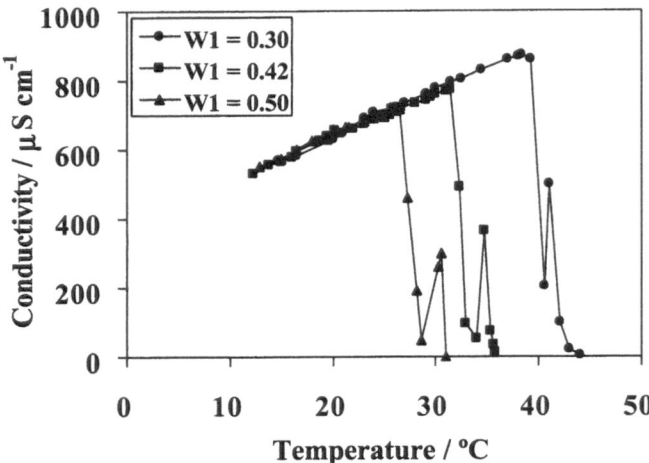

Fig. 2 Conductivity as a function of temperature for the system (aqueous solution 10^{-2} M NaCl)/$C_{16}E_4$ + $C_{12}E_6$/decane. W/O = 80:20, $S = 5$ wt%

Independently of the system, the features of the conductivity curves are similar: as the temperature increases, the conductivity initially increases slightly and then experiences an abrupt decrease to low values which is followed by an increase to intermediate values and a final decrease to very low values. The higher the W_1 value the lower the temperature at which these drastic conductivity changes occur. Similar conductivity changes were described for the water/$C_{12}E_4$/decane system [23] and were related to the phase behavior of the system. Accordingly, the first drop in conductivity can be attributed to the formation of a lamellar liquid-crystalline phase and the following slight increase in conductivity to the formation of a bicontinuous microemulsion phase or an L_3 phase. The overall conductivity changes indicate that emulsions invert from O/W to W/O via lamellar and bicontinuous phases. That is, the surfactant molecules change their affinity from water to oil or, in other words, the natural curvature of the surfactant changes from positive to negative through zero values of curvature [6, 24, 25]. In this work, T_{HLB} is taken as the mean temperature between the maximum and minimum values of conductivity. A linear relationship between T_{HLB} and W_1 for the three water/(C_iE_4 + $C_{12}E_6$)/decane systems studied, $i = 12, 14, 16$, is shown in Fig. 3. As expected, T_{HLB}

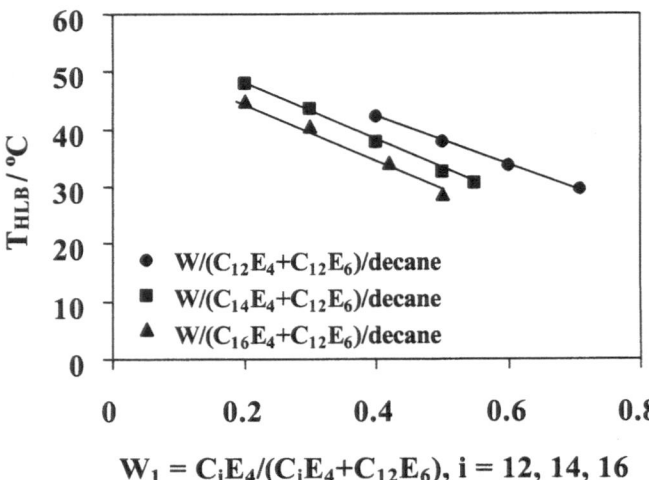

Fig. 3 Hydrophile–lipophile balance (*HLB*) temperature, T_{HLB}, as a function of surfactant mixing ratio, W_1

decreases with the increase in the lipophilicity of the surfactant mixture [6, 12]. Since in water/nonionic surfactant/oil systems the emulsions are O/W below T_{HLB} and W/O above it, the emulsions reported in this work are of the O/W type at 25 °C and at the surfactant mixing ratios considered.

Emulsification and emulsion droplet size

The emulsions were prepared according to the low-energy emulsification methods (PIT and method B) described in the Experimental section and shown schematically in Fig. 1. The droplet size of the emulsions obtained by the PIT method as a function of W_1 is shown in Fig. 4. Independently of the system and the surfactant

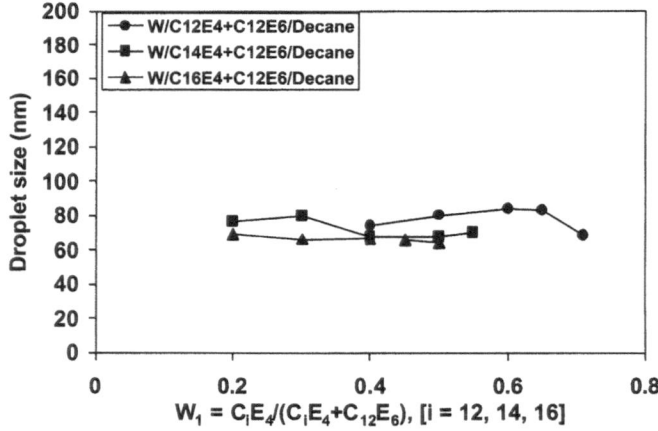

Fig. 4 Droplet size as a function of W_1 obtained by the PIT emulsification method in water/(C_iE_4 + $C_{12}E_6$)/decane systems

mixing ratio, the nano-emulsions obtained by this method have similar droplet size (between 60 and 80 nm). In contrast, the droplet size of emulsions obtained by method B depends on both the system and the surfactant mixing ratio (Fig. 5). Nevertheless, the

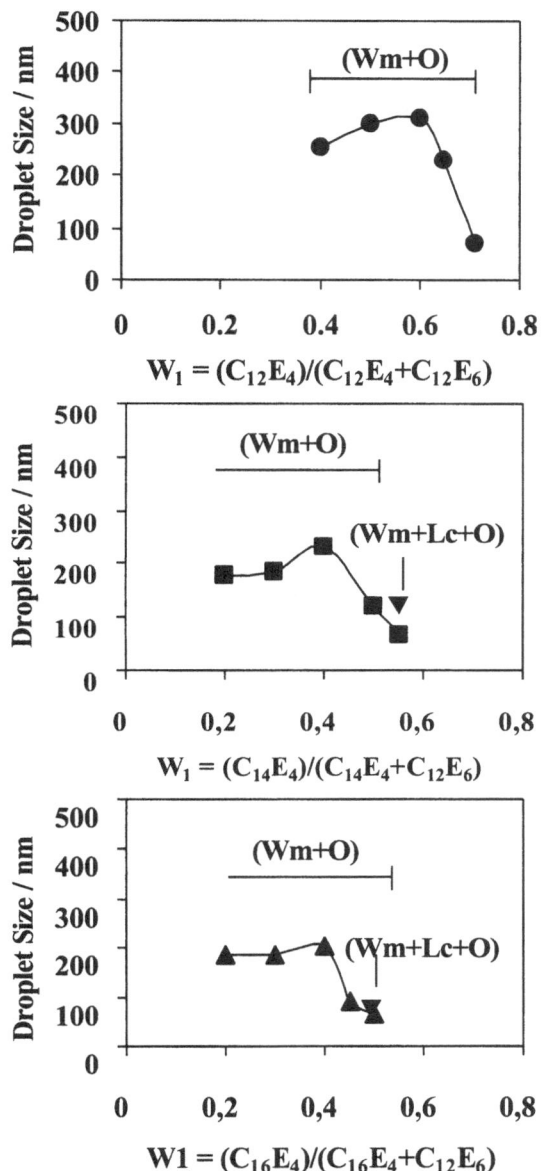

Fig. 5 Droplet size as a function of W_1 obtained by emulsification method B in water/(C_iE_4 + $C_{12}E_6$)/decane systems

tendency of the changes in droplet size versus W_1 is similar: at low W_1 (high hydrophilic surfactant content in the mixture), the droplet sizes are of the order of 150 nm; as W_1 increases, the droplet sizes also increase reaching a

maximum, after which they decrease to values around 50 nm, of the same order as those obtained by the PIT method. The equilibrium phases present in the system are also indicated in Fig. 5. Emulsions belonging to the system with the most lipophilic surfactant, $C_{12}E_4$, consist of an O/W microemulsion phase (W_m) in equilibrium with an excess oil phase (O) at all W_1 studied. However, in the other two systems, water/($C_{14}E_4$ + $C_{12}E_6$)/decane and water/($C_{16}E_4$ + $C_{12}E_6$)/decane the phase equilibrium at W_1 values where the smallest droplet sizes are obtained, consist of a lamellar liquid-crystalline phase (L_α) in equilibrium with W_m and O phases. It could be thought that the presence of the L_α phase was the cause for the smallest values of the droplet size in nano-emulsions obtained by method B. However, in the system with $C_{12}E_4$, although the droplet sizes are as small as in the other systems, the L_α phase is not present. The explanation of these results could lie in the fact that the T_{HLB} values of the compositions giving the smallest droplet sizes are very close to 25 °C, the temperature at which the emulsions are prepared.

Although the nano-emulsions obtained at compositions near T_{HLB} ($W_1 = 0.70$ in $C_{12}E_4$ + $C_{12}E_6$, $W_1 = 0.55$ in ($C_{14}E_4$ + $C_{12}E_6$) and $W_1 = 0.50$ in ($C_{16}E_4$ + $C_{12}E_6$) mixtures) showed similar droplet size, the stability of those belonging to the ($C_{12}E_4$ + $C_{12}E_6$) system was considerably lower than nano-emulsions of the other two systems; therefore, the higher stability near the HLB temperature could be due to the presence of a liquid-crystalline phase. It has been reported that emulsion stability is very low at temperatures close to the HLB temperature owing to the low interfacial tensions achieved, which enhance thermal fluctuations on the monolayers [7–11]; however, it should be noted that it is true when only liquid phases are involved in the phase equilibrium. When a liquid-crystalline phase is present, the stability is considerably enhanced at the HLB temperature. The evolution of the droplet size as a function of time allowed us to estimate the nano-emulsion stability. The radius, r, to the third power is plotted as a function of time for nano-emulsions obtained by the PIT method in water/($C_{14}E_4$ + $C_{12}E_6$)/decane and in water/($C_{16}E_4$ + $C_{12}E_6$)/decane systems at different W_1 values in Figs. 6 and 7. The linear variation of r^3 as a function of time indicates that the mechanism of instability can be attributed to Ostwald ripening. The Ostwald ripening rate obtained from the slope of the straight lines of Figs. 6 and 7 is shown in Fig. 8 as a function of W_1. The phase equilibrium is also indicated in Fig. 8. The more stable nano-emulsions are those where a liquid-crystalline phase is present. In the system containing ($C_{12}E_4$ + $C_{12}E_6$) mixtures, the Ostwald ripening rate was found to be about 2 orders of magnitude higher than that of the other two systems. Creaming followed by coalescence was also observed in that system.

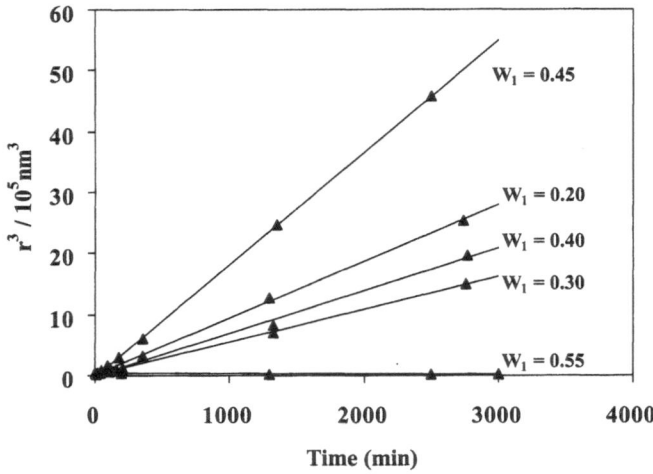

Fig. 6 Variation of r^3 with time in emulsions of the water/$(C_{14}E_4 + C_{12}E_6)$/decane. system prepared by the PIT method

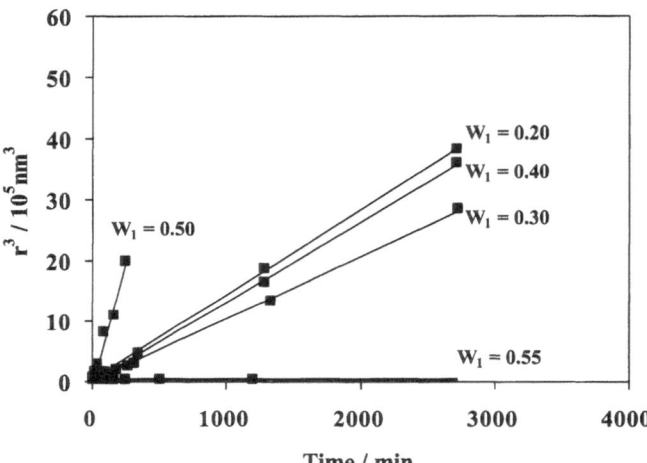

Fig. 7 Variation of r^3 with time in emulsions of the water/$(C_{16}E_4 + C_{12}E_6)$/decane system prepared by the PIT method

Conclusions

Nano-emulsions were obtained in a mixed surfactant system by two low-energy emulsification methods, PIT

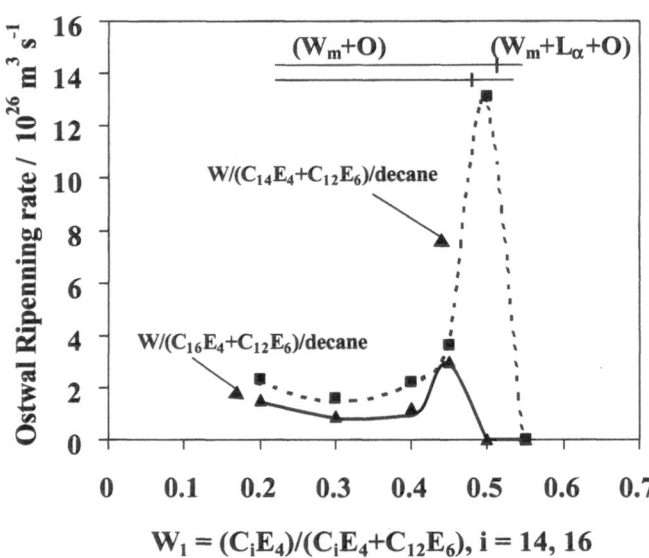

$$W_1 = (C_iE_4)/(C_iE_4 + C_{12}E_6), i = 14, 16$$

Fig. 8 Ostwald ripening rate as a function of W_1 in emulsions of the water/$(C_iE_4 + C_{12}E_6)$/decane system, $i = 12, 14, 16$, obtained by the PIT method

and addition of water to a surfactant and oil solution (method B). The droplet size of the nano-emulsions obtained by the PIT method (in the range 60–80 nm) is independent of W_1. In contrast, the droplet size in emulsification method B (in the range 60–300 nm) is highly dependent on W_1.

Near T_{HLB}, where the equilibrium phases are $(W_m + L_\alpha + O)$, nano-emulsions obtained by the two methods have similar droplet size. The most stable nano-emulsions were obtained in the vicinity of T_{HLB} where the equilibrium phases are $(W_m + L_\alpha + O)$. The main mechanism of destabilization of nano-emulsions belonging to the $(W_m + O)$ region is Ostwald ripening.

Acknowledgements The authors acknowledge financial support by CICYT (grant QUI 99-0997-CO2) and "Comissionat per a Universitats i Recerca, Generalitat de Catalunya" (grant 1999SGR-00193). A.F. acknowledges CONICIT-ULA for a Ph.D. grant.

References

1. Becher P (1965) Emulsions: theory and practice. Reinhold, New York
2. Binks BP (1998) In: Binks BP (ed) Modern aspects of emulsion science. The Royal Society of Chemistry, Cambridge, pp 1–55
3. Shinoda K, Arai H (1964) J Phys Chem 68:3485
4. Shinoda K, Saito H (1968) J Colloid Interface Sci 26:70
5. Kunieda H, Shinoda, K (1985) J Colloid Interface Sci 17:107
6. Bourrel M, Schechter R (1988) In: Bourrel M, Schechter R (eds) Microemulsions and related systems, vol 30. Dekker, New York, pp 140–148
7. Shinoda K (1967) J Colloid Interface Sci 24:4
8. Saito H, Shinoda K (1970) J Colloid Interface Sci 32:647
9. Vinatieri JE (1980) Soc Pet Eng J 20:402
10. Kunieda H, Shinoda K (1982) Bull Chem Soc 55:1777
11. Kabalnov A, Weers J (1996) Langmuir 12:1931
12. Kunieda H (1992) In: Keizo O, Mashiko A (eds) Mixed surfactant systems. Surfactant science series, vol 46. Dekker, New York, pp 235–261

13. Binks BP, Meunier J, Abillon O, Langevin D (1989) Langmuir 5:415
14. Nakajima H (1997) In: Solans C, Kunieda H (eds) Industrial applications of microemulsions, vol 66, Dekker, New York, pp 175–197
15. Lovell PA, El-Aasser MS (1997) In: Lovell PA, El-Aasser MS (eds) Emulsion polymerization and emulsion polymers. Wiley, Chichester, pp 697–722
16. Benita S (1998) In: Benita S (ed) Submicron emulsions in drug targeting and delivery. Harwood, Amsterdam, p 338
17. Taylor P, Ottewill RH (1994) Prog Colloid Polym Sci 97:199–203
18. Katsumoto Y, Ushiki H, Mendibourne B, Graciaa A, Lachaise J (2000) J Phys Condens Matter 12:3569–3583
19. Lifshitz IM, Slezov VV (1961) J Phys Chem Solids 19:35
20. Wagner C (1961) Ber Bunsenges Phys Chem 16:581
21. Kabalnov AS, Pertzov AV, Shchukin ED (1987) Colloids Surf 24:19
22. Forgiarini A, Esquena J, Gonzalez C, Solans C (2000) Prog Colloid Polym Sci 115:36–39
23. Kunieda H, Fukui Y, Uchiyama H, Solans C (1996) Langmuir 12:2136–2140
24. Shinoda K (1967) J Colloid Interface Sci 14:4–9
25. Kunieda H (1986) J Colloid Interface Sci 114:378–385

Progr Colloid Polym Sci (2001) 118: 190–195
© Springer-Verlag 2001

N. Andritsos
M. Kostoglou
A. J. Karabelas

Incipient growth of CdS films from weakly supersaturated solutions

N. Andritsos · M. Kostoglou
A. J. Karabelas (✉)
Chemical Process Engineering Research
Institute and Department of Chemical
Engineering, Aristotle University
of Thessaloniki, P.O. Box 1517
540 06 Thessaloniki, Greece
e-mail: karabaj@cperi.certh.gr
Tel.: +30-31-996201
Fax: +30-31-996209

Abstract The initial period of CdS film growth appears to strongly influence the quality of the final film. Scanning electron microscopy pictures of film grown by the flow of a supersaturated (in CdS) solution suggest that nuclei are continuously generated on the substrate and grow as discrete "surface" particles. With time, these growing particles grow and "coalesce" with neighbouring ones to create a continuous film. Similar growth patterns are also observed in the chemical bath deposition process. A simple model of the process is developed employing a "unit cell" approach, which is capable of addressing issues such as the existence of an "induction period" in film growth and its relation to supersaturation, in addition to predicting film thickness growth.

Key words Growth pattern · Film formation · Modelling · Initial growth

Introduction

CdS is widely used for junctions, buffers or passivation layers in thin-film photovoltaic devices. Among several available methods for preparing thin CdS films, the chemical bath deposition (CBD) process is considered to be quite reliable and economically attractive for large-scale applications. This process is usually based on the decomposition of thiourea in an alkaline solution containing a cadmium salt and a complexing agent such as ammonia [1–4]. Although its implementation is relatively simple, the method has some drawbacks. Firstly, only a very small percentage of cadmium (present in the bath in ionic form) is converted to a film [5]. Secondly, a rather large concentration of ammonia is required, making the application of the method in industrial scale facilities problematic. Finally, another (perhaps less serious) limitation is the contamination of the CdS films by various impurities, which may adversely affect film quality and performance [6].

The prime objective of this work was the development of an alternative to the CBD process, which may be free of the previously mentioned drawbacks, by using a simpler chemical system. The method involves wall-crystallisation of CdS on glass or glass/SnO$_2$ substrates by the flow of a solution of controlled supersaturation with respect to CdS. While pursuing our objective, it was recognised that the initial growth pattern of the CdS film (i.e. the shape and distribution of the initial nuclei formed and their growth) essentially determines the quality of the final film. Furthermore, it appears that the initial stages of film formation in the CBD process have certain similarities. Consequently, a second objective was set to simulate the evolution and growth on the substrate surface of the initial nuclei, which lead to the formation of the final film. This simulation will further aid the efforts to adequately model the CBD process [7] and other similar film growth processes.

Experimental

CdS films were grown on commercial glass and on SnO$_2$-coated glass surfaces. Separately prepared solutions of cadmium and of sulphide ions, at fixed conditions (temperature, flow rate, concentration, ionic strength), were mixed continuously to form a solution supersaturated with respect to CdS, which was then pumped through a specially designed flow cell, as shown in the schematic diagram in Fig. 1. Two Plexiglas flow cells of rectangular

Fig. 1 Schematic diagram of the experimental setup

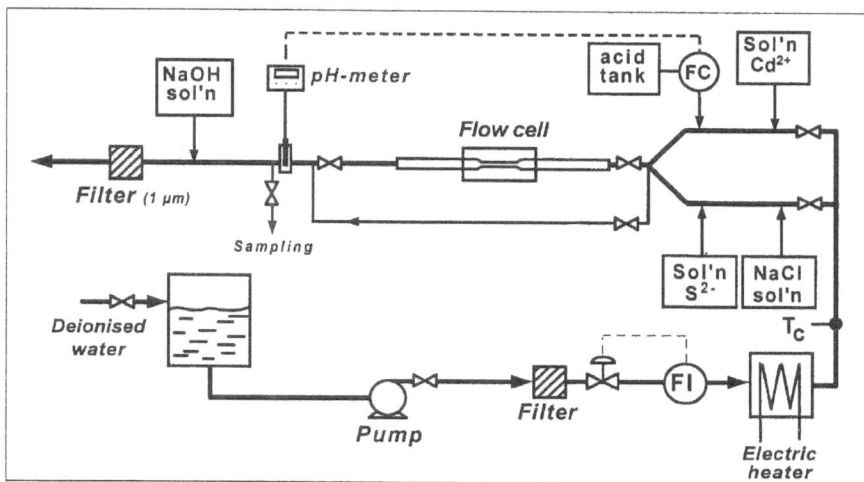

cross-section were employed; a pair of glass substrates was accommodated in each test section, forming a channel with a uniform 2-mm gap. Two sizes of substrates were used: 4.7×7.5 and 11×11 cm^2. Na$_2$S was used as the source of sulphide ions, while several cadmium salts were tested [Cd(NO$_3$)$_2$, CdCl$_2$ and Cd(CH$_3$COO)$_2$]. The sulphide solution (prepared just before each experiment) was standardised by a titrimetric method and the stock cadmium solution was standardised by inductively coupled plasma spectroscopy (PerkinElmer). Nitric acid was used for pH adjustment. In some experiments higher-ionic-strength solutions were employed by adding NaCl. In all the experiments, the supersaturated solution after passing through the flow cells was brought to a pH of about 9 by the addition of a NaOH solution, in order to precipitate CdS. The liquid stream was then passed through a system of two commercial filters (5 and 1 μm). The light absorbance of the water leaving the second filter was zero, implying that most of the particles were retained by the filter despite their small size. The Cd concentration in the effluent was always less than 0.2 mg/l.

The progress of bulk precipitation of CdS was monitored, through periodic sample withdrawal, by light absorbance measurements. The CdS-covered substrates were rinsed with distilled–deionised water after removal from the flow cell and allowed to dry under ambient conditions. Samples of CdS films were further tested as-grown and after annealing at 400 °C for about 30 min. The CdS layers and the coated glass substrates were assayed by X-ray analysis using a Siemens D500 diffractometer. The morphology of the films was examined by scanning electron microscopy (SEM) using a JEOL, JSM-840A unit. Average film thicknesses were measured using a DEKTAK 3ST profilometer. The degree of adhesion was assessed qualitatively by the "tape" test; i.e. a piece of common adhesive tape was placed on the film and then it was removed to check for attached particles. For the determination of sulphide speciation and the supersaturation ratio, S, defined as $S = ([Cd^{2+}][S^{2-}]/K_{sp})^{1/2}$, the HYDRAQL code [8] was used with $pK_{sp} = 27.8$, $pk_1 = 7.02$ and $pk_2 = 13.80$. Several CdS film samples were also grown on SnO$_2$-coated glass substrates using the currently favoured CBD process at 70 °C.

Results and discussion of CdS film development

Of the main process parameters (species concentration, pH, flow velocity, temperature) the pH appears to be the most significant, strongly affecting film development, adherence and film quality. Furthermore, in these experiments the pH is the main parameter controlling the supersaturation ratio. Using the continuous-flow process, rather uniform films can be produced on SnO$_2$/glass substrates, with thickness between 200 and 500 nm. Thicker films could also be obtained. Under the same conditions, poor quality films were produced on commercial glass surfaces. The film quality depends on the solution conditions and especially on the supersaturation ratio, S. At low S values (typically less than 5) relatively large nuclei are formed on the substrate, which tend to increase with time. Although the "grains" formed seem to adhere quite firmly on the substrate (as seen under SEM and ascertained using the tape test), the surface coverage is rather poor even for long run times. At relatively high S values ($S > 8$), the grain size is relatively small, good coverage and thinner films can be obtained, but also numerous loose particles are observed on the film and the tape can remove part of the film. Thus, the best results are obtained only within a fairly narrow range of supersaturation ratios between 5 and 8.

In view of these results, attempts were made to employ different Cd salts and increased ionic strength and to expose the substrate (for a short period – typically 1 min) to a higher supersaturation. With this technique the rapid formation of numerous CdS nuclei on the substrate is possible; these can subsequently serve as film growth sites at lower supersaturation, where bulk precipitation does not occur. Consequently, thinner CdS films were obtained, but the adherence was not as satisfactory as in runs in the range $5 < S < 8$. The various Cd salts used did not appear to have a significant effect on the "grain" size of the CdS crystals initially formed on the substrate and, consequently, on the final film thickness. An increase in the temperature tends to increase mass transfer, resulting in shorter times needed to prepare the layer and in smaller "grain" size, but the layers exhibit deterioration of adhesion characteristics. Increased ionic strength seems to improve adhesion, but at elevated temperature does not improve the layer to a sufficient degree. Finally, increasing the flow velocity (e.g. from 0.6 to 1.5 m/s) tends to improve the adhesion characteristics, but again the "grain" size remains unaffected. The addition of NaCl appears to affect the morphology of the grains, although no effect on the X-ray diffraction pattern is detected. Annealing the specimen at 400 °C for about 30 min always improves adhesion, resulting also in a somewhat darker film.

The type of glass substrate has a significant influence on the adhesion characteristics. The SnO_2/glass substrates (used as received) favour coherent and well-adhering layers, while the deposition on commercial glass surfaces (used after cleaning) leads to loosely adhering films and to film destruction even with simple water rinsing. Additionally, cracks are observed on films grown on commercial glass, which further widen with annealing.

By comparing diffractograms of an as-grown film on a SnO_2/glass substrate and of the substrate itself, peaks characteristic of the hexagonal phase α-CdS (JCPDS card no. 41-1049) are observed. This phase is usually identified in CdS films produced via the CBD process [5]. Relatively low optical transmittance was measured in both the visible and the IR parts of the spectrum,

because the CdS layers are relatively thick, exceeding 200 nm.

Scanning electron micrographs were used to examine and elucidate the incipient growth of CdS crystals on the glass substrate. The formation of a CdS layer on a SnO_2/glass substrate at various times is depicted in Fig. 2 for a supersaturation ratio of $S = 6$. Almost from the very beginning spherelike CdS grains are observed on the substrate, having a diameter smaller than about 100 nm. By tilting the scanning electron microscope stage, these crystals appear to be spherical sections (caps) with a "contact angle" smaller than 90°, reminiscent of a liquid droplet on a horizontal substrate. The initial nuclei grow with time, while new ones are also formed, at least during the initial stage when the substrate is not fully covered. On the other hand, this "contact angle" is greater than 90° in the case of a commercial glass substrate. Consequently, it appears that the shape of these initial nuclei/particles influences the adhesion characteristics of the final film.

SEM micrographs of CdS films grown via the CBD method for short times are shown in Fig. 3. Although the pictures are not very clear, the growth patterns of films from the two methods appear similar, despite the smaller size of the initial "nuclei" and the higher particle density on the substrate in the CBD method. Similar SEM pictures of films grown via the CBD method are often found in the literature [9].

Fig. 2 Scanning electron microscopy micrographs of CdS films on glass/SnO_2 obtained by a once-through flow process [Na_2S/$Cd(NO_3)_2$, pH 1.72, 0.05 N NaCl]: **a** 2 min; **b** 8 min; **c**; and **d** 45 min

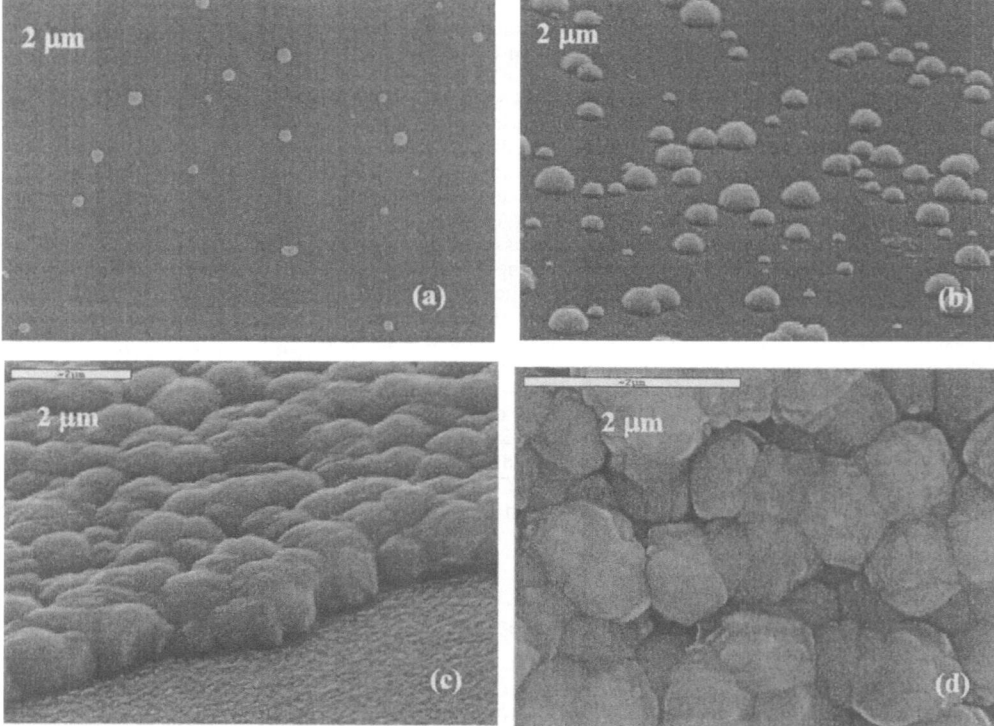

Fig. 3 CdS thin film via the chemical bath deposition process: **a** 1 min; **b** 9 min. Conditions: $[Cd^{2+}] = 0.01$ M, $[TU] = 0.014$ M, $[NH_3] = 2$ M, 70 °C

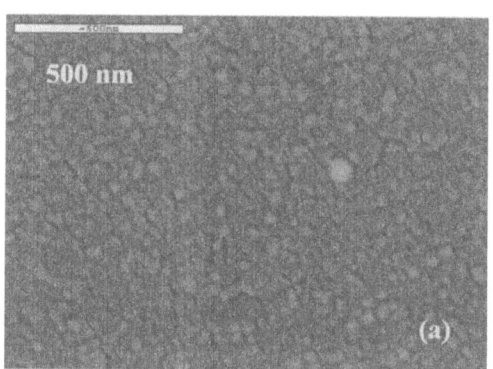
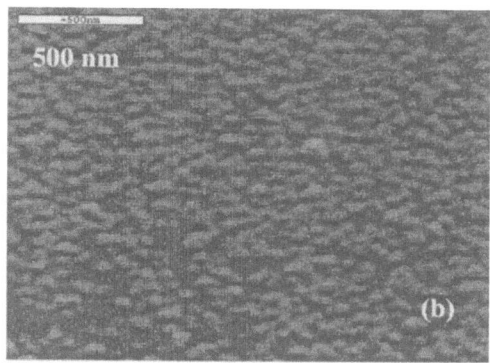

Simulation of the incipient film growth and morphology

The initial stages of film growth are very important and, as discussed previously, it appears that they strongly influence the quality of the final film. The SEM micrographs of the evolving film at several time periods reveal some very interesting features. As observed in Figs. 2 and 3, nuclei are continuously generated on the substrate and grow initially as discrete "surface" particles. This is not usually taken into account when using classical crystallisation theory (for data interpretation or modelling), which assumes a uniform film growth layer by layer [10]. With time, the "surface" particles grow and "coalesce" with each other to create a continuous film. The experiments suggest that the surface nucleation rate is a function of the supersaturation and substrate material, whereas the shape of the "surface" particles (i.e. the contact angle between the particle and the substrate) depends only on the substrate material.

A simple model of the process is developed in the present work. The fixed particles are considered to grow via a surface reaction mechanism. It is further assumed that the surface diffusion is so fast that the new mass is redistributed in order to preserve the particle spherical-cap shape with the prevailing contact angle. When the boundaries of growing neighbouring particles reach one another, the constraint on the contact angle does not apply and the particle–particle contact angle is free to evolve. The simplicity of the model is due to the spherical-cap shape of CdS, but it is expected that qualitatively similar behaviour will prevail for crystals with the same degree of isotropy (e.g. calcite rhombohedral crystals [11]). On the other hand, the initial stages of film growth may be quite different for crystals with a high degree of anisotropy, such as those of $CaSO_4$ [12]. It is interesting that the classical theories of mononuclear and polynuclear layer-by-layer growth can be obtained from the simple model by selecting appropriate values for the "contact angle", ϕ, as well as for nucleation and growth rates. In particular, the limiting case of small φ and very

large ratio of growth to nucleation rate corresponds to mononuclear growth, whereas small φ in connection with low or medium values of the ratio correspond to polynuclear growth.

In general, the surface nucleation may be time-dependent, but the study of the two extreme cases of instantaneous (site-saturated) nucleation and the constant nucleation rate is sufficient for understanding the process. The study of these two extreme cases is typical in the literature of phase transition in solids, where a problem arises similar to the present one but in three dimensions [13]. Although the experimental data (Figs. 2, 3) reveal that the constant nucleation rate mode most probably dominates, attention is focused here on the much simpler case of instantaneous nucleation. However, even this simpler case can demonstrate whether this approach is worth pursuing and is capable of explaining the experimental results, for example, the existence of an induction period.

In the case of instantaneous nucleation, it is easy to implement a "mean field" theory based on the "unit cell" concept which has been used extensively in the study of fluid mechanics of particulate systems. The total substrate area is allocated equally to the existing number of nuclei. The area "belonging" to each nucleus is assumed to be circular, with the particle evolving within, thus comprising a "unit cell". In principle, by studying the evolution of the "unit cell" one can derive the macroscopic properties of the system. The evolution of the film thickness in the unit cell and the definition of relevant parameters (contact angle, φ, and unit cell radius, α) are shown in Fig. 4. The way in which the film passes from the discrete surface–particle state to the uniform/continuous one is clear. α is related to the nuclei surface concentration, N, as $\alpha = 1/(\pi N)^{1/2}$.

By employing the previous assumptions for the growth mechanism of particles on the interface and using simple geometrical arguments, the following expressions for the evolution of the mean film thickness, h, (which is actually the experimentally obtained quantity) result.

194

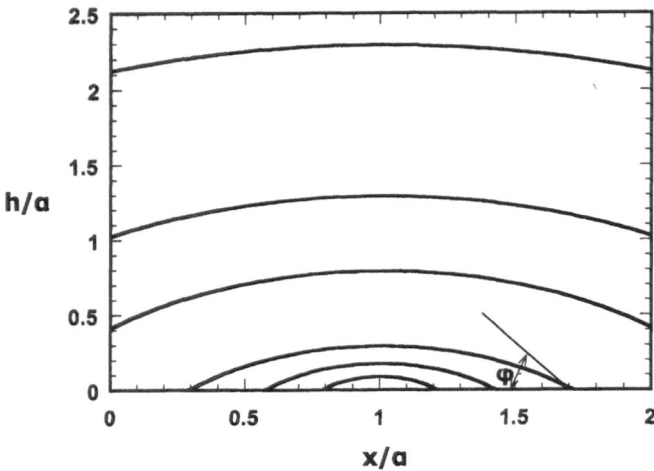

Fig. 4 Schematic of lateral and normal growth of initial nuclei. After $x = a$, growth occurs only in a direction normal to the surface

For $\varphi < \pi/2$

$$t_0 = \frac{\alpha}{k\sin\varphi}\frac{G(\varphi)}{F(\varphi)} \ ,$$

For $t \leq t_0$: $h = \frac{4}{3}\frac{(kt)^3}{\alpha^2}\frac{[G(\varphi)]^3}{[F(\varphi)]^2} \ ,$

For $t > t_0$: $h = \frac{2}{3\alpha^2}[R^3 - (R^2 - \alpha^2)^{1.5}] - \frac{\alpha}{\tan\varphi} \ ,$

with $R = \frac{\alpha}{\sin\varphi} + k(t - t_0) \ .$

For $\varphi > \pi/2$

$$t_0 = \frac{\alpha}{k}\frac{G(\varphi)}{F(\varphi)} \ ,$$

For $t \leq t_0$: $h = \frac{4}{3}\frac{(kt)^3}{\alpha^2}\frac{[G(\varphi)]^3}{[F(\varphi)]^2} \ ,$

For $t > t_0$: $h = \frac{2}{3\alpha^2}[R^3 - (R^2 - \alpha^2)^{1.5}]$
$$- \alpha\cos\varphi + \frac{\alpha}{3}\cos^3\varphi \ ,$$

with $R = \alpha + k(t - t_0)$

The functions F and G are defined as follows:

$$F(\varphi) = 1 - \frac{1}{4}(1 + \cos\varphi)^2(2 - \cos\varphi) \ ,$$

$$G(\varphi) = \frac{1}{2}(1 - \cos\varphi) \ .$$

In order to assess the reliability of the "unit cell" approach, a direct Monte Carlo simulation of the process for the case $\varphi = \pi/2$ was made. The main difference between the two approaches is that the unit cell is

assigned a specified distance between the nuclei, whereas the direct simulation corresponds to a random distribution of this distance. However, the unit cell approach may be more appropriate for cases where interactions between nuclei exist. For example, in diffusion-dominated growth, the concentration field around a growing particle prohibits the generation of new nuclei. The comparison between the two approaches is shown in Fig. 5. At a first glance, it appears that the unit cell approach is incapable of adequately simulating the film growth process; furthermore, by proper selection of the area corresponding to each nucleus (size of the unit cell) perfect matching can be achieved. It is finally concluded that the unit cell model gives the correct result by assigning a nucleus surface density 40% smaller than the real one. A more important conclusion, however, is that (apart from the need for an apparent nucleus density) the unit cell approach provides the correct qualitative behaviour. This is shown in Fig. 6, where the evolution of the film thickness for various values of the contact angle is plotted. The existence of an "induction time" in the film growth is more pronounced in the case of large contact angles (poor film–substrate adherence). For small contact angles (good adherence) the film growth pattern tends to be perfectly linear as the classical theory predicts. In Fig. 7 a comparison is made between the unit cell approach and a set of experimental data of Oladeji and Chow (curve a in Fig. 4 in Ref. [4]) in which the existence of an induction time is obvious. The growth rate, k, is estimated from the linear part of the curve and the value for N is 12 nuclei/μm^2. The qualitative agreement between theory and experiment is very good, but a greater induction time is needed for the model to perfectly match the data. This is expected since in reality the nucleation rate is nearly constant and it can be shown

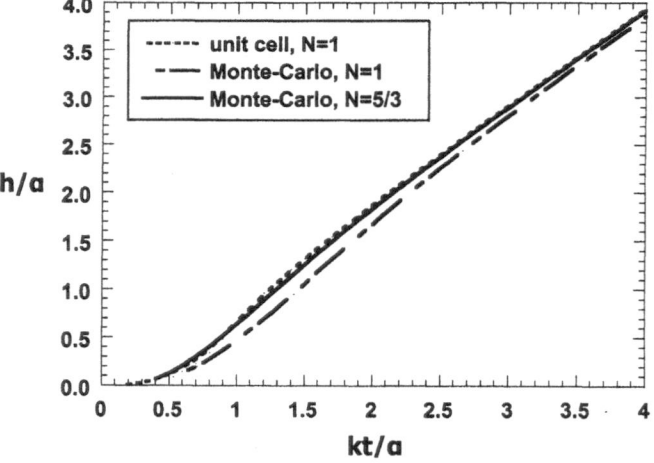

Fig. 5 Comparison between "unit cell" model and Monte Carlo simulation for $\phi = \pi/2$

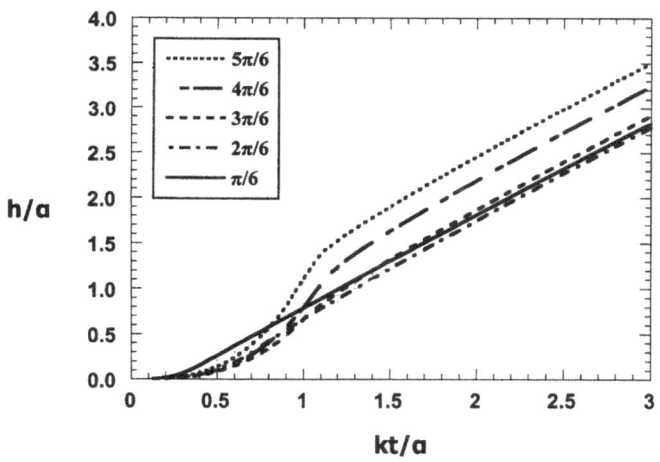

Fig. 6 Film growth evolution for various values of the contact angle

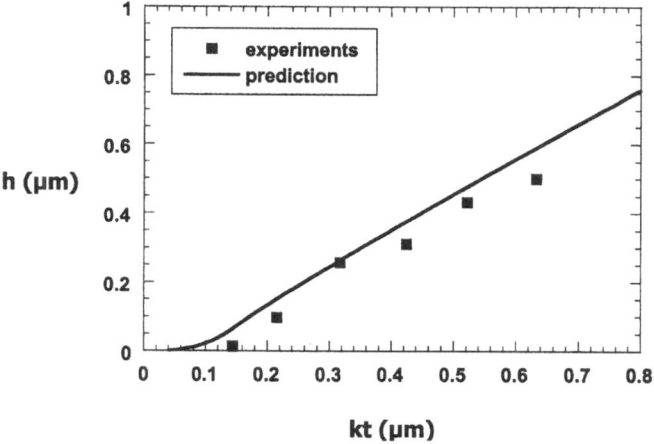

Fig. 7 Comparison between experimental results of CdS film growth via the chemical bath deposition process [4] and "unit cell" model predictions for $\varphi = 2\pi/6$

that the constant nucleation case is associated with induction times longer than the instantaneous one.

Conclusions

With a once-through flow process, it is shown that it is possible to produce good-quality CdS films having a thickness larger than 200 nm. The main parameter affecting the deposition characteristics is the supersaturation ratio. The best results (concerning coverage and adherence of the layer) are obtained for supersaturation ratios between 5 and 8. As in other film growth processes, the nuclei appear to be continuously generated on the substrate, to grow as discrete particles and to coalesce with each other. A simple model is developed, which is capable of addressing key issues, such as the existence of an induction time in film growth and its relation to supersaturation, which cannot be predicted by using classical approaches. The same approach could be extended to include more realistic models of transient nucleation and more sophisticated (than "particle in cell") spatial interaction models between the "surface" particles. The ultimate objective of this approach is to incorporate the structural surface growth submodel in a general kinetic model [7] in order to obtain information on film quality in addition to film thickness.

Acknowledgements Financial support by the European Commission under contract JOR3-CT97-0124 is gratefully acknowledged.

References

1. Ortega-Borges R, Lincot D (1993) J Electrochem Soc 140:3464
2. Doña JM, Herrero J (1997) J Electrochem Soc 144:4081
3. O'Brien P, McAleese J (1998) J Mater Chem 8:2309
4. Oladeji IO, Chow LJ (1997) Electrochem Soc 144:2342
5. Boyle DS, Bayer A, Heinrich MR, Robe O, O'Brien P (2000) Thin Solid Films 361–362:150
6. Webb JD, Keane J, Ribelin R, Gedvilas L, Swartzlander A, Ramanathan K, Albin DS, Noufi A (1998) Report NREL/CP-520-25287
7. Kostoglou M, Andritsos N, Karabelas AJ (2000) Ind Eng Chem Res 39:3272
8. Papelis C, Hayes KF, Leckie JO (1988) HYDRAQL. Technical report no 306. Stanford University
9. Martinez MA, Guillen C, Herrero J (1998) Appl Surf Sci 136:8
10. Dirksen JA, Ring TA (1991) Chem Eng Sci 46:2389
11. Andritsos N, Karabelas AJ, Koutsoukos PG (1997) Langmuir 13:2873
12. Sheikholeslami R (2000) Heat Transfer Eng 21:17
13. Lorenz B (1989) Acta Metall 10:2689

Progr Colloid Polym Sci (2001) 118:196–201
© Springer-Verlag 2001

E. Duggan
E. Waghorne

Effect of addition of chitosan on rheological properties of acidified milk gels

E. Duggan · E. Waghorne (✉)
Chemistry Department
University College Dublin
Belfield, Dublin 4 Ireland
e-mail: earle.waghorne@ucd.ie
Tel.: +353-1-7062132
Fax: +353-1-7062127

Abstract A study was made of the effect of addition of chitosan on the rheology of acidified milk gels. The milk gels were made from skimmed milk powder and were acidified at 43 °C until pH 4.5 was reached. Glucono-δ-lactone or two starter cultures, differing in the viscosity of the final product, one culture producing exopolysaccharide were used. Set- and stirred-type milk gels were studied, both with and without chitosan. Dynamic low amplitude oscillatory rheological mea-surements were made during acidification, following gelation, and after 2-days storage at 5 °C. Dynamic rheology showed interesting differences between the elastic moduli of those milk gels with chitosan added and those prepared without. The syneresis and the effective viscosity after 2 days were compared for gels with or without chitosan.

Key words Yoghurt · Chitosan · Rheology

Introduction

Fermented milk products are produced throughout the world, with yoghurt being the most popular. The popularity of these products is due in part to various nutritional benefits that have been associated with them. In yoghurt production, the culture, usually a mixture of *Lactobacillus delbruckii* subsp. *bulgaricus* and *Streptococcus salivarius* subsp. *thermophilus*, transforms lactose into lactic acid. The resultant decrease in pH lowers the net negative charge of casein micelles and dissolves the colloidal calcium phosphate associated with the micelles. This colloidal destabilisation of the micelles leads to aggregation, leading to the formation of a gel.

Two different cultures were used in this study. Both are mixtures of *L. delbruckii* subsp. *bulgaricus* and *S. salivarius* subsp. *thermophilus*. One culture is the more traditional type while the second culture produces an exopolysaccharide (EPS). This culture is described as a "ropy" culture. In general, the use of a ropy culture enhances the viscosity of yoghurt, influences gel strength and can prevent gel fracture and wheying off.

In commercial yoghurt production two types are found; set, where the yoghurt is fermented in the retail cups and cooled without disturbance of the gel, and stirred, where the milk is fermented in vats and then stirred, thus breaking the gel, and packaged.

The lack of reproducibility introduced by the use of bacterial fermentations has led many researchers to use alternative methods for acidification to study the nature of acid milk gels. A method frequently used is the hydrolysis of glucono-δ-lactone (GDL) to gluconic acid [1, 2]. The rate of acidification is different between milk acidified with GDL and bacterial cultures. GDL is rapidly hydrolysed to gluconic acid, whereas with cultures the pH does not change very much initially, but decreases quickly later in the fermentation. The final pH attained in GDL-induced gels is a function of the initial GDL concentration, whereas bacteria can continue to produce acid until a very low pH (e.g. below 4.0) is attained. However, in practice, bacterial activity can be stopped by rapid cooling when the required pH is reached.

In recent years much work has been done to improve the texture of fermented milk products. Whey-separation, or the appearance of liquid (whey) on the surface of a milk gel, is a common and unwanted sight in fermented milk products. Syneresis is defined as shrinkage of a gel causing an expulsion of liquid. The addition of milk solids has been shown to improve the texture of yoghurts [3, 4]. Besides this, work has focused on the use of thickeners and stabilisers to increase viscosity and reduce syneresis. κ-Carrageenan and pectins have been shown to improve the texture in yoghurt [5, 6]. More recently work has been done on the addition of gelatin [7], where it was found that the yoghurts with added gelatin exhibited more solidlike behaviour than those without.

Chitosan, a polysaccharide derived from chitin found in the exoskeletons of crustaceans and fungi, is a heteropolymer of $\beta(1-4)$-linked glucosamine and N-acetyl-D-glucosamine units. This polymer has been shown to exhibit interesting association and gelling properties in aqueous solution [8, 9]. Recently attention has been focused on chitosan as a dietary fibre. Also it exhibits anticholesterolemic properties and a capacity to bind to fatty acids and bile acids [10]. In this study chitosan was modified to be water-soluble. Chitosan needs conditions of pH < 6.3 to be soluble (milk typically above 6.3). To this end, the chitosan was hydrobrominated to be soluble at the milk pH.

This study focused on the rheological properties of acidified milk gels in the presence of different concentrations of chitosan hydrobromide. The milk gels were acidified chemically using GDL and in the more traditional bacterial method. Two cultures were used: a traditional type (YC 460) and a high-molecular-weight EPS-producing culture (YC 191). Gelation of both milk gel types was followed by pH and rheological measurement. At the end pH (4.5) the yoghurts were stirred and stored for 2 days, whereafter they were subjected to rheological examination, and the syneresis and the effective viscosity were measured. The GDL milk gels were monitored only during acidification (not studied after 2 days).

Materials and methods

Preparation of chitosan hydrobromide

Chitosan hydrobromide was prepared by the method of Domszy and Roberts [11]. Chitosan (Aldrich, Milwaukee, Wis., USA) (low molecular weight, 75–85% deacetylation) was dissolved in 0.2 M HBr and 9 M HBr was added with vigorous stirring to precipitate the hydrobromide salt. The resultant slurry was centrifuged (5000 rpm) for 30 min and the supernatant decanted. The chitosan hydrobromide was filtered off, washed with methanol until the filtrate was neutral to litmus, then washed with several portions of ether and air-dried. When dry it was slurried in methanol for 10 min and filtered. This was repeated three times, then the product was washed with ether and dried in a vacuum desiccator.

Preparation of GDL-induced milk gels

The milk was reconstituted from skimmed milk powder (Glanbia, Waterford, Ireland) at a 14% (w/w) concentration of solids in distilled water. For all the tests the same batch of skimmed milk powder was used. The reconstituted milk was heat-treated for 20 min at 85 °C, then cooled.

Chitosan hydrobromide was allowed to dissolve in the milk overnight at concentrations of 1 and 3 wt%. The milk was heated to 43 °C. GDL (2.1 wt%) was added to the samples while stirring. The samples were stirred for 5 min, then the pH was monitored throughout acidification, while maintaining the temperature at 43 °C. When the pH reached 4.5 the samples were cooled to 5 °C.

Preparation of yoghurt

The procedure for yoghurt had to be modified slightly as it was prepared on a larger scale. As it was difficult to see if the chitosan was dissolved fully, it was predissolved and the solution was added to the milk.

Chitosan hydrobromide was dissolved in distilled water and heated to 85 °C for 20 min, then allowed to cool to 43 °C. The milk was reconstituted from skimmed milk powder in distilled water to give a final concentration of 14% solids (after addition of chitosan solution). The milk was heated to 85 °C for 20 min in 4 l culture vessels and 100 ml bottles, then allowed to cool to 43 °C. The chitosan solution was added to the milk at concentrations of 0.05, 0.1, 0.2% (w/v) and stirred for 10 min. The culture vessels (at 43 °C) were inoculated with the cultures consisting of *S. thermophilus* and *L. bulgarius* (YC 460, YC 191, Chr. Hansen, Horsholm, Denmark) at a concentration of 0.02%. When the yoghurt reached a pH of 4.5, the culture vessels were removed from the waterbath and the yoghurt was stirred for 2 min at 500 rpm and for 3 min at 450 rpm, then cooled to 10 °C, put into pots and stored at 5 °C.

Rheology

Rheological measurement of gelation

Rheological measurements were carried out using a Rheometric Scientific rheometer (SR 2000). The couette was pre-set to the fermentation temperature (43 °C). Inoculated sample (20 ml) was poured into the couette. The yoghurt samples were taken 30 min after bacterial inoculation, whereas the GDL-inoculated samples were taken after 5 min stirring after addition of GDL. The gelation was monitored using low-amplitude oscillation until pH 4.5 was reached. At this stage the gels were of the set type.

A test sequence was run to examine the gel once pH 4.5 was reached. This comprised

1. Dynamic temperature ramp: 0.16 Hz, 0.5% strain, temperature rate of 0.05 °C/min from 43 to 5 °C.
2. "Mini" strain sweep: 0.16 Hz, 0.01–0.5 strain.
3. Frequency sweep: 0.5% strain, 0.005–10 Hz.
4. "Full" strain sweep: 0.16 Hz, 0.01–20% strain.

The tests were at low-oscillation amplitude, thus not harming the gel, apart from the last strain sweep, which measured the yield strain of the gel.

Rheological measurement after 2 days (yoghurt)

The couette was preset to 5 °C. Yoghurt (20 ml) was poured into the couette. A test sequence was run to examine the stirred-type gel after 2-days storage at 5 °C:

1. Time sweep: 0.16 Hz, 0.5% strain, 60 min.
2. "Mini" strain sweep: 0.16 Hz, 0.01–0.5% strain.
3. Frequency sweep: 0.5% strain, 0.005–10 Hz.
4. "Full" strain sweep: 0.16 Hz, 0.01–20% strain.

The test sequence was repeated after the yoghurt had been kept at 20 °C for 5 h to examine the dependence of the gel strength on temperature.

Viscosity by the funnel flow test

The method used is adapted from the Posthumous method. Three pots of yoghurt were mixed together, stirred gently and kept at 20 °C for 5 h. The yoghurt was poured into the calibrated funnel, with the outlet closed. The outlet was released and the time for the yoghurt to pass between two markers was measured. The test was repeated.

Syneresis by centrifugation

Four pots of yoghurt were mixed together and stirred gently. The yoghurt was poured into four preweighed centrifuge tubes. The tubes were centrifuged at 1400g for 20 min at 6 °C. The supernatant (separated whey) was suctioned off and the tubes reweighed. The weight to weight percentage syneresis was then calculated.

Results and discussion

The acidification of the milk gels was monitored by the pH. The chosen end point of the acidification process was pH 4.5. At this stage both milk gel types, i.e. chemically and bacterially acidified gels, were immersed in cold water (5 °C) to stop the acidification. The yoghurt, however, was stirred before cooling to break the gel structure and then stored for 2 days to allow any recovery of the gel structure. Gel formation was monitored at 43 °C by dynamic rheology during acidification, giving G', G''–time curves. Once pH 4.5 was reached in the bulk gel, the gel (still in Couette unstirred) was subjected to a set of rheological tests to test the strength of the gel, which was the set type at this stage. Initially the gel was cooled to 5 °C, then strain and frequency sweeps were carried out. The final test, called a "full" strain sweep, tested the yield strain to break the gel.

The stirred yoghurts were stored at 5 °C. These were then characterised as to syneresis, viscosity and rheological properties.

GDL-induced gels

A GDL concentration of 2.1 wt% was chosen to imitate yoghurt formation, taking 4–5 h to reach pH 4.5. Addition of chitosan did not affect the final pH reached nor the time taken for it to be reached. The gel formation was followed by measuring the dynamic moduli (G' and G'') as a function of time (Fig. 1).

Gelling began immediately both with and without added chitosan. Adding 0.1% chitosan hydrobromide made little difference to the gel strength as indicated by the elastic modulus, G'. However 0.3% chitosan hydrobro-
mide lowered the gel strength considerably. Cooling the gels from the fermentation temperature to 5 °C resulted in an increase in G' in all the gels, although the increase was most pronounced with the chitosan-added gels. This can be seen from the frequency sweeps after cooling (Fig. 2), where G' of the 0.1% chitosan product exceeded that of the chitosan-free product by an order of magnitude.

Yoghurt gel formation

Two yoghurt cultures were examined, one a traditional commercial type (YC 460) and the second an EPS-producing culture (YC 191). Typically YC 191 takes longer to reach the chosen endpoint of pH 4.5 than YC 460 (above 7 h versus about 4.5 h). For YC 460 the pH drops very gradually in the beginning stage. After a lag phase, the rate of the pH change increases. There is a slight plateau where the pH levels off, then drops steadily again, slowing slightly towards the end at pH 4.5. Addition of chitosan hydrobromide to yoghurt YC 460 marginally lowered the initial pH of the milk (6.3 to 6.2). It also caused a lengthening of the fermentation time, with the pH reducing at a much slower rate than for the standard yoghurt (Fig. 3). Overall, the fermentation time rises with increasing chitosan concentration, up to a threefold time increase with 0.2% chitosan hydrobromide. The same trend was seen with YC 191, i.e. chitosan did not change the pH profile, but lengthened it considerably (not shown).

The gel formation was followed by measuring the dynamic moduli (G' and G'') as a function of time. Added chitosan hydrobromide had a pronounced effect on the gelation profile of the yoghurt. The formation of the gel network began much later in the chitosan-added yoghurt than in the standard yoghurt, as can be seen in Fig. 4. G' rose rapidly at the onset of gel formation. After the rapid increase G' levelled off and increased at a much slower rate until pH 4.5 was reached (Fig. 4a). Addition of chitosan caused the onset of gel formation to occur later, but overall the gelation profile was similar to the standard product. For YC 191 the same trend can be seen, with chitosan lengthening the fermentation time by a factor of 3 for the product with 0.2% added chitosan. There is a difference with the 0.05% chitosan. The rapid G' increase does not follow the other chitosan concentrations as with YC 460, but stands alone between the standard YC 191 and the yoghurts with higher chitosan concentration added (Fig. 4b).

The yoghurts were cooled in the rheometer and the dynamic moduli followed. For both YC 460 and YC 191, G' and G'' increased with decreasing temperature. The G' values were higher for 0.2% chitosan hydrobromide, with 0.05% lowest. G' followed the same trend (0.2 > 0.1 > 0.005) in frequency sweeps and strain sweeps for both YC 460 and YC 191.

After 2-days storage

After 2-days storage at 5 °C, the yoghurts were examined using similar dynamic low-amplitude-oscillatory rheological measurements as those following gelation. In addition the estimate of the viscosity was examined, whereby the time taken for a fixed volume of yoghurt to pass through a funnel is a measure of the viscosity of the

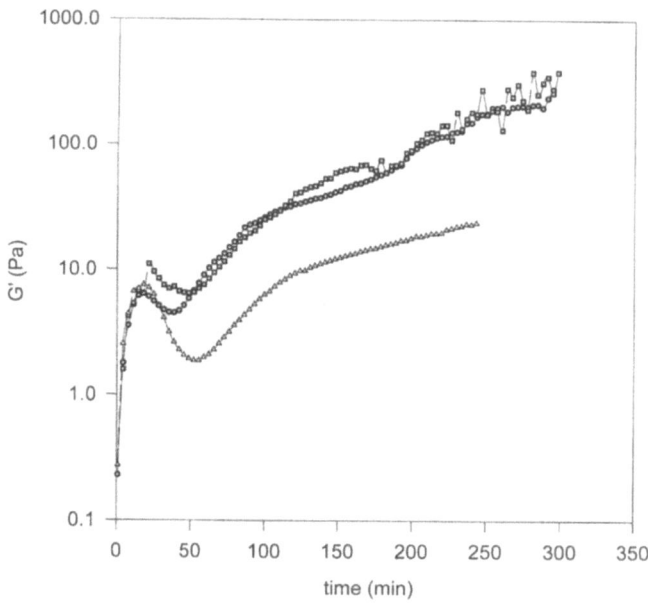

Fig. 1 Fermentation profile of 2.1 wt% glucono-δ-lactone (*GDL*) milk gel: no added chitosan (*circles*), 0.1% chitosan hydrobromide (*squares*), 0.3% chitosan hydrobromide (*triangles*). All gelation was monitored at 0.01 Hz and 0.05 Pa

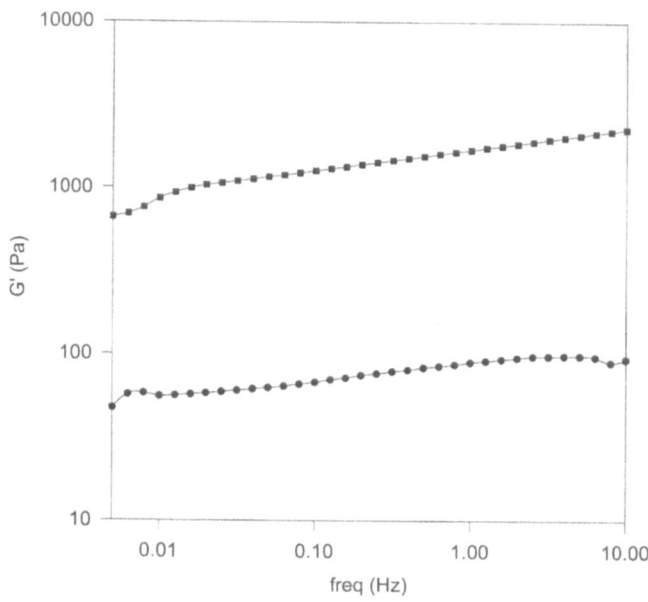

Fig. 2 Frequency sweep of 2.1 wt% GDL milk gel: no added chitosan (*circles*), 0.1% chitosan hydrobromide (*squares*). Frequency sweep from 0.005 to 10 Hz at 0.5% strain

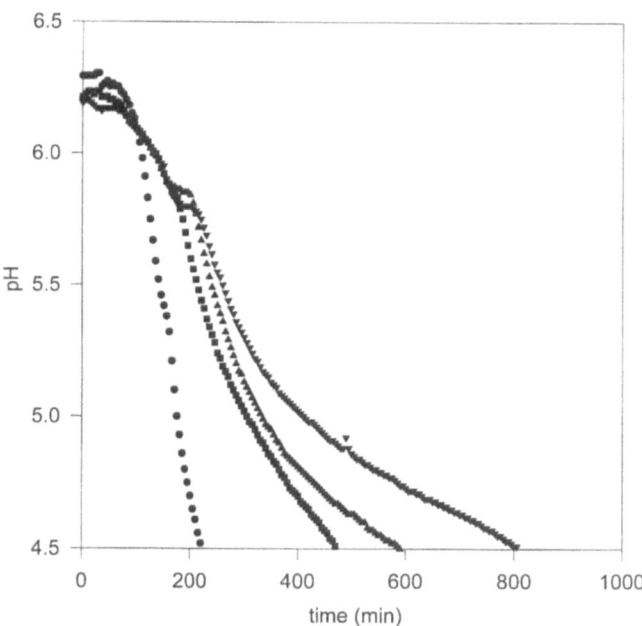

Fig. 3 pH profile of YC 460 following fermentation to pH 4.5: no added chitosan (*circles*), 0.05% chitosan hydrobromide (*squares*), 0.1% chitosan hydrobromide (*up triangles*), 0.2% chitosan hydrobromide (*down triangles*)

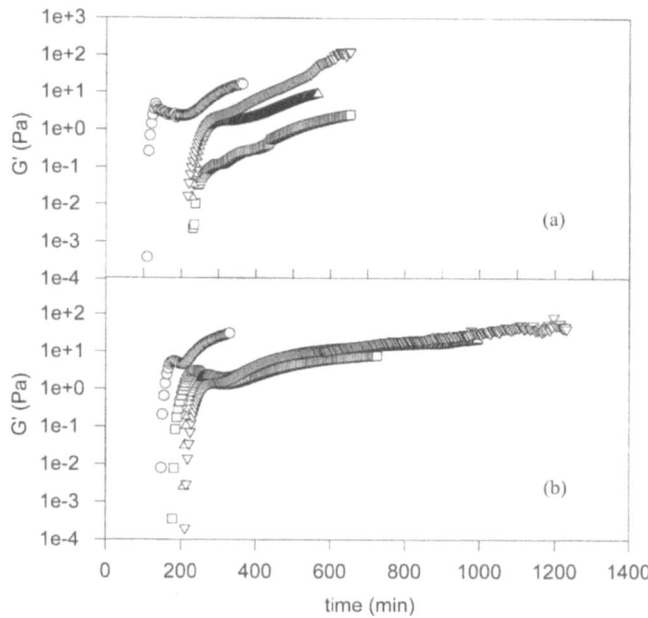

Fig. 4 Gelation profiles following G' of **a** YC 460 and **b** YC 191: no added chitosan (*circles*), 0.05% chitosan hydrobromide (*squares*), 0.1% chitosan hydrobromide (*up triangles*), 0.2% chitosan hydrobromide (*down triangles*). All gelation was monitored at 0.01 Hz and 0.05 Pa

Table 1 Funnel flowtimes and syneresis of yoghurts YC 460 and YC 191 after 2-days storage

Chitosan (%)	Funnel flowtime (s)		Syneresis (%)	
	YC 460[a]	YC 191[a]	YC 460[b]	YC 191[b]
0	42	127	39	41
0.05	48	48	44	45
0.1	34	40	45	47
0.2	19	25	50	51

[a] Standard deviation typically less than 0.3 (within one production batch). Replicates typically within 2 s for different productions of yoghurt

[b] Standard deviation typically less than 0.2 (within one production batch). Replicates typically within 2% for different productions of yoghurt

sample. Because of the high-molecular-weight EPS produced by the ropy culture, YC 191, its flowtimes are approximately 3 times that of YC 460 (130 versus 40 s). Adding 0.05 wt% chitosan hydrobromide to YC 460 caused an increase in the flowtime, but each increase in the chitosan hydrobromide concentration lowered the flowtime thereafter (Table 1). A more striking difference was seen with YC 191 by the addition of chitosan hydrobromide. The flowtimes were lowered threefold at all chitosan concentrations, reaching those of YC 460. This implies that the ropy nature of the yoghurt was counteracted by the addition of chitosan.

The rheological measurements were done in conjunction with the viscosity measurements, i.e. the yoghurt was examined directly out of storage at 5 °C and again after 5 h at 20 °C. After 5 h the yoghurt was examined in the couette at 5 °C, so the time sweep included the cooling of the yoghurt from 20 to 5 °C. In general, the frequency dependence of the dynamic moduli of the different yoghurt gels was similar. The moduli increased with increasing frequency between 0.005 and 10 Hz. G' was substantially higher than G'' in all cases. Comparing the yoghurts, the elastic modulus is highest for the standard yoghurt, with G' decreasing with each addition of chitosan hydrobromide. This is seen in both YC 460 and YC 191.

G' of YC 460 of all chitosan concentrations is shown as a function of strain in Fig. 5. The four yoghurts showed linear behaviour in the range between 0.01 and 1.5%. After 1.5% strain, a slight decrease in G' was seen, while G'' remained fairly constant, indicating partial breakdown of the elastic structure. A rapid decrease at 3% (standard YC 460) or 4% (yoghurts with chitosan) was then seen in G'. G' and G'' crossed over, indicating the full elastic structure of the material had broken down. The yoghurt now becomes a predominantly viscous rather than an elastic material. This behaviour was also seen with YC 191.

The susceptibility to syneresis for these yoghurts was established by high-speed centrifugation, which gives an

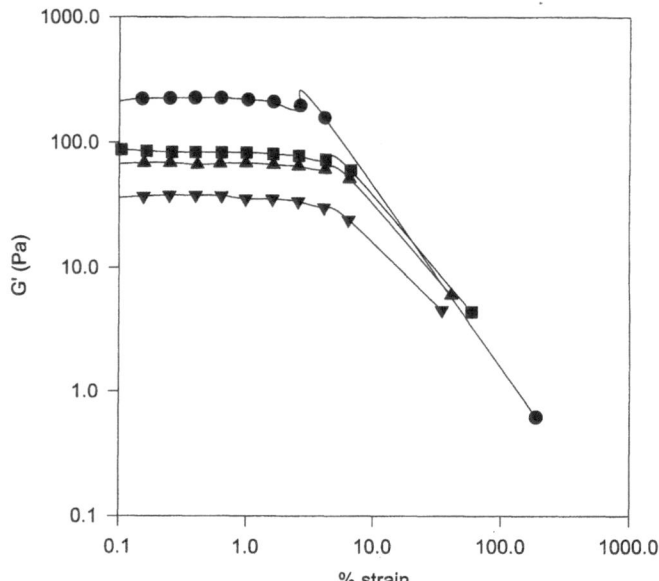

Fig. 5 Strain sweeps of YC 460 after 2-days storage at 5 °C: no added chitosan (*circles*), 0.05% chitosan hydrobromide (*squares*), 0.1% chitosan hydrobromide (*up triangles*), 0.2% chitosan hydrobromide (*down triangles*). Tests carried out at 5 °C at 0.16 Hz, from 0.01 to 20% strain

indication of the spontaneous syneresis which would be seen over long time periods. This syneresis was expressed as the percentage of the weight of removed supernatant (whey) relative to the original weight of the yoghurt. Typically, a nonropy culture is less susceptible to syneresis, i.e. YC 191 > YC 460. Addition of chitosan hydrobromide caused the syneresis to increase in both cultures, with YC 191 staying consistently higher than YC 460 (Table 1).

Discussion

Addition of chitosan to acidified milk gels has different effects, depending on the method of acidification. Chemically acidified milk gels, i.e. induced by GDL, behave as expected when adding a gelling polysaccharide. Increasing the chitosan concentration increases the gel strength as seen with G'; however acidifying with bacteria proves to be very different. The fermentation time lengthens considerably, as seen in the pH and rheological profiles. The frequency dependence of GDL and bacteria-induced gels also differs. Adding chitosan increases G' for GDL-induced gels, whereas G' decreases with increasing chitosan concentration. It is believed the reason for the differences is due to the bacteria themselves. Bacteria cell walls are generally negatively charged. Chitosan, being a cationic polysaccharide, could interact with the bacteria. The chemically acidified gels have no such negative charge for the chitosan to bind to, thus no lengthening of the fermentation time, and the chitosan can just act as a gelling agent.

Future work will include examining these yoghurts using electron microscopy. The effect of adding chitosan on isolated bacterial colonies grown on agar is currently being investigated.

Acknowledgements This work was supported by FAIR grant no. CT97 3098. E.D. acknowledges financial assistance from Chivers through the Irish-American Partnership.

References

1. Harwalker VR, Kalab M, Emmons DB (1977) Milchwissenschaft 32:400
2. Lucey JA, van Vliet T, Grolle K, Guerts T, Walstra P (1997) Int Dairy J 7:381
3. Harwalker VR, Kalab M (1986) Food Microstruct 5:287
4. Mistry VV, Hassan HN (1992) J Dairy Sci 75:947
5. Xu S, Stanley DW, Goff HD, Davison VJ, Le Maguer M (1992) J Food Sci 57:96
6. Ramaswamy HS, Basak S (1992) J Food Sci 57:357
7. Fiszmann SM, Lluch MA, Salvador A (1999) Int Dairy J 9:895
8. Nystrom B, Kjoniksen A, Iversen C (1999) Adv Colloid Interface Sci 79:81
9. Mucha M (1997) Macromol Chem Phys 198:471
10. Muzzarelli RAA (1996) Carbohydr Polym 29:309
11. Domszy JG, Roberts GAF (1985) Makromol Chem 186:1671

Progr Colloid Polym Sci (2001) 118: 202–207
© Springer-Verlag 2001

RHEOLOGY

R. Biehl
T. Palberg

Real-space imaging and motion analysis in sheared colloidal crystals

R. Biehl · T. Palberg (✉)
Institut für Physik der Universität Mainz
Staudinger Weg 7, 55099, Mainz, Germany
e-mail: thomas.palberg@uni-mainz.de

Abstract Microscopic imaging of single particles is a powerful tool to investigate the local structure of colloidal suspensions. For single-particle identification with high-resolution microscopy the resolution power is limited by refraction to roughly the wavelength of light. In this case the depth of sharpness is on a scale of less than this limit. For this reason the simultaneous observation of particles in two or more layers of a colloidal crystal seems to be impossible. We report a method with which we can image more than one particle layer in dilute colloidal suspensions with preserved resolution. The analysis of the images obtained, in particular for the investigation of crystal layer motion in sheared colloidal crystals, is discussed.

Key words Colloidal suspensions Particle identification Microscopy · Depth of sharpness · Shear mechanism

Introduction

Colloidal suspensions show a wide range of different structures, like face-centred-cubic (fcc) and body-centred-cubic (bcc) crystal structures or a fluid structure in equilibrium [1] or a triangular layer structure (also named a hexagonal layer) for sheared suspensions [2, 3]. These structures are studied by static light scattering, neutron scattering [3] or small-angle X-ray scattering [4] as well as by Bragg microscopy [5], polarisation microscopy [6] or high-resolution microscopy [7, 8] with single-particle resolution. For the scattering techniques the observation takes place in the corresponding reciprocal lattice and gives a description of the structure in a volume of about 1 mm^3, depending on the scattering technique and ray optics. In this way scattering techniques yield an ensemble average over the observed volume. On the other hand, microscopic images contain real-space information, like crystal morphology for low-resolution microscopy or individual particle positions for high-resolution microscopy. The disadvantage of microscopy is the more or less 2D character of the resulting images caused by the small depth of sharpness.

We describe here a technique to overcome the problem of the 2D images by combining inverse ultra-microscopy with a specially adjusted cover glass correction. This yields images, which include information collected from several micrometres in the z-direction. To avoid problems of multiple scattering and back-ground scattering we used an optical plate–plate shear cell of variable but small gap width $d = 10$–1800 μm. For the same reason we decided to work with dilute suspensions with typical particle densities of $n \leq 1$ μm^{-1} only. The resulting images of different structures occurring in sheared suspensions are discussed and the image analysis procedure to interpret the overall structure is described in detail to demonstrate the capabilities of this technique. We begin, however with a short description of the particles used and the preparation technique.

Experimental

Preparation

The colloidal suspension used consists of polystyrene spheres of nominal diameter $2a = 301$ nm (IDC, Portland, USA, batch

10-66-58) in water. We carefully characterised these and obtained a static light scattering radius of $a_{SLS} = (155.5 \pm 1.2)$ nm [9]. The titratable number of surface charges was $N = 2.3 \times 10^4$ and the effective charge from conductivity was $Z^* = 1980$ [10, 11]. All sample conditioning follows procedures recently described in more detail [2]. The actual measuring cell was integrated into a closed tubing system. During conditioning, the suspension was pumped peristaltically through a set of devices allowing precise adjustment and in situ control of the interaction parameters, particle number density, n, and concentration of added salt, c. We further note that the concentration of residual impurities, in general, was negligibly small, $c_B < 10^{-7}$ mol l^{-1}, and that the contamination with airborne carbonate can be kept below 5×10^{-7} mol l^{-1} h^{-1} [10]. At typical conditions of particle concentration $n \approx 1$ μm^{-3} and salt concentration $c = 5 \times 10^{-6}$ mol l^{-1} residual uncertainties are of the order of 1 and 5%, respectively. We added H_2CO_3 by contaminating the suspension with airborne CO_2 [12].

The suspension was observed within an optical plate–plate shear cell of variable gap width (10–1800 μm) mounted on an inverted microscope (Leica DM IRB, Leitz, Wetzlar, Germany). The lower plate was a 1-mm-thick quartz plate with a diameter of 60 mm. The upper plate was an 18-mm-thick quartz plate with a special design able to be assembled in a rotating bearing and to be sealed by a Viton O-ring connected to the aforementioned closed tubing system. Both have a planarity of less than $\lambda/4$ in the observation volume with a 32-mm diameter. In this way the suspension can be sheared with different shear rates dependent on the rotation velocity and the gap width. A complete description is given elsewhere [13].

The aperture of the dry, long distance 63× objective (PL Fluotar L 63 × /0.7 corr PH2 ∞ /0.1–1.3/C; Leitz) used here collects light under angles $\Theta \leq 44.4°$. It covers practically the whole range of plate–plate distances adjustable in the shear cell. Additionally, the objective allows an adaptation of the thickness of the cover glass used in the range from 0.1 to 1.3 mm for a standard cover glass. The resulting images were recorded digitally with a charge-coupled-device (CCD) camera (CV-M10, Jai, Copenhagen, Denmark) and stored as 8 bit grey level images in a computer for analysis.

For high-resolution imaging we used a special kind of ultramicroscopic illumination. Conventional ultramicroscopic images where taken with a dark-field illumination in the sense that the illuminating light passes the objective outside the collecting aperture of the objective and eliminating in this way scattered light of zero order [14]. On the basis of the special design of the upper plate mounted in the bearing and the high aperture of high-resolution objectives we had to modify this. Dark-field illumination is based on the fading out of nonscattered light of zero order. Illuminating the sample only from a restricted angular range and fading out this light by a central stop in the back focal plane of the objective also can eliminate zero order light. Note that parallel incoming light is collected in one point in the back focal plane; thus, the size of the central stop has to be adjusted to the aperture of the illuminating light. Usually the back focal plane is not directly accessible because of the compact type of construction of the high-quality objectives. We overcome this problem by using a central stop in front of the objective. In our case this was done by a special additional cover glass (thickness 0.17 mm, species bk7, Schott) with a silver coating of 1.3-mm diameter in the entrance pupil of the objective, which has a diameter of 4 mm. Thus, we collect light, for the depiction in air, under angles from 20.3° to 44.4° corresponding to an aperture for the central stop of 0.26 and 0.7 for the objective.

To reduce the background intensity from multiple scattering and particles beyond the depth of sharpness the distance between the plates was adjusted to a small gap of 28 μm.

Microscopic depth of sharpness

For standard microscopy, the resolution is limited by refraction to $l = 0.61\lambda/A$ and the depth of sharpness is limited to $b = n\lambda/2A^2$. The aperture $A = n \sin \Theta$ (n is the refractive index, Θ is the exposed angle) is typical in the range of 1 for high-resolution imaging. For this aperture and a wavelength in the visible region around 0.55 μm, as an example, the resolution limit is about 0.335 μm and the depth of sharpness is 0.275 μm, both in air. If we want to study colloidal suspensions we have to choose a particle concentration with a mean particle distance significantly above the resolution limit, for example 1 μm^{-3}. Because of the depth of sharpness, typically smaller than the resolution, microscopic imaging is like a 2D picture of a small range in the z-direction. To examine 3D structures, like the triangular sliding layers in sheared colloidal suspensions or a crystal lattice, we have to expand the depth of sharpness to roughly 2 or 3 times the mean particle distance. Note that the main information gathered for our particles is the particle position.

Our way to increase the depth of sharpness works as follows. The adjustable cover glass correction of our objective is calculated for a cover glass made of glass k5 (Schott, Germany), which is similar to the standard glass bk7. If we use a different glass, like quartz glass, we have to pay attention to the change in the refractive index, $n(\lambda)$. The offset, δ, for a beam, with angle α to the vertical on the surface of the plate (in the following all angles are measured from the normal to the surfaces), going through a plate of thickness, d, (Fig. 1) is

$$\delta = \frac{d \tan \alpha - d \tan\left[\arcsin\left(\frac{\sin(\alpha)}{n(\lambda)}\right)\right]}{\tan\left[\arcsin\left(\frac{\sin(\alpha)}{n(\lambda)}\right)\right]} . \tag{1}$$

If we assume that for the regular cover glass k5 the objective collects all rays with a starting position at a point in the focal plane and an angle $\gamma < \gamma_{aperture}$ in one image point, in front of the objective these rays have an angle of $\alpha = \arcsin(n \sin \gamma)$. In other words, Eq. (1) gives the angle and wavelength-dependent offset, which is corrected by the objective. If the adjusted cover glass correction and the cover glass used do not match, the difference between both gives the remaining offset, which is also angle and wavelength-dependent. In principle, this is zero for the right cover glass and the correctly adjusted thickness. The resulting offset is shown for a cover glass of 1-mm thickness made of quartz glass and different settings of the cover glass thickness in Fig. 2. For each α a different object plane gives a contribution to the same image point. When imaging a 2D surface, each of these object planes, which is not equivalent to the surface, contributes to a blurred image. In our case of scattering particles these planes contribute only to the

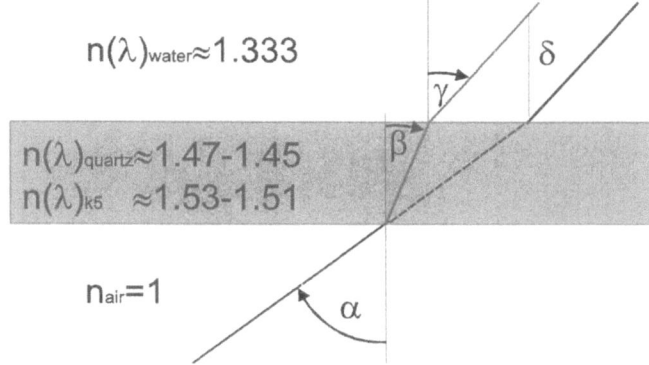

Fig. 1 Pathway of a ray from the objective (*below*, air) through the cover glass into the suspension (*above*, water refractive index). δ is the resulting offset and α, β and γ are the angles between the ray and the surface normal. The refractive indices for quartz glass and k5 (Schott) for 400 nm and 700 nm are indicated

Fig. 2 The difference between the offset caused by a parallel plate of 1-mm quartz and a plate of thickness d of glass k5 for different angles of aperture. The *grey areas* show the regions contributing to the image. The angular limitations are due to the central stop and the aperture of the objective. Further limitations are given by the wavelength of transmitted light (*straight line* 400 nm, *broken line* 700 nm). The height of the *grey areas* gives the effective focal deep indicated at the left of the bars

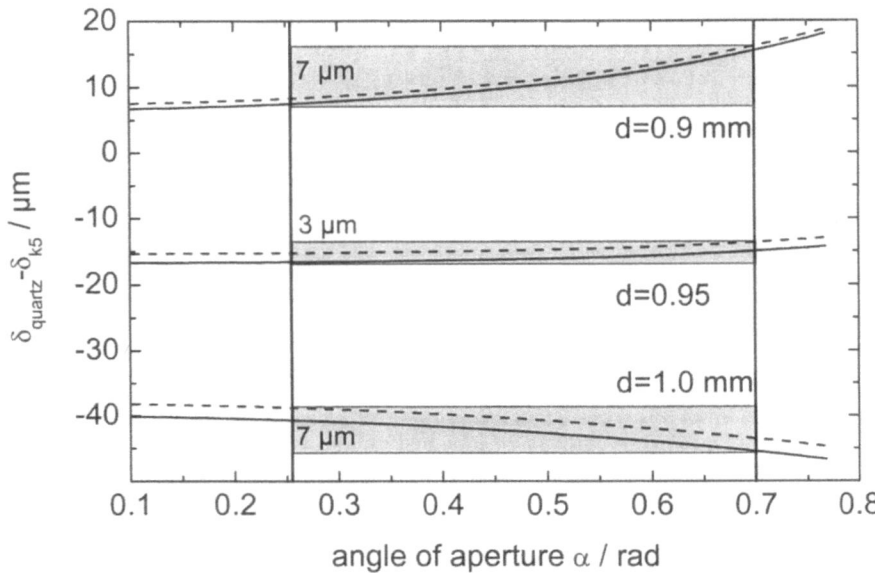

Fig. 2 The difference between the offset caused by a parallel plate of 1-mm quartz and a plate of thickness d of glass k5 for different angles of aperture. The *grey areas* show the regions contributing to the image. The angular limitations are due to the central stop and the aperture of the objective. Further limitations are given by the wavelength of transmitted light (*straight line* 400 nm, *broken line* 700 nm). The height of the *grey areas* gives the effective focal deep indicated at the left of the bars

image, if they coincide with a particle position. If they do not coincide, they do not blur the image because there is only nonscattering water at these positions. This leads to an increased depth of sharpness due to the incorrect cover glass correction, in the examples shown in the range of some microns at the cost of a small decrease in resolution. Even for the best matching at $d = 0.963$ mm the depth of sharpness is about 2 μm (not shown). As an example a magnified image of a bcc crystal at the lower plate of our shear cell is shown in Fig. 3a. The points indicated reflect the particles of the (1 0 1) plane positioned parallel to the wall. The particles visible on the connecting lines correspond to the particles in the second layer parallel to the (1 0 1) plane with a distance of $a/\sqrt{2} = 1.27\ \mu$m as the minimum focal depth. As a second example a magnified image of a stack of triangular layers after cessation of shear is shown (Fig. 3b). The particles circled belong to the same layer and in the clearance between these, the particles of the next layer are visible. Notice that we cannot distinguish between the upper and lower layers without gathering further information. This will have consequences to the image analysis.

Image analysis procedure

The images were taken with a conventional CCD camera with a typical exposure time of 2 ms below the time a single particle needs

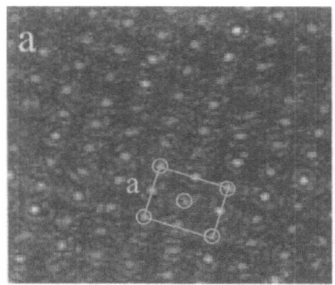 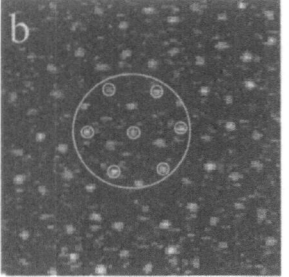

Fig. 3 a Image of an equilibrium body-centered-cubic crystal. The points indicated with connecting lines represent the (1 0 1) plane positioned parallel to the glass plate. The particles on the line belong to the next upper layer. **b** Triangular layers with particles circled and the next layer in the clearance

to diffuse a distance of its own radius of 150 nm. For demonstration the suspension was prepared at a number density of $n = 0.14\ \mu$m^{-3} and a typical particle distance of 1.9 μm.

The analysis starts with a 3×3-pixel Gaussian filter for noise reduction and a 20×20-pixel local background correction to equalise the background level. Particle identification follows the idea that a particle has an intensity above the underlying noise and that the particle centre should have a maximum intensity in the neighbourhood. Accordingly the filter first calculates in a neighbourhood the difference between minimum and maximum intensities. If this is larger than the noise level and the central pixel is equal to the maximum, then the central pixel is accepted as a particle location. For this procedure the two parameters filter size and noise level have to be chosen. For the noise level we first calculate the overall histogram for the prefiltered image with a mean grey value and standard deviation. Because of the prefiltering these calculated values are good parameters for the whole image. For the best results we set the noise level to a value of twice the standard deviation. The Filter size is set to 5×5 pixel according to a pixel resolution of 0.13 μm and a resolution limit of 0.48 μm for our objective. A sequence illustrating this procedure is shown in Fig. 4. Figure 4c shows that some particles are indicated by more than one pixel. This is due to a small signal-to-noise ratio for the images of about 1:10. This leads to neighbouring pixels with the same values, which are both accepted as a particle, if they are a local maximum. Figure 4d demonstrates the quality of our procedure. We see that most of the particles are recognised, but still there are some intensity distributions, which are regarded as noise but can be a particle as well.

To reveal the overall structure we now have to compute the position correlation, $G_{xy\Sigma z}(x, y)$, of the particles recognised. To do this, for each particle identified its 100×100-pixel neighbourhood, with the particle in the centre, is taken and at the position of the identified particles inside a counter is increased. Afterwards the number of neighbourhoods that have a pixel at this position weights the counting to consider the smaller neighbourhood of particles at the image border. In this manner we get a 100×100-pixel area representing the overall structure. Notice that for the reason of undistinguishable particles these position correlations have a point symmetric character. To reduce noise, we sum over a complete sequence of 37 images. An example is given for a shear rate of 0.9 Hz and a shear direction along the tortuous lines in Fig. 5 to illustrate the characteristics. At the origin a total count of

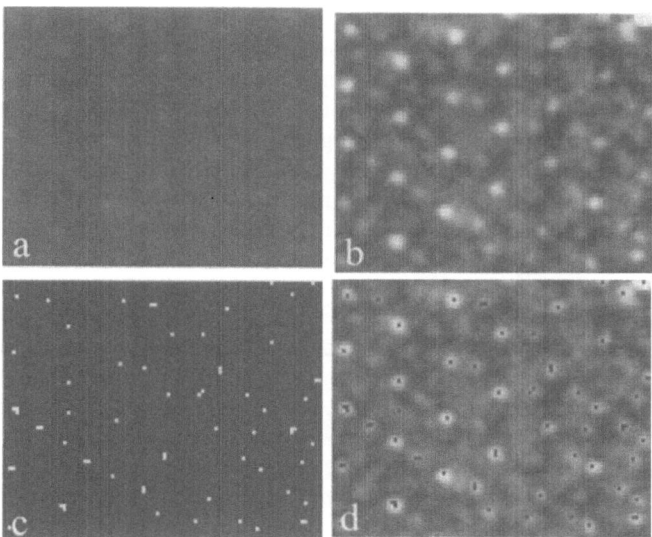

Fig. 4 Sequence of image analysis: **a** original image; **b** after 3×3 Gaussian filter and local background correction; **c** result of particle identification; and **d** comparison of improved original from **b** and inverted **c**. Most of the particles from the original image are identified. For the example given the efficiency is estimated by comparing the respective numbers of particles identified in the layers to the total number of particles present (assuming a defect-free structure). For the first layer (brightest particles) nearly all the particles were found. For the second layer the efficiency is about 50% and for the third layer about 30% in this example. For the overall efficiency, see text

270 000 is found. The distribution around the central position is due to the particles, which produce double or triple pixels, and give a kind of broadening for all other positions. Analysis of this distribution around the origin yields a total number of about 184 000 particles, corresponding to about 5000 identified particles per image (ippi). About 40% of the particles produce pixel distributions with more than one pixel. The overall triangular

structure is clearly visible with a counting of about 800 ippi in the maximum and a mean distance of 2.17 μm, corresponding to 1830 particles in a plane of 7500-μm^2 image size for the triangular structure. Thus, $G_{xy\Sigma z}(x, y)$ represents the relative particle positions of nearly three complete layers per image. Integrating the counting in a single triangular peak yields the same counting as at the origin. The same result is obtained if we integrate the counting in the triangle between the origin and two neighbouring triangular peaks excluding the peaks. Both results are due to the fact that each central particle is also a next neighbour or a next layer particle. The tortuous lines between the triangular positions correspond to the positions of a second layer under or above the central particle layer. In this way these lines are in a statistical meaning equivalent to the common way one layer moves relative to the other. The relative values on such a line correspond to the probability to find a particle at this position.

The black ring around the origin is due to the filter size of 5×5, which suppresses the detection of particles inside. Inside the black ring we find a minimum of fewer than 10 ippi, demonstrating the suppression of the particle identification of nearby particles by the filter. The intensity points in the shear direction from the origin, next outside this filter area, evolve through particles from a third layer moving relative to the others. There are two possibilities for a third layer to move conserving the possible stacking sequences ABA and ABC for hexagonal-close-packed (hcp) and fcc crystallites. For the ABA stacking sequence the third layer moves along the connecting line between the central particle and the next intraplane particle in the shear direction. For the ABC stacking sequence let us define the particle position in the centre of the left, middle triangle, seen from the central particle, as B. Position C is in the centre of the triangle below. In this case the movement follows the line from position C trough the central particle to the opposite triangle centre position. Both movements are broadened and overlap in our projection in the position of the aforementioned intensity points. As both stacking sequences are random the counting adds in the overlapping region. This yields a local maximum at these positions.

To clarify the last points the radial average computed from the image in Fig. 5 is shown in Fig. 6. Near $r = 0$ μm the central peak is visible with a decrease due to the oppression of particle identification by the filter. The following increase is due to the particles

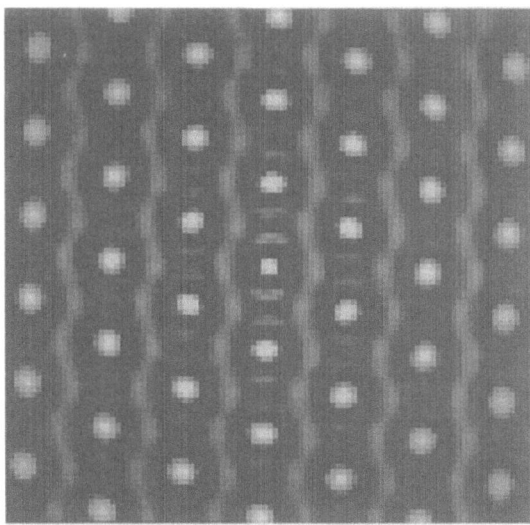

Fig. 5 Result of the image analysis and the accumulation of all the particle neighbourhoods as a nonnormalised $G_{xy\Sigma z}(x, y)$. A detailed description is found in the text

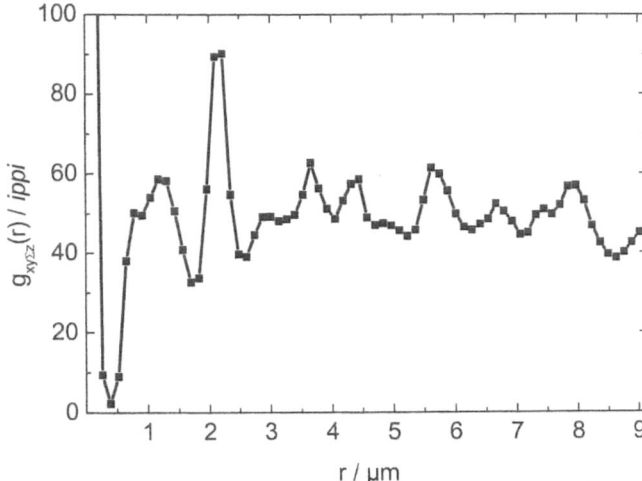

Fig. 6 Nonnormalised radial average from Fig. 5, $G_{xy\Sigma z}(r)$, representing the projection of a radial $g(r)$, taken within the range of sharpness, onto the xy-plane. The first peak, at 1 μm, corresponds to particle positions in the adjacent layer. The peak at 2 μm belongs to the next neighbours in the plane

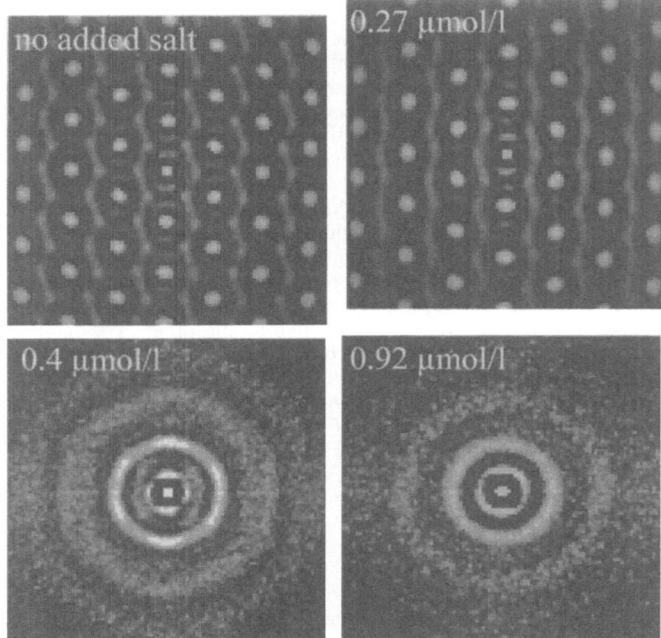

Fig. 7 Nonnormalised $G_{xy\Sigma z}(x, y)$ for different salt concentrations: no added salt, 0.27 μmol/l, 0.4 μmol/l, 0.92 μmol/l. We added H_2CO_3 by contaminating the suspension with CO_2. A detailed description is found in the text

situated in other triangular layers; therefore, we call it the interplane region. The next increase to a peak is caused by the triangular structure inside a layer; therefore, called the intraplane peak. Owing to the smearing by the averaging, the subsequent structures are not as clearly attributed to the regions mentioned. Anyhow, for a clear triangular structure, like here, the second and third-nearest-neighbour peaks are visible, and as for fluid structures a slowly varying slope is found.

Results

$G_{xy\Sigma z}(x, y)$ images for a shear rate of 0.89 Hz directly at the lower plate are shown in Fig. 7 for different salt concentrations. The number density is $n = 0.15$ μm^{-3}, resulting in an equilibrium bcc crystal up to an added salt concentration of 0.125 μmol/l. For all images the shear direction is equal from the lower to the upper image border along the tortuous lines of the first images. The first image without added salt exhibits a clear triangular structure with a long-range order typical for crystals. This demonstrates the crystal-like order inside the single triangular planes. The second outstanding attributes are the zigzag lines in the interplane region between the triangular peaks. The minimum counting in the straight segment is about 135 ippi over a background counting of 50 ippi. At the corners a higher counting of 220 ippi is obtained. This is interpreted in a statistical meaning as a higher probability to find a particle at the corner. For triangular planes sliding over each other this means a clearly visible zigzag motion with registered planes in the

middle of the triangular intraplane peaks. This position corresponds to the position of the next plane of an equilibrium fcc or hcp crystal. The second image with an added salt concentration of 0.27 μmol/l H_2CO_3 in the equilibrium fluid-phase region displays, in principle, the same structure as the first. The triangular peaks are the same, demonstrating the shear-induced intraplane triangular structure in the equilibrium fluid phase. The zigzag pattern with accumulation points is still visible. The corners of the lines in the interplane region are not as sharp as before and also the accumulation points are not as bright as before. In the next image, for a salt concentration of 0.4 μmol/l H_2CO_3, a completely different structure is observed. The triangular long-range order has vanished and the whole pattern has a more fluidlike character. As deviations from a complete fluidlike pattern the angular appearance of the outer fluid ring and the weak maxima on the first fluid ring are recognized. The weak maxima on the first ring have the same position as the triangular peaks in the first two images and represent the first stage of an evolving triangular structure. In the interplane region inside the first fluid ring a leftover from the tortuous lines is found with the same direction as before. Although there is a dominating fluidlike structure the last point indicates a still existing interplane coupling between the still existing layers. The counting around the filter-dominated region now has a component perpendicular to the shear direction. This is due to the fact that for a high interplane coupling of triangular layers a particle in a third layer above the central particle of the first layer is restricted to a small string in the shear direction. This results in the covering of this central third plane particle by the filter. For a lower coupling the string is broadened and so the counting of these particles for higher salt concentrations

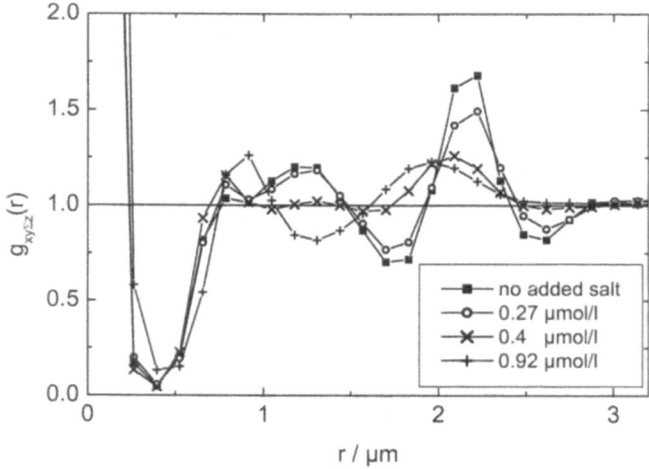

Fig. 8 Radial averages from Fig. 7. The averages are divided by the average for values with $r > 3$ μm, yielding a normalised $g_{xy\Sigma z}(r)$. A detailed description is found in the text

rises. In the last image, for a salt concentration of 0.92 μmol/l H_2CO_3, the pattern shows a complete fluidlike appearance. The outer fluid rings have a lower counting than in the third image and the first fluid ring has no additional structure, pointing to a complete loss of any triangular intraplane short-range order. Again in the interplane region the third-plane particle counting is raised but there is no counting visible which can be attributed to a correlated second layer.

The images in Fig. 7 represent 2D particle correlations with absolute counting. From that we calculate the radial average $G_{xy\Sigma z}(r)$ and divide it by the average counting for $r > 3$ μm to compare different structures. The resulting normalised $g_{xy\Sigma z}(r)$ with the main characteristics from the origin to next plane behind the first peak is shown in Fig. 8. At the origin the high counting from the central particle with a strong decrease due to the filter is seen. Outside the filter-influenced region the third-layer particle forms a small maximum, followed by the second-layer particles with a salt-concentration-dependent course. The course shows a maximum for the high interplane coupling and a minimum for lower coupling in the fluidlike case. The first peak, however, has a salt-concentration dependence. With higher salt concentration the maximum position changes to a lower distance and the maximum value decreases.

Here it should be mentioned that owing to the more or less 3D character of our particle detection, the slope of the complete curve is influenced by the 3D $g(r)$ and the projection of a bar of about two or three particle distances onto the xy-plane. This leads to an additional broadening of all peaks and fluid rings. Additionally the minima in $g(r)$ were overlapped by positions at a different depth and because of this were not as deep as seen in a real $g(r)$.

Conclusions

In our plate–plate shear cell with variable gap width colloidal suspensions were observed with a high-resolution microscope. A specially adapted inverted ultramicroscopic illumination for high-contrast images together with an adjusted mismatch of cover glass correction yields a 2D projection of 3D particle positions in a colloidal suspension under shear. The images were analysed by a procedure of particle identification and accumulation of particle neighbourhoods. This procedure provides particle correlation of next-neighbourhood particles not only with 2D character. The cover-glass-correction mismatch responsible for the depiction of a, some microns thick, gap between the two rheometer plates yields additional information on particle positions of different height inside the cell.

We presented the first real-space observations of the complete path a triangular layer moves relative to the next triangular layer. The result confirms the sliding mechanism discussed elsewhere [15]. As a preliminary scenario we observe with the addition of extra salt the change from a zigzag motion with accumulation points at the corners to less pronounced corners and accumulation points inside the fluid phase right next to the crystalline-phase boundary. For higher salt concentration, for example, deeper in the fluid phase, we find a collapse of triangular long-range order and an evolving fluidlike order. Anyhow some triangular short-range order still exists even if the fluid like long-range order is dominant and interplane coupling still exist. At the final point a completely fluid structure is reached. The collapse of the triangular order does not coincide with the phase-boundary position of the equilibrium colloidal suspension. Here a shear-induced ordering in the equilibrium fluid phase is found.

A detailed study with more data of the shear mechanism in colloidal suspensions for different shear rates and salt concentrations in real space will be presented in the future.

Acknowledgements We gratefully acknowledge financial support by the Deutsche Forschungsgemeinschaft (grant Pa459/7-3) and the Materialwissenschaftliches Forschungszentrum Mainz.

References

1. Sirota EB, Ou-Yand HD, Sinha SK, Chaikin PM, Axe JD, Fujii Y (1989) Phys Rev Lett 62:524
2. Dux C, Musa S, Reus V, Versmold H (1998) J Chem Phys 109:2556
3. Versmold H (1995) Phys Rev Lett 75:763
4. Konishi T, Ise N (1998) Phys Rev B 57:2655
5. Maaroufi MR, Stipp A, Palberg T (1998) Prog Colloid Polym Sci 110:83
6. Imhof A, van Blaaderen A, Dhont JKG (1994) Langmuir 10:3477
7. van Winkle DH, Murray CA (1986) Phys Rev A 34:562
8. Weiss JA, Oxtoby DW, Grier DG (1995) J Chem Phys 103:1180
9. Garbow N, Müller J, Schätzel K, Palberg T (1997) Physica A 235:291
10. Evers M, Garbow N, Hessinger D, Palberg T (1998) Phys Rev E 57:6774
11. Hessinger D, Evers M, Palberg T (1999) Phys Rev E (accepted)
12. Wette P, Schöpe HJ, Biehl R (to be published)
13. Biehl R, Palberg T (2000) Prog Colloid Polym Sci 115:300
14. Siedentopf H, Zsigmondy R (1903) Ann Physi 10:1
15. Loose W, Ackerson BJ (1994) J Chem Phys 101:7211

Progr Colloid Polym Sci (2001) 118: 208–215
© Springer-Verlag 2001

RHEOLOGY

Hasan Mousa
Wim Agterof
Jorrit Mellema

Theoretical and experimental investigation of the coalescence efficiency of droplets in simple shear flow

H. Mousa (✉)
Jordan University of Science
and Technology
Department of Chemical Engineering
P.O. Box 3030, Irbid 22110, Jordan
e-mail: akras@just.edu.jo

W. Agterof
Unilever Research Laboratorim
Vlaardingen
Olivier van Noortlaan 120
3133 AT Vlaardingen, The Netherlands
e-mail: wim.agterof@unilever.com

J. Mellema
University of Twente,
Department of Applied Physics
P.O. Box 217,
7500 AE Enschede, The Netherlands
e-mail: j.mellema@tn.utwente.nl

Abstract The coalescence efficiency of two Newtonian droplets submerged in a Newtonian fluid subjected to a simple shear flow was investigated experimentally and theoretically. The experimental investigation was based on observing collisions between two droplets under a microscope. The theoretical investigation considered three drainage models: immobile, partially mobile and mobile interfaces. Both the experimental results and the theoretical analysis showed that a critical approach angle exists below which the colliding droplets separate. Above this critical angle the collision leads to coalescence. Knowledge of the critical angle permits calculation of the coalescence efficiency. The dependence of the coalescence efficiency on various dimensionless groups such as the flow number, the capillary number and the viscosity ratio was studied. The theoretical analysis indicated that the coalescence efficiency decreases as the capillary number and the flow number increase. The experimental results showed that the coalescence efficiency goes through a minimum as the value of the flow number increases. The discrepancy between the experimental and the theoretical results was attributed to some mechanism that enhances coalescence and that is not accounted for in the equation used for the critical thickness for film rupture. Both the experimental and the theoretical results indicated that the coalescence efficiency decreases as the viscosity ratio decreases.

Key words Coalescence efficiency · Droplets · Emulsion · Simple shear · Stability

Introduction

Consider two droplets in the same shear plane (xy plane) and being subjected to a simple shear flow as can be seen in Fig. 1. The line of the centers makes an angle α with the y-axis. The approaching droplet moves in a straight line until it starts to feel the presence of the central droplet (interaction zone). The film that forms between the droplets at this stage will continue to drain owing to the hydrodynamic force and the van der Waals attraction forces. The latter become more important as the droplets get closer to each other. The rate of film drainage depends on the mobility of the interfaces. For clean interfaces, the mobility depends on the ratio of the dispersed-phase viscosity to the continuous-phase viscosity. The rate of drainage for immobile, mobile and partially mobile films is given by the following equations:

$$-\frac{dh'}{dt'} = \frac{8.68}{3} h' \sin(2\alpha) + \frac{1}{3h'\mathrm{Fl}\left(1+\frac{h'}{4}\right)^2\left(1+\frac{h'}{2}\right)^3},$$

immobile rigid interfaces   (1)

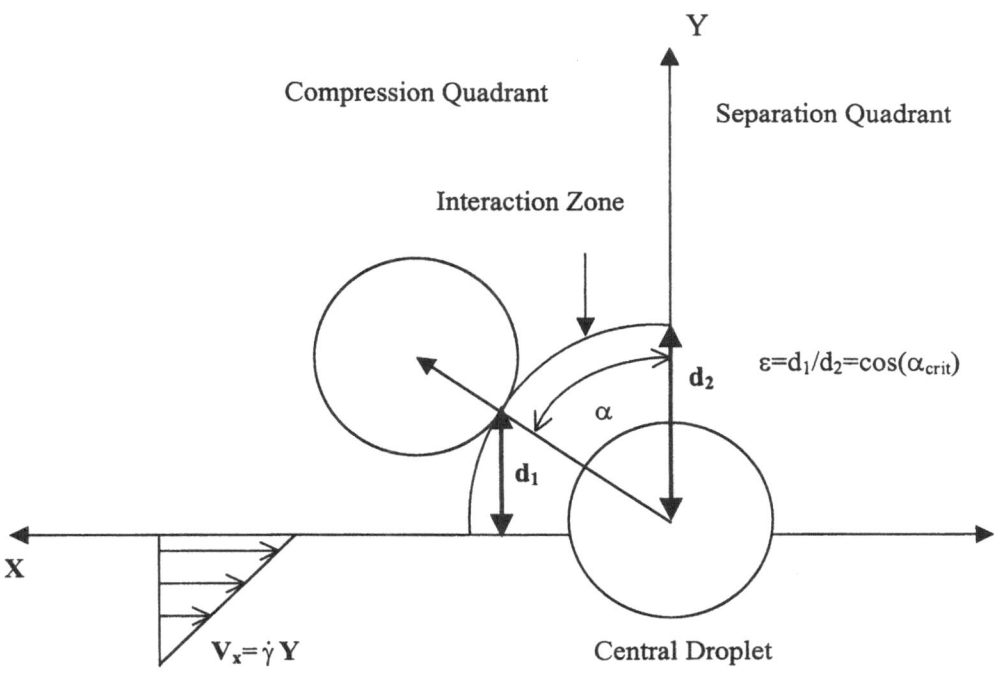

Fig. 1 A droplet approaching the central droplet in a simple shear flow. The approaching droplet will move in a straight line until it reaches the interaction zone, where its motion is influenced by the central droplet. The arc shown in the figure represents the interaction zone. The line of centers makes an angle, α, with the line perpendicular to the flow direction, y-axis. A critical approach angle, α_{crit}, exists such that droplets approaching at $\alpha > \alpha_{crit}$ will coalesce with the central one; those approaching at $\alpha < \alpha_{crit}$ will not. The coalescence efficiency, ε, is $d_1/d_2 = \cos(\alpha_{crit})$

$$-\frac{dh'}{dt'} = \frac{8h'^3}{13\Omega^2 \sin(2\alpha) + \frac{3\Omega^2 \xi}{h'^2 Fl}},$$

immobile deformable interfaces (2)

IC: $t' = 0$, $\alpha = \alpha_{approach}$ and $h' = h'_0 = 0.5$ [1–3], (3)

$$\frac{dh'}{dt'} = \frac{\frac{8.68}{3}\sin(2\alpha) + \frac{\xi}{Fl} - \frac{2}{\Omega}}{\left(\frac{3}{h'} - \frac{1}{h' + (1-a')^2}\right)}, \text{ mobile interfaces}$$ (4)

IC: $t' = 0$, $\alpha = \alpha_{approach}$ and $h' = h'_0 = 0.037$[4], (5)

$$-\frac{dh'}{dt'} = \frac{4h'^2}{\lambda}\left(\frac{2}{\Omega^3\left[4.34\left(\frac{2/3+\lambda}{1+\lambda}\right)\sin(2\alpha) + \frac{\xi}{Fl}\right]}\right)^{1/2},$$

partially mobile (6)

IC: $t' = 0$, $\alpha = \alpha_{approach}$ and $h' = h'_0 = 2(\Omega_d)^{2/3}$[5]. (7)

The change in the angle α with time was calculated from [4]

$$0.4t' = \left[\arctan\left(\frac{\tan(\alpha_{approach})}{2}\right) - \arctan\left(\frac{\tan(\alpha)}{2}\right)\right].$$ (8)

In these equations $h' = h/R_{eqv}$ is the dimensionless film thickness, $R_{eqv} = \frac{2R_1R_2}{(R_1+R_2)}$ is the equivalent radius, R_1 and R_2 are the radii of the colliding droplets, $a' = a/R_{eqv}$ is the

dimensionless radius of the film, $t' = t\dot\gamma$ is the dimensionless time, $\dot\gamma$ is the shear rate, $Fl = 6\pi\mu_c\dot\gamma R^3_{eqv}/A$ is the flow number, A is the Hamaker constant, $\Omega = \dot\gamma R^3_{eqv}\mu_c/\sigma$ is the capillary number, σ is the interfacial tension, ω is a lengthy expression that depends on h' and a' and $\lambda = \mu_d/\mu_c$ is the viscosity ratio, where μ_d and μ_c are the viscosities of the dispersed-phase and the continuous-phase viscosities.

The coalescence efficiency was determined by solving the equations subjected to the initial conditions shown by Eqs. (3), (5) and (7) for the immobile, mobile and partially mobile models, respectively. For immobile interfaces the criterion for coalescence and separation was based on the change in the film thickness with time as the two droplets approach each other. If a decrease followed by an increase in the film thickness with time occurs, the droplets are considered to be separating after collision. If, however, the film thickness continues to decrease, the droplets are considered to be coalescing. For mobile and partially mobile interfaces a critical film thickness, h'_{crit}, given by the following equation was used as a criterion for coalescence [6]:

$$h'_{crit} = \left(\frac{3\Omega}{4Fl}\right)^{1/3}.$$ (9)

The calculations showed that there is a critical approach angle, α_{crit}, below which the colliding droplets separate and above which coalescence takes place. From geometrical analysis the coalescence efficiency, ε, is related to the α_{crit} by the following equation (see Fig. 1):

$$\varepsilon = \cos(\alpha_{crit}).$$ (10)

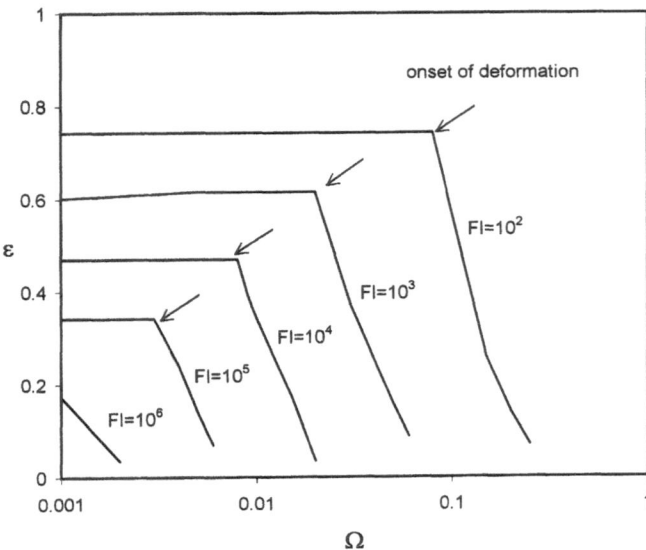

Fig. 2 ε versus Ω for various values of the flow number, Fl, for immobile interfaces. The *arrows* indicate the onset of deformation

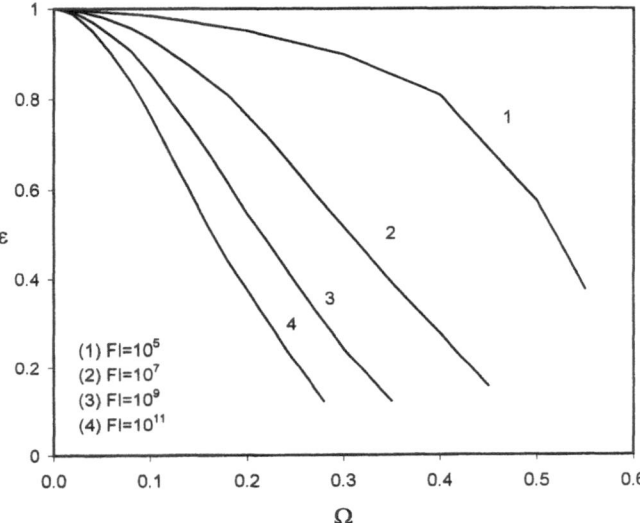

Fig. 3 ε versus Ω for various values of Fl for mobile interfaces

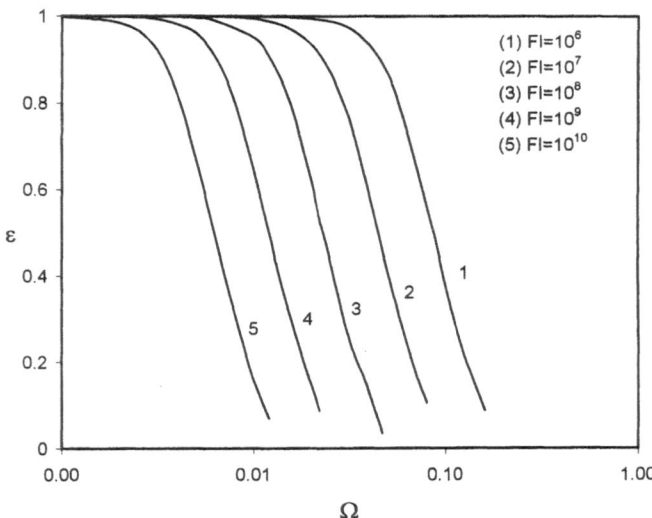

Fig. 4 ε versus Ω for various values of Fl for partially mobile interfaces. $\lambda = 1.0$

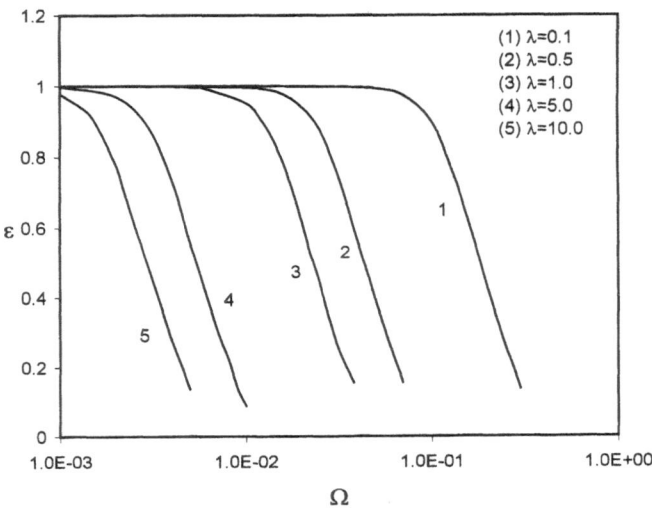

Fig. 5 ε versus Ω for various values of λ for partially mobile interfaces. $Fl = 1.0 \times 10^8$

The coalescence efficiency predicted using these models is shown in Figs. 2, 3, 4 and 5, where it can be seen that the coalescence efficiency decreases as the values of the flow number and the capillary number increase. Moreover the coalescence efficiency decreases as the viscosity ratio, λ, increases.

Experimental

Experimental setup

A Couette apparatus was used to carry out the experiments. Full details of the apparatus is given in Ref. [7]. A schematic representation of the Couette apparatus used is shown in Fig. 6. It consists of two concentric cylinders, each one is driven by a separate motor, the speed of which is controlled by a speed control unit. The speed control units are operated with a personal computer so that a shear rate ranging between 0.03 and 91 s^{-1} can be generated. The software is written such that the rotational speed of the cylinders can either be increased or decreased at small or large steps as desired. This helps in controlling the speed of the colliding droplets or holding one droplet at the stagnation point while the second one is approaching. The software also allows an immediate halt of the motion of the cylinders. The two cylinders can be rotated in a counterclockwise or a clockwise fashion. The inner cylinder is made from aluminum and its radius and height are 4 and 5 cm respectively. The outer cylinder is made from precision-pore glass and its radius and height are both 5 cm. A video camera is mounted perpendicular to the shear plane (along the vorticity

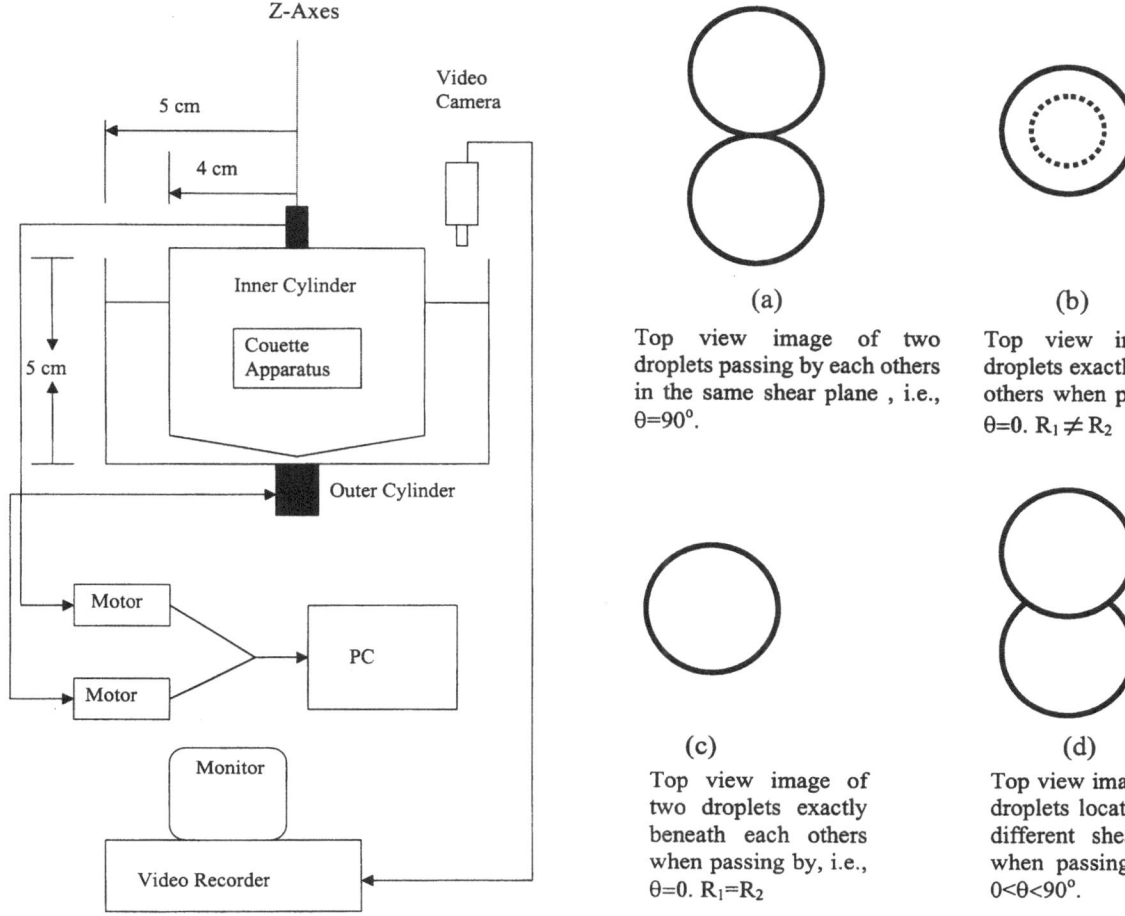

Z-Axes

Fig. 6 Schematic diagram of the Couette apparatus used

(a) Top view image of two droplets passing by each others in the same shear plane , i.e., $\theta=90°$.

(b) Top view image of two droplets exactly beneath each others when passing by, i.e., $\theta=0$. $R_1 \neq R_2$

(c) Top view image of two droplets exactly beneath each others when passing by, i.e., $\theta=0$. $R_1=R_2$

(d) Top view image of two droplets located in two different shear planes when passing by, i.e., $0<\theta<90°$.

Fig. 7 Top view image of two droplets passing each other

z-axis) and is connected to a black and white monitor and to a video cassette recorder to record images for further analysis. The video camera is mounted on a stand that allows the camera to rotate in order to follow the droplets when necessary. The stand also allows the camera to move horizontally in the xy plane to the right or to the left as well as in forward and backward directions. The camera can also move up and down for focusing purposes. The objective lens of the camera magnifies the images 105 times as seen on the monitor screen. In all the experiments, observations were made along the z-axis (Fig. 6); however the glass outer cylinder allows observation from beneath and from the side of the cylinder when necessary.

Material

Silicone oil (Baysilon from Roland Chemie, The Netherlands) of the viscosity 1, 3.1, 3.5, 3.7 and 4.9 Pas was used as a continuous media. Glycerin (Merck, Germany) was used as a dispersed phase. The viscosities of glycerin (0.9 Pas) and silicone oil were measured using a Haake RheoStress RS 150. The interfacial tension was 28 mNm^{-1}.

Experimental procedure

The Couette apparatus was filled with silicone oil of the desired viscosity. Enough time was allowed for any air bubbles that may

have formed to disappear. The shear rate was increased to the required value without allowing the cylinders to rotate. This was achieved by switching off the speed controllers. When the required shear rate was reached a glycerin droplet was injected via a disposable syringe. After the droplet had sunk under the surface of the oil, the speed controllers were switched on and this caused the cylinders to rotate immediately at the previously chosen speeds. After few seconds the rotation of the cylinders was suddenly stopped. This procedure causes the large glycerin droplet to break into smaller ones and guarantees that the droplets are in the same shear plane. The sizes of the droplets generated can be varied by applying higher or lower shear rates or by waiting for a longer time before the cylinders are stopped. The size of the droplets generated is also a function of the viscosity of the continuous medium [8, 9].

The droplets produced by this procedures are in the same shear plane, which makes an angle, θ, of 90° with the vorticity axes; however the angle α varies from 0 to 90° (Fig. 1). To check that the droplets were in the same plane, the camera was occasionally tilted by 90° so that it could look through the shear plane. Such a test revealed that the droplets were aligned in the same shear plane and hence all the experiments were performed at $\theta = 90°$. This point was checked further by watching the droplets passing each other, i.e. when $\alpha = 0$. When the two droplets are in the same shear plane ($\theta = 90°$) the top view image is similar to that shown in Fig. 7a. If the radii are not equal, one sees a small droplet inside a big one (Fig. 7b). If the droplets are exactly beneath each other ($\theta = 0$) the top view image resembles Fig. 7c if the radii are equal. When

Table 1 The measured values of R_1, R_2, $\alpha_{approach}$, $\alpha_{capture}$, $t_{contact}$ and the calculated values of R_{eqv}, Fl and Ω. The system is glycerin droplets in silicone oil (3.5 Pas). The applied shear rate is 0.14 s^{-1}. $A = 5 \times 10^{-21}$ J and $\sigma = 0.028$ Nm^{-1} were used to calculate Fl and Ω

R_1 (μm)	R_2 (μm)	R_{eqv} (μm)	q	$\alpha_{approach}$	$t_{contact}$ (s)	$\alpha_{capture}$	Fl $\times 10^{-8}$	$\Omega \times 10^3$
94	90	92	0.96	81	40.2	−34	14.6	1.64
89	84	86	0.94	78	34.8	−37	12	1.53
90	86	88	0.96	77	–	–	12.8	1.57
94	90	92	0.96	68	–	–	14.6	1.64
94	86	90	0.92	65	–	–	13.6	1.60
103	94	98	0.91	54	–	–	17.7	1.74

$0° < \theta < 90°$ the top view image will look similar to that presented in Fig. 7d. In our experiments only images similar to that represented in Fig. 7a were analyzed; the others were ignored.

Once the small droplets had formed the cylinders were allowed to rotate in opposite directions at low shear rate and this allowed the droplets to collide. All the experiments were performed at a shear rate of 0.14 s^{-1} and were recorded on videotapes for analysis. The analysis includes measuring the approach angle, $\alpha_{approach}$ and the radii of the colliding droplets, R_1 and R_2. When coalescence takes place the contact time, $t_{contact}$, and the capture angle, $\alpha_{capture}$, are also measured. The measurements of $\alpha_{approach}$ and t are started when the distance between the surfaces of the two droplets is 0.5 times the equivalent radius, i.e. $h' = 0.5$ [6]. The colliding droplets are grouped in classes according to R_{eqv}. Particles within 10 μm are considered to be one class. The resulting classes are further grouped according to the size ratio, $q = R_1/R_2$, such that the width of each class is 0.1. Hence, by knowing the critical approach angle, the coalescence efficiency for that class of droplets is calculated from Eq. (10). An example of such results is given in Table 1, where it can be seen that coalescence takes place when $\alpha_{approach} \geq 78°$. For values of $\alpha_{approach} \leq 77°$ no coalescence takes place; therefore α_{crit} is $77.5 \pm 0.5°$, which gives an efficiency of 0.22 ± 0.01 for $85 \ \mu$m $\leq R_{eqv} \leq 95 \ \mu$m and $0.9 \leq q \leq 1$.

Experimental results and discussion

Measured values of the coalescence efficiency as a function of Fl for glycerin droplets in silicone oil (3.1 Pas) are shown in Fig. 8. As can be seen the coalescence efficiency decreases with Fl then starts to increase. A comparison with the efficiency predicted from mobile, partially mobile and immobile drainage can also be seen in Fig. 8. None of the models predict the trend observed experimentally. Previous work on this subject investigating coagulation of solid particles (e.g. [10, 11]) and coalescence of drops in simple shear flow [6, 12–14] showed that ε is a decreasing function of Fl. The same results were found by studying gravity-driven coalescence of slightly deformable drops [15]. On the other hand, the study of the coalescence of drops in stirred dispersions showed that the coalescence efficiency increases with the radius [16–18]. Our results seem to agree qualitatively with the first group of research for Fl $< \sim 3 \times 10^9$; above this value of Fl they seems to agree with the second group of research. Taking repulsion and attraction forces into account van de Ven and Mason [10], van de Ven [11] and Adler [19] studied theoretically the coagulation of solid particles in simple shear flow.

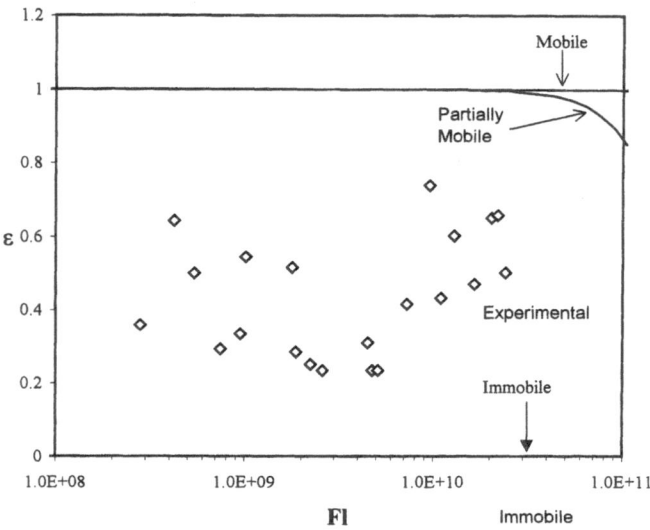

Fig. 8 Experimentally measured values of ε versus Fl and theoretically calculated ones from mobile, partially mobile and immobile interface models for glycerin/silicone oil (3.1 Pas)

They found that the efficiency decreases as Fl increases until a minimum is reached after which it starts to increase, a result similar to that observed in Fig. 8. Note that $C_A = 1/6$Fl is used in the work reported in Refs. [10, 11, 19]. The competition between the attractive and the repulsive forces causes the decrease followed by an increase in the efficiency. The same observation was found experimentally by Mousa and van de Ven [20], where Fl was changed by altering the shear rate applied to the emulsion. In the system studied here, where silicone oil is the continuous phase, the repulsive forces are unimportant and the previous argument cannot explain the experimental results. Recently Chesters and Bazhlekov [21] found theoretically that in the presence of a nondiffusing surfactant at the interface, dimple formation takes place as the droplets approach each other. Dimple formation is further enhanced when the droplets are large. In this situation rim coalescence rather than nose coalescence takes place. This leads to a higher coalescence efficiency. To investigate if dimple formation is the reason behind the trend shown in Fig. 8 videotaped images of the largest droplets ($R \sim 250 \ \mu$m) encountering each other were examined. The images revealed that no

Fig. 9 Calculated values of h'_{crit} necessary to predict the coalescence efficiency for the experimental data shown in Fig. 4 using the immobile, mobile and the partially mobile interface models. The values of h'_{crit} calculated from Eq. (9) are also shown

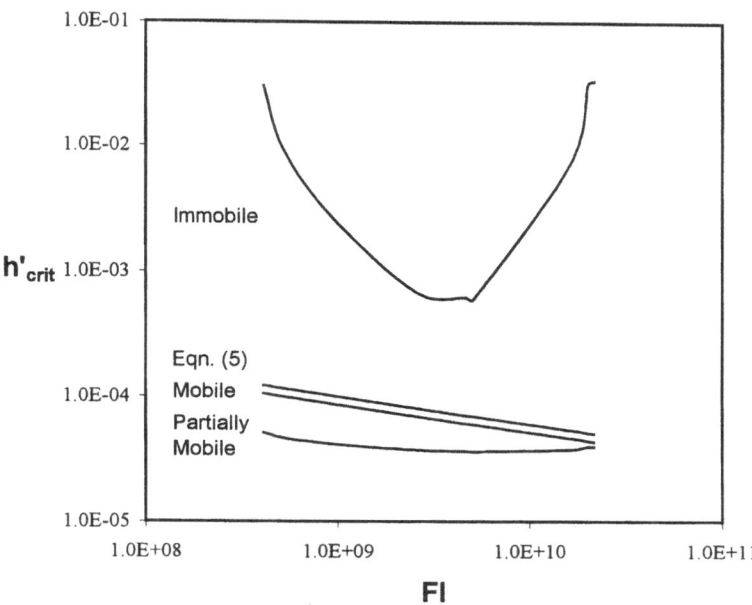

deformation takes place during the whole period of the encounter. This indicates that dimple formation is unlikely to occur and hence the trend shown in Fig. 8 cannot be explained by dimple formation.

The drainage models indicate that the coalescence efficiency can theoretically be changed by altering h'_{crit}. In other words h'_{crit} is the only variable that theoretically controls the coalescence efficiency for a given system (constant Fl and Ω). A different value for the critical film thickness results in a different value for the coalescence efficiency. This required an expression for the critical film thickness different from Eq. (9) to be assumed. In order to find such an expression, the values of h'_{crit} necessary to predict the measured coalescence efficiency, shown in

Table 2 A list of the measured values of R_1, R_2, $\alpha_{approach}$, $\alpha_{capture}$, $t_{contact}$ and the calculated ones using partially mobile and mobile models. The values of h'_{crit} calculated from Eq. (9) were used as a criterion for coalescence. The system is glycerin droplets in silicone oil ($\mu_c = 3.1$ Pas) subjected to a shear rate $\dot{\gamma} = 0.14$ s^{-1}

Experimental					Theoretical			
					Partially mobile		Mobile	
R_1 (μm)	R_2 (μm)	$\alpha_{approach}$	$t_{contact}$ (s)	$\alpha_{capture}$	$t_{contact}$ (s)	$\alpha_{capture}$	$T_{contact}$ (s)	$\alpha_{capture}$
219	248	83	43.3	22	1.60	80	0.36	82
209	209	47	6.8	−6	2.37	37	0.31	46
209	267	52	7.4	−4	3.10	40	0.35	51
219	167	80	17.6	55	1.20	78	0.28	79
200	276	81	28.3	51	1.90	78	0.35	80
119	124	75	18.0	15	0.53	74	0.17	75
124	124	81	36.8	−42	0.44	80	0.18	81
129	124	79	23.6	3	0.50	78	0.18	79
100	100	71	22.6	−15	0.38	70	0.14	71
176	62	76	18.9	5	0.04	75	0.13	76
219	171	66	10.9	27	1.76	62	0.29	65
200	176	77	18.1	38	1.29	75	0.28	76
200	229	60	8.7	19	2.38	53	0.32	59
238	271	66	9.0	32	3.28	57	0.38	65
210	214	60	7.4	28	2.36	53	0.32	59
310	241	52	8.2	−4	4.15	35	0.41	50
276	276	50	5.0	45	4.30	31	0.42	48
276	310	72	6.8	60	4.10	62	0.45	71
276	241	51	3.1	31	3.71	36	0.39	50
76	76	73	19.8	25	0.20	73	0.10	73
71	62	70	7.2	36	0.16	70	0.09	70
76	71	54	10.0	−13	0.24	53	0.10	54
86	67	65	16.3	−5	0.23	64	0.10	65
171	152	82	47.0	−18	0.74	81	0.74	82
247	195	71	5.8	65	2.15	66	0.33	70

Table 3 The measured values of R_1, R_2, $\alpha_{approach}$, $\alpha_{capture}$, $t_{contact}$ and the calculated ones using immobile, partially mobile and mobile models. The values of h'_{crit} were obtained from Fig. 9 and were used as a criterion for coalescence. The system is glycerin droplets in silicone oil ($\mu_c = 3.1$ Pas) subjected to a shear rate $\dot{\gamma} = 0.14$ s^{-1}

Experimental					Theoretical					
					Immobile		Partially mobile		Mobile	
R_1 (µm)	R_2 (µm)	$\alpha_{approach}$	$t_{contact}$ (s)	$\alpha_{capture}$	$t_{contact}$ (s)	$\alpha_{capture}$	$t_{contact}$ (s)	$\alpha_{capture}$	$t_{contact}$ (s)	$\alpha_{capture}$
219	248	83	43.3	22	11.9	57	7.5	66	23.7	0
209	209	47	6.8	−6	−	−	−	−	8.7	0
209	267	52	7.4	−4	−	−	9.5	4	10.2	0
219	167	80	17.6	55	18.7	21	19.1	18	22.0	0
200	276	81	28.3	51	11.1	56	8.0	64	22.6	0
119	124	75	18.0	15	19	0	−	−	19.3	0
124	124	81	36.8	−42	22.6	0	19.2	21	22.6	0
129	124	79	23.6	3	21.4	0	19.6	11	21.4	0
100	100	71	22.6	−15	17.1	1	−	−	17.2	0
176	62	76	18.9	5	17.6	14	17	17	19.8	0
219	171	66	10.9	27	15	0	−	−	15.1	0
200	176	77	18.1	38	18.4	12	20.1	1	20.3	0
200	229	60	8.7	19	12	0	−	−	12.7	0
238	271	66	9.0	32	7.6	42	8.5	38	15.0	0
210	214	60	7.4	28	12.7	0	−	−	12.7	0
310	241	52	8.2	−4	4.3	34	−	−	10.2	0
276	276	50	5.0	45	4.2	31	−	−	9.6	0
276	310	72	6.8	60	5.4	59	−	−	17.6	0
276	241	51	3.1	31	5	29	8.3	10	9.9	0
76	76	73	19.8	25	13.4	29	14.5	23	18.2	0
71	62	70	7.2	36	10.9	34	11.3	32	16.8	0
76	71	54	10.0	−13	−	−	−	−	10.7	0
86	67	65	16.3	−5	14.2	3	14.5	1	14.6	0
171	152	82	47.0	−18	23.1	0.3	18.5	28	23.2	0
247	195	71	5.8	65	11.1	36	12.2	30	17.3	0

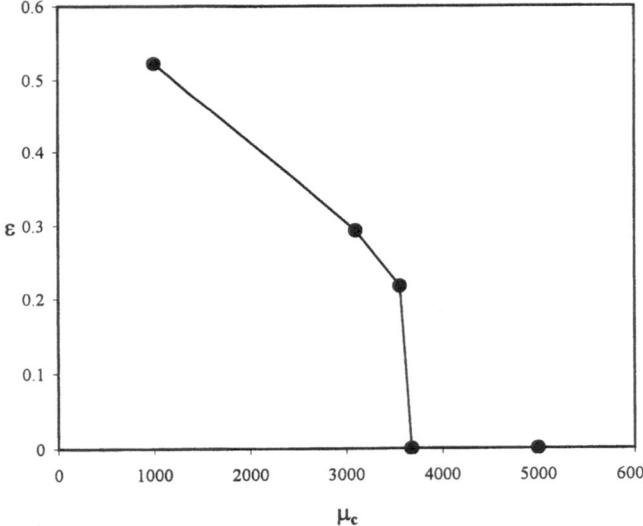

Fig. 10 Experimentally measured values of ε versus silicone oil viscosity at Fl $= 1.2 \times 10^9$ and $q = 0.95 \pm 0.05$

Fig. 8, were calculated using the drainage models. The results of such calculations are portrayed in Fig. 9, which shows, except for the mobile interface, a decrease

followed by an increase in h'_{crit} with Fl. It should be pointed out that the interfaces studied here cannot be mobile ones since the viscosity ratio tends to zero for mobile interfaces. Figure 9 also shows the values of h'_{crit} calculated from Eq. (9). The results of Fig. 9 elucidate the influence of film mobility on coalescence through h'_{crit}. This is in accordance with the results of Klaseboer [22]. The dependence of h'_{crit} on Fl for the immobile and the partially mobile films is best represented by the following equation:

$$h'_{crit} = \Omega^{1/3}\left(\frac{a}{Fl^b} + cFl^d\right), \qquad (11)$$

where a, b, c and d are fitting parameters. This equation indicates a decrease followed by an increase in h'_{crit} as Fl increases. The increase in h'_{crit}, represented by the second term, is perhaps due to some mechanism that enhances coalescence. It is observed experimentally that the distance, h, between two approaching droplets fluctuates when they come close to each other. The results also showed that the fluctuation is larger when the droplets are bigger. This fluctuation could arise from some perturbations that take place at the interface creating some kind of waves or instabilities enhancing coalescence. However this point still needs more investigation.

Theoretically predicted values of $\alpha_{capture}$ and $t'_{contact}$ from the partially mobile and the mobile models, where h'_{crit} calculated from Eq. (9) was used as a criterion for coalescence, are compared to the experimentally measured ones in Table 2 for an oil viscosity of 3.1. A poor agreement between the experimental results and the theoretical predictions is obtained. The immobile interface model (not shown in Table 2) predicts no coalescence. The mobile and the partially mobile interface models predict immediate coalescence between the droplets. The disagreement between the experimental and the theoretical values provides another opportunity to check the validity of Eq. (11) or the form of h'_{crit} versus Fl shown in Fig. 9. To do so the calculations were repeated using values of h'_{crit} obtained from Fig. 9 as a criterion for coalescence. The experimentally measured $\alpha_{capture}$ and $t'_{contact}$ and the predicted ones are listed in Table 3. Tables 2 and 3 clearly illustrate that the dependence of h'_{crit} on Fl in the manner shown in Fig. 9 provides better results than Eq. (9). The contact time predicted was in excellent agreement with the experimental values. The theoretically predicted values of $\alpha_{capture}$ differ slightly from the experimental ones; however, since the measured angles are highly influenced by the time, within experimental error the deviation is plausible. It can also be seen from Table 3 that the mobile interface model predicts zero capture angles. This implies that the interfaces of the droplets are either immobile or partially mobile. All the models were unable to predict negative capture angles. It should be noted that the interfaces were assumed to be clean and devoid of contaminants in the calculations. In practice this is very hard to achieve in spite of all the precautions and provisions one makes. The presence of a very minute amount of contaminant at the interface can have a tremendous effect on the drainage process [22]. An attempt was made to correlate the capture angle to the system parameters, such as the radii of the colliding droplets or Fl, the radii ratio and the approach angle, but the attempt was not successful. A similar observation was also made by Allan and Mason [23].

The effect of the continuous-phase viscosity on the coalescence efficiency can be seen in Fig. 10. As expected ε increases as μ_c decreases (or λ increases) since the film drains more easily when its viscosity is lower. The results are in agreement with those found elsewhere [17, 24–26].

Conclusions

The following conclusions can be made from the results:

1. The coalescence efficiency versus Fl goes through a minimum.
2. Experimental evidence indicates that the critical film thickness decreases then increases as Fl becomes larger, a point that still needs further investigation.
3. The coalescence efficiency increases as the viscosity of the continuous phase decreases.

References

1. Groeneweg F, van Voorst Vader F, Agterof WGM (1993) Chem Eng Sci 48:229
2. Arp PA, Mason SG (1976) Can J Chem 54:3769
3. Guido S, Simeone M (1998) J Fluid Mech 357:1
4. Jaeger PT, Janssen JJM, Groeneweg F, Agterof WGM (1994) Colloids Surf 85:255
5. Abid S, Chesters AK (1994) Int J Multiphase Flow 20:613
6. Chesters AK (1991) Trans Inst Chem Eng 69:259
7. de Bruijn RA (1989) PhD thesis. Eindhoven University, Eindhoven, The Netherlands
8. Tomotika S (1935) Proc R Soc Lond Ser A 153:302
9. Mikami T, Cox RG, Mason SG (1975) Int J Multiphase Flow 2:113
10. van de Ven TGM, Mason SG (1977) Colloid Polym Sci 255:468
11. van de Ven TGM (1982) Adv Colloid Interface Sci 17:105
12. Patlazhan SA, Lindt JT (1996) J Rheol 40:1095
13. Wang H, Zinchenko AZ, Davis RH (1994) J Fluid Mech 265:161
14. Brazier-Smith PR, Jennings SG, Latham J (1972) Proc R Soc Lond Ser A 326:393
15. Rother MA, Zinchenko AZ, Davis RH (1997) J Fluid Mech 346:117
16. Wright H, Ramkrishna D (1994) AIChE J 40:767
17. Kumar S, Kumar R, Gandhi KS (1993) Chem Eng Sci 48:2025
18. Muralidhar R, Ramkrishna D (1986) Ind Eng Chem Fundam 25:554
19. Adler PM (1981) J Colloid Interface Sci 83:106
20. Mousa H, van de Ven TGM (1991) Colloids Surf 60:39
21. Chesters AK, Bazhlekov IB (2000) J Colloid Interface Sci 230:229
22. Klaseboer E (1998) PhD thesis. Institute National Polytechnique de Toulouse
23. Allan RS, Mason SG (1962) J Colloid Interface Sci 17:383
24. Bazhlekov IB, Chesters AK, van de Vosse FN (2000) Int J Multiphase Flow 26:445
25. MacKay, Mason SG (1963) Can J Chem Eng 41:203
26. Zhang X, Davis RH (1991) J Fluid Mech 230:479

Progr Colloid Polym Sci (2001) 118: 216–220
© Springer-Verlag 2001

Rheological properties of highly concentrated fluorinated water-in-oil emulsions

Valéry G. Babak
Anne Langenfeld
Nathalie Fa
Marie José Stébé

V. G. Babak
A. Nesmeyanov Institute of
Organoelement Compounds
Russian Academy of Sciences
28, Vavilova str., Moscow 117813
Russia

V. G. Babak · A. Langenfeld · N. Fa
M. J. Stébé (✉)
Equipe Physico-Chimie des Colloïdes,
UMR N°7565 CNRS-Université
H. Poincaré, Faculté des Sciences, B.P. 239
54506 Vandoeuvre les Nancy cédex
France
e-mail: stebe@lesoc.uhp-nancy.fr
Tel.: +33-383-912369
Fax: +33-383-912532

Abstract A new rheological model for highly concentrated emulsions containing 77–98% of the dispersion phase is suggested which relates the macroscopic functional properties of these systems (elasticity modulus, yield stress and yield strain) to the microscopic physicochemical parameters (droplet size, interfacial tension, surface forces acting in thin liquid films, specific surface of these films, adhesion force between the droplets, their deformability, etc.). Whereas Princen's model describes only the effect of the capillary pressure and the volume fraction of the internal phase on the elasticity modulus of such emulsions, the new model also predicts the effect of the adhesion free energy between the droplets (or the contact angle between the droplets) on the elasticity modulus and the yield stress and strain of these emulsions. The model proposed is applied to explain and systematize the effect of physicochemical parameters on the rheological properties of highly concentrated fluorinated water-in-oil emulsions.

Key words Concentrated emulsion · Reverse emulsion · Fluorinated emulsion · Rheological model

Introduction

The rheological behavior of highly concentrated emulsions, or high-internal-phase-ratio emulsions [1], is strongly influenced by the viscoelastic properties of liquid droplets and by the contact interaction between them. At relatively low volume fractions, φ_v, of the dispersed phase, the droplets are almost spherical in shape; however, with increasing φ_v, the droplets are distorted and the structure of these emulsions becomes similar to that of ordinary foams. Consequently, these emulsions are also called "biliquid foams" [2, 3]. The ensemble of compressed and interacting liquid droplets forms a gel-like structure that manifests viscoelastic behavior and plasticity owing to the high specific interfacial area and the adhesion between the droplets. This justifies these systems to be designed as gel–emulsions [4–6].

The gel–emulsions (with φ_v of the order of about 0.8–0.99) may be obtained spontaneously by slow mechanical agitation from micellar solutions (microemulsions) of nonionic surfactants [7–9]. The relatively low interfacial tension, $\sigma_0 \leq 1$ mNm^{-1}, which is typical for these systems, on the one hand, and the high resistance to rupture of microscopic emulsion films stabilized by the nonionic surfactants, on the other, favor the formation of high φ_v without applying a remarkable osmotic pressure (Fig. 1). This explains the exceptional stability and the high rheological parameters (elasticity, viscosity, yield stress and strain) of fluorinated water-in-oil (w/o) gel–emulsions [10–12].

The attraction (the adhesion force, f^*) between liquid droplets may influence the character of the dependence of the shear-storage elasticity modulus, G', the yield stress, τ^*, and the yield strain, ε_{max}, of highly concentrated emulsions on the physicochemical parameters, such as the interfacial tension, σ_0, the mean droplet radius, R, φ_v of the dispersed phase, etc. Although the proportionality of G' to the ratio σ_0/R holds for all known theoretical or semiempirical relationships [13–17], the dependence of

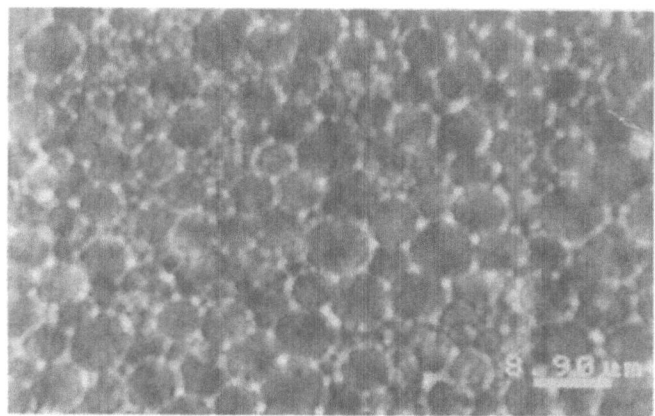

Fig. 1 Highly concentrated fluorinated emulsion (system: $C_8^F E_2/$ perfluorodecalin/water; $\varphi_v = 0,95$; $r_0 = $ perfluorodecalin/$C_8^F E_2 = 5$

G' on φ_v is sometimes controversial and is the subject of discussion [18–20]. The well-known semiempirical formula of Princen [14],

$$G' = C\left(\frac{\sigma_0}{R}\right)\varphi_v^{1/3}(\varphi_v - \varphi_c) \quad \varphi_c < \varphi_v < 1 \;, \tag{1}$$

with $C = 1.77$ and $\varphi_c = 0.74$, which was verified experimentally for different types of highly concentrated polydisperse emulsions [21–24], is in agreement with the limit $G'_{max} \cong 0.5(\sigma_0/R)$ as $\varphi_v \rightarrow 1$ predicted for the elasticity modulus of highly concentrated emulsions which behave as dry foams [13, 15].

As far as the plastic properties of these emulsions and their deformability are concerned, there is no convenient model which is able to explain the effect of physicochemical factors (such as droplet size, φ_v, hydrophile–lipophile balance of the surfactants, temperature, etc.) on the values of τ^* and ε_{max}. For example, the higher values of G', τ^* and ε_{max} of fluorinated emulsions with respect to the hydrogenated systems are currently not well understood. In spite of the presumably important role which could be played by the attractive forces in the viscoelastic and plastic

behavior of highly concentrated emulsions [16, 25], no attempts have been made to quantify and describe this influence in terms of the thermodynamics of the surface forces and the microrheology.

In this communication we present some preliminary results concerning the thermodynamic study of the effect of the attractive forces acting in thin liquid films between liquid droplets on the visco-elasto-plastic properties of highly concentrated adhesive emulsions.

Materials and methods

The fluorinated oil used in the w/o emulsions is perfluorodecalin purchased from Interchim and used as received. The nonionic fluorinated surfactants of the chemical formula C_mF_{2m+1}–$C_2H_4SC_2H_4$–$(OC_2H_4)_n$–OH (noted $C_m^F E_n$) were synthesized and characterized as described in Refs. [8, 10]. The surfactants $C_8^F E_2$, $C_6^F E_2$ and $C_6^F E_3$ were used to prepare these emulsions.

The emulsions with a water volume fraction of $\varphi_v = 0.95$ and an oil-to-surfactant mass ratio of $r_0 = 5$ were prepared by mechanical stirring as described in Ref. [10]. The mean droplet radius of these emulsions was determined by microscopy.

The rheological measurements (determination of G', τ^* and ε_{max}) were performed using a stress-controlled Carri-Med CSL500 rheometer with a cone/plate geometry as described in Ref. [10]. A spinning drop (Texas University) instrument, modified for temperature regulation, was used in order to perform the interfacial measurements.

Model for elasticity modulus of highly concentrated emulsions

The known rheological models explain the dependence of G' of highly concentrated emulsions on the ratio σ_0/R; however, these models cannot explain the effect of physicochemical factors on τ^* and ε_{max} of these emulsions.

The suggested rheological model for highly concentrated emulsions is based on the following assumptions (Fig. 2).

1. The macroscopic quantity, the tensile stress, σ_{zz}, acting inside an emulsion, on a microscopic level is a

Fig. 2 Scheme illustrating the tensile deformation of **a** a model emulsion and **b** an emulsion droplet

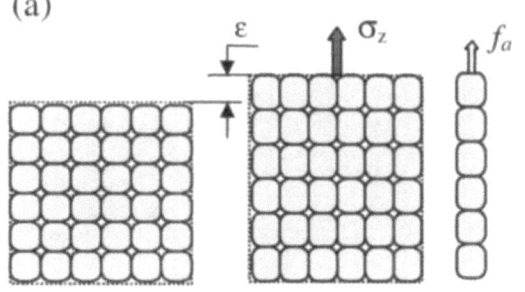

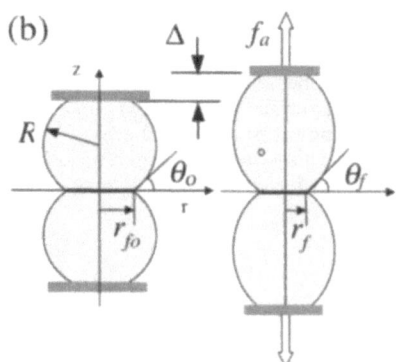

nonhomogeneous discrete quantity which is localized in the contact regions between the droplets and may be represented as an ensemble of forces, f_a, acting between the droplets. As a first approximation, σ_{zz} is assumed to be uniformly distributed over all the contacts between the droplets forming a gel-like network

$$\sigma_{zz} = n_\Sigma f_a , \tag{2}$$

where n_Σ is the number of droplets in a unit area of the layer.

2. The deformation of each emulsion droplet is elastic (i.e. obeys Hooke's law):

$$f_a = K\Delta \quad \text{and} \quad \sigma_{zz} = E\varepsilon , \tag{3}$$

where K is the "rigidity" coefficient of the droplet (considered as a spring) and $\varepsilon = \Delta/2R$ is the tensile strain.

It has been shown [26] that for small deformations of the droplets, K is a linear function of the interfacial tension, σ_0,

$$K \cong 2\pi\sigma_0 . \tag{4}$$

By using an appropriate expression for n_Σ, for example, as in Princen's model,

$$n_\Sigma = S(\varphi_v)/(2R)^2 , \tag{5}$$

with $S(\varphi_v) = C\varphi_v^{1/3}(\varphi_v - \varphi_c)$, where $C = 1.77$ and $\varphi_c = 0.74$ for polydisperse emulsions, and taking into account the relationship $E \cong 3G'$ [27, 28], one obtains

$$G' = 1.77\left(\frac{\sigma_0}{R}\right)\varphi_v^{1/3}(\varphi_v - \varphi_c) . \tag{6}$$

Therefore, the model proposed describes correctly the observed experimental dependence $G' \sim (\sigma_0/R)$ although it does not rely upon the proportionality of G' to the capillary pressure, P_c, in the droplets, which is also proportional to the ratio σ_0/R. The coincidence of the numerical constants in Eqs. (1) and (6) is surprising taking into account that rather rough approximations were made to obtain this expression.

Expressions of yield stress and yield strain for highly concentrated emulsions

The tensile yield stress, σ_b, of a highly concentrated emulsion is proportional to a mean adhesion force, f_a^*, between two emulsion droplets,

$$\sigma_b = n_\Sigma f_a^* . \tag{7}$$

From the elaborate theory of the adhesion of fluid particles in liquid media [29] it follows that

$$f_a^* \cong \pi R \Delta_{ad}\Omega . \tag{8}$$

The adhesion free energy, $\Delta_{ad}\Omega$, may be estimated theoretically on the basis of the appropriate disjoining pressure isotherm, $\Pi(h)$, for a thin liquid film [30] (Fig. 3),

$$\Delta_{ad}\Omega \cong \int_{H_f}^\infty \Pi(h)\mathrm{d}h , \tag{9}$$

where H_f is the liquid film thickness.

$\Delta_{ad}\Omega$ may be also estimated experimentally from the measurements of the interfacial tension, σ_0, and the contact angle, θ_f, between the droplets (Fig. 3) as [30, 31]

$$\Delta_{ad}\Omega = 2\sigma_0(\cos\theta_f - 1) - \tau/r_f , \tag{10}$$

where $\tau \sim 10^{-11}$ N is the line tension and r_f is the radius of the liquid film.

From Eqs. (5), (7), (8) and (9) one obtains

$$\sigma_b \cong \tau^* \cong |f_a^*|\frac{S(\varphi_v)}{(2R)^2} \cong \frac{\pi}{4}\frac{|\Delta_{ad}\Omega|}{R}S(\varphi_v) , \tag{11}$$

i.e. σ_b may be predicted from the measurements of the radius of the droplet, σ_0 and θ_f. By neglecting the term τ/r_f in Eq. (10) and expanding $\cos\theta_f$ in a series, one obtains $\Delta_{ad}\Omega \cong -\sigma_0\sin^2\theta_f$ and finally

$$\sigma_b \cong \tau^* \cong \frac{\sigma_0}{R}\sin^2\theta_f S(\varphi_v) . \tag{12}$$

The significance of the result obtained consists in the possibility of predicting σ_b of gel–emulsions from a measurement of θ_f in thin liquid films (Eq. 10) or from the theoretical estimation of $\Delta_{ad}\Omega$ according to Eq. (9). Inversely, from the measurement of σ_b, one may estimate

Fig. 3 Scheme illustrating the effect of surface forces (the disjoining pressure) on the profile of a thin liquid film between two emulsion droplets

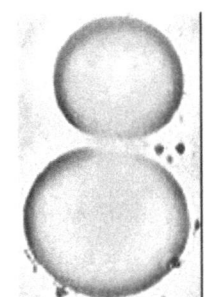
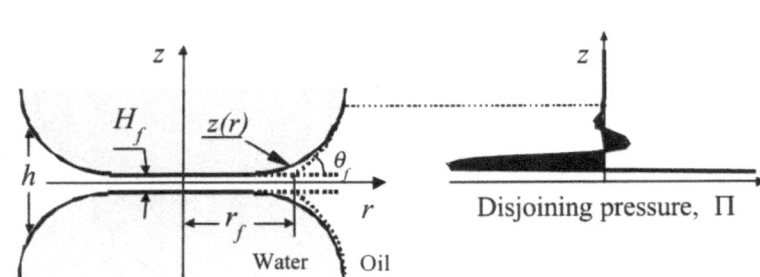

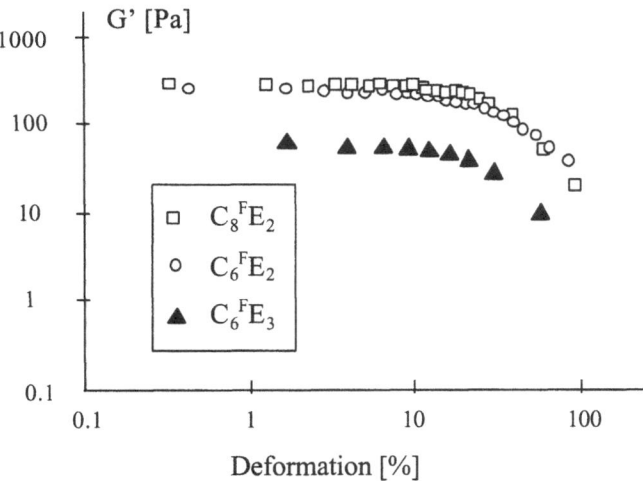

Fig. 4 Effect of the molecular structure of fluorinated surfactants on the elasticity of gel–emulsions

the value of $\Delta_{ad}\Omega$ and proceed to the systematization of the effect of surface forces on the rheological properties of these emulsions.

ε_{max} of gel–emulsions may be expressed using Eqs. (3), (4) and (8),

$$\varepsilon_{max} = \frac{f_a^*}{4\pi R\sigma_0} \cong \frac{1}{4}\frac{\Delta_{ad}\Omega}{\sigma_0} \cong -\frac{1}{4}\sin^2\theta_f, \qquad (13)$$

where the negative sign of ε_{max} is conventionally attributed to extension strains. Therefore, by using thermodynamic considerations, ε_{max} may be predicted from measurements of θ_f in emulsion films. It should be pointed out that the macroscopic parameter of gel–emulsions, ε_{max}, is identified as the ratio (the "adhesiveness" parameter [26]) $v_{ad} = \Delta_{ad}\Omega/\sigma_0$, which characterizes the surface interaction in microscopic liquid films.

Application of the suggested model to fluorinated gel–emulsions

Rheological measurements on the highly concentrated fluorinated emulsions (Fig. 4, Table 1) show that G' of gel–emulsions is very sensitive to the hydrophilicity of the

surfactants used to stabilize these systems and decreases remarkably from about 250 Pa to about 100 Pa with increasing length of the hydrophilic chain of the surfactants by only one oxythylene group. At the same time, the region of linear deformation (ε_{max}) of these gel–emulsions undergoes a sharp decrease from 14 to 9%. It should be pointed out that the mean radius of the emulsion droplet was almost the same in all cases (2 μm).

The general trend of a diminishing G' corresponds to the decrease in σ_0, whereas the increase of the oxythylene chain length induces an increasing phase inversion temperature [32] of the system. However, the relative values of the decreases in G' is much greater than the corresponding decrease in σ_0 (or σ_0/R). This lets us assume that other factors (e.g. $\Delta_{ad}\Omega$ acting between the droplets in thin liquid films) could contribute to the decrease in G'. The sharp decrease in τ^* and ε_{max}, which are very sensitive to $\Delta_{ad}\Omega$ according to Eqs. (11) and (13), confirms this assumption. The mechanism of the decrease in $\Delta_{ad}\Omega$ with an increase in the hydrophilicity of a surfactant molecule is not understood for the moment. By using thermodynamic considerations, this may signify the increase of the film thickness, H_{max}, and consequently the decrease in θ_f in the film. An alternative explanation consists of decreasing the depletion forces (which join droplets) provided from the osmotic pressure of micelles in the oil phase. It should be noted that the observed decrease in ε_{max} with temperature (Fig. 5) may be rationally explained only by assuming that $\Delta_{ad}\Omega$ also decreases with temperature.

Conclusion

The elaborate rheological model of highly concentrated emulsions correlates well with Princen's model explaining the proportionality of G' of these emulsions to the ratio σ_o/R. As far as the yield rheological parameters

Table 1 Effect of the molecular structure of the fluorinated surfactants on the rheological properties of water-in-oil emulsions

Surfactant	Hydrophile–lipophile balance (Griffin)	Interfacial tension (mN/m)	Storage modulus (Pa)	Yield strain (%)	Yield stress (Pa)
$C_8^F E_2$	3.4	0.8	250	14	35
$C_6^F E_2$	4.1	0.6	270	9	24
$C_6^F E_3$	5.4	0.4	80	3	3

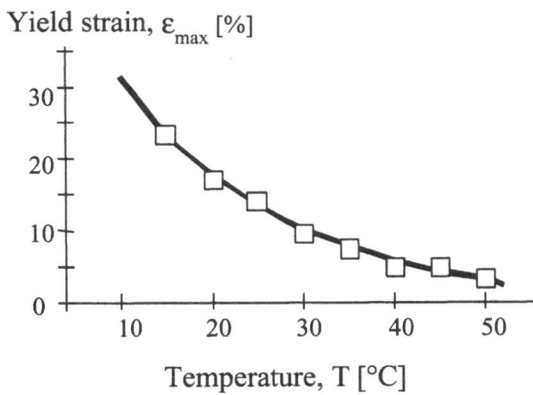

Fig. 5 Effect of the temperature on the yield strain. ε_{max} of a gel–emulsions stabilized by $C_8^F E_2$

($\tau*$ and ε_{max}) are concerned, the model predicts the following dependences

$$\varepsilon_{max} \sim \Delta_{ad}\Omega/\sigma_0 \sim \sin^2 \theta_f$$

and

$$\tau* \sim \Delta_{ad}\Omega R.$$

The model is proposed based on the assumption that the surface forces acting in thin liquid films between the emulsion droplets play an important role in the rheological behavior of gel–emulsions. This model permits the explanation (or at least the systematization) of the effect of different physicochemical parameters on the rheological properties of gel–emulsions.

References

1. Lissant KJ (1966) J Colloid Interface Sci 22:462
2. Princen HM (1987) Langmuir 3:36
3. Sonneville-Aubrun O, Bergeron V, Gulik-Krzywicki T, Jönsson B, Wennerström H, Linder P, Cabane B (2000) Langmuir 16:1566
4. Pons R, Erra P, Solans C, Ravey JC, Stébé M-J (1993) J Phys Chem 97:2320
5. Ravey JC, Stébé M-J, Sauvage S (1994) Colloids Surf 91:237
6. Pons R, Carrera J, Erra P, Kunieda H, Solans C (1994) Colloids Surf 91:259
7. Ravey JC, Stébé M-J (1989) Physica B 394:156
8. Ravey JC, Stébé M-J (1990) Prog Colloid Polymer Sci 82:218
9. Kunieda H, Fukui Y, Uchiyama H, Solans C (1996) Langmuir 12:2136
10. (a) Langenfeld A, Lequeux F, Stébé MJ, Schmitt V (1998) Langmuir 14:6030; (b) Langenfeld A, Schmitt V, Stébé MJ (1999) J Colloid Interface Sci 218:522–528
11. Caldero G, Garcia-Celma MJ, Solans C, Stébé M-J, Ravey JC, Rocca S, Pons R (1998) Langmuir 14:1580
12. Rocca S, Muller S, Stébé MJ (1999) J Controlled Release 61:251
13. Derjaguin BV (1933) Kolloid 64:1
14. Princen HM. (1986) Langmuir 2:519
15. Stamenovic D (1991) J Colloid Interface Sci 145:255
16. Mason TG, Krall AH, Gang H, Bibette J, Weitz DA (1996) In: Becher P (ed) Encyclopedia of emulsion technology. Dekker, New York, p 299
17. Mason TG, Lacasse MD, Grest GS, Levine D, Bibette J, Weitz DA (1997) Phys Rev E 56:3150
18. Mason TG, Bibette J, Weitz DA (1996) J Colloid Interface Sci 179:439
19. Mason TG (1999) Curr Opin Colloid Interface Sci 4:231
20. Hemar Y, Horne DS (2000) Langmuir 16:3050
21. Ebert G, Platz G, Rehage H (1988) Ber Bunsenges Phys Chem 92:1158
22. Jager-Léger N, Tranchant JF, Alard V, Vu C, Tchoreloff PC, Grossiord JL (1998) Rheol Acta 37:129
23. Pons R, Taylor P, Tadros TF (1997) Colloid Polym Sci 275:769
24. Taylor P (1996) Colloid Polym Sci 274:1061
25. Manoj P, Watson AD, Hibberd DJ, Fillery-Travis AJ, Robins MM (1998) J Colloid Interface Sci 207:294
26. Babak VG, Stébé M-J, to be published
27. Landau LD, Lifshitz EM (1959) Theory of elasticity. Pergamon, London
28. Buzza DMA, Cates ME (1994) Langmuir 10:4503
29. Babak VG (1993) Russ Chem Rev 62:703
30. De Feijter JA, Vrij A (1972) J Electroanal Chem Interfacial Electrochem 37:9
31. Babak VG (1992) Russ Chem Rev 61:975
32. Shinoda K, Friberg S (Eds) (1986) Emulsions and Solubilization, Wiley, New York

Progr Colloid Polym Sci (2001) 118: 221–225
© Springer-Verlag 2001

RHEOLOGY

Are particle gels "glasses"?

G. Foffi
E. Zaccarelli
P. Tartaglia
F. Sciortino
K. A. Dawson

G. Foffi (✉) · E. Zaccarelli
K. A. Dawson
Irish Centre for Colloid Science and
Biomaterials, Department of Chemistry
University College Dublin,
Belfield, Dublin 4, Ireland
e-mail: giuseppe@fiachra.ucd.ie
Tel.: +353-1-7062418
Fax: +353-1-7062415

P. Tartaglia · F. Sciortino
Dipartimento di Fisica,
Università di Roma La Sapienza,
Rome, Italy

P. Tartaglia · F. Sciortino
Istituto Nazionale di Fisica della Materia
Unità di Roma La Sapienza, Piazzale
Aldo Moro 2, 00185 Rome, Italy

Abstract We propose an analogy
between a new type of glass, recently
found within the mode coupling
theory framework, and a particle
gel, experimentally observed in colloidal suspensions where the particles have attractive interactions. We
report the study of a colloidal system
model, made of particles with hard
core interacting via an attractive
square-well potential. The well-
width has a range much shorter than
the particle diameter. We find new
phenomena in the temperature-
composition plane related to the
width of the attractive interactions,
namely a re-entrant behaviour in the
'phase' diagram and a coexistence
line between two types of glasses.
One has been identified as the commonly studied colloidal glass and the
other as a new type, the 'attractive'
glass, that can be viewed as a particle
gel. The coexistence line terminates
at an end-point, named A_3, after
which the gel and the colloidal glass
become indistinguishable. We also
show characteristic features of the
normalised density correlators, for
the gel at a relatively low density and
close to this singularity point, where
the gel and the colloidal glass start to
coexist. For the latter it is remarkable to note that the density correlators show a logarithmic time
decay.

Key words Colloidal systems ·
Square-well potential Glass
transition · Particle gels ·
Nonergodic systems

Introduction

Systems of colloidal particles have long been of both
practical and scientific importance, and there has been a
considerable growth of knowledge in recent years in the
area of dense colloidal systems. However, particle gels
and the process of gelation itself have not been much
studied at a very fundamental level, despite the practical
importance of this part of the field. The reason is
probably that there are few theoretical approaches within
which to rationalise the information. We present here a
new way of viewing these systems, based on developments already well established in glass theory.

For colloidal particle systems with short-range attractions, it was quite natural to argue by analogy to all
of the phenomena present for molecules. Thus, we expect
to find liquids, gases, and crystals, and perhaps some sort
of dense but imperfectly packed or glassy state. This
perspective, whilst quite reasonable in many regards,
does not take into account the great difference in energy
scales and length scales between molecules and colloidal
particles. The hard core of a colloidal particle might be
on the scale of a micron, and the attraction, tunable using
a variety of solvents and other additives, might have a
range of only a few percent of this [1, 2, 3]. Typically,
though not always, the repulsion is very hard and short-
ranged, being mainly derived from the material properties of the particle itself. In addition, the flexibility in the
use of solvents and additives may lead to very strong
effective attractions, and this combined with the large

mass of the particle means that attractive interactions might easily overcome the tendency of the particles to disperse, even if the true entropic interaction balance is more favourable to a dispersion. So, far from being a rare occurrence, much effort is devoted to preventing colloidal particles from "collapsing" into a condensed phase, a precipitate, or a "gel", the result usually considered to be poorly characterised, or characterisable. Thus, whilst there are clear analogies between molecular behaviour and colloidal systems, we may also need to look again more carefully at the nature and prominence of all the phenomena, and we may expect new features to emerge and a change in the relative importance of existing phenomena. In fact, as we will show, the result of such a reexamination may well lead us to be able to systematise colloidal phenomena long considered to be inconvenient, rather than scientifically interesting.

We chose as a model of attractive colloidal particles the square-well potential. Such a choice was driven by two reasons: firstly, the square-well potential is a good approximation for many colloidal systems (i.e. colloids with depletion interaction, grafted colloids, etc. and, secondly, it possesses many relevant characteristics of a vast range of potentials i.e. hard-core repulsion and short-range interaction). This model has been widely used to describe interacting colloidal solutions and, to some extent, phenomena such as depletion interaction and grafted colloids are well represented by it [3–7].

Our idea is that the glass theoretical framework, in particular the mode-coupling theory (MCT) [8], that has been introduced and extensively used to describe the glass transition for both simple liquids and colloidal systems, can also be extended to describe other nonergodic states of matter, such as particle gels. The new results recently found for a Yukawa [9, 10] and a square-well potential [11] for an 'attractive glass', that will be discussed later in detail led us to propose an analogy between this nonergodic state and a gel. In brief, we propose that the ergodic state (fluid) can be considered to be a sol phase, the "repulsive" glass, also discussed later, is the commonly studied colloidal glass (hard-sphere type), and the "attractive" glass is a particle gel. Thus, what follows will be discussed in these terms.

Mode-coupling theory

We have already alluded to the fact that we use glass theory to study the square-well potential. The reason is that we shall be looking for transitions to a nonergodic state that could represent the process of gelation. The most practical theory to use in this respect is MCT, which has previously been shown to describe colloidal glasses driven by packing forces or pure repulsive interactions and, indeed, has been found to be in very good agreement with experiments [12]. The theory is, in outline, as follows.

The MCT equations of motions for the normalised density correlators, $\Phi_q(t) = \langle \rho_q^*(t)\rho_q \rangle / \langle |\rho_q|^2 \rangle$, for a colloidal system [13] are

$$\tau_q \dot{\phi}_q(t) + \phi_q(t) + \int_0^t m_q(t-t')\dot{\phi}_q(t')\mathrm{d}t' = 0 , \tag{1}$$

where $\tau_q = \mathrm{v}S(q)/(vq)^2$, with $S(q)$ being the static structure factor, v the thermal velocity, and the approximation of the instantaneous friction is the constant v. The kernel $m_{t\,q}$ is given as $m_q(t) = \mathscr{F}_q(\{\Phi_k(t)\})$, where the mode-coupling functional $\mathscr{F}_q$ is,

$$\mathscr{F}_q(\{f_k\}) = \frac{1}{2} \int \frac{\mathrm{d}^3 k}{(2\pi)^3} V_{\vec{q},\vec{k}} f_k f_{|\vec{q}-\vec{k}|} , \tag{2}$$

$$V_{\vec{q},\vec{k}} \equiv S_q S_k S_{|\vec{q}-\vec{k}|} \frac{n}{q^4} [\vec{q}\cdot\vec{k}c_k + \vec{q}\cdot(\vec{q}-\vec{k})c_{|\vec{q}-\vec{k}|}]^2 , \tag{3}$$

where $c_q = (1 - S_q^{-1})/\rho$ is the Fourier-transformed direct correlation function. In order to locate and characterise the gel phase we define the nonergodicity parameter (or Edwards–Anderson parameter) as the long-time limit of the density–density correlation function, $f_q = \lim_{t\to\infty} \Phi_q(t)$. It is clear that if the system is ergodic the correlation function will decay to zero after a certain time; in contrast when the system is in a nonergodic regime, the density fluctuations will not be able to relax and, consequently, the function f_q will have a finite value. Indeed, the fact that from dynamic light scattering $\Phi_q(\infty)$ is nonzero, is a frequent observation as the gel transition is crossed. We numerically solved Eq. (1) on a grid of 2000 equally spaced q values extending up to $q\sigma = 72$; f_q is obtained by an iterative solution of the bifurcation equation,

$$\frac{f_q}{1-f_q} = \mathscr{F}_q(\{f_k\}) , \tag{4}$$

that corresponds to the long time limit of Eq. (1). From Eq. (4) is evident that $f_q = 0$ is always a solution but it is not always a stable one: at the transition a new solution $f_q \neq 0$ emerges owing to the formation of a nonergodic phase. The most striking feature of MCT is the fact that it can produce dynamics using as an input only static quantities (i.e. static structure factor and number density). Therefore it is possible, by only providing the structure factor of the system, to locate the ergodic–nonergodic transitions. This brief exploration of the MCT results is not by any means exhaustive and for greater insight we suggest the reader explore the literature [8, 14].

Results

For the interaction of the colloidal particles we have chose a square-well potential,

$$\beta V(r) = \begin{cases} \infty & 0 < r < \sigma \\ \beta u_o & \sigma < r < \sigma + \Delta \\ 0 & \sigma + \Delta < r \end{cases}, \qquad (5)$$

where $\beta = (k_B T)^{-1}$, where k_B is Boltzmann's constant. In order to obtain the structure factors we solved the Percus–Yevick equation for this model; details of this procedure can be found in Ref. [11]. In contrast to the hard-sphere case, where the volume fraction, ϕ, is the only control parameter, the ratio $u_0/k_B T$ between the well depth and the temperature and the well width, $\Delta\phi$, are important control parameters as well. In what follows will make use, in order to characterise the width of the attraction, of the adimensional parameter $\epsilon = \Delta/(\Delta + \sigma)$. We remind the reader that for polymer-induced depletion interactions the well width is controlled by the size of the polymer and the well depth increases with polymer concentration.

The "phase" diagram for the case $\epsilon = 0.03$ in the thermodynamic plane (ϕ, T) is represented in Figs. 1 and 2; the lower density region is reproduced in Fig. 1, whereas the high density regime is shown in Fig. 2. In Fig. 1 we present the sol–gel (fluid–attractive glass) transition line together with the underlying gas–liquid spinodal. It is remarkable to note that the gel transition line in this region lies above the coexistence curve. This type of behaviour may indeed have been observed experimentally for effective potentials with a very narrow range of attractions [1, 5]. We believe that in this region the attractive gel is a space-spanning structure of particles strongly attached to each other; however, we caution the reader that on the left-hand side of the critical point and in its vicinity, we cannot accept blindly the MCT results. Further work is necessary to elucidate this region. Some of these concerns have been addressed in

Ref. [15] while discussing the number of bonds present in these gel states.

In Fig. 2 firstly we note that for $T \to \infty$ we recover the glass-transition packing fraction, already found in MCT, for hard spheres (i.e. $\phi_g \simeq 0.516$). In this case the kinetic arrest is understandable in terms of a cage effect, i.e. each particle is trapped in a cage formed by its near neighbours and consequently the system is frozen. This is what we call "repulsive" glass, commonly manifested as the typical colloidal glass [12]. By decreasing the temperature, the attractions start to be relevant and the packing fraction at which the system freezes becomes larger than the hard-sphere one. This unusual reentrant behaviour continues up to a certain temperature where the line joins another branch of the transition line. The latter line extends towards lower densities and along it the arrest is due to the attraction, i.e. at low temperatures the transition is driven by the fact that the particles tend to stick together. We have named this second state the "attractive" glass or particle gel. The process of the arrest, glass or gel formation, is then driven by repulsion along the "vertical" line and by attraction on the "horizontal" one. The vertical repulsive line is explained by the fact that glassification is driven by the hard core, which lacks any energy scale, whilst the attractive line, being fairly horizontal, implies that there is a single, fairly well characterized energy scale that drives the gel formation. These simple observations essentially determine much of the shape of the phase diagram as a vertical line at roughly $\phi \simeq 0.52\%$ and a horizontal line at the characteristic energy (temperature) scale of gelation. An interesting feature to emphasise is the presence of a characteristic reentrant behaviour, corresponding to the presence of a liquid phase between the two glassy phases.

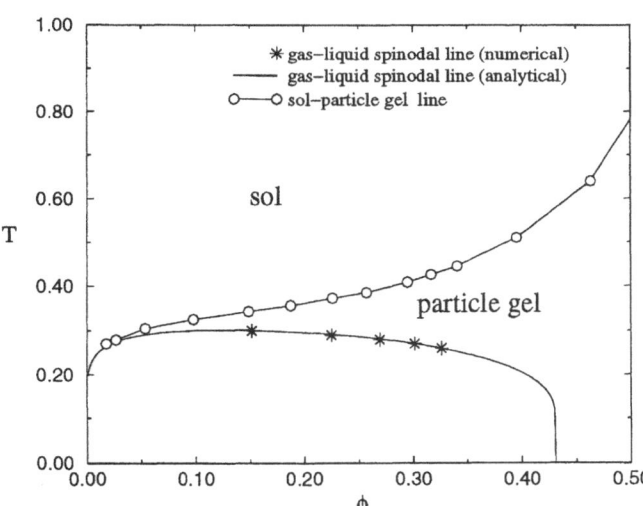

Fig. 1 Sol–gel transition line at low packing fraction for $\epsilon = 3\%$

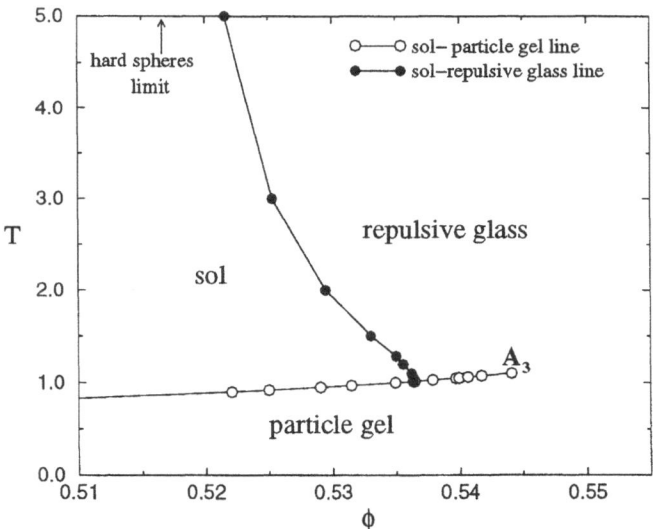

Fig. 2 Same as Fig. 1 for high packing fraction. A_3 is the endpoint of the gel–repulsive glass transion line

This behaviour is encountered in MCT calculations for different kinds of attractive potentials and was shown to be strictly related to the choice of a very narrow well width [9, 16, 11].

Another important feature emerging from Fig. 2 is the presence of a transition line between the two nonergodic states. Crossing it, the system passes from a repulsive glass to a gel abruptly, as the nonergodicity parameter shows a discontinuous behaviour. Thus, this line corresponds to the remarkable phenomenon of a gel–repulsive glass coexistence. It would be interesting to have unambiguous confirmation of this phenomenon, in the light of the current predictions [6].

The "order parameter" of the transition is the nonergodicity parameter. It shows very different behaviour in the two types of arrested states. For a repulsive glass, it remains almost unchanged with temperature, whereas for a gel it varies consistently, and in particular its range increases by decreasing the temperature. The range of f_q is related to the localisation length of the particles [17]. Thus, while for a repulsive glass this length (i.e. the size of the cage) remains almost the same with varying temperature, for the attractive gel the particles become more and more localised, strengthening the attractions between them. The latter phenomenon is the indication that in the gel the particle arrest is due to the formation of bonds between particles at close distance. Indeed, the bond formation has been seen as an important issue for colloidal aggregation and it has been studied within MCT [9, 10, 15]. It is important to note that the glass–glass transition line presents an end

point (labeled A_3 in MCT notation) after which the two gels become indistinguishable (i.e. f_q varies continuously).

The behaviour of the phase diagram on varying the range of the attractions has been studied in Ref. [11]. Here, we limit ourselves to report the most important features of that study. Firstly, the typical reentrant behaviour that have discussed for $\epsilon = 3\%$ tends to vanish, increasing the attractive range. A second and more important feature is related to the behaviour of the gel–repulsive glass transition line. For values of ϵ roughly between 3 and 4% this line shrinks and eventually, for a certain value (i.e. $\epsilon \simeq 4.11\%$), the end point A_3 touches the sol–gel transition line, giving origin to a very peculiar point, corresponding to a higher-order singularity in MCT and is thus referred as the A_4 point [11].

The importance of finding these singularities A_3 and A_4 lies in the fact that in the proximity of them MCT predicts for the intermediate scattering function, $\Phi_q(t)$, instead of the typical two-step relaxation scenario [8] a very peculiar logarithmic decay.

For $\epsilon \geq 4.11\%$ the A_4 singularity disappears and the transition between the gel and a repulsive glass becomes continuous along the transition curve.

The behaviour of $\Phi_q(t)$ at the constant packing fraction value $\Phi = 0.340349$ is shown in Fig. 3. The data correspond to the wave vector $q = 10.0$, with the temperature being varied as indicated in the figure. In the phase diagram (Fig. 1), the case represented can be found to be far both from the underlying critical point for the gas–liquid transition and from the singularity A_3.

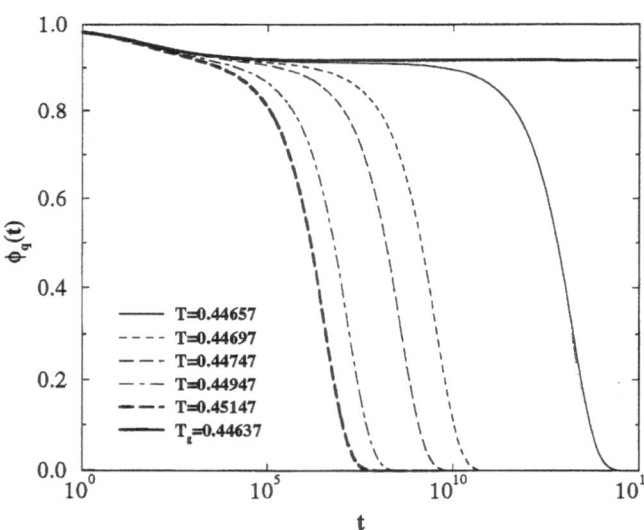

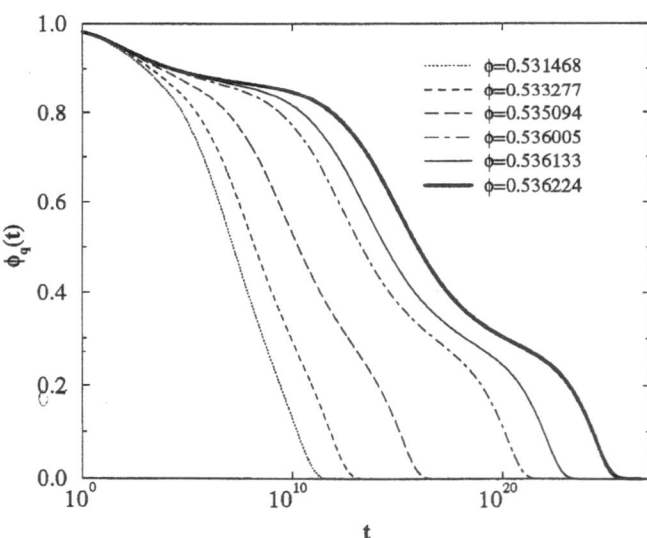

Fig. 3 Correlators $\Phi_q(t)$ for $q = 10.0$ at constant packing fraction $\Phi = 0.340349$ for various temperatures, close to and at the sol–particle gel transition. A typical two-step relaxation is developed the closer the system is to the transition temperature, T_g, finally approaching a nonzero plateau at T_g itself

Fig. 4 Correlators $\Phi_q(t)$ for $q = 10.0$ at constant temperature $T = 1.0132$ for various packing fractions, close to and at the matching point between the sol–particle gel line and the sol–repulsive glass line. A logarithmic decay is observed for various decades in time, and approaching the transition ($\Phi_g = 0.536270$), the presence of the two different types of arrested states is also eviden

Thus, it represents a typical two-step relaxation function, as predicted from MCT [8], for a sol–particle gel transition which is more evident the closer the system is to the transition temperature $T_g = 0.44637$.

In contrast to the previous case, we show the behaviour of the intermediate scattering function close to the singularity A_3 (Fig. 2). To see the relaxation dynamics, we approach from the fluid side at constant temperature $T = 1.0132$, corresponding to the crossing point of the sol–repulsive glass line with the sol–particle gel one. Here, we vary the packing fraction, as reported in the figure, and we always refer to the wave vector $q = 10.0$. It is possible to observe a logarithmic decay of $\Phi_q(t)$ over a number of decades in time (e.g. at $\Phi = 0.531468$ it holds for about 5 decades). There may be an early indication of such behaviour in recent work [18]. Then, going closer and closer to the transition, the presence of two different types of arrested states starts to become evident, as the correlators seem to develop a double-plateau structure.

Conclusions

We have sought to introduce the reader to the broad developments that are taking place in connecting the traditional science of nonergodic systems, as glasses, to the world of disordered soft matter, and in particular particle gels. It would appear that the methods used in glass theory are also applicable to soft interactions where attractions dominate the loss of ergodicity. If this proposition turns out to be correct, it seems likely that we can interpret particle gels as glasses of a novel type, the so-called attractive glass. The ramifications of this connection are extensive and most have yet to be appreciated. At the most simplistic level, we now have the machinery to calculate the "phase" or state diagrams of the system, their dynamics and transport coefficients and other properties. Also, we can expect to find in experiments all of the traditional phenomena of the two-step relaxation as we approach gelation, as illustrated in Fig. 3 and, possibly, of the peculiar logarithmic decay predicted within our theoretical model (Fig. 4).

References

1. Verduin H, Dhont JKG (1995) J Colloid Interface Sci 172:425
2. Gast AP, Hall CK, Russel WB (1983) J Colloid Interface Sci 96:251
3. Lekkerkerker HNW, Poon WCK, Pusey PN, Stroobants A, Warren PB (1992) Europhys Lett 20:559
4. Ilett SM, Orrock A, Poon WCK, Pusey PN (1995) Phys Rev E 51:1344
5. Lekkerkerker HNW, Dhont JKG, Verduin H, Smits C, van Duijneveldt JS (1995) Physica A 213:18
6. Pusey PN, Pirie AD, Poon WCK (1993) Physica A 201:322
7. Bolhuis P, Hagen M, Frenkel D (1995) Phys Rev E 50:4880
8. Götze W (1991) In: Hansen JP, Levesqueand D, Zinn-Justin J (eds) Liquids, freezing and glass transition. Les Houches Session LI, 1989. North-Holland, Amsterdam, p 287; (b) Götze W, Sjögren L (1992) Rep Prog Phys 55:241
9. Berghenholtz J, Fuchs M (1999) Phys Rev E 59:5706
10. Berghenholtz J, Fuchs M, Voigtmann T (1999) J Phys Condens Matter 12:40
11. Dawson KA, Foffi G, Fuchs M, Götze W, Sciortino F, Sperl M, Tartaglia P, Voigtmann T, Zaccarelli E (2001) Phys Rev E 63:114
12. (a) van Megen W, Underwood SM, Pusey PN (1991) Phys Rev Lett 67:1586; (b) van Megen W, Underwood SM (1993) Phys Rev Lett 70:2766; van Megen W, Underwood SM (1994) 72:1773
13. Franosch T, Fuchs M, Götze W, Mayr MR, Singh AP (1997) Phys Rev E 55:7153
14. Götze W (1999) J Phys Condens Matter 11:A1
15. Zaccarelli E, Foffi G, Tartaglia P, Sciortino F, Dawson KA (2001) Phys Rev E 63:315
16. (a) Fabbian L, Götze W, Sciortino F, Thiery F, Tartaglia P (1999) Phys Rev E 59:R1347 (b) Fabbian L, Götze W, Sciortino F, Thiery F, Tartaglia P (1999) Phys Rev E 60:2430
17. Fuchs M, Götze W, Mayr MR (1998) Phys Rev E 58:3384
18. Mallamace F, Gambadauro P, Micali N, Tartaglia P, Liao C, Chen S-H (2000) Phys Rev Lett 84:5431

Progr Colloid Polym Sci (2001) 118: 226–231
© Springer-Verlag 2001

D. M. McLoughlin
A. V. Gorelov
K. A. Dawson

Influence of cationic surfactants on DNA conformation

D. M. McLoughlin · A. V. Gorelov (✉)
K. A. Dawson
Irish Centre for Colloid Science and
Biomaterials, Department of Chemistry
University College Dublin, Belfield
Dublin 4, Ireland
e-mail: gorelov@pop3.ucd.ie
Tel.: +353-1-7162417
Fax: +353-1-7162127

A. V. Gorelov
Institute of Theoretical and Experimental
Biophysics, Pushchino, Russia

Abstract In previous work the binding of the cationic surfactant dodecyltrimethylammonium bromide to DNA was studied. The original work has been extended to include data for the decyltrimethylammonium ion. Additionally, while previously considerations of the complex structure were done within the framework of the dynamics of rigid rods, in the present case we have extended this analysis to include a wide range of three-dimensional configurations. Furthermore, the secondary structure of the DNA within the complex is taken into account. Examination of secondary structural changes on surfactant binding indicates that there are no significant changes in the DNA secondary structure. Consideration of the hydrodynamic properties of the surfactant–DNA complex and extension of the experimental data to include decyltrimethylammonium along with application of hydrodynamic modelling allowed us to exclude highly bent or folded complex conformations. The magnitude of the DNA diffusion coefficient decrease on surfactant binding was measured for surfactant molecules of two different tail lengths. The data showed that a rod covered in a single surfactant layer can provide a simple explanation for the difference in the magnitude of the decrease between the two surfactants. In order to account for the observed ratio of 0.8 surfactants per DNA phosphate observed on completion of the first-stage binding, the surfactant headgroups should be located close to the DNA surface, within the condensation volume. This would leave the tail groups projecting outwards, with lateral hydrophobic association between the tails.

Key words Dodecyltrimethylammonium · Decyltrimethylammonium · DNA–surfactant complex

Introduction

In previous work the binding of the cationic surfactant dodecyltrimethylammonium (DTA^+) to DNA was studied using a combination of techniques, including dynamic light scattering, electrophoresis and an ion-selective electrode [1]. In that study, short DNA fragments were used because the relation between the rodlike structure and the dynamics of short DNA fragments has been investigated extensively both theoretically and experimentally [2–6] and thus it was possible to interpret the diffusion results within the framework of the dynamics of rodlike molecules in dilute solution. This allowed a direct study of the complex structure. Dodecyltrimethylammonium bromide (DTAB) was used because the DNA remained monomolecularly dispersed after binding of surfactant, allowing the complex to be examined using solution techniques.

The original work has been extended to include binding data for decyltrimethylammonium ion ($DeTA^+$). Additionally, while previous considerations of the complex structure were made within the framework of the dynamics of rigid rods, in the present case we have extended this analysis to include a wide range of three-dimensional configurations generated using bead modelling, and which include bent and compacted conformations. Furthermore, the secondary structure of the DNA within the complex is taken into account.

Experimental

Purification of reagents

All buffer salts used were analytical grade. Surfactant and DNA were purified as described in a previous publication [1]. The preparation and size distribution characterisation of small DNA fragments were described previously [11].

Light scattering measurements

Measurements of the z-average diffusion coefficient were done using a Malvern 4700 with an argon ion laser source. The experimental setup and analysis procedures have been described previously [1]. Prior to a run, the samples were diluted with the same buffer to a DNA concentration of 0.9 mM DNA phosphates (double the final concentration). Decyltrimethylammonium bromide (DeTAB) and DTAB solutions were made up to double their final concentration. The two solutions were transferred to plastic Eppendorf tubes and centrifuged at 10,000g for 2 h. The DNA solution was then filtered into a preweighed light scattering cell through a 0.2-μm filter. Then, an equal volume of surfactant solution was added with mixing in a similar manner. The tube was then sealed and the solution was mixed using a vortex mixer. All weighing was done on a balance with an accuracy of $\pm 2 \times 10^{-4}$ g, and exact final surfactant concentrations could be determined from the weights obtained. This procedure eliminated artefacts due to aggregation, which could occur if concentrated surfactant was titrated into the cell. The samples were run at final surfactant concentrations between 0 and 10 mM. The DNA concentration was 0.45 mM. The measurements were conducted at 20 °C.

Circular dichroism spectroscopy

Circular dichroism (CD) spectra were run on a Jasco 720 CD spectrometer; the scan rates were 10 nm/min. A 0.15 mM DNA solution was made up, and aliquots of a 100 mM DTAB solution were added using a Hamilton microsyringe with a Chaney adaptor. The solution was stirred during and after the additions. Absorbance measurements were taken concurrently with the CD measurements to correct for the dilution owing to addition of the surfactant solution.

Calculation of the diffusion coefficient for different models

A suite of FORTRAN subroutines for the calculation of hydrodynamic properties of rigid macromolecules in solution has been made available in the public domain [7]. A bead model, composed of an arbitrary number of spherical elements of any size represents the macromolecule. The input data is a set of coordinates of the beads and their radii. The calculations are based on the Kirkwood–Riseman theory, but were extended to arrays of subunits of unequal size.

Using this program, it is possible to calculate translational diffusion coefficients for an array of three-dimensional conformations. By calculating the diffusion coefficients for a wide range of hypothetical DNA–surfactant complex structures and comparing the calculated values to the values obtained experimentally from light scattering it is possible to make deductions about the nature of the final complex.

The modelling of DNA rods as straight strings of beads carries an inherent error as DNA is more closely represented by a cylindrical model; thus, HYDRO was used to determine relative changes in the diffusion coefficients on binding of surfactant. Firstly, the diffusion coefficient was calculated for a 238 base pair Na–DNA rod, modelled as a string of beads, each with a diameter of 2.0 nm, which is in agreement with recent measurements of the effective hydrodynamic diameter of DNA [8, 9]. This served as a reference point, as any calculated diffusion coefficients for the surfactant–DNA complex must give diffusion values lower than for the bare rod, in order to be consistent with the experimental data. Then, calculations were carried out for the DNA complex in a variety of three-dimensional conformations.

For quantitative comparison with experiment an analytical expression for the diffusion of cylindrical rods of length L and diameter d was derived using a subunit approach by Tirado and et al. [3]: the authors considered spheres arranged into rings, and these rings were then stacked to form cylinders. The diffusion coefficient was then calculated for the whole assembly, on the basis of Kirkwood–Riseman theory, with extrapolation to the shell model limit and with correction for end effects. The expression for translational diffusion (D_τ) obtained is

$$D_\tau = \frac{k_B T}{3\pi\eta L}[\ln(p) + \gamma], \tag{1}$$

where k_B is the Boltzmann constant, T is the absolute temperature, η is solvent viscosity, $p = L/d$ and γ, the end effect correction, is expressed by

$$\gamma = 0.316 + 0.578/p + 0.050/p^2. \tag{2}$$

The relation is valid over the range $2 < p < 30$. The effective hydrodynamic diameter, d, was taken as 2.0 nm. L was calculated by multiplying the number of DNA base pairs by a rise per base pair of 0.34 nm, which is characteristic of B-form DNA [10].

Results and discussion

Effect of surfactant binding on DNA secondary structure

In order to determine whether binding of surfactant influences the helical structure, the CD spectrum of DNA between 220 and 320 nm was measured as a function of DTAB concentration. Under the conditions of the experiment the critical aggregation concentration (cac) is known to be to be about 0.9 mM from electrophoretic measurements [11]. Measurements show that below the cac the DNA is in the B-form. However, on commencement of first-stage binding, there is a small change in the intensity of the 275 nm band. This change in magnitude at 275 nm has been directly related to a change in the average rotation per base pair [12]. Correlating the change in $\Delta\varepsilon$ with the change in the average rotation, using the data of Baase and Johnson gives the result presented in Fig. 1.

On binding there is an increase in rotation, up to a maximum of $+0.4°$. Assuming that the starting B-form DNA without surfactant has 10.4 base pairs per turn

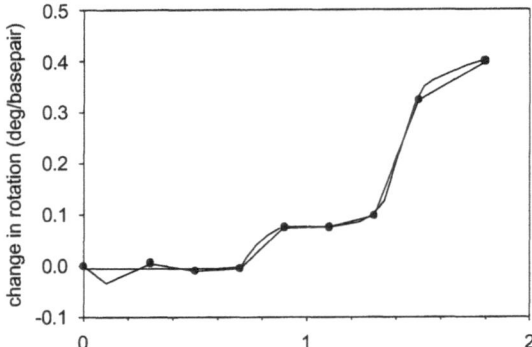

Fig. 1 Change in average rotation of the helix as a function of dodecyltrimethylammonium bromide (*DTAB*) concentration. The DNA concentration was 0.15 mM. The buffer conditions were 10 mM NaBr, 5 mM *N*-(2-hydroxyethyl)piperazine-*N'*-ethanesulfonic acid (*HEPES*) and 2 mM etylenediaminetetraacetic acid (*EDTA*)

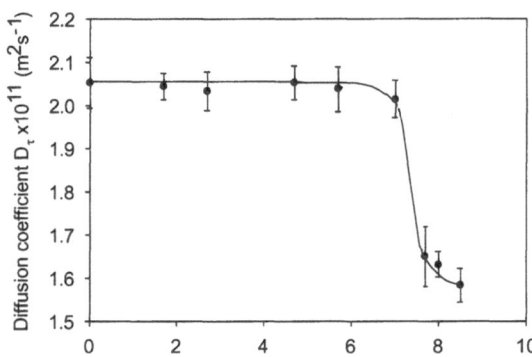

Fig. 2 Diffusion coefficient as a function of DeTAB concentration. The DNA concentration was 0.45 mM. The temperature was 20 °C. The buffer conditions were 10 mM NaBr, 5 mM HEPES and 2 mM EDTA

[13], giving a turn per base pair of 34.6°, this decreases to 10.3 base pairs per turn at 1.8 mM DTAB; however, aggregation occurs close to this surfactant concentration, so this value must be treated with caution. The spectral changes between 0.9 and 1.3 mM are certainly not the result of aggregation, as electrophoresis experiments carried out under identical conditions have shown that the DNA complex is monomolecular under these conditions [11]. They show there is a change in rotation of only +0.1° per base pair. Thus, there are no changes in DNA length on binding great enough to influence its hydrodynamic properties. This is an important point when determining the hydrodynamic properties of the surfactant–DNA complex, as any length changes in the DNA on surfactant binding will have to be taken into account when interpreting the data.

Diffusion coefficient of the DeTA$^+$ ion with DNA and comparison with data for DTA$^+$

The diffusion coefficient of DNA as a function of DeTAB concentration is presented in Fig. 2. In the case of DeTA$^+$ binding the diffusion coefficient for the uncomplexed Na–DNA was found to be 2.05×10^{-11} m^2/s. The concentration used in the titration experiment was 0.45 mM, consequently variations with DNA concentration were neglected and it was considered that the measured diffusion coefficients (D_{app}) obtained were equal to the infinite dilution values (D_τ) for practical purposes of calculations. An excellent agreement was obtained with the theoretical value for the translational diffusion of a rigid rod (2.02×10^{-11} m^2/s). Above 7.8 mM added DeTAB, corresponding to the cac, as measured by electrophoretic mobility [11], there was a sharp decrease in the diffusion coefficient to 1.58×10^{-11} m^2/s.

For DTA$^+$, measurements were made under identical conditions to those used previously in the binding study of Gorelov and et al. [1]. Thus, referring to the binding isotherm, diffusion coefficient measurements were taken at 0 mM surfactant (corresponding to bare DNA) and 1.2 mM free DTA$^+$ (corresponding to first-stage bound, unaggregated DTA–DNA). Binding resulted in a decrease in te diffusion coefficient from 2.66×10^{-11} to 1.98×10^{-11} m^2/s.

Consideration of the surfactant–DNA complex structure

We considered the following surfactant–DNA complex structures (Fig. 3)

1. Wrapped micelle model, where the surfactant is bound to the DNA in micellar form and there is some wrapping of the DNA around the micelles.
2. Surface micelle model, where the surfactant is bound to DNA in micellar form, but with no DNA bending
3. Surfactant-covered rod model, where there is no DNA bending and where the surfactant headgroups are located close to the DNA surface and there is lateral association between tailgroups.

Firstly, we calculated the diffusion coefficient of the DNA rod complexed with DeTA$^+$ micelles, in a conformation where the surfactant is bound in a micellar form and the DNA rod is bent around the micelles. It is still not clear whether the micellar aggregation number should be higher, lower, or unchanged from that for the free micelles. Thus, to start with, the micellar aggregation number was taken to be the same as in free micelles. Taking an aggregation number of 36 for DeTAB [14] around 11 micelles are required for a 238 base pair fragment (476 phosphates) to give binding of 0.8 surfactants per phosphate. The micellar radius was taken as

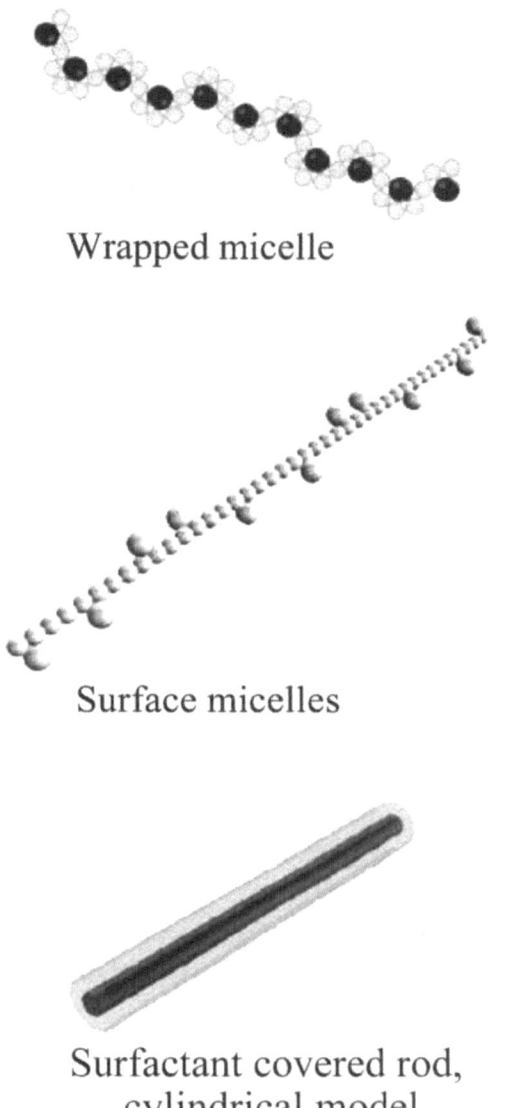

Wrapped micelle

Surface micelles

Surfactant covered rod,
cylindrical model

Fig. 3 Diagrammatic representation of different hypothetical surfactant–DNA complex structure

Table 1 Comparison of diffusion coefficients calculated using the bead model for a surfactant–DNA complex in different conformations with reference to a Na–DNA rod, represented as a string of beads. The surfactants used were decyltrimethylammonium (*DeTA*) and dodecyltrimethylammonium (*DTA*)

Model type	Calculated D_τ ($\times 10^{11}$ m²/s)
Na–DNA rod	2.2
DeTA–DNA, compact sphere	5.25
DeTA–DNA, wrapped micelles	2.17
DeTA–DNA, surface micelles	1.84
DTA–DNA, surface micelles	1.89
DeTA, surfactant-covered rod	1.76
DTA–DNA, surfactant-covered rod	1.71

changes in the effective length of the complex. As the diffusion coefficient dependence on length for rigid rods is a logarithm/linear dependence, while the diameter only appears as a logarithmic term, any changes in the effective length will cause dramatic deviations in the diffusion coefficient from that expected for a rodlike configuration. We can also exclude a compact globular conformation similar to that which has been observed in the case of long-chain DNA [15]. The diffusion coefficient of the complex in a hypothetical conformation where the DNA and surfactant in the complex were condensed in a closely packed sphere was calculated. The volume of this sphere was estimated by using intrinsic molar volume data for DNA [16] and assuming that the density of the decyl chains in the complex is the same as pure decane. The radius of the sphere was converted to a diffusion coefficient by application of the Stokes–Einstein equation. The calculated diffusion value (5.25×10^{-11} m²/s) is much higher than that calculated for the bare DNA in a rodlike conformation. This is in contradiction to the experimental data and excludes a compact structure. Thus, there is no compaction for DNA fragments with length close to the persistence length. That there is no bending is not unexpected as electrophoretic measurements have shown no significant changes in the effective charge of the complex on binding surfactant [11].

Next we considered the case where no DNA bending occurred. In analogy to surfactant binding a two-dimensional planar surface we can distinguish two types of surfactant aggregate structure. For planar surfaces binding has been proposed to occur by formation of "hemi-micelles" [17]. The reverse-orientation model [18] describes these hemi-micelles as laterally associated surfactants with charged headgroups interacting with the charged surface and with tailgroups pointing outwards towards the solvent. An alternative model, called the surface micelle model, proposes that the surfactant is bound to the surface in a form similar to free micelles [19].

Thus, diffusion coefficients were calculated for both surface-bound micelles and a surfactant-covered rod and

1.7 nm. The calculated diffusion value for the DeTA–DNA complex in this conformation was 2.17×10^{-11} m²/s (Table 1), which is close to that for the Na–DNA rod (2.20×10^{-11} m²/s). This is at variance with what has been obtained experimentally, where a decrease in the diffusion coefficient of more than 20% was seen. Thus, it is unlikely that the DNA adopts this type of configuration. Of course, the wrapped micelle model can have a number of different structures. For example, the aggregation number can be lower or higher, leading to more or fewer bends; however, no matter what the conformation, there will be a number of bends in the polyion, in order to stabilise as many of the bound micelles as possible. Hence, there will be large

were compared with the experimental data. Firstly, the diffusion coefficient was calculated for surfactant bound to DNA in micellar form. This micellar binding has been proposed as more favourable, as it does not involve the projection of hydrophobic tails into the solvent. In common with the case for the wrapped micelle model, to start with the micellar aggregation number was first taken to be the same as for free micelles. In general for surfactant binding to polyelectrolytes of high charge density the aggregation number of the resulting micelles is usually similar to or greater than that for free micelles [20]. Also the observed highly cooperative binding suggests a high aggregation number. The diffusion coefficient for the 238 base pair DNA rod, with DeTA micelles bound on the surface, was calculated to be 1.84×10^{-11} m^2/s. The same calculation was then carried out for a DTA–DNA complex with the same structure. As the aggregation number (55) is larger than in the case of DeTAB [14], the number of bound micelles will be lower. Also comparison of the two calculations allows an assessment of the influence of aggregation number on the hydrodynamic properties. The radius of the DTA$^+$ micelles was taken to be 2.0 nm [21]. The calculated diffusion coefficients for the DeTA–DNA and DTA–DNA complexes are 1.84×10^{-11} and 1.89×10^{-11} m^2/s, respectively (Table 1). Thus, if the micelle model is correct, the decrease in the diffusion coefficient of the DNA on surfactant binding will actually become smaller on substitution of the larger DTA$^+$ for DeTA$^+$. Again this is in contrast to the experimental data. This result also rules out micelles of very large aggregation number, as this will lead to even smaller changes in the diffusion coefficient on binding.

However, while the previous analysis rules out large bound micelles it does not discriminate between a surfactant-covered rod and a covered rod with micelles of small aggregation number. These two are difficult to distinguish from the consideration of their hydrodynamic properties alone; however, surfactant-specific electrode measurements have shown that after first-stage binding there are 0.8 surfactants per DNA phosphate. This corresponds closely to the 0.76 condensed counterions per DNA charge, which can be calculated from the Manning theory [22]. This suggests that the sodium ions within the condensation volume have been replaced by surfactant. While there is no exact theoretical expression describing the condensation volume, continuum electrostatic, Monte Carlo and molecular dynamic simulations have shown that condensed counterions are localised in a volume less that 2 nm from the central DNA axis [23]. As the DTA$^+$ tailgroup alone is 1.7 nm it is difficult to see how even small micelles can fully occupy the condensation layer. Similarly this rules out "admicelles" [24] where some surfactant headgroups are pointing outwards from the DNA.

For quantitative comparisons with experimental data the bead model is not suitable, as DNA molecules are more closely represented by a cylindrical geometry. Thus, to make direct comparisons with experimentally derived diffusion coefficients the diffusion coefficient for a cylindrical rigid rod was used. In addition, in order to account for polydispersity and differences in molecular weight between samples, the size distribution of the DNA used in the dynamic light scattering experiments was determined using capillary gel electrophoresis [11]. For the DTA$^+$ study the DNA weight-average molecular weight was 113,520 g/mol and the polydispersity was 1.18, while in the case of DeTA$^+$ these were 149,820 g/mol and 1.27, respectively. The actual size distributions were used as the input for the diffusion coefficient equation (Eq. 1) and the calculated z-average diffusion coefficients of the DNA complexes were obtained.

As shown in the first section, binding of surfactant does not lead to significant length changes in the DNA. If there is no DNA bending and if the surfactant is considered to be evenly distributed on the DNA surface, binding of surfactant can be considered to increase the effective diameter of the DNA rod. Using the equation for the diffusion coefficient for a rigid rod (Eq. 1) and treating the diameter as a variable gives the best fit of the calculated diffusion coefficient to the experimental values for the case where the effective diameter is 4.8 nm in the case of DeTA-DNA and 5.4 nm for DTA–DNA. The calculated diffusion coefficients of DeTA$^+$- and DTA$^+$-covered DNA are compared to the experimental values in Table 2. Subtracting the effective diameter of DNA and dividing by 2 gives the effective thickness of the surfactant layer. This is 1.4 and 1.7 nm for DeTA$^+$ and DTA$^+$, respectively.

The Tanford relation links the number of carbons in the surfactant tailgroup to the length of the hydrocarbon chain of the alkyltrimethylammonium ion in a micelle, l [25]

$$l \approx (0.15 + 0.1265\, N_c)\ \text{nm}, \tag{3}$$

Table 2 Comparison of experimental DNA translational diffusion coefficient with that calculated for a DNA rod modelled as a rigid cylinder. *Top* Comparison of experimental translation diffusion with that calculated for a rigid cylinder for Na–DNA and DeTA–DNA. *Bottom* Comparison of experimental translation diffusion with that calculated for a rigid cylinder for Na–DNA and DTA–DNA

	Calculated D_τ ($\times 10^{11}$ m^2/s)	Experimental D_τ ($\times 10^{11}$ m^2/s)
Na–DNA	2.02 ± 0.02	2.05 ± 0.07
DeTA–DNA	1.63 ± 0.03	1.58 ± 0.03
	Calculated D_τ ($\times 10^{11}$ m^2/s)	Experimental D_τ ($\times 10^{11}$ m^2/s)
Na–DNA	2.62 ± 0.04	2.66 ± 0.02
DTA–DNA	1.96 ± 0.03	1.98 ± 0.02

where N_c is the number of carbon atoms in the hydrocarbon chain of the surfactant. For N_c equal to 10(DeTA$^+$), the value of l corresponds to 1.4 nm and for N_c equal to 12(DTA$^+$) it is 1.7 nm. These two values are identical to those determined for the surfactant layer thickness in the surfactant–DNA complex. Thus, the increment of the diameter increase on binding of surfactant, both in the case of DeTAB and DTAB, is close to that expected for the size of a single surfactant layer as estimated using the Tanford relation. A possible structure which would fit with the experimental findings would be a surfactant-covered rod, where the surfactant headgroups are located close to the DNA surface, within the condensation volume, and the tailgroups projecting outwards, with lateral hydrophobic association between the tails.

Conclusion

Consideration of the hydrodynamic properties of the DTA–DNA complex and extension of the experimental data to include DeTAB along with application of hydrodynamic modelling allows us to exclude highly bent or folded complex conformations. Examination of the secondary structural changes on binding indicates

that there are no significant changes in the DNA length on binding. Binding of surfactant in the form of micelles of aggregation numbers larger or similar to those of free micelles is also ruled out by hydrodynamic considerations; however, the possibility that surfactant binds to the DNA as micelles of small aggregation number cannot be ruled out. The magnitude of the DNA diffusion coefficient decrease on surfactant binding was measured for surfactant molecules of two different tail lengths (DeTA$^+$ and DTA$^+$). The data showed that a rod covered in a single surfactant layer can provide a simple explanation for the difference in the magnitude of the decrease between the two surfactants. In order to account for the observed ratio of 0.8 surfactants per DNA phosphate observed on completion of first-stage binding, the surfactant headgroups should be located close to the DNA surface, within the condensation volume. This would leave the tailgroups projecting outwards, with lateral hydrophobic association between the tails.

Acknowledgements The authors acknowledge Professor Boyd of Queen's University Belfast for allowing us access to the CD spectrometer. We would also like to thank all our colleagues in the Irish Centre for Colloid Science and Biomaterials for helpful discussions. We thank J.-C. Jacquier for carrying out the DNA characterisation by capillary gel electrophoresis.

References

1. Gorelov AV, Kudryashov ED, Jacquier J-C, McLoughlin DM, Dawson KA (1998) Physica A 249:216
2. Broesrma S (1960) J Chem Phys 32:1626
3. Tirado MM, Martinez CL Garcia de la Torre J (1984) J Chem Phys. 81:2047
4. Elias JG, Eden D (1981) Biopolymers 20:2369
5. Goinga HT, Pecora R (1991) Macromolecules 24:6128
6. Ferrari ME, Bloomfield VA (1992) Macromolecules 25:5266
7. Garcia de la Torre J, Navarro S, Lopez-Martinez MC, Diaz FG, Lopez Cascales JJ (1994) Biophys J 67:530
8. Garcia de la Torre J, Navarro S, Lopez Martinez MC (1994) Biophys J 66:1573
9. Eimer W, Williamson JR, Boxer SG, Pecora R (1990) Biochemistry 29:799
10. Cantor CR, Schimmel PR (1980) Biophysical chemistry part 1: conformation of biological macromolecules. Freeman, New York, p 179
11. Jacquier J-C, Gorelov AV, McLoughlin DM, Dawson KA (1998) J Chromatogr A 817:263
12. Baase WA, Johnson WC (1979) Nucleic Acids Res 6:797
13. Wang AHJ, Quigley GJ, Kolpak FJ, Crawford JL, van Boom JH, van der Marel G, Rich A (1979) Nature 282:680
14. Phillips JN (1955) J Chem Soc Faraday Trans 51:561
15. Mel'nikov SM, Sergeyev VG, Yoshikawa K (1995) J Am Chem Soc 117:2401
16. Buckin VA, Kankiya D, Sarvazyan AP, Uedaira H (1989) Nucleic Acids Res 17:4189
17. Gaudin AM, Fuerstenau DW (1955) Trans AIME 202:958
18. Somasundaran P, Fuerstenau DW (1966) J Phys Chem 70:90
19. Gu T, Huang Z (1989) Colloids Surf 40:71
20. Hansson P, Lindman B (1996) Curr Opin Colloid Interface Sci 1:604
21. Hayter JB (1985) In: Degiorgio V, Corti M (eds) Physics of amphiphiles, micelles, vesicles and microemulsions. North Holland, Amsterdam, pp 59–94
22. Manning GS (1978) Q Rev Biophys 11:179
23. Young MA, Jayaram B, Beveridge DL (1997) J Am Chem Soc 119:59
24. Yeskie MA, Harwell JH (1988) J Phys Chem 92:2346
25. Tanford C (1980) The hydrophobic effect – formation of micelles and biological membranes. Wiley, New York

Progr Colloid Polym Sci (2001) 118: 232–237
© Springer-Verlag 2001

U.-S. Jeng
T.-L. Lin
T.-S. Chang
H.-Y. Lee
C.-H. Hsu
Y.-W. Hsieh
T. Canteenwala
L. Y. Chiang

Comparison of the aggregation behavior of water-soluble hexa(sulfobutyl)fullerenes and polyhydroxylated fullerenes for their free-radical scavenging activity

U.-S. Jeng · T.-L. Lin (✉) · T.-S. Chang
Department of Engineering and System
Science, National Tsing Hua University
Hsinchu 30043, Taiwan

H.-Y. Lee · C.-H. Hsu · Y.-W. Hseih
Synchrotron Radiation Research Center
Hsinchu, 30043, Taiwan

T. Canteenwala · L. Y. Chiang
Center for Condensed Matter Sciences
National Taiwan University
Taipei 10617, Taiwan
e-mail: tllin@mx.nthu.edu.tw
Fax: +886-3-5728445

Abstract We study the aggregation behavior of two highly water-soluble fullerene derivatives, hexa(sulfobutyl)fullerenes (FC$_4$S) and polyhydroxylated fullerenes (fullerenols) [C$_{60}$(OH)$_{18}$] using small-angle x-ray scattering. We found that FC$_4$S forms spheroidal aggregates having a similar radius of gyration $R_g \approx 19$ Å, in a wide concentration range from 0.4 to 26 mM in water solutions, whereas the mean sizes observed for C$_{60}$(OH)$_{18}$ aggregates in water solutions grow nearly two-fold from $R_g = 20$ Å to $R_g = 40$ Å as the concentration increases from 0.7 to 50 mM. The implication of the structural differences between the two fullerene derivatives on their free-radical scavenging activity is discussed.

Key words Small-angle X-ray scattering · Water-soluble fullerene derivatives · Aggregation

Introduction

Fullernols, C$_{60}$(OH)$_{18}$, of high water solubility and a moderate electron affinity were reported to be potent free-radical scavengers in biological systems [1]. In an effort to modify the electron affinity of the C$_{60}$ cage of fullerene derivatives for a better free-radical scavenging efficiency, e, Chi et al. [2] synthesized the sodium salt of hexa(sulfobutyl)fullerenes, FC$_4$S, which has a number of addends per C$_{60}$ cage differing much from that of C$_{60}$(OH)$_{18}$. In a xanthine/xanthine oxidase enzymatic system of superoxide radicals (O$_2^-$), Chi et al. [2] compared the free-radical scavenging activity for these two fullerene derivatives. It was found that for a dose level of 50 μM FC$_4$S has an O$_2^-$ suppression efficiency of 60%, comparable to the 59% for C$_{60}$(OH)$_{18}$[2]. When the dose level increased to 100 μM, FC$_4$S showed a higher radical suppression efficiency of 96% than the 70% for C$_{60}$(OH)$_{18}$. This concentration-induced difference in eliminating superoxide radical species implies that there should be a correlation between the aggregation behavior and the free-radical scavenging efficiency. In a similar case, Guldi et al. [3, 4] attributed the low free-electron (or radical) affinity observed for C$_{60}$(COONa)$_2$ to the formation of aggregates in aqueous solutions.

Using small-angle X-ray scattering (SAXS) and small-angle neutron scattering we have observed and reported previously the aggregation behavior for FC$_4$S [5] as well as C$_{60}$(OH)$_{18}$ [6] in a concentration range from 3 to 50 mM. Nevertheless, this concentration range is much higher than the biochemistry concentration range (10–100 μM) used for the free-radical scavenging activity measurement. To understand better the possible correlation between the aggregation behavior and the scavenging efficiency, we extend our study on FC$_4$S and C$_{60}$(OH)$_{18}$ to a concentration level of 350 μM using SAXS of a high photon flux from synchrotron radiation.

Experimental

Water-soluble fullerenols, C$_{60}$(OH)$_{18}$ (Fig. 1), fullerene oxide with 18 hydroxy groups on average, were synthesized following the steps reported previously [7]. The sodium salt of hexa(sulfobutyl)fullerene-ness, C$_{60}$(CH$_2$CH$_2$CH$_2$CH$_2$SO$_3$Na)$_6$, a C$_{60}$ cage covalently bonded with six ionic sulphonate arms (Fig. 2), were synthesized using the procedures detailed in Ref. [5].

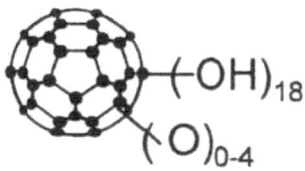

Fig. 1 Scheme for polyhydroxylated fullerenes, $C_{60}(OH)_{18}$

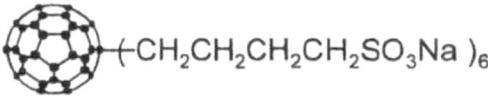

Fig. 2 Scheme for hexa(sulfobutyl)fullerenes, FC_4S

SAXS measurements for FC_4S aqueous solutions of higher concentrations of 3.2, 6.4, 12.8, and 25.6 mM were performed on the 8-m SAXS instrument at Tsing-Hua University, Taiwan. SAXS data for water solutions of 6.1, 12.2, 24.4, 48.8 and 122 mM $C_{60}(OH)_{18}$ were also collected using the same instrument. The instrument and measurement details were described in Refs. [5, 6, 8].

For low sample concentrations, the SAXS data for water solutions of 0.35, 0.7 and 1.8 mM FC_4S as well as 0.7 and 1.8 mM $C_{60}(OH)_{18}$ were measured using the SAXS setup on the BL17B wiggler beamline at the synchrotron radiation research center (SRRC) in Taiwan. The SRRC has an electron storage ring

operated at an electron energy of 1.5 GeV and a maximum beam current of 200 mA. The incident radiation quanta, focussed vertically by a mirror and monochromated by a Si (111) double-crystal monochromator for an energy of 8 keV, were then collimated by two sets of slits, 0.4×0.4 and 0.4×0.6 mm² for the first and second set, separated by 1.0 m. A linear detector was positioned 1.1 m (vacuum path) after the sample position, covering a Q range from 0.01 to 0.12 Å⁻¹. Here, Q is scattering wavevector of X-rays. A higher Q range can be easily achieved by moving the detector sideways from the direct beam position.

Results

The SAXS data measured for FC_4S solutions of concentrations from 3.2 to 25.6 mM are shown in Fig. 3. Using an ellipsoid model [9], we can fit (curves) all the scattering data in the higher-Q region adequately with a major axis $a \approx 29$ Å and a minor axis $b \approx 22$ Å for the FC_4S aggregates. The corresponding radius of gyration $R_g = [(a^2 + 2b^2)/5]^{1/2} = 19$ Å for the aggregates, is consistent with 20 ± 1 Å obtained from the Guinier approximation [9]. For these higher concentrations, the fitting curves deviate from the data in the low-Q region owing to the neglect of the interaction effect between aggregates. The interaction effect, which diminishes gradually as the concentration decreases, was discussed

Fig. 3 The in-house small-angle X-ray scattering (*SAXS*) data for FC_4S water solutions are fitted (*curves*) using an ellipsoid-like model for the form factor of FC_4S aggregates. For higher-concentration cases, the data deviate gradually from the fitting curves in the low-Q region owing to the neglect of the interparticle interaction effect in the fitting model

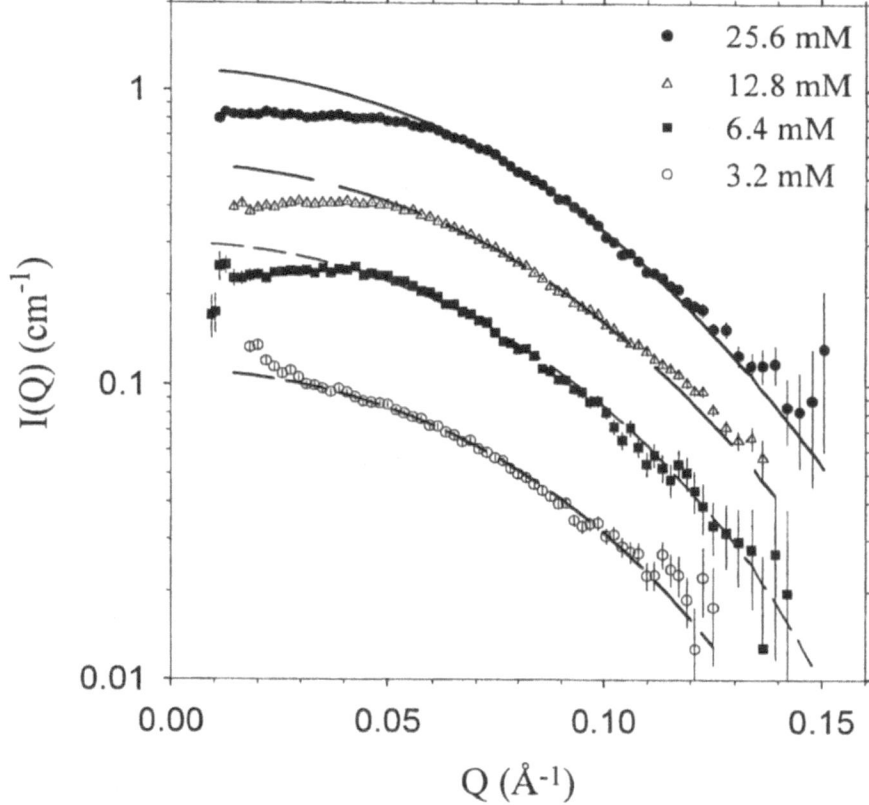

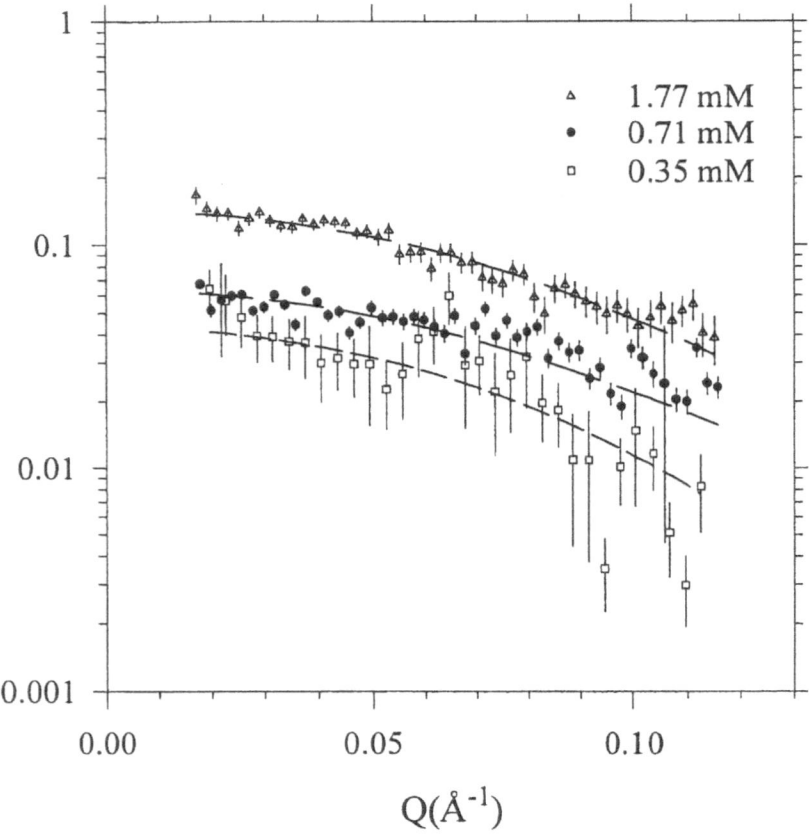

Fig. 4 Synchrotron radiation SAXS data for FC$_4$S water solutions of low concentrations. The data were fitted (*curves*) using the same ellipsoid-like model

in Ref. [5]. In the present study, we concentrate on the data in the higher-Q region, where the form factor of the aggregates dominates [9], for determining the size and shape of the aggregates.

The SAXS data taken by synchrotron X-rays for lower FC$_4$S concentrations of 1.8, 0.71, and 0.35 mM are shown in Fig. 4. Using the same ellipsoidal model of similar sizes as used for the higher concentrations, we can fit (curves in Fig. 4) the data adequately. We summarize the fitting results for FC$_4$S in Table 1. From the result, we conclude that the aggregation size and shape of FC$_4$S do not change appreciably with concentration in the concentration range studied.

The SAXS data measured for water solutions of low C$_{60}$(OH)$_{18}$ concentrations of 1.77 and 0.7 mM are shown in Fig. 5. These data can be fitted (solid and dashed curves) reasonably well using the same fractal model for the C$_{60}$(OH)$_{18}$ aggregates as that used for the higher-concentration data measured previously [6]. In the fractal model, the C$_{60}$(OH)$_{18}$ fractal aggregates, constructed by primary particles of fullerenols in a self-similar geometrical arrangement, are characterized by a pair correlation function

$$g(r) \propto r^{D-d} \exp(-r/\xi) , \qquad (1)$$

where D is the fractal dimension, d the Euclidean dimension, and ξ the correlation length [10]. The mass of the fractal objects, M, increases with their linear dimension, r, in the form of $M \propto r^D$. We can deduce an average $R_g = [D(D+1)/2]^{1/2}\xi$ for the aggregates.[1]

The SAXS results for C$_{60}$(OH)$_{18}$ are summarized in Table 2. From the result we can see C$_{60}$(OH)$_{18}$ aggregates grow continuously to nearly twice the size as the concentration increases from 0.7 to 50 mM, with a common fractal dimension of 2.5. Indeed, for concentration higher than 3 mM, traces of large C$_{60}$(OH)$_{18}$ aggregates can be detected by dynamic light scattering [11].

We compare the two different aggregation characteristics of FC$_4$S and C$_{60}$(OH)$_{18}$ in Fig. 6. For FC$_4$S, the aggregates remain relatively the same size in the concentration range studied; therefore, the total aggregates surface area, S_{tot}, for the FC$_4$S aggregates in solution is proportional to the concentration, c, whereas for the C$_{60}$(OH)$_{18}$ aggregates $S_{tot} = n_p S = cS/N \propto cR_g^{-1/2}$. Here, $S \propto R_g^2$ is the smooth surface area of one aggregate and $n_p = c/N$ is the number density of the aggregates having a fractal dimension of 2.5 and an aggregation number

---
[1] There is a factor of 2 missing from the same relation in Ref. [6]

Table 1 Summary of the small-angle X-ray scattering (*SAXS*) results for FC$_4$S water solutions. c is the sample concentration, R_g the radius of gyration, and a and b the major and minor axes for the ellipsoid-like aggregates

c (mM)	a (Å)	b (Å)	R_g (Å)
0.35	30.9	20.8	19.1
0.71	30.1	19.9	18.4
1.77	29.7	19.8	18.5
3.2	27.1	21.7	18.3
6.4	27.7	21.9	18.6
12.8	29.1	21.4	18.8
25.6	29.3	21.5	18.9

Table 2 Summary of the SAXS results for C$_{60}$(OH)$_{18}$ water solutions. ξ is the correlation length

c (mM)	ξ (Å)	R_g (Å)
0.71	9.5	20.2
1.77	12.5	26.5
6.1	13.4	28.3
12.2	15.0	31.7
24.4	17.6	37.2
48.8	20.9	44.2

Fig. 5 Synchrotron radiation SAXS data for C$_{60}$(OH)$_{18}$ water solutions of low concentrations. The data were fitted (*curves*) using a fractal model for the aggregates

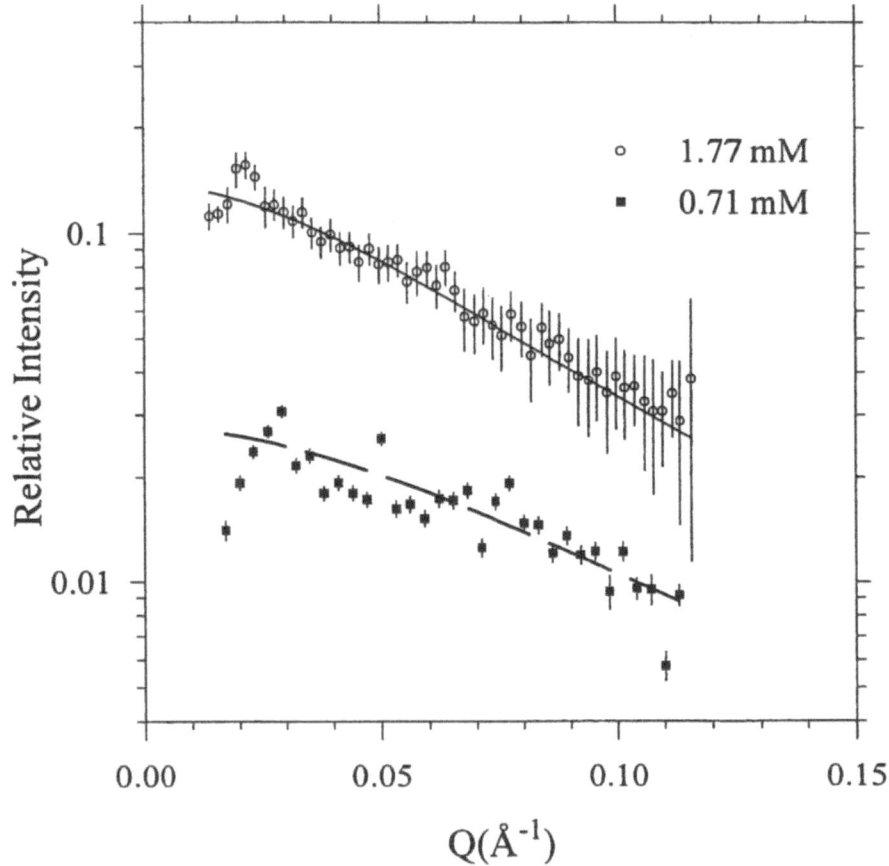

$N \propto M \propto R_g^{2.5}$ [12]^2. Furthermore, since R_g grows roughly linearly with c (see the dashed line in Fig. 6), we have $S_{tot} \propto c^{1/2}$ for C$_{60}$(OH)$_{18}$ aggregates. This implies a slower growth rate of S_{tot} with concentration, compared with $S_{tot} \propto c$ for FC$_4$S aggregates.

Assuming the free-radical scavenging efficiency, e, is related to the available C$_{60}$ cage surfaces of the fullerene

derivatives, namely, the unblocked olefinic moieties of the fullerene derivatives as indicated in Ref. [13], we may expect e for FC$_4$S grows faster with concentration than that for C$_{60}$(OH)$_{18}$, on the basis of the previous discussion of aggregate surface area. This may explain why FC$_4$S has a similar O$_2^-$ suppression efficiency as that for C$_{60}$(OH)$_{18}$ at a lower dose of 50 μM, but has a significantly better efficiency when the dose is doubled to 100 μM as reported in Ref. [2].

Assuming a linear relation of $e \propto S_{tot}$, we have $e \propto c^{1/2}$ for the C$_{60}$(OH)$_{18}$ aggregates, since $S_{tot} \propto c^{1/2}$ (discussed earlier). Using this relation, we tried to explain

---

^{2} We consider only the smooth surface of the fractal aggregate of a characteriser surface $S \propto R,^{2.5}$ [13]. Since the additional surface area inside the fractal aggregate may be blocked to radical species owing to the fractal morphology and contribute insignificantly to the colision cross-section for radical scavenging activity

Fig. 6 Radii of gyration, R_g, for FC$_4$S (*circles*) and C$_{60}$(OH)$_{18}$ (*squares*) at different concentrations fitted with *straight lines*

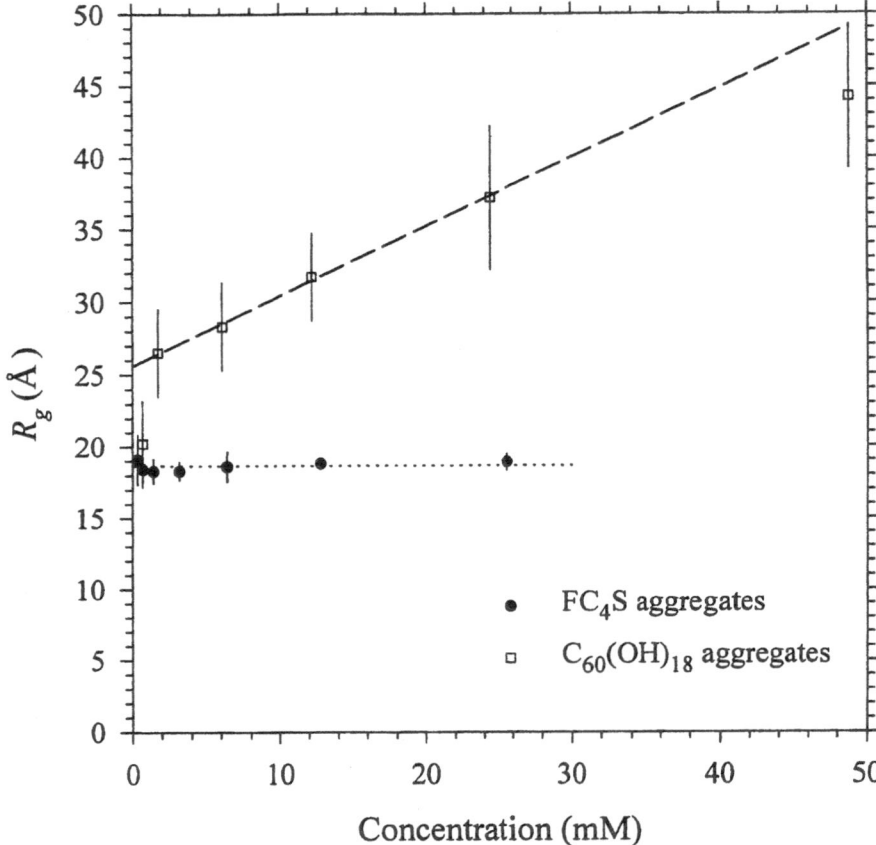

Fig. 7 Free-radical scavenging efficiency data, measured for C$_{60}$(OH)$_{18}$ in a xanthine/xanthine oxidase enzymatic system of superoxide radicals (O$_2^-$), fitted using the relation $e \propto c^{1/2}$ (*dashed line*) The data shown here are redrawn from Ref. [1]

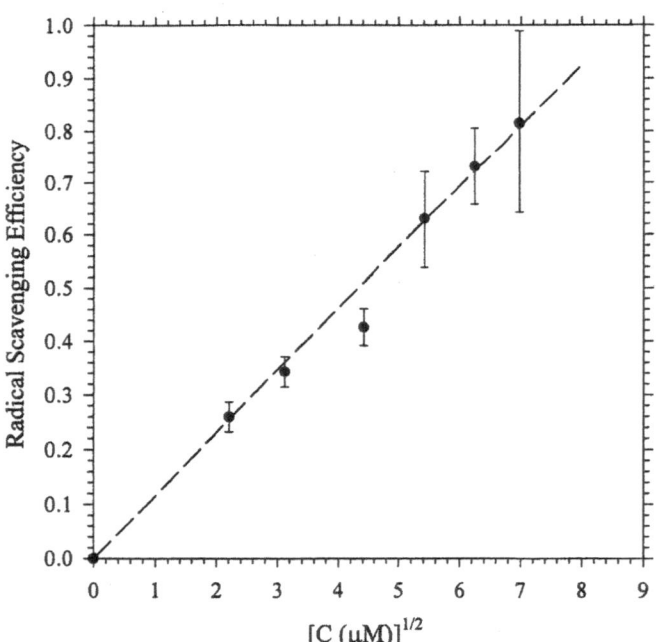

the radical scavenging efficiency data measured in Ref. [1] for C$_{60}$(OH)$_{18}$. The result (see the dashed line in Fig. 7) suggests that the radical scavenging efficiency for C$_{60}$(OH)$_{18}$ is indeed approximately proportional to $c^{1/2}$.

The previous discussion for the relation between the aggregation behavior and the radical scavenging activity is based on the SAXS results measured for sample solutions of concentrations slightly higher than the biochemistry concentrations. Limited by the photon flux of SAXS, we do not have the mean aggregation sizes for FC$_4$S and C$_{60}$(OH)$_{18}$ in the concentrations range from 50 to 100 μM, where the superoxide-radical eliminating efficiencies were measured and compared. We also neglected the possible effect of the critical aggregation concentration (cac) on the radical scavenging efficiency. Nevertheless, the cac value for either FC$_4$S or C$_{60}$(OH)$_{18}$ seems to be rather low, as suggested by the measurable SAXS intensity for low sample concentrations. This is consistent with the low cac values, 0.9 ± 0.7 [5] and 2.0 ± 2.6 mM [6], extracted from the data of higher concentrations for FC$_4$S and C$_{60}$(OH)$_{18}$, respectively.

Conclusion

We have extended our SAXS study on the aggregation behavior of FC$_4$S and C$_{60}$(OH)$_{18}$ to low sample concentrations. In the wide concentration range investigated, FC$_4$S forms ellipsoidal aggregates of similar size and shape, while C$_{60}$(OH)$_{18}$ forms fractal aggregates that grow with increasing concentration. We have compared the two fundamentally different aggregation behaviors of FC$_4$S and C$_{60}$(OH)$_{18}$ and proposed a relation between the total aggregate surface area and the radical scavenging activity. It seems that their concentration-dependent radical scavenging efficiencies can be explained qualitatively on the basis of their aggregation behavior.

Acknowledgements We thank C.-Y. Liu, J.-M. Juang and C.-H. Chang for the essential technical support in the SAXS setup at SRRC. This work was supported by the National Science Council, grant NSC 89-2113-M-007-018.

References

1. Chiang LY, Lou FJ, Lin JT (1995) J Chem Soc Chem Commun 1283
2. Chi Y, Bhonsle JB, Canteenwala T, Huang JP, Shica J, Chen BJ, Chiang LY (1998) Chem Lett 465
3. Guldi DM, Hungerbuhler H, Asmus KD (1995) J Phys Chem 99:13487
4. Guldi DM, Hungerbuhler H, Asmus KD (1997) J Phys Chem 101:1783
5. Jeng U, Lin TL, Tsao CS, Lee CH, Canteenwala T, Wang LY, Chiang LY, Han CC (1999) J Phys Chem B 103:1059
6. Jeng U, Liu WJ, Lin TL, Wang LY, Chiang LY (1999) Fullerene Sci Technol 7:599
7. Chiang LY, Upasani RB, Swirczewski JW (1992) J Am Chem Soc 114:10154
8. Linliu K, Chen SA, Yu TL, Lin TL, Lee CH, Kai JJ, Chang SL, Lin JS (1995) J Polym Res 2:63
9. Chen SH (1986) Annu Rev Phys Chem 37:351
10. Freltoft T, Kjems JK, Sinha SK (1986) Phys Rev B 33:269
11. Mohan H, Palit DK, Mittal JP, Chiang LY, Asmus KD, Guldi DM (1998) J Chem Soc Faraday Trans 94:359
12. Martin JE, Hurd AJ (1987) J App Crystallog 20:61
13. Yu C, Bhonsle JB, Wang LY, Lin JG, Chen BJ, Chiang LY (1997) Fullerene Sci Technol 5:1407

Progr Colloid Polym Sci (2001) 118: 238–242
© Springer-Verlag 2001

Pierandrea Lo Nostro
Giulia Capuzzi
Emiliano Fratini
Luigi Dei
Piero Baglioni

Modulation of interfacial properties of functionalized calixarenes

P. Lo Nostro (✉) · G. Capuzzi
E. Fratini · L. Dei · P. Baglioni
CSGI and Department of Chemistry
University of Florence
Via Gino Capponi
9, 50121 Florence, Italy
e-mail: pln@csgi.unifi.it
Tel.: +39-055-2757575
Fax: +39-055-240865

Abstract Calix[n]arenes and their amphiphilic derivatives act as strong ligands, by selectively binding cations and small organic molecules (such as chloroform, toluene, fullerene). The presence of a cavity and of specific electron donor sites results in their ability to form stable host–guest systems with electron-acceptor molecules. We report our studies on the interfacial properties of amphiphilic calixarenes, both at the liquid/liquid and at the air/liquid interfaces, and their selective complexation of alkaline cations and small organic molecules, by performing Langmuir isotherms and small-angle neutron scattering experiments. Changing the guest species, the functional groups, and the experimental conditions (such as temperature, concentration, and ionic strength) can efficiently modulate the interfacial properties of these amphiphilic ligands.

Key words Calix[n]arene(s) · Monolayer(s) · Host–guest · Inclusion compound(s) · Small-angle neutron scattering

Introduction

Calix[n]arenes are cavity-shaped cyclic molecules made up of n phenol units linked through alkylidene bonds (Fig. 1). R (lower rim) can either be a hydrogen atom, an alkyl chain (usually 1–8 carbons), or even a crown ether ring, bridging different phenolic units across the calixarene molecule. R_1 (upper rim) can be H, an alkyl chain, OH, SH, COOH, COOR, and so forth.

Calixarenes, as well as cyclodextrins and crown ethers are considered to be the third generation of supramolecules [1–3]. The peculiarity of these compounds resides both in their host–guest properties and in the powerful selectivity towards different ions and small organic molecules. These cavity-bearing macrocycles have already been tested for setting up liquid membranes and ion-selective electrodes in order to detect and separate alkali ions [4].

Calixarenes show a particular binding efficiency towards alkali- and alkaline-earth metal ions, depending on their stereochemical conformation [1–3]. The rate of complexation depends on the nature of the ligand, of the guest species, of the solvent, and on the desolvation of the ligand sites prior to complexation. The Corey–Pauling–Koltun (CPK) models provide a useful tool for predicting both the conformational and geometrical properties (volume, cross-section, and cavity dimension) of these macromolecular systems. CPK models of several calix[n]arenes indicate that the aryl units have a certain degree of flexibility, depending on the number of aryl units (n), because of the interconversion between different structures, as illustrated in Fig. 2 in the case of calix[8]arene [7–9]. The presence of crown ether rings in the lower rim dramatically decreases the flexibility of the whole molecule [7–9].

The nature and the position of the substituting groups in calix[n]arenes strongly affect their selectivity in binding ions and organic molecules. As a matter of fact, when $R_1 = C(CH_3)_3$, the thermodynamic stability of calix[4]arene/alkali-ion complexes shows a sequence of binding efficiency for alkali ions in solution that decreases in the order partial cone > 1,3-alternate > cone [7–9]. Unlikely, the 1,3-alternate conformation is favored for unsubstituted calixarenes, which results in an increased

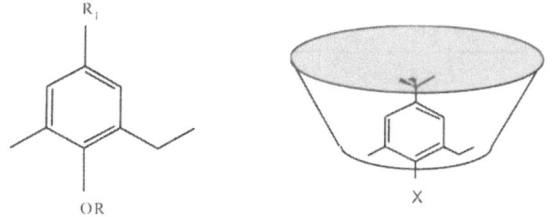

Fig. 1 Phenolic unit showing the R and R_1 residues in calix[n]arenes and the typical cone-shaped conformation of calix[n]arenes

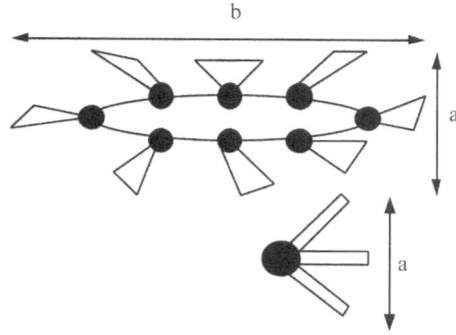

Fig. 2 Calix[8]arene pleated loop conformation and ring flexibility

complexation efficiency and ion selectivity in the order 1,3-alternate > partial cone > cone [7–9].

For calix[n]arenes with $n = 6$ and 8, these effects are less significant, as the increased size of the cavity and the ion selectivity of these hosts are not strongly affected by the different conformation adopted by the whole macrocycle. In these cases the ion or molecule complexation basically depends on the size of the ligand's cavity and of the guest molecule.

Since calix[n]arenes are mostly water-insoluble, they form stable Langmuir films at the air/water interface. The surface properties of such monolayers basically depend on the nature of the ligand, the spreading solvent, the subphase, temperature, and pH [5, 6, 10–12]. The formation of very stable complexes between calix[n]arenes and alkali ions has been observed and investigated, and the selectivity of different ligands towards various ions has been ascertained.

In solution, macrocyclic ligands such as calix[n]arenes may be of great help in understanding cation–micelle interactions. In this perspective, small-angle neutron scattering (SANS) experiments can be performed to determine the micelle–macrocyclic cage interactions at the micellar interface [11, 12]. In fact, by depicting a micelle as a compact particle formed by a hydrophobic core (surfactant's aliphatic tails) and a hydrophilic shell (polar headgroups, solvent molecules, and counterions), the [macrocyclic–ion]$^+$ complexes will tightly stick to the hydrophilic layer, and an increment in the ligand-to-

surfactant molar ratio will modify the surfactant's self-assembling behavior and will result in significant changes in the geometrical micellar parameters.

We report a concise review of our studies on the ion complexation carried out by some calix[n]arene derivatives, in spreading monolayers at the air/water interface and in micellar dispersions of cesium and potassium dodecyl sulfate.

Materials and methods

The calix[n]arenes were either a kind gift from A. Pochini, R. Ungaro and coworkers (University of Parma, Italy) (HXAM, CARB-CAL6, C_3CAL6, C_3CAL5, and C_8CAL6) or were purchased from Acros (Belgium) (CAL6 and CAL8) and were used as received.

For monolayer experiments we followed procedures reported earlier [5, 6, 10, 11]. The complexation properties of each ligand were studied as a function of the subphase composition by using 1 M water solutions of Na$^+$, K$^+$, and Cs$^+$ and 0.1 M water solutions of C(NH$_2$)$_3^+$ (GNDT). In the case of calix[8]arene the 1:1 complex with C$_{60}$ fullerene was also investigated.

SANS experiments were performed according to procedures described in previous work [11, 12]. In each sample, the surfactant concentration (cesium or potassium dodecyl sulfate in deuterium oxide) was kept constant (1% w/w), while the calixarene-to-surfactant molar ratio was scaled up.

Results and discussion

The different calix[n]arene-based ligands used in these studies are shown in Figs. 3, 4, and 5.

CAL6, HXAM, and CAL8 produced stable monolayers, while C_3CAL6, C_3CAL5, and C_8CAL6 were added to CsDS or KDS solutions in D$_2$O for SANS experiments. All the experimental results are shown in the tables.

Spreading π/A isotherms provided the limit area (A_{lim}, Å^{2}molecule^{-1}), the collapse pressure (π_{coll}, mNm^{-1}), the compressibility modulus (C_s^{-1}, mNm^{-1}),

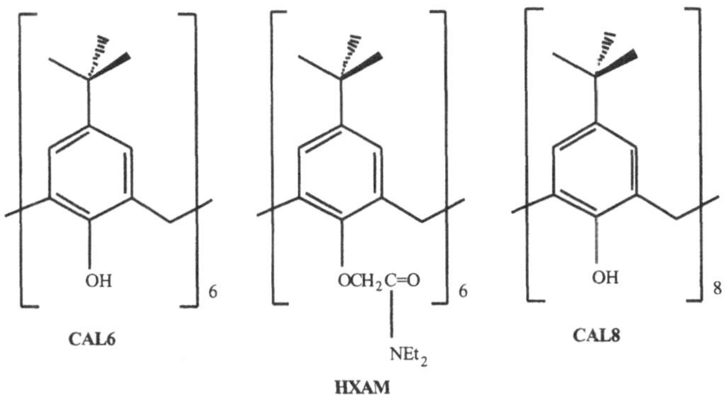

Fig. 3 Chemical structure of CAL6, HXAM, and CAL8

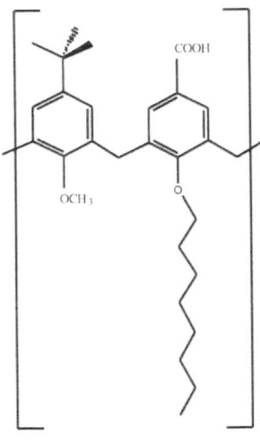

Fig. 4 Chemical structure of CARB–CAL6

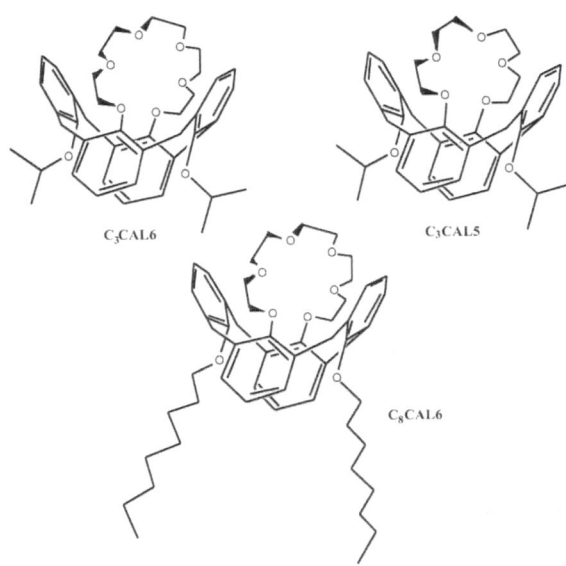

Fig. 5 Chemical structure of the crown ether based calix[4]arenes (C₃CAL6, C₃CAL5, and C₈CAL6)

Table 1 A_{lim} (20 °C) and ΔG_{salt} (25 °C) values for CAL6, HXAM, and CARB–CAL6

Subphase	CAL6		HXAM		CARB–CAL6	
	A_{lim}	ΔG_{salt}	A_{lim}	ΔG_{salt}	A_{lim}	ΔG_{salt}
Water	180	–	258	–	192	–
HCl (pH 3)	185	–	258	–	–	–
1 M NaCl	190	–0.5	304	2.0	–	–
1 M KCl	195	1.5	311	3.8	210	1.3
1 M CsCl	210	2.5	339	9.9	187	–0.3
0.1 M C(NH₂)₃⁺	165	–2.0	318	6.0	–	–

Table 2 A_{lim} (20 °C) and ΔG_{salt} (25 °C) values for crown ether based calixarenes

Subphase	C₃CAL6		C₃CAL5		C₈CAL6	
	A_{lim}	ΔG_{salt}	A_{lim}	ΔG_{salt}	A_{lim}	ΔG_{salt}
Water	100	–	122	–	93	–
1 M KCl	102	0.0	160	3.1	96	0.7
1 M CsCl	111	1.1	133	1.2	101	1.3

Table 3 A_{lim} and C_s^{-1} (20 °C) values for CAL8 and the C₆₀–CAL8 complex. A_{lim} (± 3, Å²molecule⁻¹); $\Delta G_{salt} = \int_0^\pi (A_{salt} - A_{water}) d\pi$ (± 0.2, kJ/mol); C_s^{-1} (± 10, mNm⁻¹)

Solvent	CAL8		C₆₀–CAL8	
	A_{lim}	C_s^{-1}	A_{lim}	C_s^{-1}
CHCl₃	68	110		
C₆H₅CH₃	87	95		
C₆H₆	108	90		
CH₂Cl₂	167	70		
CCl₄	75	110	59	149

and the Gibbs' free-energy change of spreading (ΔG_{salt}, kJmol⁻¹). A_{lim} was evaluated as the x-axis intercept of the tangent line in the condensed-phase region of the spreading isotherm; π_{coll} was determined as the maximum surface pressure before the collapse of the film; ΔG_{salt} was calculated for a specific π range, according to the formula $\Delta G_{salt} = \int_0^\pi (A_{salt} - A_{water}) d\pi$, where A_{salt} and A_{water} are the monolayer areas between 0 and π for monolayers spread over a salt solution and pure water, respectively [5]. The data for A_{lim} and C_s^{-1} determined at 20 °C and for ΔG_{salt} calculated at 25 °C are reported in Tables 1, 2, and 3 for different ligands and different subphases. CAL6 and HXAM form stable monolayers

at the air/water interface at all temperatures. The A_{lim} values indicate that CAL6 molecules are arranged in a perpendicular orientation at the interface, optimizing the hydrophobic interactions between interdigitated *tert*-butyl tails and the hydrogen bonding between facing lower rims of adjacent molecules (Fig. 6). The A_{lim} and ΔG_{salt} values show that at the air/water interface, CAL6 forms stable complexes with alkali ions, with the selectivity decreasing in the order Cs⁺ > K⁺ > Na⁺ > GND⁺. The same perpendicular orientation of the ligand molecules spread at the air/water interface was found for HXAM, but the trend of the selectivity towards ions is Cs⁺ > GND⁺ > K⁺ > Na⁺. Owing to the strength of Cs⁺ and GND⁺ complexes with CAL6 and HXAM and to strong electrostatic repulsions between complexed ligand molecules, these ions dramatically affect confor-

Table 4 Fitting parameters extracted from small-angle neutron scattering experiments

	C₃CAL6/CsDS		C₃CAL5/KDS		C₈CAL6/CsDS	
	0%	5%	0%	5%	0%	5%
Aggregation number	110	127	87	95	110	278
Ionization (%)	13.6	8.0	26.0	19.0	13.6	4.5
Shell thickness	5.0	5.5	4.7	4.7	5.0	7.0
Short axis	18	18	17	18	18	20
Axial ratio	1.6	1.9	1.5	1.5	1.6	3.0
Average diameter	52	56	49	51	52	73
Hydration	7.5	8.2	8.4	8.2	7.5	7.5

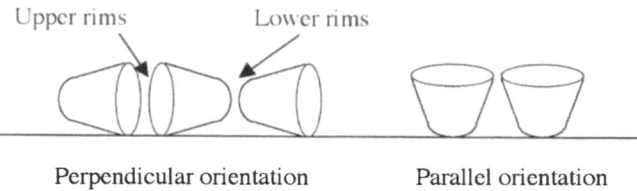

Fig. 6 Perpendicular and parallel orientations in calix[n]arenes at the air/water interface

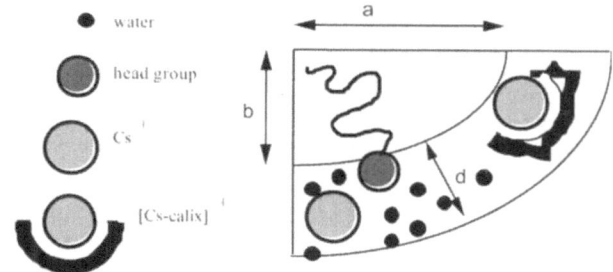

Fig. 7 Model for the micellar structure as used in the small-angle neutron scattering fitting program

mational changes in CAL6 and HXAM and interfere in the ligand's parallel-to-perpendicular orientation transitions. Regarding CAL8 (Table 3), the A_{lim} values indicate that this ligand does not adopt a cone-shaped conformation (Fig. 1) at the air/water interface. In fact the cross-sections calculated according to the CPK models are 322 and 140 $Å^2$molecule^{-1} for the pleated loop conformation (Fig. 2) in the parallel and perpendicular molecular orientation at the air/water interface, respectively (Fig. 6). For the cone-shaped conformations much higher values would be expected. The compressibility modulus values, $C_s^{-1} = -A(\partial\pi/\partial A)_T$, range between 70 and 110 mNm^{-1}, indicating that CAL8 monolayers are in the liquid-expanded phase with fluidlike hydrophobic chains in the whole A range investigated [10]. These findings indicate that the larger the ring, the more flexible the whole calixarene molecule. Another confirmation of the existence of the pleated loop conformation is the change in A_{lim} with the nature of the spreading solvent, as a result of the different interactions established between the ligand and the solvent molecules. CAL8 forms stable inclusions complexed with C_{60} fullerene, as shown by the lowering in A_{lim} and the increment in the C_s^{-1} values. The formation of the C_{60}–CAL8 complex was also confirmed by means of UV–vis spectroscopy performed on Langmuir–Blodgett films and on collapsed material transferred onto quartz plates. CARB–CAL6 monolayers indicate the presence of strong interactions in the case of the complex with K$^+$ ions. Preliminary results from monolayer and black lipid membranes experiments suggest the formation of dimers through hydrophobic interactions and hydrogen bonding. Finally, in

the case of crown ether based calix[4]arenes (C₃CAL6, C₃CAL5, and C₈CAL6), the π/A isotherms show that these ligands are selective for either potassium or cesium depending on the crown ether cavity size. The same ligands were added to potassium or cesium dodecyl sulfate micellar solutions in D₂O in order to investigate the modification induced on the micellar interface charge by the ligand's complexation by SANS experiments. The absolute intensity, corrected for cell by cell, the background scattering, transmissions, and detector efficiency, and calibrated for absolute intensity with H₂O, was calculated with fitting programs, and provided the parameters listed in Table 4. The fitting program was set up by depicting the micelle structure as sketched in Fig. 7, as already mentioned in the Introduction.

Conclusions

We reported a summary of our work on some calix[n]arene derivatives, with different cavity size and selectivity towards specific ions. According to our experiments, the conformations of calix[n]arenes depend on the number of aryl units and on the kind of substituents on the upper and lower rims. Cone-shaped conformations were found for CAL6 and its amide derivative (HXAM), while CAL8 possesses a pleated-loop conformation. Calix[4]arenes with crown ether bridges, which force the ligand molecule into a 1,3-alternate conformation, were found to specifically bind potassium and cesium ions. This evidence was used to monitor the interface charge variation in the micellar dispersion by addition of proper amounts of crown ether based calix[4]arenes.

Acknowledgements The authors are greatly thankful to Andrea Pochini, Rocco Ungaro, Arturo Arduini, and Alessandro Casnati for providing the calixarenes. MURST and CSGI are acknowledged for partial financial support.

References

1. Gutsche CD (1989) In: Stoddard JF (ed) Calixarenes. The Royal Society of Chemistry, Cambridge, chapter 1
2. Shinkai S (1993) Tetrahedron 49:8933
3. Böhmer V (1995) Angew Chem Int Ed Engl 34:713
4. Perrin R, Harris S (1990) In: Vicens J, Böhmer V (eds) Calixarenes. A versatile class of macrocyclic compounds. Kluwer, Dordrecht, pp 235–259
5. Dei L, Casnati A, Lo Nostro P, Baglioni P (1995) Langmuir 11:1268
6. Lo Nostro P, Casnati A, Bossoletti L, Baglioni P (1996) Colloids Surf A 116:203
7. Ungaro R, Casnati A, Ugozzoli F, Pochini A, Dozol JF, Hill C, Rouquette H (1994) Angew Chem Int Ed Engl 33:1506
8. Casnati A, Pochini A, Ungaro R, Ugozzoli F, Arnaud F, Fanni S, Schwing MJ, Egberink RJM, de Jong F, Reinhoudt DN (1995) J Am Chem Soc 117:2767
9. Casnati A, Pochini A, Ungaro R, Bocchi C, Ugozzoli F, Schwing MJ, Egberink RJM, Struijk H, Lugtemberg R, de Jong F, Reinhoudt DN (1996) Chem Eur J 2:436
10. Dei L, Lo Nostro P, Capuzzi G, Baglioni P (1998) Langmuir 14:4143
11. Capuzzi G, Fratini E, Pini F, Baglioni P, Casnati, A, Teixeira J (2000) Langmuir 16:188
12. Capuzzi G, Fratini F, Dei L, Lo Nostro P, Casnati A, Gilles R, Baglioni P (2000) Colloids Surf A 167:105

Progr Colloid Polym Sci (2001) 118: 243–247
© Springer-Verlag 2001

Mark E. Hayes
Alexander V. Gorelov
Kenneth A. Dawson

DNA-induced fusion of phosphatidylcholine vesicles

M. E. Hayes · A. V. Gorelov (✉)
K. A. Dawson
Irish Centre for Colloid Science
and Biomaterials,
Department of Chemistry
University College Dublin, Belfield
Dublin 4, Ireland
e-mail: gorelov@pop3.ucd.ie
Tel.: +353-1-7162417
Fax: +353-1-7162127

A. V. Gorelov
Institute of Theoretical and Experimental
Biophysics, Pushchino, Russia

Abstract We show that unilamellar vesicles composed of zwitterionic lipids such as dipalmitoylphosphatidylcholine can fuse in the presence of DNA. Ca^{2+} ions are required to induce fusion. Nonradiative energy transfer was used to monitor the fusion process. The dependence of the extent and kinetics of fusion on the mode of preparation, the size of the vesicles, calcium concentration and lipid concentration was investigated. Small unilamellar vesicles of size less than 50 nm fuse rapidly, whereas vesicles larger than 100 nm will not fuse.

Key words DNA–lipid complexes · Lipid mixing assay · Vesicle fusion

Introduction

Gene therapy is a new medical treatment with the potential to cure diseases at the molecular level. It is becoming a more realistic goal with the rapid advances being made in molecular biology. The most important obstacle in this new technology is in achieving efficient delivery of DNA in vivo [1]. Cationic vesicles composed of synthetic and/or natural components are, at present, the most popular choice of nonviral genetic delivery vehicle. They are receiving much interest from the scientific community and are also available commercially for transfection of tissue cultures. Natural phospholipids offer a distinct advantage over cationic lipids in that they lack cytotoxicity. Zwitterionic phospholipids do not complex spontaneously with DNA like cationic lipids, but their interaction can be mediated by the addition of divalent cations [2]. It was shown previously that DNA is able to induce polymorphic phase transitions in lecithins in the presence of calcium cations [3]. Also, the number of lipid molecules interacting with DNA in the complex was estimated from temperature-scanning ultrasound velocimetry [4].

In the present work, we looked at the interaction of small unilamellar vesicles (SUV) and large unilamellar vesicles (LUV) composed of dipalmitoylphosphatidylcholine (DPPC) with short fragments of DNA. We have recently found in preliminary small-angle X-ray scattering experiments that DNA intercalates between the lamellar membranes of DPPC, after mixing DNA with SUV in the presence of calcium. The formation of such a highly organized structure implies that lipid vesicles should break during complex formation. The purpose of the present work is to show that lipid mixing occurred during this complex formation. A lipid mixing assay was employed which used nonradiative energy transfer (NRET) between two fluorescent membrane probes to monitor lipid mixing. We propose a two-wavelength excitation method, where we express the efficiency of transfer, E, as the ratio of sensitized emission to acceptor emission when the acceptor is excited directly. We show the effect of calcium and lipid concentration on the extent of fusion and also show that vesicle size determines whether fusion occurs or not.

Materials and methods

DPPC, N-(7-nitro-2,1,3-benzoxadiazol-4-yl)dipalmitoylphosphatidylethanolamine (NBD-PE) and N-(lissamine rhodamine B sulfonyl)dipalmitoylphosphatidylethanolamine (Rh-PE) were purchased from Avanti Polar Lipids (Alabaster, Ala., USA) and were used without further purification. Appropriate amounts of DPPC was weighed out and dissolved in freshly distilled chloroform. To this, appropriate volumes of NBD-PE and Rh-PE from stock CHCl₃/CH₃OH (3:1 v/v) solutions were added for the labelled vesicle preparation. The resulting solution was mixed well and dried with a stream of N₂ and dried under vacuum overnight.

The films of lipid (labelled and unlabelled) were hydrated with 5 mM *N*-(2-hydroxyethyl)piperazine-*N'*-ethanesulfonic acid (HEPES, Sigma) buffer at pH 7.5 for 1 h at room temperature and then a further 1 h at 50 °C vortexing intermittently. LUV of diameter 100 nm were prepared by passing this multilamellar suspension through a miniextruder (Avanti Polar Lipids) with two stacked (Nucleopore) polycarbonate membranes (100-nm pore size), a total of 11 times at 50 °C. The lipid concentration was determined by the Stewart assay [5]. SUV were prepared from the same multilamellar suspension diluted to 1 g/l, by sonication in a temperature-controlled ultrasonic bath (Cole-Parmer 8890) for 1 h under an Ar atmosphere at 50 °C. The resulting SUV suspension gave vesicle sizes between 30–50 nm in diameter which were measured by dynamic light scattering and the vesicles were tested for degradation products by a standard Bligh–Dryer extraction and subsequent thin-layer chromatography [6]. At this low level of ultrasonic intensity there was no appearance of degradation products. All fluorescence spectra were taken with a PerkinElmer LS-50 luminescence spectrometer with a stirring cell holder thermostated by an external water bath. The concentrations of both fluorescence probes were determined spectrophotometrically using a Helios UV–vis spectrometer using the quoted extinction coefficients in methanol [7].

Calf thymus DNA from Sigma was used in the current work. Purification was carried out using the standard chloroform/phenol extraction procedure [8]. Short fragments were prepared by sonication of the DNA solution after purification, at a concentration of 2 mg/cm³, in 50 mM sodium citrate buffer (trisodium dihydrate from Aldrich) and 2 mM ethylenediaminetetraacetate (disodium salt dihydrate from Sigma) at pH 7.0. This procedure was continued for 12 h on ice under a nitrogen atmosphere using a VC 50 ultrasonic processor (Sonics and Materials, Danbury, Conn., USA) at low intensity. The DNA was dialysed exhaustively against 5 mM HEPES, pH 7.5 buffer using Spectra/Por 1 membrane tubing (Spectrum Medical Industries, Calif., USA) which had a molecular-weight cutoff of 6–8,000 Da. All the water used was distilled and purified using a Milli-Q RG apparatus from Millipore (Ireland).

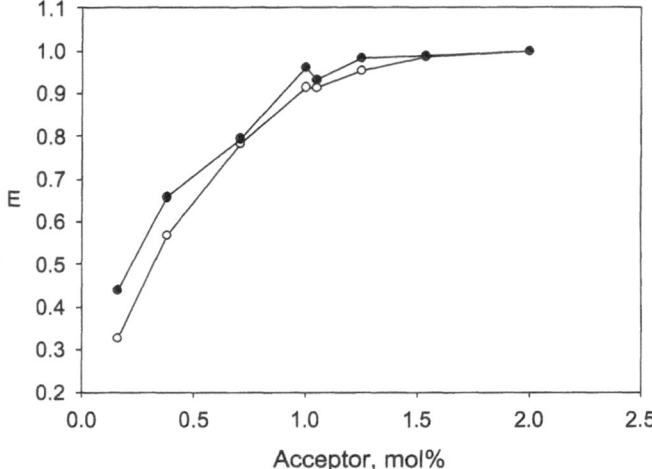

Fig. 1 The dependence of E on the surface density of the acceptor, measured at two temperatures. The *closed circles* are at 25 °C (gel phase) and the *open circles* are at 50 °C (fluid phase)

Results

We used a lipid mixing assay based on NRET to elucidate the mechanism and rate of complex formation [9, 10]. This approach relies upon the interactions which occur between two fluorophores in close proximity when the emission band of one (the energy donor, NBD-PE) overlaps with the absorption band of another (the energy acceptor, Rh-PE). The method used here was to incorporate both probes in one population of vesicles with an initially high efficiency of transfer. When such vesicles fuse with an unlabeled population, in the presence of DNA and calcium, the probes effectively dilute within the newly formed membrane. As a result, their average distance increases with a reduction in the efficiency of transfer according to Förster's equation for transfer efficiency as a function of the distance between the donor and acceptor [11]. At low surface densities of Rh-PE (less than 0.7 mol% with respect to total lipid) the efficiency of transfer is nearly linearly related to the ratio of acceptor lipid to total lipid in the vesicle membrane (Fig. 1). In order to calculate E, essentially the quantum yield (or emission intensity) of the donor must be measured in the presence and absence of acceptor molecules. A common procedure is to solubilize the vesicles by adding detergents such as Triton X-100 using the relationship

$$E = 1 - F/F_0 \ , \tag{1}$$

where F is the fluorescence intensity at the donor emission maximum (530 nm) and F_0 is the intensity at the same wavelength in the presence of Triton X-100, with correction for donor quenching and dilution [10].

As the emission of the donor is measured, any changes in the donor quantum yield during complex formation can give rise to a signal which could be interpreted as lipid mixing.

In the present system, it was found that the fluorescence spectrum of the labelled vesicles was affected by the addition of DNA and calcium without any unlabelled vesicles present. So, we developed a method to calculate E which is independent of the quantum yield changes that occur during complex formation. The emission intensity from a fluorescent molecule excited at λ_e and detected at λ can be expressed as [12]

$$F(\lambda) = \varepsilon(\lambda_e)\varphi f(\lambda)c I_0 k \ , \tag{2}$$

where $\varepsilon(\lambda_e)$ is the extinction coefficient at the excitation wavelength, φ is the quantum yield, $f(\lambda)$ is the fraction of total emission that occurs at the wavelength λ, c is the concentration of absorbing molecules and I_0 is the intensity of the incident light. The instrumental constant k depends on the characteristics of the spectrometer and the geometry of the experiment. In the following expressions, λ_1 and λ_2 refer to the wavelengths of the maximum absorption of donor and acceptor. λ_3 and λ_4 refer to donor and acceptor emission maxima, respectively. First, we calculate the fluorescence intensity at λ_4, $F_{\lambda 1}^{\lambda 4}$, with excitation at λ_1. $F_{\lambda 1}^{\lambda 4}$ contains the sensitized emission from the acceptor, emission from the acceptor when excited at λ_1 and the contribution of donor emission at λ_4.

$$F_{\lambda 1}^{\lambda 4} = I_0 k \left[\varepsilon_D^{\lambda 1} c_D E \varphi_A + \varepsilon_A^{\lambda 1} c_A \varphi_A + f_D^{\lambda 4} \varepsilon_D^{\lambda 1} c_D (1-E) \varphi_D \right] , \tag{3}$$

where $\varepsilon_D^{\lambda 1}$ and $\varepsilon_A^{\lambda 1}$ are the extinction coefficients of donor and acceptor at λ_1. c_D and c_A are the concentrations of donor and acceptor and φ_D and φ_A are the quantum yields of the donor and acceptor. $f_D^{\lambda 4}$ is the emission of the donor at λ_4 relative to λ_3. $F_{\lambda 1}^{\lambda 3}$ expresses the fluorescence intensity with excitation at λ_1 and observation at λ_3. It contains the terms for direct excitation of the donor and the contribution of acceptor emission at λ_3. $f_A^{\lambda 3}$ is the acceptor emission at λ_3 relative to λ_4.

$$F_{\lambda 1}^{\lambda 3} = I_0 k \left[\varepsilon_D^{\lambda 1} c_D (1-E) \varphi_D + f_A^{\lambda 3} \varepsilon_D c_D E \varphi_A \right] \tag{4}$$

$F_{\lambda 2}^{\lambda 4}$ expresses the fluorescence intensity when exciting at λ_2 and observing at λ_4 and contains only the direct excitation term of the acceptor.

$$F_{\lambda 2}^{\lambda 4} = I_0 k \varepsilon_A^{\lambda 2} c_A \varphi_A \tag{5}$$

E can be calculated by using Eqs. (3), (4) and (5) to yield

$$E = \frac{\varepsilon_A^{\lambda 2} c_A}{\varepsilon_D^{\lambda 1} c_D} \left\{ \frac{F_{\lambda 1}^{\lambda 4}}{F_{\lambda 2}^{\lambda 4}} - \frac{\varepsilon_A^{\lambda 1}}{\varepsilon_A^{\lambda 2}} - f_D^{\lambda 4} \left[\frac{F_{\lambda 1}^{\lambda 3}}{F_{\lambda 1}^{\lambda 4}} + f_A^{\lambda 3} \left(\frac{F_{\lambda 1}^{\lambda 4}}{F_{\lambda 2}^{\lambda 4}} \right) \right] \right\} , \tag{6}$$

where $\varepsilon_A^{\lambda 2} c_A / \varepsilon_D^{\lambda 1} c_D$ is the ratio of acceptor absorbance to the donor absorbance of the sample. In the derivation of Eq. (6) we relied on the fact that the $I_0 k$ term in Eq. (2) stays constant when the excitation wavelength is changed from λ_1 to λ_2. This depends on the corrections for the spectral characteristics of the light source and the detector. Also, the ratio of the acceptor absorbance to the donor absorbance is difficult to determine accurately in vesicles owing to the strong scattering contribution to the spectra. So, Eq. (6) is exact for the proportionality factor which has to be measured for each particular vesicle preparation and spectrometer setting.

We obtained the proportionality factor in two ways. Firstly, at high acceptor concentration the efficiency of transfer should be 1. A chloroform solution of DPPC with a relatively high acceptor (2 mol%) and donor (1 mol%) concentration was prepared and subsequent additions of DPPC were made, effectively diluting the probes by unlabelled lipid. From each sample, vesicles were prepared as described before. The proportionality factor was calculated by setting E for the sample with the highest probe concentration to 1. The ratio of the probe concentrations remained constant in all the other

samples, so the same proportionality factor can be used to calculate E. The dependence of E on the surface density of the acceptor is shown in Fig. 1. E is independent of the surface density of the donor [10]. The dependence of E on the surface density of the acceptor obtained this way is similar to the dependence calculated from Eq. (1) for the same probes [10]. In the other way, the proportionality factor was calculated using Eq. (1) for each vesicle preparation and then the changes in E that occurred upon the addition of DNA and Ca^{2+} were calculated using Eq. (6). The advantage of Eq. (6) is that changes in E calculated this way are less susceptible to the light scattering contribution to fluorescence that occurs when aggregates are formed. Also, Eq. (6) takes into account any changes in the quantum yield of the probes that occur upon addition of other molecules to the system. Thus, the changes in E during the fusion experiments are purely a result of probe dilution during lipid mixing.

DNA-induced fusion of SUV

The stability of SUV and LUV in the absence and presence of DNA and Ca^{2+} with time are shown in Fig. 2. It shows how SUV are susceptible to slow spontaneous fusion in the absence of DNA and Ca^{2+}, which is well known from the literature [13]. However, in the case of SUV when DNA and Ca^{2+} were added, E rapidly decreased. A two-stage process for SUV fusion is shown. Initially, there was fast lipid mixing, which was

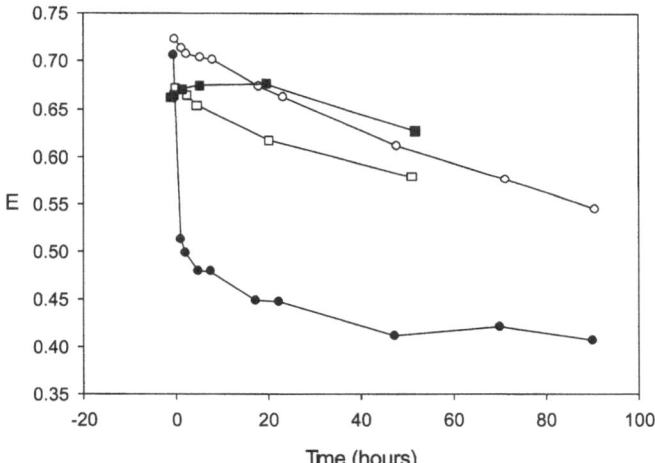

Fig. 2 The fusion of small unilamellar vesicles (*SUV*) (*closed circles*) and large unilamellar vesicles (*LUV*) (*closed squares*) in the presence of DNA and Ca^{2+} as a function of time. Also shown is the spontaneous fusion of SUV (*open circles*) and LUV (*open squares*) in the absence of DNA and Ca^{2+}. The DNA phosphate-to-lipid molar ratio was 1:4 and the ratio of labeled to unlabeled vesicles was also 1:4 with both types of vesicles. The calcium concentration was 10 mM. The total lipid concentration was 100 μM. Measurements were taken at 25 °C

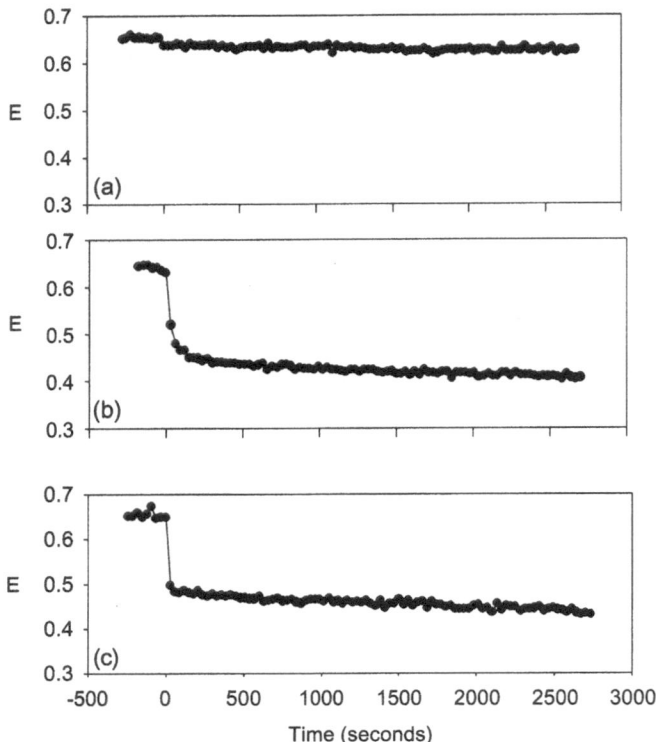

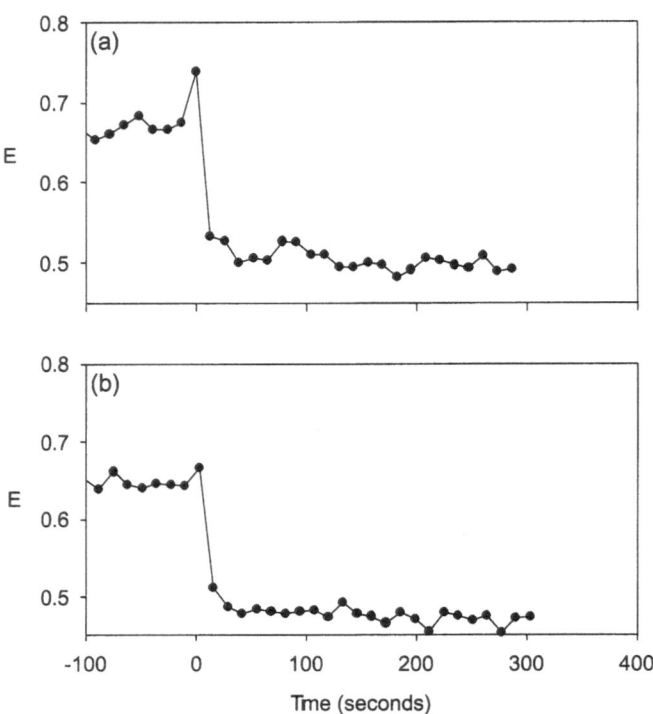

Fig. 3a–c The dependence of fusion on Ca^{2+} concentration. **a, b** and **c** are 0.1, 1 and 10 mM, respectively. The DNA phosphate-to-lipid molar ratio was 1:4 and the ratio of labelled to unlabelled vesicles was 1:4. The graphs initially show the vesicles and DNA present with Ca^{2+} added at time zero. The total lipid concentration was 100 μM. Measurements were taken at 25 °C

Fig. 4 The dependence of fusion on the total lipid concentration: **a** 25 μM and **b** 100 μM. The DNA-to-lipid molar ratio was kept constant at 1:4 and the ratio of labelled to unlabelled vesicles was also kept constant at 1:4. Ca^{2+} (10 mM) was added at time zero. Measurements were taken at 25 °C

faster than the time resolution of our measurements (about 13 s), followed by a slower decrease in the transfer efficiency over 90 h. Using the dependence of E on the acceptor concentration at 25 °C (Fig. 1) we can calculate the extent of fusion taking into account that the initial ratio of labeled to unlabeled vesicles is 1:4. Fusion of 100% corresponds to the case where all the lipids are mixed, so the acceptor concentration in the bilayer decreases by 5 times. The extent of fusion calculated for the rapid first stage is 65% and after 90 h the extent of fusion is 82%. LUV were relatively stable over time and the presence of DNA and Ca^{2+} causes LUV aggregation without any indication of fusion. So, we can conclude that it is fusion upon the interaction of DNA and Ca^{2+} which causes lipid mixing in the case of SUV. We can exclude, for example, probe transfer between adjacent bilayers in DNA lipid aggregates.

Calcium concentration dependence

The dependence of fusion on Ca^{2+} concentration is shown in Fig. 3. With 0.1 mM there is no noticeable change in E. Above 1 mM we observe rapid changes in E

followed by a slower phase. The extent of fusion above 1 mM was 65–75% for the concentration range studied (0.1–50 mM) after 45 min. The stock vesicle suspension used in the Ca^{2+}-dependence experiments had an extent of spontaneous fusion over the entire time course of the experiments of less than 5%.

Vesicle concentration dependence

The dependence of the extent of fusion of SUV on lipid concentration in the presence of DNA and Ca^{2+} for 25 and 100 μM lipid is shown in Fig. 4. The rate of fusion in the concentration range studied (25–200 μM) was in all cases faster than the time resolution of the measurements, even at 25 μM, which was as low as the sensitivity of the spectrometer would allow. The extent of fusion was approximately the same for all the lipid concentrations studied.

Discussion

Vesicles of all sizes will eventually fuse; this is primarily due to the reduction in the free energy owing to the loss of stress from the curved surfaces. Generally electrostatic, steric and hydration forces create a sufficiently strong repulsive barrier to prevent membranes coming into close contact. These forces are believed to maintain the colloidal stability of the vesicle suspension under physiologically relevant conditions [14]. Consequently, to initiate vesicles aggregation one must diminish the intermembrane repulsion. Lowering the number of charges or decreasing the number of water binding sites on membranes or condensing their interfaces lowers the lipid bilayer repulsion and facilitates vesicle aggregation.

We showed that low concentrations of Ca^{2+} in the presence of DNA are required to initiate lipid mixing between labeled and unlabeled SUV. As lipid mixing does not occur during transient vesicle collisions [10] but only during vesicle aggregation that result in fusion, it is therefore DNA which acts in bringing the vesicles together so that fusion may occur.

Calcium binds strongly to negatively charged phospholipids such as phosphatidylserine [15] and phosphatidylglycerol [16]. Indeed, calcium-promoted fusion has been extensively studied in vesicles composed of phosphatidylserine or cardiolipin with phosphatidylcholine [17, 18]. Calcium binding to zwitterionic lipids such as phosphatidylcholines is rather weak. It was found that the ratio of bound Ca^{2+} to DPPC liposomes is 0.03–0.08 per DPPC molecule as determined by different techniques [19, 20]. So, Ca^{2+} alone actually stabilizes zwitterionic vesicle suspensions by adding this slight positive charge to the surface [19]. It therefore must be the interaction of the vesicles with DNA mediated by Ca^{2+} which leads to lipid mixing. Figure 3 shows that the critical concentration where Ca^{2+} induces fusion is between 0.1 and 1 mM. This compares with our previous findings where in this region of Ca^{2+} concentrations DNA partitioned into a reversed micellar phase of phosphatidylcholines in cyclohexane [21].

LUV aggregate but do not fuse in the presence of Ca^{2+} and DNA. This resistance of LUV to fusion was observed previously in the poly(ethylene glycol) (PEG) mediated fusion of DPPC vesicles. SUV were found to fuse rapidly but LUV (even when severely dehydrated) did not fuse unless the bilayer was disrupted in some way [13]. The highly curved membranes of SUV increase their propensity for fusion. The comparison between our vesicle fusion with Ca^{2+} and DNA with the PEG-mediated fusion of DPPC shows similar results in that vesicles smaller than 70 nm in diameter were readily fused, whereas vesicles greater than 90 nm in diameter did not fuse [13]. So this raises an interesting question as to the structural events that occur during the DNA–Ca^{2+}-induced fusion of DPPC SUV. Lipid mixing can be brought about by DNA-induced aggregation of the vesicles, followed by close approach of the curved membranes, leading to fusion. Additionally, SUV can be ruptured upon the interaction with DNA in the presence of Ca^{2+}. In the aggregates, ruptured SUV can reseal with other vesicles, causing lipid mixing. This question will be the focus of further study.

Acknowledgements The authors wish to thank their colleagues in the Irish Centre for Colloid Science and Biomaterials for their helpful discussions and A.G. would also like to acknowledge very useful discussions with D.P. Kharakoz.

References

1. Bally MB, Harvie P, Wong F, Kong S, Wasan EK, Reimer DL (1999) Adv Drug Deliv Rev 38:291–315
2. Gruzdev AD, Khramtsov VV, Weiner LM, Budker VG (1982) FEBS Lett 137:227–230
3. Tarahovsky YS, Khusainova RS, Gorelov AV, Nicolaeva TI, Deev AA, Dawson KA, Ivanitsky GR (1996) FEBS Lett 390:133–136
4. Kharakoz DP, Khusainova RS, Gorelov AV, Dawson KA (1999) FEBS Lett 446:27–29
5. New RRC (ed) (1990) Liposomes: a practical approach. IRL press, p 108
6. New RRC (ed) (1990) Liposomes: a practical approach. IRL press, pp 109–256
7. Haugland RP (ed) (1996) Molecular probes. Handbook of fluorescent probes and research chemicals. p 304
8. Wallace DM (1987) Methods Enzymol 152:33–48
9. Stryer L (1978) Annu Rev Biochem 47:819–846
10. Struck DK, Hoekstra D, Pagano RE (1981) Biochemistry 20:4093–4099
11. Förster T (1949) Z Naturforsch A 4:321–327
12. Cantor CR, Schimmel PR (1980) Biophysical Chemistry. WH Freeman Company, USA
13. Lentz BR (1994) Chem Phys Lipids 73:91–106
14. Cevc G, Marsh D (1987) Phospholipid Bilayers, Physical Principles and Models, Wiley Interscience, USA
15. Roux M, Bloom M (1991) Biophys J 60:38–44
16. Macdonald P, Seelig J (1987) Biochemistry 26:1231–1240
17. Wilschut J, Nir S, Scholma J, Hoekstra D (1985) Biochemistry 24:4630–4636
18. Wilschut J, Scholma J, Bental M, Hoekstra D, Nir S (1985) Biochim Biophys Acta 821:45–55
19. Satoh K (1995) Biochim Biophys Acta 1239:239–248
20. Lis LJ, Lis WT, Parsegian VA, Rand RP (1981) Biochemistry 20:1771–1777
21. Gorelov AV, Hayes ME, Wehling A, Dawson KA (1998) Nuovo Cimento 20:2401–2408

Progr Colloid Polym Sci (2001) 118: 248–250
© Springer-Verlag 2001

D. Letellier
V. Cabuil

Mineralization of lipidic tubules

D. Letellier · V. Cabuil (✉)
Laboratoire des Liquides Ioniques et
Interfaces Chargées – Colloïdes
Magnétiques, Université Paris 6, Bât. F,
Case 63, 75252 Paris cédex 05, France

Abstract We describe the use of
supramolecular assemblies of phospholipidic molecules as a support for
the surface-mediated synthesis of
inorganic nanoparticles. Diacetylenic phospholipids such as $DC_{8,9}PC$
are known to form hollow tubular
microstructures reffered to as tubules. Mineralization of $DC_{8,9}PC$
tubules is obtained by nucleation
and growth of an iron oxyhydroxide
on their surface. This leads to
tubular organic–inorganic hybrid
microstructures.

Key words Self-assemblies ·
Tubules · Phospholipids ·
Mineralization

Introduction

We report the use of self-assemblies of amphiphilic
molecules as a support for the precipitation of anisotropic iron oxide nanoparticles. The organic template is a
hollow cylinder made of tensioactive molecules. These
molecules are phospholipids, namely diacetylenic phospholipid ($DC_{n,m}PC$). This phospholipid forms bilayers in
water which have the specificity to roll themselves up into
open-ended hollow tubes at room temperature, giving
rise to so-called lipidic tubules (Figs. 1, 2) [1]. Tubule
formation is driven by a reversible first-order phase
transition from a intralamellar chain-melted L_α phase to
a chain-frozen $L_{\beta*}$ phase [2], the tubule phase being the
one in which the hydrocarbon chains are highly ordered
[3]. The $DC_{8,9}PC$ chain-melting transition temperature is
approximately 43 °C [4]. Above this temperature
$DC_{8,9}PC$ forms liposomes in aqueous solutions.

Their rigidity, hollowness, high aspect ratio and
resistance to a wide range of pH and ionic strength
make the tubules suitable for technical applications and
model bioinorganic template synthesis [5]. It was shown
that lipidic tubules could be metallized by electroless
deposition of copper [1] or permalloy [6] and mineralized
by chemical precipitation of metal carbonates [7] or
coated with pre-formed silica particles [8]. Archibald and
Mann [9] described the coating of galactocerebroside
tubules with iron.

Despite their interesting morphology (high aspect
ratio and hollowness) and their chemical resistance,
$DC_{8,9}PC$ tubules have not been used yet for the template
precipitation of ferric species. After studying the interactions between iron oxide nanoparticles and $DC_{8,9}PC$
tubules [10], our purpose is now to perform the
mineralization of $DC_{8,9}PC$ tubules that are easier to
prepare than the galactocerebroside ones. We shall refer
to mineralization as to the surface-specific polycondensation of mono- or oligomeric species present in the bulk
solution.

Experimental

The tubular lipidic microstructure were prepared with 1,2-
bis(10,12-tricosadiynoyl)-*sn*-glycero-3-phosphocholine ($DC_{8,9}PC$)
obtained from Avanti Polar Lipids and used as purchased. In
order to get tubules, we used the so-called precipitation method
firstly described by Georger et al. [4], which consists of the
solubilization of the lipid in an appropriate solvent and the further
precipitation by addition of water. Water was added to a solution
of $DC_{8,9}PC$ in ethanol, the water/ethanol ratio being about 40:60
(v:v) and the final concentration of lipids being approximately 1 g/
l. A flocculent precipitate of tubules was observed. This precipitate
was centrifuged at 6000 rmp for 15 min and the resulting pellet was
dispersed in distilled water. The tubules obtained had a constant
diameter of 0.5 μm and were multiamellar. Their length ranged
from a few microns to tens of microns.

According to Schnur and coworkers [11] the self-assembly
process of $DC_{8,9}PC$ tubules begins with the formation of helical

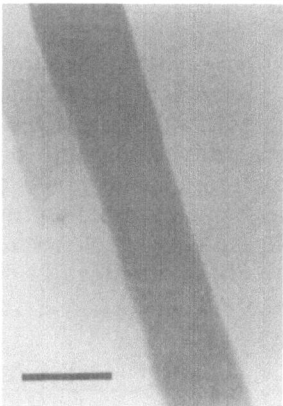

Fig. 1 Transmission electron microscopy (*TEM*) picture of a DC$_{8,9}$PC tubule colored with ammonium molybdate. The ribbons fused to form a smooth surface. The *bar* represents 500 nm

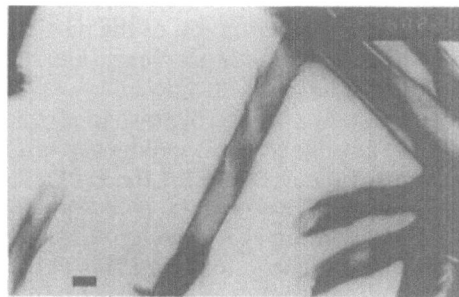

Fig. 2 TEM picture of DC$_{8,9}$PC tubules colored with uranyl acetate. Several overlapping open ribbons form an edged tubule. The *bar* represents 500 nm

ribbons that fuse into cylinderical tubules to reduce their edge energy. We established that the tubules' precipitation rate allows some control of the proportion of ribbons. Rapid precipitation leads to cylindrical structures constituted of overlapping wound ribbons; we call them edged tubules for they are lipidic tubules that bear large numbers of bilayer edges exposed to the solvent. Fused ribbons obtained by slow precipitation are refered to as smooth tubules.

Mineralization of the tubules was performed by slow hydrolysis of an aqueous solution of iron(III) chloride in the presence of DC$_{8,9}$PC tubules. An aqueous dispersion of the tubules was thus prepared, then centrifuged. The heavy phase of the tubules (around 10 g/l) was then dispersed in an aqueous solution of iron chloride. On leaving the FeCl$_3$/DC$_{8,9}$PC tubule mixture at room temperature for aging, the hydrolysis of the ferric species drives the polycondensation process. It gives rise to a yellow-brown precipitate. The effects of the aging time, of the reactant concentrations and of the nature of the tubules (edged or smooth) were investigated. The ferric species concentration ranged from 10^{-4} to 1 mol^{-1} and the tubule concentration ranged from 0.1 to 20 gl^{-1}.

Optical microscopy and transmission electron microscopy (TEM) were performed in order to confirm the effective mineralization of the DC$_{8,9}$PC tubules. The nature of the inorganic species deposited on the tubules was indentified by electron scattering. TEM and electron scattering were performed using a JEOL 100CXII top-entry UHR electron microscope on samples deposited on carbon films and air-dried. Tubules without ferric precipitate were colored by negative stain with an aqueous solution of ammonium molybdate or uranylacetare (2 wt%). Mineralized tubules were observed without coloration.

Results and discussion

TEM photographs of the structures obtained are shown in Figs. 3 and 4. Two types of objects are observed: mineralized tubules (Fig. 3a), which are covered with iron oxyhydroxide microcrystals, and decorated tubules (Fig. 4), which bear discrete particles. These structure are never found together in the same sample. Moreover, the formation of one structure rather than the other only depends on the nature of the DC$_{8,9}$PC tubules.

The mineralized tubules are discernible by optical microscopy: they appear light brown and thicker than the uncoated tubules (Fig. 3b). Smooth tubules lead to large leaflike microcrystals that covers the entire surface of the tubules giving the latter a hairy shape (Fig. 3). The inorganic phase was identified by selected-area electron diffraction as lepidocrocite γ-FeOOH. The thickness of the inorganic coating ranged from 0.1 to 0.3 μm in the samples observed. Edged tubules exhibit small rice-bean-shaped particles that are very specifically deposited on the DC$_{8,9}$PC ribbon edges (Fig. 4). The iron oxyhydroxide particles were identified as poorly ordered akaganeite β-FeOOH and were about 70-nm long and 20-nm wide , on average. In our case the particles are monodisperse in size and their binding to the ribbon edges is obvious on the TEM pictures.

Fig. 3 a TEM picture of a mineralised DC$_{8,9}$PC tubule, obtained with a smooth tubule, exhibiting leaflike iron oxyhydroxide (γ-FeOOH) particles covering the entire surface of the tubule. The *bar* represents 100 nm. **b** Optical microscopy picture of mineralized tubules (indicated by *arrows*). The *bar* represents 10 μm

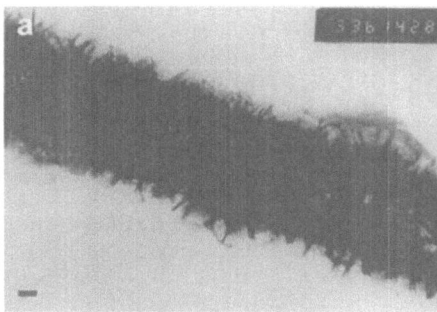
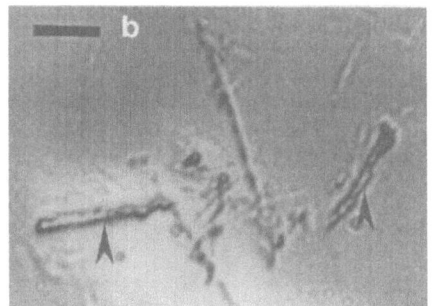

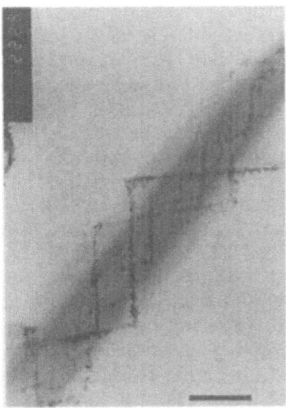

Fig. 4 TEM picture of a decorated $DC_{8,9}PC$ tubule, obtained with an edged tubule, exhibiting rice-shaped iron oxyhydroxide (β-FeOOH) particles bound on the $DC_{8,9}PC$ ribbon edges. The *bar* represents 500 nm

Very few particles not bonded to a tubule are observed in the samples. As it has been established [12] that metal ions bind to phospholipidic bilayers, it is reasonable to assume that the reaction of condensation of the ferric species is confined on the tubule's surface. The surface of the tubule favors the nucleation of the solid phase. In both cases the precipitated inorganic phase is a usual product of iron(III) hydrolysis in acidic media [13]. γ-FeOOH is the spontaneously occurring phase in the same conditions in the absence of tubules.

We shall point out that the structure of the solid phase depends on the binding location on the tubules: their surface favors the formation of the spontaneously occurring compound, whereas the bilayer edges induce the formation of another species (β-FeOOH). The decoration pattern on the edged tubules is comparable to the gold particles deposited on tubules by Burkett and Mann [14]. It points out the specific structure of the edges of tubules that induce specific precipitation of Fe(III): at those $DC_{8,9}PC$ bilayer edges the hydrophobic chains are exposed to the solvent. This conformation is not favorable but George et al. [4] mention that it can be stabilized by the presence of remaining alcohol in the sample in the neighborhood of the edges. In the scope of this hypothesis, the acetylenic groups of the chains are accessible and can form π bonds with ferric ions, implying site-specific nucleation and growth of particles on the ribbon edges. This selectivity reveals the high chemical reactivity of the ribbon edges compared to the $DC_{8,9}PC$ bilayer surface.

Conclusion

It is possible to coat lipidic tubules with a mineral phase composed of iron oxyhydroxide by hydrolysis of ferric chloride in the presence of $DC_{8,9}PC$ tubules. Owing to the binding of the ferric cations to the surface and to the edges of the bilayers, the nucleation of the particles takes place on the tubules and leads to composite structures with high aspect ratio.

The selective binding of rice-bean-shaped β-FeOOH nanoparticles on the bilayer ribbon edges is interpreted as a consequence of the exposure of the tubule's ribbon edges to the solvent. These edges constituted by the hydrophobic chain contain acetylenic groups which may form π bonds with the ferric ions and this explains the specific nucleation and growth of particles on the bilayer edges of the tubules. A further challenge is to transform the β-FeOOH particles precipitated on the tubules into the magnetite Fe_3O_4 in order to confer magnetic properties to the coated tubules.

Acknowledgements We thank Michel Lavergne from the service de Microscopie Electronique du Groupement Régional de Mesures Physiques, who obtained the electron microscopy pictures presented in this work.

References

1. Schnur JM, Price R, Schoen P, Yager P, Calvert JM, Singh A (1987) Thin Solid Films 152:181
2. Chappell JS, Yager P (1991) Chem Phys Lipids 58:253
3. (a) Schoen PE, Yager P (1985) J Polym Sci Polym Phys Ed 23:2203; (b) Rudolph AS, Burke TG (1907) Biophys Acta 902:349; (c) Schoen PE, Yager P, Sheridan JP, Price RR, Schnur JM, Singh A, Rhodes DG, Blechener SL (1987) Mol Cryst Liq Cryst 153:357; (d) Sheridan JP (1988) NRL Memorandum Rep 5975
4. Georger JH, Singh A, Price RR, Schnur JM, Yager P, Schoen PE (1987) J Am Chem Soc 109:6169
5. Schnur JM (1993) Science 262:1669
6. Krebs JJ, Rubinstein R, Lubitz P, Harford MZ, Baral S, Shashidhar R, Ho YS, Chow GM, Qadri S (1991) J Appl Phys 70:6404
7. Chappel JS, Yagar P (1992) J Mater Sci Lett 11:633
8. Baral S, Schoen P (1993) Chem Mater 5:154
9. Archibald DD, Mann S (1993) Nature 364:430
10. Letellier D, Sandre O, Ménager C, Cabuil V, Lavergne M (1997) Mater Sci Eng C5:153
11. (a) Schnur JM; Ratna BR, Selinger JV, Singh A, Jyothi G, Easwaran KRK (1994) Science 264:945; (b) Selinger JV, MacKintosh FC, Schnur JM (1996) Phys Rev E 53:3804
12. Akustu H, Seeling J (1981) Biochemistry 20:7366
13. Misawa T (1973) Corros Sci 13:659
14. Burkett S, Mann S (1996) Chem Commun 321

Progr Colloid Polym Sci (2001) 118: 251–255
© Springer-Verlag 2001

Dieter Gräbner
Takashi Matsuo
Ernst Hoinkis
Christine Thunig
Heinz Hoffmann

Vesicular precipitates from surfactant/cosurfactant mixtures

D. Gräbner (✉) · C. Thunig
H. Hoffmann
Lehrstuhl für Physikalische Chemie I
Universität Bayreuth, 95440 Bayreuth
Germany
e-mail: dieter.graebner@uni-bayreuth.de
Tel.: +49-921-552773
Fax: +49-921-552780

E. Hoinkis
Hahn-Meitner-Institut, Glienicker Strasse
100, 14109 Berlin, Germany

T. Matsuo
Tokyo Research Laboratories
Kao Corporation, 2-1-3 Bunka
Sumida-ku, Tokyo 131, Japan

Abstract A new anionic surfactant lauryl amidomethylsulfate ($LAMS^-$) was investigated. The aggregation behaviour was determined by small-angle neutron scattering (SANS) measurements. The Na-salt micelles are highly charged, while the Ca-salt micelles are almost uncharged. Na-LAMS (100 mM) solutions in the presence of 100 mM $CaCl_2$ undergo several phase transformations with increasing n-hexanol concentration. We found the expected micellar L_1 phase and a lamellar phase, but also a novel phase: a white precipitate is formed at the bottom of the sample. With increasing n-hexanol concentration, the precipitate dissolves into a liquid-crystalline L_α phase. Investigation by freeze–fracture transmission electron microscopy, light micros-

copy and SANS shows that the precipitate consists of agglomerated polydisperse multilamellar vesicles. The bilayer thickness is about 20 Å and is independent of the composition, whereas the interlamellar distance is strikingly linked to the concentrations of cosurfactant (surfactant/cosurfactant ratio) and electrolyte. With increasing cosurfactant content, the bilayers become less rigid and resulting thermal undulations force the membranes apart until a common L_α phase is formed. This transition is an example of a bonding–nonbonding transition of membranes.

Key words Micelles · Vesicles · Membrane flexibility · Undulations · Bonding–nonbonding transitions

Introduction

Surfactants (amphiphiles) are known to form a wide variety of aggregates (e.g. globular or rodlike micelles, bilayer membranes, vesicles, etc.) in aqueous solutions [1]. Their shape and size depend on the surfactant concentration and can also be heavily influenced by cosurfactants (mostly medium-chain-length alcohols) that join these aggregates [2, 3]. Ionic amphiphiles are additionally affected by electrolytes (new counterions, Coulombic shielding).

We report here a new anionic surfactant, lauryl amidomethylsulfate (LAMS), with Na and Ca counterions.

We investigated binary, ternary and quaternary systems consisting of aqueous solutions of the Na and Ca salts, Na-LAMS and Ca-LAMS, solutions of Ca-LAMS with hexanol and solutions of Na-LAMS, $CaCl_2$ and hexanol.

Materials and methods

Substances

The surfactants Na-LAMS and Ca-LAMS were synthesized and purified according to the method described in Ref. [4], $CaCl_2$ was from Merck, purchased as $CaCl_2 \cdot 6H_2O$ and dried at about 120 °C until no further decrease in weight was observed. n-Hexanol was from Fluka, purum quality.

Experimental techniques

Small-angle neutron scattering measurements

The small-angle neutron scattering (SANS) experiments on the micellar solutions of Na-LAMS and Ca-LAMS were performed in Risoe, Denmark. The SANS experiments on the vesicular systems (ternary and quaternary) were performed at the Hahn-Meitner Institute in Berlin [5]. The raw data were corrected for background and cuvette scattering; normalization was done in the conventional manner.

Experimental results

The concentrations of surfactant salt with bivalent cations always refer to the surfactant; thus, 100 mM Ca-LAMS means 100 mM $LAMS^-$ and 50 mM Ca^{2+}.

Binary systems: M-LAMS in aqueous solution

SANS shows that the aggregation behaviours of Na-LAMS and Ca-LAMS as a function of concentration are totally different, owing to the considerably higher degree of counterion dissociation of Na^+ compared to Ca^{2+}. Whereas Na-LAMS yields increasingly sharp peaks with increasing concentration, the spectra of Ca-LAMS do not even show a maximum. This means that Na-LAMS micelles are increasingly ordered with rising concentration as a result of strong electrostatic repulsion, indicating that the micelles are highly charged. Ca-LAMS spectra show the opposite behaviour: there are no peaks, not even maxima in the spectra; the micelles are practically uncharged.

The Ca-LAMS/hexanol/water ternary system

Phase diagram and freeze–fracture transmission electron microscopy investigation

Anionic surfactants in aqueous solutions tend to from micellar L_1 phases only; in the presence of cosurfactants such as *n*-alcohols they also form lamellar L_α phases at high concentrations where bilayer distances are small [6]. Diluting these phases does not lead to a swelling of the lamellar phase, but causes a phase transition from lamellar to micellar (L_1).

Obviously, this is connected to the surface charge density of the bilayers, as mixing charged anionic and uncharged surfactant decreases the total surfactant concentration needed to obtain a lamellar phase [6] and adding charged surfactant ends the swelling of a lamellar phase formed by uncharged surfactant [7].

On the other hand, combinations of uncharged surfactants and cosurfactants often form lamellar phases at very low total concentrations, i.e. they form highly

swollen systems. Dodecyldimethylaminoxide and *n*-hexanol in water, for examle, form lamellar phases at 1 wt% total concentration [8]. With tetradecyldimethylaminoxide, similar phases were found [9, 10].

This phenomenon of increasing tendency towards swelling of lamellar phases with decreasing charge density is counterintuitive. One would expect the contrary, i.e. increasing swelling with increasing charge density, because the strong electrostatic repulsion of like-charged membranes should help keep the membranes at a distance.

This is not the case; there seems to be an attractive force that stops swelling and increases with increasing charge density. Such attractive forces have been postulated [11]; Poisson–Boltzmann theory does not predict this force [12], whereas Monte Carlo simulations [13] predict reduced repulsion in the presence of high-valency counterions and in some cases the interaction even becomes attractive.

More recent theoretical attempts to explain the attractive forces between like-charged particles [14] were claimed to be false by others [15]. As far as we know, this problem has no solution so far.

Replacing monovalent by divalent counterions reduces the surface charge density because of the higher degree of counterion condensation and, therefore, according to the previously mentioned experimental findings, should yield swollen lamellar phases. Maciejewska et al. [16] and Khan and coworkers [17–19] searched for swollen lamellar phases in the $Ca(DS)_2$ [and $Mg(DS)_2$], decanol and water system, but found none. Similarly, Sein et al. [20] investigated sulfonate-based surfactants with bivalent counterions and also found no swollen lamellar phases.

More recently, however, extended swollen lamellar phases in $Ca(DS)_2$/*n*-alcohol/water systems were found with *n*-alcohols from pentanol to octanol [21]. Here, we present results on a surfactant that forms lamellar phases with a broad variety of bilayer distances from collapsed to strongly swollen upon addition of *n*-hexanol.

The phase diagram (25 °C) of this system is simple (Fig. 1). The increase in the hexanol concentration transforms the single-phase L_1 into a two-phase L_1/L_α system and further into a single L_α phase, which occurs at a total concentration (surfactant plus cosurfactant) as low as 3.5 wt%. A freeze–fracture transmission electron microscopy (FF-TEM) image (Fig. 2) that was taken of the L_α phase of the two-phase 100 mM Ca-LAMS/1.2 vol% (96 mM) hexanol (in a mixture of 80% water and 20% glycerol to avoid formation of ice crystals during preparation) L_1/L_α system shows the microstructure of this L_α phase (glycerol did not change the appearance of the sample). Polydisperse multilamellar vesicles with a broad distribution of sizes, shapes (see the peanut-shaped vesicle in Fig. 2, top middle) and number of layers can be seen. The well-

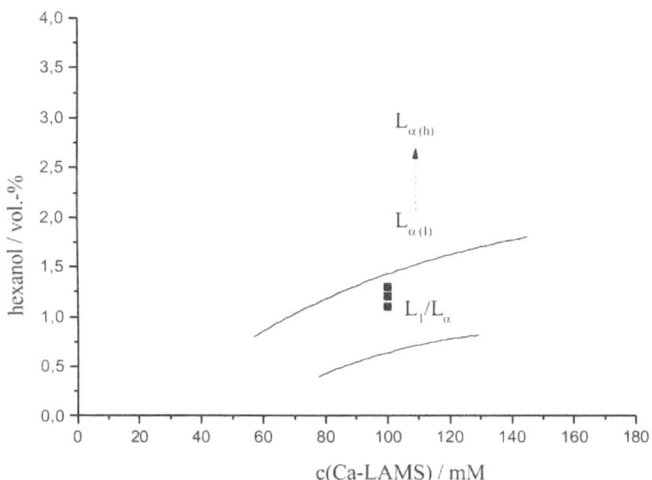

Fig. 1 Phase diagram of calcium lauryl amidomethylsulfate (*Ca-LAMS*) with *n*-hexanol at 25 °C. The *squares* mark the composition of samples that were investigated by small-angle neutron scattering (*SANS*)

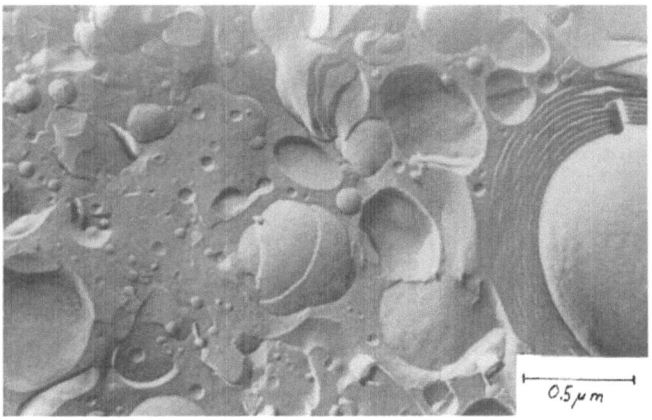

Fig. 2 Freeze–fracture transmission electron microscopy image (*FF-TEM*) of the L$_\alpha$ phase of 100 mM Ca-LAMS/1.2 vol% *n*-hexanol. Note that this sample is in the two-phase region (L$_1$/L$_\alpha$); the FF-TEM image was taken of the L$_\alpha$ phase only

defined intervesicular spacing of the membranes as can best be seen in the broken-up vesicle (Fig. 2 right) is about 240 Å.

SANS experiments

SANS experiments were done on 100 mM Ca-LAMS (Fig. 3) with 1.1 (88 mM), 1.2 (96 mM) and 1.3 vol% (0.10 M) hexanol. All these samples are biphasic L$_1$/L$_\alpha$, (see phase diagram, Fig. 1, where the overall compositions of the samples investigated are marked by filled-squares). Only the L$_\alpha$ phase was used for the SANS experiment.

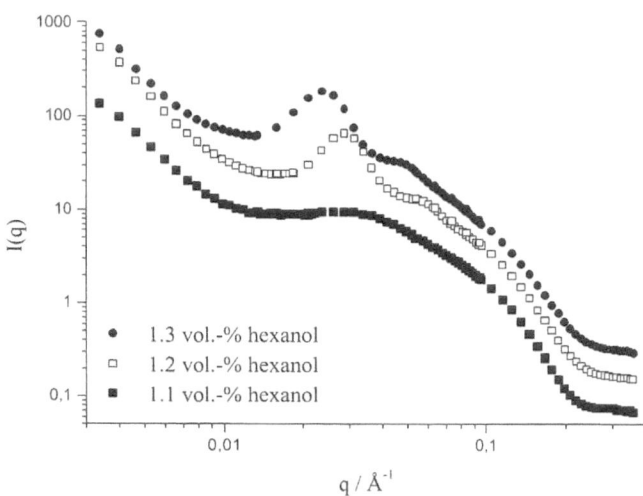

Fig. 3 SANS spectra of 100 mM Ca-LAMS with different amounts of *n*-hexanol at 25 °C; for each increase in cosurfactant concentration, the intensities were doubled to render the spectra clear and nonoverlapping. Note that here all three samples were in the two-phase region (L$_1$/L$_\alpha$); only the L$_\alpha$ phase was looked at

Three peaks were found, indicating repeat distances of 210 Å (very broad and weak), 220 Å (compared to 240 Å from the FF-TEM investigation) and 260 Å for 1.1, 1.2 and 1.3 vol%, respectively.

In the high-q region, the intensity is dominated by the form factor of a planar lamella, which has an $(lq)^{-4}$ dependence, where l is the lamellar thickness; we fitted that part of the spectrum to get the value for l from the fit function. We found the lamellar thickness to be surprisingly independent of the system. Both in the ternary and in the quaternary system, D was 20–25 Å, independent of surfactant, cosurfactant and (in the quaternary system) CaCl$_2$ concentration.

The Ca-LAMS/CaCl$_2$/hexanol quaternary system

Phase diagram

This phase diagram (Fig. 4) was established at 30 °C with the Na-LAMS concentration kept constant at 100 mM. The L$_\alpha$ phase in the L$_1$/L$_\alpha$ two-phase region again has a cottonlike appearance and looks very much like a solid precipitate, which is denoted p.

Similar precipitates have been found in catanionic systems with [22] and without [23] additional electrolyte where the cationic and anionic components mix to form a bilayer with low surface charge density which swell upon a change in the ionic strength or increasing surface charge density.

At 50 mM CaCl$_2$, a precipitate is found between 1.3 (0.10 M) and 1.7 vol% (0.14 M) cosurfactant; at

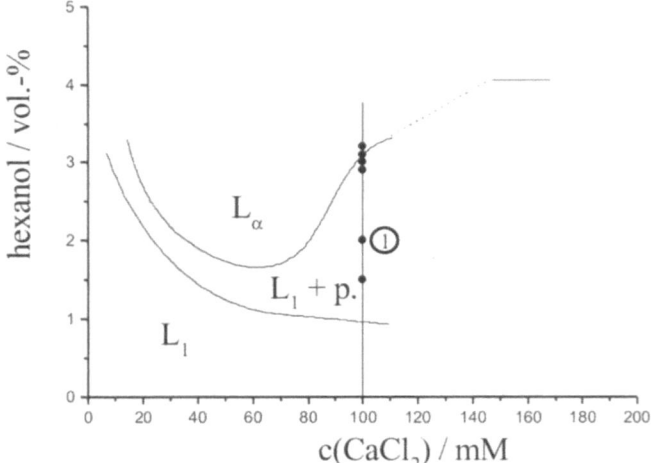

Fig. 4 Phase diagram of 100 mM Na-LAMS/CaCl₂/*n*-hexanol at 30 °C The *filled circles* mark the composition of samples that were investigated by SANS

100 mM CaCl₂, it is found between 1.0 (80 mM) and about 3.0 vol% (0.24 M) hexanol. Along path 1 (100 mM CaCl₂, varying cosurfactant concentration), at about 1.1 vol% (88 mM) hexanol, a precipitate is found. The volume fraction of the precipitate increases only moderately from 0 to about 25% upon an increase in cosurfactant from 1.1 to about 2.9 vol% (0.23 M). Upon a slight further increase, the precipitate swells into a common L_α phase.

At 160 mM CaCl₂, no common L_α phase could be observed upon increasing the hexanol concentration. Obviously, very high CaCl₂ concentrations keep the membranes stiff enough to prevent thermal undulations.

SANS experiments

The spectra dealt with in this section were mostly taken from the collapsed vesicle state (the precipitate), coexisting with an L_1 phase; some are from samples that are single-phase L_α; this information is included in the figure captions. The overall compositions of the samples of which the SANS spectra are shown are marked in the phase diagram (Fig. 4) by circles.

At 100 mM Na-LAMS/100 mM CaCl₂, the SANS spectra (Figs. 5) show peaks until 2.9 vol% (0.23 M) and flattening shoulders at 3.0 (0.24 M) and 3.1 vol% (0.25 M). At 3.2 vol% (0.26 M), only a continuous decline is observed. The repeat distances first stay nearly constant upon cosurfactant increase [45 Å at 1.5 vol% (0.12 M), 46 Å at 2.0 vol% (0.16 M)]. At 2.9 vol%, it reaches 66 Å and at 3.0 and 3.1 vol%, 75 Å is observed. Smaller repeat distances with bivalent ions than with monovalent ions have been found in other binary and ternary systems [24, 25] and explained by smaller

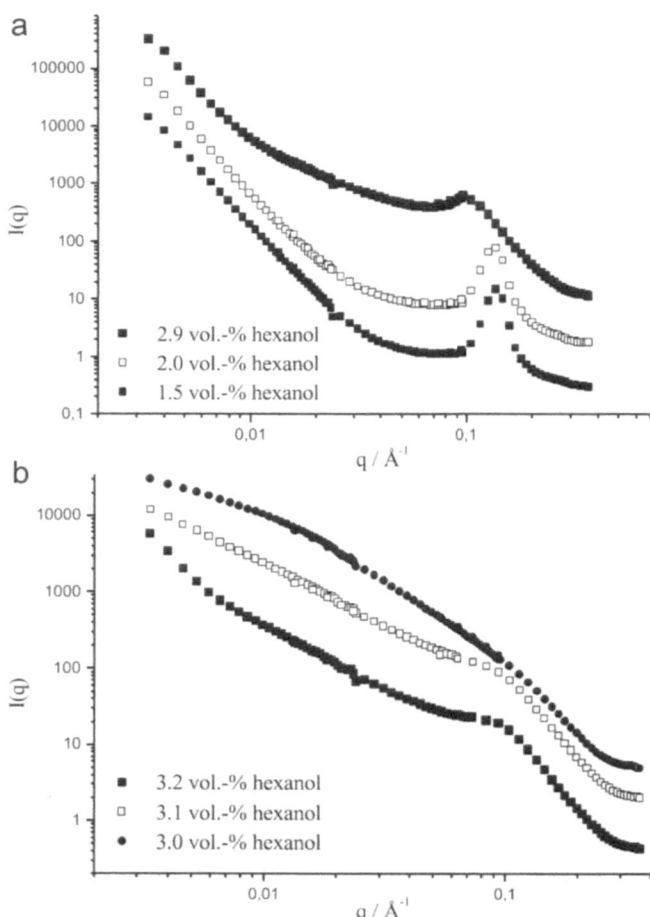

Fig. 5 a SANS spectra of the precipitate from samples with 100 mM Na-LAMS/100 mM CaCl₂ with different amounts of *n*-hexanol at 30 °C; for each increase in cosurfactant concentration, the intensities were multiplied by 5 to render the spectra clear and nonoverlapping. These samples are in the two-phase region (L_1/precipitate). **b** SANS spectra of 100 mM Na-LAMS/100 mM CaCl₂ with different amounts of *n*-hexanol at 30 °C; for each increase in the cosurfactant concentration, the intensities were multiplied by 5 to render the spectra clear and nonoverlapping. These samples are single-phase L_α

electrostatic repulsion between lamella with smaller charge density.

Conclusions

Together with cosurfactant *n*-hexanol, Ca-LAMS forms an L_α phase. FF-TEM shows that this L_α phase consists of multilamellar vesicles of broad distribution of size, shape and number of lamellae per vesicle. SANS spectra show that the repeat distance in the two-phase region increases upon an increase in the cosurfactant concentration and the cosurfactant/surfactant ratio; also, the distribution of the repeat distance becomes broader.

Table 1 Repeat distances (interlamellar spacing plus membrane thickness) in angstroms in the calcium lauryl amidomethylsulfate (*Ca-LAMS*)/*n*-hexanol and the Ca-LAMS/CaCl₂/*n*-hexanol system from small angle neutron scattering spectra

n-Hexanol (vol%)	70 mM Ca-LAMS	100 mM Ca-LAMS	100 mM Na-LAMS + 50 mM CaCl₂	100 mM Na-LAMS + 100 mM CaCl₂
1.1	570	210		
1.2	780 (shoulder)	220		
1.3	780 (shoulder)	260		
1.4			290	
1.5				45
1.6			No peak	
1.7			No peak	
2.0				46
2.9				66
3.0				75 (shoulder)
3.1				75 (shoulder)
3.2				No peak

Na-LAMS together with CaCl₂ forms a precipitate of vesicles upon addition of hexanol. This precipitate is found in a wide range of cosurfactant concentrations when the CaCl₂ concentration is high. Whereas the L_α phase from Ca-LAMS/hexanol has repeat distances varying about linearly with cosurfactant concentration in the L_1/L_α two-phase region, the precipitate in the quaternary system shows distances almost independent of hexanol concentration as long as the latter is below about 2.9 vol% (at 100 mM CaCl₂); above this value, a sudden swelling of the precipitate into an L_α phase occurs. Softening of the membrane by rising cosurfactant content allows thermal undulations to force the mem-branes apart. This is an example of a bonding–nonbonding transition of membranes.

As 100 mM Ca-LAMS/1.3 vol% hexanol has a repeat distance of 260 Å, while 100 mM Na-LAMS/50 mM CaCl₂ (which is the same as 100 mM Ca-LAMS/100 mM NaCl)/1.4 vol% has a repeat distance of 290 Å, the effect of NaCl as an electrolyte is negligible. In contrast, increasing the CaCl₂ concentration by 50 mM in order to obtain 100 mM Na-LAMS/100 mM CaCl₂ (including a small rise in hexanol from 1.4 to 1.5 vol%) causes the repeat distance to shrink to 45 Å, 20% of its former value. This indicates that the effect of Ca^{2+} is not only Coulombic.

References

1. Laughlin PG (1994) The aqueous phase behaviour of surfactants. Academic, London
2. Hoffmann H (1994) Adv Mater 6:116
3. Hoffmann H (1994) Ber Bunsenges Phys Chem 98:1433
4. Mizushima H, Matsuo T, Satoh N, Hoffmann H, Graebner D (1999) Langmuir 15:6664
5. Keiderling U, Wiedenmann A (1995) Physica B 213:895
6. Hoffmann H, Illner J-C (1995) Tenside Surfactants Deterg 32:318
7. Douglas CB, Kaler EW (1994) J Chem Soc Faraday Trans 90:471
8. Platz G, Thunig C, Hoffmann H (1992) Ber Bunsenges Phys Chem 96:667
9. Hoffmann H, Thunig C, Schmiedel P, Munkert U (1994) Langmuir 10:3972
10. Hoffmann H, Munkert U, Thunig C, Valiente M (1994) J Colloid Interface Sci 163:217
11. Levine S, Hall DG (1992) Langmuir 8:1090
12. Wennerström H, Jönsson B, Linse P (1982) J Chem Phys 76:4665
13. Guldbrand L, Jönsson B, Wennerström H (1984) J Phys Chem 80:2221
14. Grier DG (1998) Nature 393:621
15. Sader JE, Chan D (1999) J Colloid Interface Sci 213:268
16. Maciejewska D, Khan A, Lindman B (1986) Colloid Polym Sci 264:909
17. Khan A, Lindman B, Shinoda K (1989) J Colloid Interface Sci 128:396
18. Khan A, Fontell K, Lindman B (1984) Colloids Surf 11:401
19. Khan A, Fontell K, Lindblom G, Lindman B (1982) J Phys Chem 86:4266
20. Sein A, Engberts JBFN, van der Linden E, van de Pas JC (1986) Langmuir 12:2913
21. Hornfeck U (1998) PhD thesis. University of Bayreuth, Germany
22. Horbaschek K, Hoffmann H, Thunig C (2000) J Colloid Interface Sci 206:439
23. Jingcheng H, Hoffmann H, Horbaschek K (2000) J Phys Chem 104:2781
24. Helfrich W (1994) J Phys Condens Matter A 79:6
25. Khan A, Fontell K, Lindman B (1984) J Colloid Interface Sci 101:193

Progr Colloid Polym Sci (2001) 118: 256–259
© Springer-Verlag 2001

R. Marinov
S. Panayotova
A. Derzhanski

The phase behavior of the Triton X-114–water system and the hydrophile–lipophile balance theory

R. Marinov (✉) · A. Derzhanski
Institute of Solid State Physics
72 Tzarigradsko Chaussee Blvd.
1784 Sofia, Bulgaria
e-mail: marinov@scientist.com

S. Panayotova
Department of Physics, Higher Institute
of Food and Flavor Industries
26 Maritza Blvd., Plovdiv 4000
Bulgaria

Abstract The existence of two lyotropic liquid-crystal phases (micelles and lamellar) of the Triton X-114–water binary system has already been determined by several different methods [Walsh MF (1980) *PhD thesis. Salford University*; Heusch R (1984) *Ber Bunsenges Phys Chem* 88:1094]. We observed changes in the phase transitions of the system studied upon addition of octanol, butanol, glucose or fructose. A semiquantitative explanation of the experimental results is given based on the hydrophile–lipophile balance theory.

Key words Triton X-114–water system · Micellar and lamellar liquid-crystal phases Hydrophile–lipophile balance · Phase transition changes in the presence of an alcohol or a sugar

Introduction

The aggregation of amphiphilic molecules in a solvent is governed by the balance between the interactions with the solvent molecules and between the amphiphilic molecules themselves. Several types of models have been developed to describe the self-assembly: some authors use simple geometrical considerations [1–3], while others use thermodynamic descriptions or numerical calculations based on mean-field approximations.

The hydrophile–lipophile balance (HLB) theory was originally developed for classification of the surfactants used to stabilize different emulsions [6]. Davies and Rideal [4] related the HLB number of a surfactant to the relative coalescence rate of oil-in-water and water-in-oil emulsions. Thus, they succeeded in giving the empirical HLB values a fundamental significance in terms of free energies, and a relation was obtained between the HLB number and the distribution of a substance between oil and water phases under certain conditions.

We have adopted the approximation of Davies and Rideal in order to characterize semiquantitatively the influence of different additives on the phase transition of a surfactant–water system. The surfactant–water system is considered to consist of two phases – a hydrophilic and a hydrophobic one. The hydrophobic phase is in fact the hydrocarbon interior of the surfactant (Triton in our case) micelles. The hydrophilic surfactant heads together with the water are regarded as the hydrophilic phase. The influence of the additives is analysed in terms of the redistribution between the hydrophobic and hydrophilic phases calculated from the respective HLB numbers. There is good agreement between the theoretical predictions based on the HLB theory and the experimentally observed trends. The same approach can be applied to predict the changes in the phase transitions in the presence of additives in many other binary systems as well.

Experimental

The Triton X-114–water system was studied in the temperature region 290–313 K; the concentrations of Triton were 51–54 wt% and concentrations of the additives were 1 and 2 wt%. Samples with glucose and fructose were prepared by dissolving the sugar in water and subsequently adding Triton X-114. The alcohols were initially dissolved in the surfactant (Triton X-114) and after that the necessary quantity of water was added. The phase transitions

Fig. 1a, b Changes in the phase diagram of the Triton X-114–water system in the presence of 1% octanol or butanol. (*M*-micellar phase, *L*-lamellar phase)

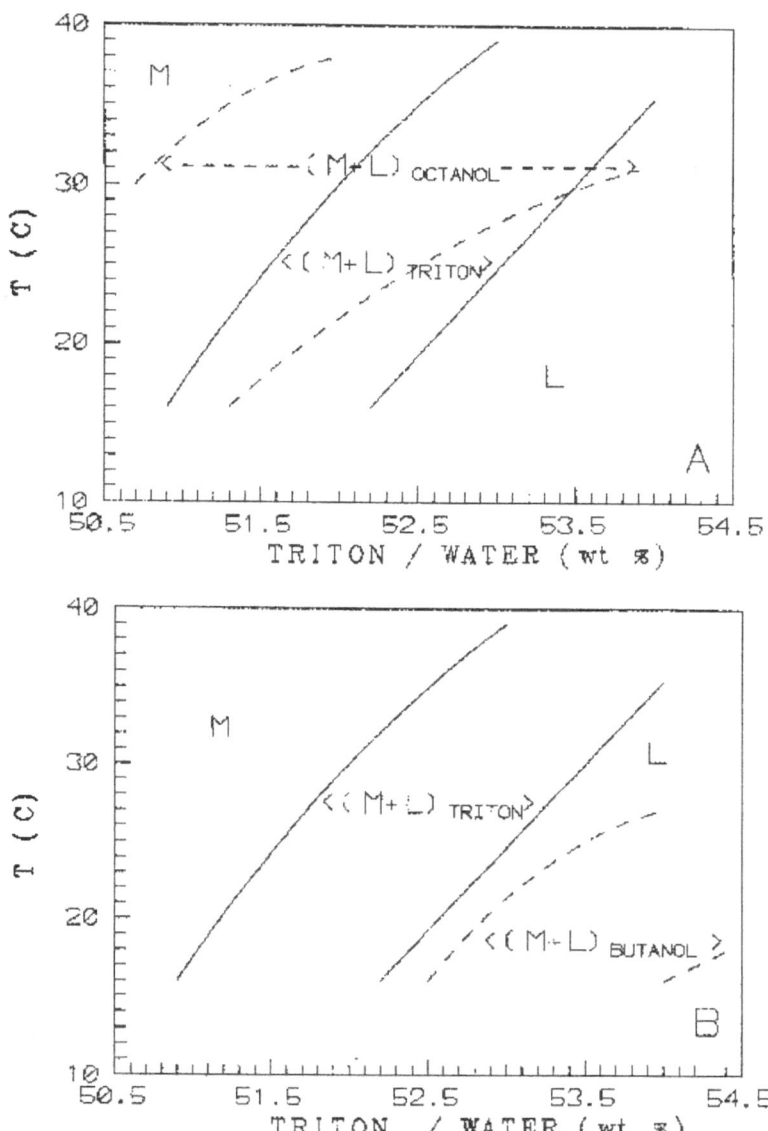

were determined by optical observations of the samples between crossed polarizers.

Results and discussion

The experimentally observed changes in the micellar-to-lamellar liquid crystal phase transition of the Triton X-114–water system in the presence of several additives are presented in Figs. 1 and 2.

At the same temperature the addition of octanol to the system leads to decrease in the surfactant concentration at which the transition occurs (Fig. 1), while the addition of the much shorter butanol leads to an increase in the concentration at which the transition occurs (Fig. 1). At

the same temperatures, the addition of sugars, either fructose (Fig. 2) or glucose (Fig. 2) shifts the transition toward higher concentration of Triton.

All the additives used changed the phase behavior of the Triton X-114–water system. This change is most probably due to incorporation of the additives in the surfactant layers and changing of the spontaneous curvature.

To quantify the influence of the additives used we applied HLB theory in order to calculate the redistribution of the additives between the hydrophilic and hydrophobic phases of the system.

We calculated the corresponding HLB values of the constituents (Table 1). As shown in Ref. [4] for a component with a given HLB value the relation of the

Fig. 2a, b Changes in the phase diagram of the Triton X-114–water system in the presence of 1% fructose or glucose

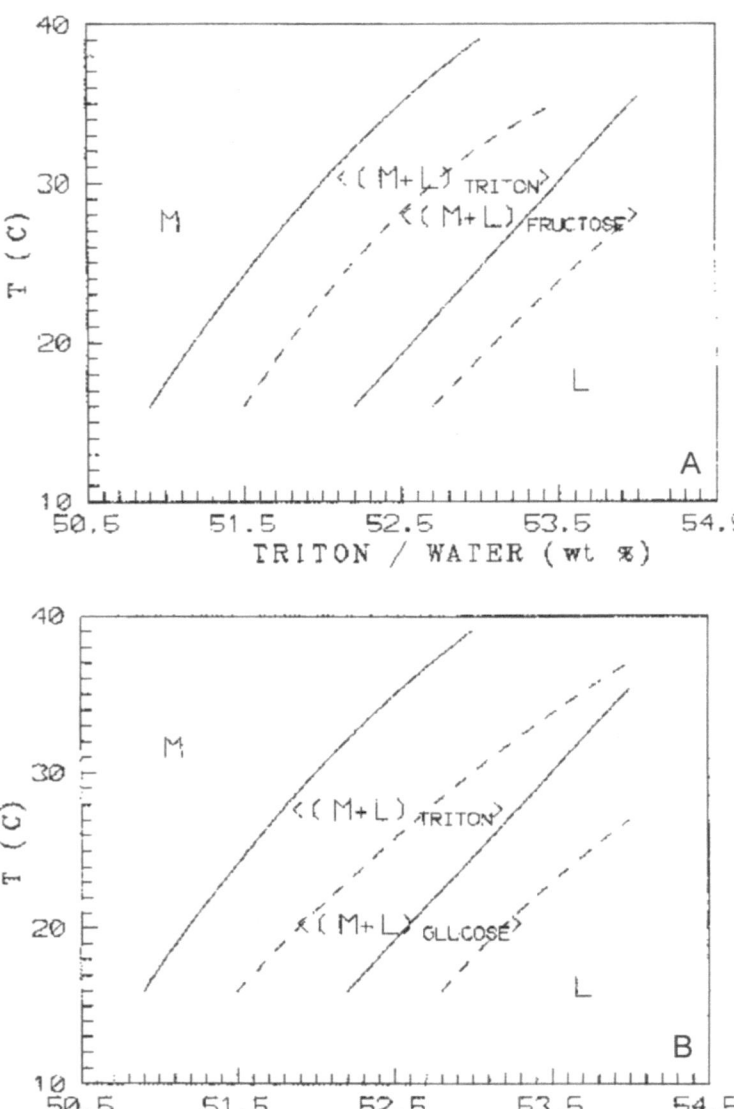

concentrations c_w and c_o in water (the hydrophilic phase in our consideration) and oil (the hydrophilic phase) solutions in equilibrium is

$$HLB - 7 = 0.36\ln(c_w/c_o) \tag{1}$$

Using the HLB values form Table 1 and Eq. (1) we calculated the redistribution of the additives between the hydrophillic and hydrophobic phases of the system. The relations obtained are shown in Table 2. They show that fructose and glucose are entirely in the water phase, which is as expected. The octanol concentration in the oil phase (in the interior of the micelles) is about 200 times greater in comparison with its concentration in the water phase, which is also not surprising. However, the butanol molecules redistribute equally between the hydrocarbon and polar surfactant parts.

Table 1 Hydrophile–lipophile balance number of the substances used

Octanol	5.1
Butanol	7.0
Fructose	16.5
Glucose	16.5

Table 2 Calculated relations between the equilibrium concentrations of the additives used in the so-called water, c_w, and the oil, c_o, phases

Octanol	$c_w^{oct} = 0.0051\, c_o^{oct}$
Butanol	$c_w^{but} = c_o^{but}$
Fructose	$c_w^{fr} \approx 3 \times 10^{11}\, c_o^{fr}$
Glucose	$c_w^{gl} \approx 3 \times 10^{11}\, c_o^{gl}$

The presence of sugars (fructose and glucose) shifts the micellar-to-lamellar phase transition toward higher concentrations of Triton. This shift signifies an effective increase in the hydrophile part of the Triton micelles by the presence of fructose and glucose. The theoretical considerations, using the HLB theory, also predict sugars to be entirely in the water phase (oxyethylene layer).

Addition of octanol shifts the micellar-to-lamellar phase transition toward lower concentrations of Triton. Octanol molecules should incorporate mostly between the hydrocarbon surfactant chains to realize such a change. The HLB considerations also predict that octanol is mostly distributed in the oil phase (hydrocarbon chain layer).

In contrast to octanol, the presence of butanol is observed to shift the micellar-to-lamellar phase transition toward higher concentrations of Triton. The butanol effect is even greater than that of the very hydrophilic glucose and fructose. Estrada-Alexanders et al. [5] have proposed that the oxyethylene chains screen the hydrophobic interactions between butanol and water molecules. The HLB theory predicts equal concentrations of the butanol molecules between the water and oil phases; however, because of the screening effect it is possible that the butanol molecules are preferably in the oxyethylene layer. This hypothesis has been confirmed recently by small-angle neutron scattering experiments [6]. Thus, butanol effectively expands the hydrophilic surfactant parts and shifts the phase transition to higher concentrations of Triton (which is observed experimentally).

Acknowledgements This work was supported by Bulgarian National Fund "Scientific Studies" under project F581/1995.

References

1. Israelachvili JN (1991) Intermolecular and surface forces. Academic, New York
2. Marinov R, Zheliaskova A, Derzhanski A (1996) J Dispersion Sci Technol 17:591
3. Petrov AG, Derzhanski A (1976) J Phys Suppl 47:155
4. Davies JT, Rideal EK (1963) Interfacial phenomena. Academic, New York, pp 371–382
5. Estrada-Alexanders AF, Remis R, Guzman F (1999) Langmuir 15:1879
6. Möller A, Lang P, Findenegg GH, Keiderling U (1998) J Phys Chem B 102:8958

Progr Colloid Polym Sci (2001) 118: 260–265
© Springer-Verlag 2001

P. Wette
H.-J. Schöpe
T. Palberg

Properties of mixed colloidal crystals

P. Wette · H.-J. Schöpe · T. Palberg (✉)
Institut für Physik der Universität Mainz
Staudinger Weg 7
55099 Mainz, Germany
e-mail: thomas.palberg@uni-mainz.de

Abstract We prepared colloidal crystals from aqueous suspensions of spherical, charged polystyrene spheres under deionised conditions. Using a home-built multipurpose light scattering apparatus we measured the static structure factor, the static shear modulus and the intermediate scattering function. In addition we also monitored also the conductivity as a function of the composition and the particle number density. For the mixture investigated the data are well described assuming the formation of randomly substituted body-centered-cubic crystals.

Key words Colloids · Charged spheres · Mixtures · Light scattering

Introduction

Colloidal crystals are the subject of intense research for several reasons. First, owing to their specific length and time scales their properties are readily accessible by various forms of light scattering and microscopy. Second, a number of interesting questions of fundamental interest in condensed matter physics can be addressed in a well-defined model systems. This concerns both equilibrium properties (like diffusive dynamics, structure or phase behaviour), response to external fields (like electrokinetic properties or sedimentation) and situations far from equilibrium (like the glass transition, shear-melting or crystallisation kinetics). Third, however, important applications may also result from a better understanding of the generic properties. Optical-band-gap materials are one of these.

Recently focus has shifted from single-component suspensions to mixtures. The most prominent examples deal with hard-sphere or hard-sphere-polymer mixtures of large size difference, where phase separation is observed owing to attractive terms in the potential of the mean force [1, 2]. Much less is known about charged mixtures. Only a few reports exist on the phase behaviour [3, 4, 5], and fewer on even other properties, like shear rigidity [6], diffusion [7, 8] or crystallisation kinetics.

We report on a mixture of two species of polystyrene latex spheres in deionised aqueous suspension. Their size ratio is $\Gamma = 0.85$. Such a small value should not induce an entropically driven instability. Nevertheless, the phase behaviour is not immediately obvious. The formation of alloy structures of glasses or of precipitats can be expected. Alloys may either be compositionally ordered (e.g. of NaCl structure) or disordered (e.g. Cu/Au). We conducted a number of different experiments to comprehensively characterise the resulting solids. Besides the static structure factor, $S(q)$, we also measured the conductivity, σ, and the static shear modulus, G. The former is very sensitive to the formation of precipitates; the latter has been shown to yield valuable information about the local crystal structure [9]. Finally, we also report some of the first dynamic light scattering measurements on polycrystalline samples. Their interpretation can be qualitative only, since a rigorous scattering theory of such nonergodic and inhomogeneous materials is still demanded. The combination of our results shows that at low-to-intermediate particle concentrations randomly substituted alloys of body-centered-cubic (bcc) structure are formed.

Table 1 Properties of the pure components. Sample name, source, nominal diameter, $2a_{nom}$, effective charges, Z^*_σ and Z^*_G, from conductivity and shear modulus measurements, shear modulus, G, at the particle number density used for the mixture as well as the number densities at freezing and melting are given

Sample	Source	$2a_{nom}$	Z^*_σ	Z^*_G	G at $n = 1\ \mu m^{-3}$	n_F [μm^{-3}]	n_M [μm^{-3}]
PS85	IDC Batch No 767.1	85 nm	530 ± 32	350 ± 20	0.786 Pa	3.8 ± 0.5	4.4 ± 0.5
PS100	Banys Lab. Batch No. 3512	100 nm	530 ± 38	327 ± 10	0.8 Pa	4.0 ± 0.5	7.0 ± 0.5

Experimental

Two species of commercially available polystyrene latex spheres were investigated. The particle parameters are compiled in Table 1. All the suspensions were prepared from diluted and precleaned stock suspensions of approximately 1% packing fraction. The suspensions were thoroughly deionised using advanced continuous conditioning methods. Details of this have recently been given elsewhere [10]. Both individual species were comprehensively characterised before conducting experiments on the mixtures.

Both pure samples crystalline at low particle number densities in a bcc structure. Measurements of $S(q)$ were further used to calibrate measurements of the conductivity, σ. The particle number density dependence of σ was observed to be strictly linear and independent of the phase state of the suspension. The data for thoroughly deionised suspensions are well described by $\sigma = neZ^*_\sigma(\mu_p + \mu_+) + \sigma_B$, with the elementary charge e, the particle and proton mobilities μ_p and μ_+ and the background conductivity σ_B (mainly stemming from dissociated water and ionic impurities). The effectively transported charge, Z^*_σ, was determined as a fit parameter. It is tabulated in Table 1. The shear modulus, G, is in good agreement with theoretical fits yielding an effective shear modulus charge of Z^*_G as the only free parameter. We note that Z^*_G is observed to be somewhat smaller than the effectively transported charge.

Mixtures were prepared at fixed particle number density under thoroughly deionised conditions. The background concentration of ionic impurities was below 5×10^{-7} M. Under these conditions the samples were completely solidified. Colloidal solids, however, show very low yield moduli and may easily be shear-molten by mechanical treatment. Since for some properties the sample morphology may have a severe influence on the results, it was of great importance to leave the samples undisturbed once crystallised. To this end all measurements were performed using a recently reported multipurpose light scattering apparatus. In this static structure, the shear modulus and the dynamic structure factors may be measured quasi simultaneously on the same crystalline sample. The apparatus is equipped with a double-arm goniometer. It uses two counterpropagating illumination/detection schemes for static and dynamic light scattering and a third independently adjustable scheme for the torsional resonance detection. Details have recently been given elsewhere [11].

Properties of mixed colloidal crystals

Our results are given in Figs. 1, 2, 3 and 4.

The conductivity varies linearly with the composition. It is well described (solid line) using an extension of Hessinger's conductivity model of independent ion migration with exchange of icons across the plane of shear [12]

$$\sigma = ne\left(XZ^*_{\sigma,1}(\mu_1 + \mu^+_H) + (1 - X)Z^*_{\sigma,2}(\mu_2 + \mu^+_H)\right) + \sigma_B \ ,$$

where X is the fraction of component 1, e is the elementary charge and μ_i is the independently measurable electrophoretic mobility of particle species i. The effective charges, $Z^*_{\sigma,i}$, used in the fit were taken from the single-component measurements and the background conductivity indicates a background concentration of impurities of some 0.5 μM.

The structure of the mixtures stays bcc independent of composition. Note the constancy of the position of the (1 1 0) Bragg reflection, indicating constant particle density.

Together with the conductivity data this rules out any phase separation or correlation between particle positions. In the first case the conductivity should show deviations from the theoretical expectations; in the latter case the appearance of superstructure peaks is expected. This demonstrates the formation of an alloy with a random distribution of particles on the bcc lattice sites.

This can be checked via G. Here a prediction of Lindsay and Chaikin [13] exists for the shear modulus of a randomly substituted alloy:

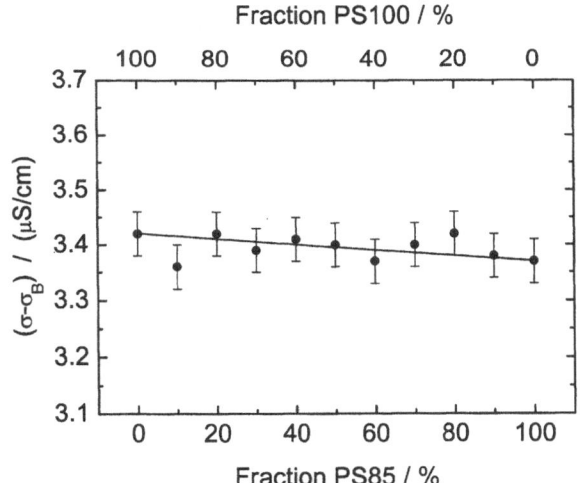

Fig. 1 Background-corrected conductivity, $\sigma - \sigma_B$, of the PS85/PS100 mixture as a function of composition

$$G_{bcc} = f_A \frac{4}{9} n \frac{X^2 \tilde{Z}_1^2 + 2X(1-X)\tilde{Z}_1 \tilde{Z}_2 + (1-X)^2 \tilde{Z}_2^2}{4\pi\varepsilon\varepsilon_0}$$
$$\times \frac{\exp(-\kappa d)}{d} \kappa^2 d^2 \ ,$$

where f_A is a known factor for orientational averaging in polycrystalline materials and $\varepsilon\varepsilon_0$ is the dielectric permittivity of the suspension.

$$\tilde{Z}_i = Z^*_{G,i} \frac{\exp(\kappa a_i)}{1+\kappa a_i} ,$$

$$\kappa^2 = \frac{e^2}{\varepsilon\varepsilon_0} [2N_A 1000c + n(XZ^*_{G,1} + (1-X)Z^*_{G,2})] \ ,$$

where a_i and $Z^*_{G,i}$ are the radius and shear modulus effective charge of the ith component, respectively. In Fig. 2 the upper line gives the prediction as calculated using the single-component shear modulus charges and the lower line is a prediction based on the single-component measurements in this run. The discrepancy is attributed to the small background concentration of impurities present. More important, however, the composition dependence is correctly described.

Finally, we report the dynamics in our polycrystalline alloys. The samples are nonergodic as the whole phase space is not explored during the time of a measurement. In this case the Siegert relation may not be employed to calculate the desired field autocorrelation functions, $g^{(1)}(q, \tau)$, or intermediate scattering functions, $f(q, \tau)$, from the measured intensity autocorrelation functions, $g^{(2)}(q, \tau)$. Instead the method of Pusey and van Megan [14] was employed using the relation

$$g_E^{(1)}(q, \tau) = f(q, \tau)$$
$$= 1 + \frac{I_t}{I_E} \left(\sqrt{g_t^{(2)}(q, \tau) - g_t^{(2)}(q, 0) - 1} + 1 \right) \ ,$$

where I is the average scattered intensity and the indices t and E refer to time and ensemble averages, respectively. All the measurements reported in Fig. 4 were taken at the

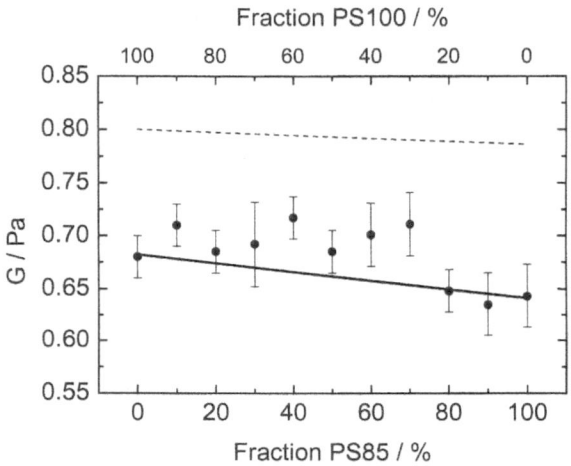

Fig. 3 Shear moduli of the PS85/PS100 mixture as a function of composition

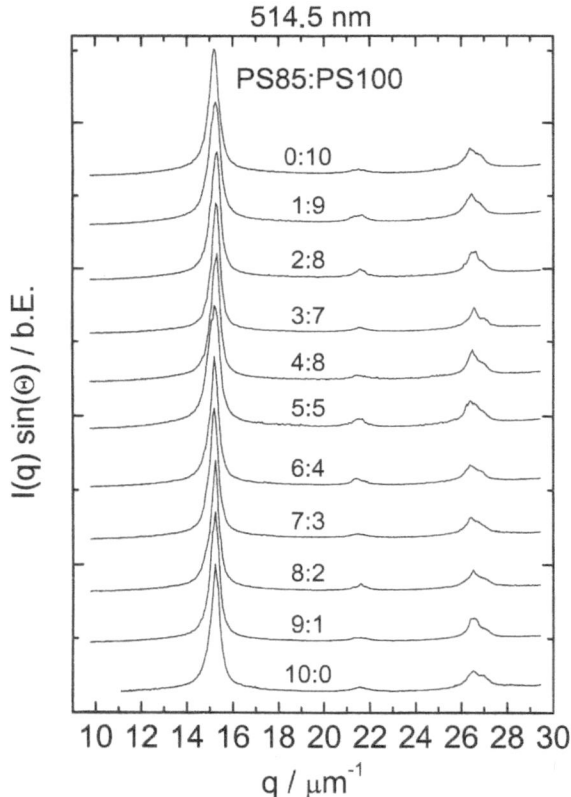

Fig. 2 Angle-corrected scattered intensity of the PS85/PS100 mixture as a function of composition

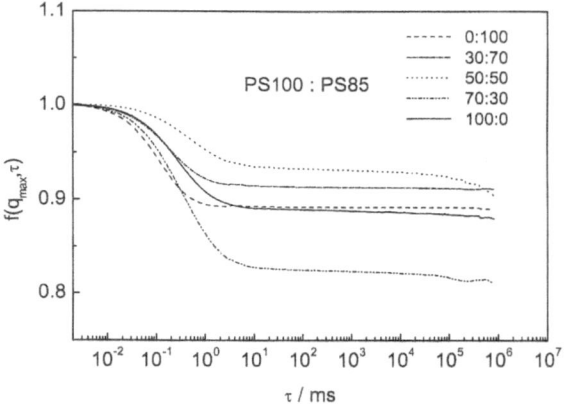

Fig. 4 Intermediate scattering functions of the PS85/PS100 mixture as a function of composition

maximum of the (1 1 0) Bragg reflection. They thus correspond to the dynamics on the scale of the nearest-neighbour distance. Two relaxation time scales can be discriminated. The first is located at about $\tau = (10^{-4} - 10^{-3})$ s. $f(q, \tau)$ drops from its initial value to plateau values between 0.83 and 0.93. The second decay is best visible for the 50:50 mixture and is located at times longer than 10^3 s. In the present case the time window was restricted by sample times of 1 h (the thermal stability of the setup was 5×10^4 s or less as tested using a ground quartz glass sample).

The interpretation of these data faces some difficulties. First, in addition to being nonergodic the samples are inhomogeneous as well. Not much is known about the structure of the grain boundaries. One may, however envisage there being two short time processes present. According to the cage model, the first corresponds to short time excursions of particles about their places on lattice sites or in grain boundaries. In addition, the grain boundaries may be disordered and possibly less dense and thus allow grain boundary diffusion. Assuming the boundaries posses fluids order one would expect the time scale of this process to be close to the structural relaxation observed in the melt. A comparison of three $f(q, \tau)$ as measured in poly(tetrafluoroethylene) particles

in a water glycerol mixed solvent sample is given in Fig. 5. This was prepared at coexistence, in the polycrystalline state and in an aged glassy state. The figure illustrates that at coexistence, diffusion within the crystalline cage and the spatially distinct diffusion in the fluid occupy the same time window. Clearly, without more detailed information about the sample morphologies the origin of the first decay in Fig. 4 cannot be assigned unequivocally. Further the height of the plateau plus the amplitude of the first decay in the crystal should be connected to the amount of crystalline material present. A rigorous theoretical treatment of this situation is still missing. Nevertheless we checked for systematic trends in the plateau heights of the mixed samples. Figure 6 shows that no definite dependence of this plateau value on composition is observed.

Two further pieces of morphological information may be extracted from the static structure factors. First, the width of the principal maximum, Δq, is correlated with the crystallite size:

$$L = \frac{2\pi K_{hkl}}{\Delta q} \, ,$$

where K_{hkl} is the Scherrer constant, which is of the order of 1 [15] and L is the edge length of the crystals, which are assumed to be cubes. The result of this evaluation is shown in Fig. 7. The crystallite sizes are of the order of some 15 μm. There is some scatter but no definite dependence on sample composition. In general the crystallite size is determined by the nucleation rate density, J, and the crystal growth velocity, v. The crystals stop growing once they intersect. In our case the nucleation rate densities are of the order of some $(10^{14} - 10^{15})$ m^{-3} s^{-1}. Solidification is thus nucleation dominated and rather small crystals result. J, however is

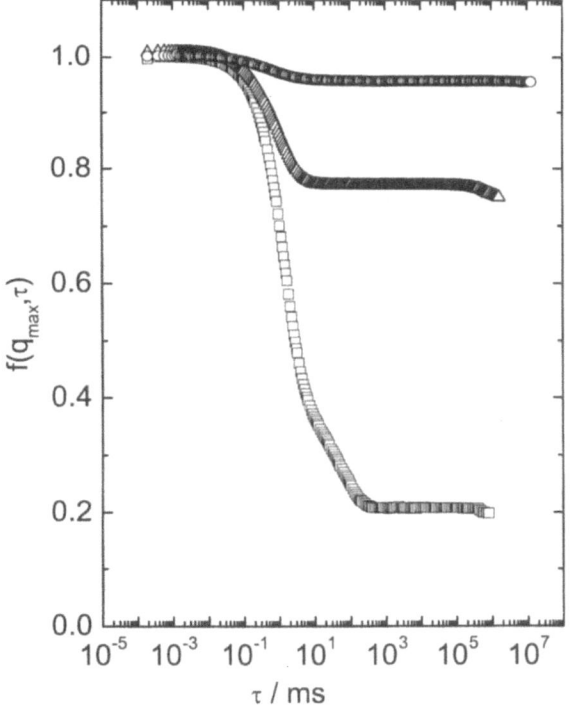

Fig. 5 Comparison of representative intermediate scattering functions, $f(q, \tau)$, taken on deionised samples of poly (tetrafluoroethylene) particles in a water/glycerol mixed solvent. From *bottom* to *top* the curves correspond to a sample at fluid/crystal coexistence, a polycrystalline sample and a glassy sample aged for about 1 year

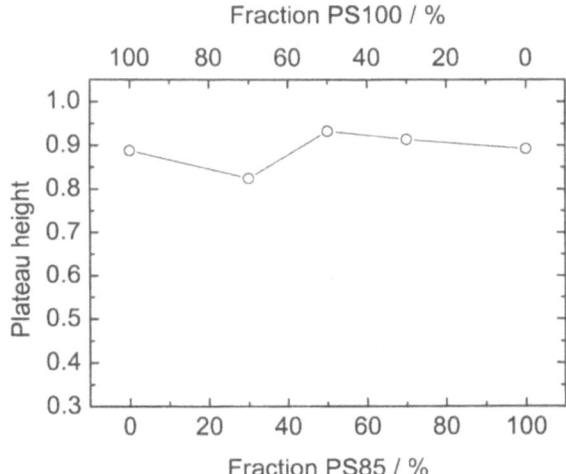

Fig. 6 Plateau heights of $f(q, t)$ for the PS85/PS100 mixture as a function of composition

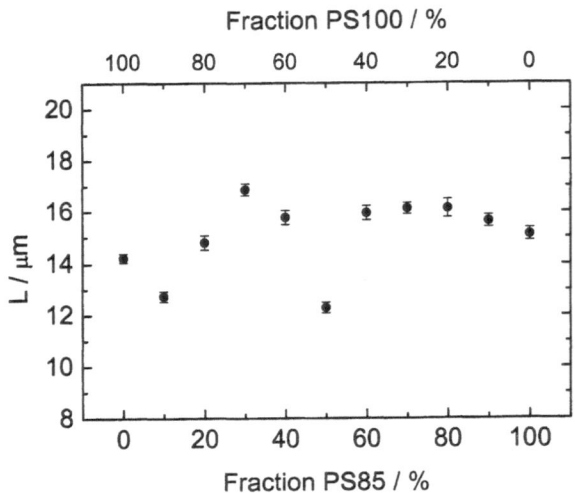

Fig. 7 Crystallite sizes, L, for the PS85/PS100 mixture as a function of composition

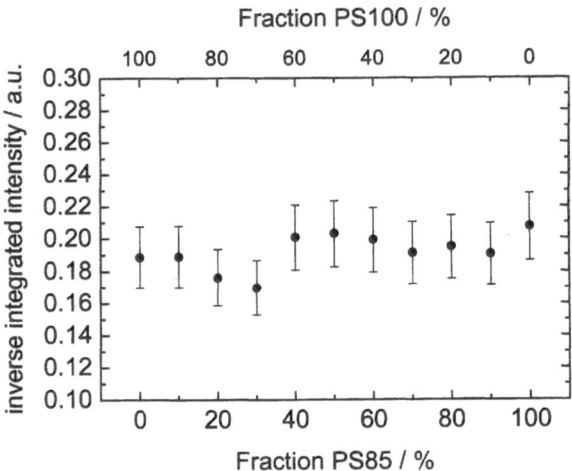

Fig. 8 Inverse integrated intensity of the background corrected (1 1 0) Bragg reflection for the PS85/PS100 mixture as a function of composition

very sensitive to minor changes in the chemical potential difference between the melt and the solid. The scatter of the data is therefore attributed to minor changes in the deionisation state.

The second piece of information can be gained via the integrated intensity of the first peak. This is a measure of the square of the number of lattice planes scattering in a given direction and thus of the amount of crystalline material. Further the fraction of the grain boundary material can be estimated. It is given as $(I_{SC}-I_{PC})/I_{PC}$. Here, the indices PC and SC refer to the measurements on polycrystalline material and a reference measurement on a defect-free single crystal. The latter is not yet available for the present case.

Alternatively we use the inverse of the integrated intensity as a qualitative and relative estimate. This is shown in Fig. 8. Again no definite dependence on composition is observed.

Conclusions

We performed a comprehensive characterisation of a mixture of charged spherical colloids of size ratio $\Gamma = 0.85$ in deionised aqueous suspensions. At constant particle number density all the samples were observed to crystallise completely. For polycrystalline solids, the static structure and the variation of shear modulus and conductivity with composition were found to be compatible with predictions for bcc alloys with randomly distributed composition. By comparing Figs. 6, 7 and 8 we can further state that the plateau heights, the crystal sizes and the amount of grain boundary material are independent of composition.

Two comments are appropriate. First, our data confirm our expectations. In other words, we were lucky to investigate an uncomplicated textbook case. On the other hand, it was exactly this that allowed the test and demonstration of the versatility and accuracy of the methods employed. It also allowed a discussion of the urgently needed extensions of dynamic scattering theory. Second, not all samples behave in such an uncomplicated way. For example, recent measurements of Hunt and Zukoski [6] indicated the formation of A/B phase-separated crystals with significant consequences for the elastic properties. Okubo and Fujita [5] reported the formation of compositionally ordered alloys. The multipurpose apparatus is well suited to precisely measure several intricately connected properties of colloidal solids quasisimultaneously and without the need of mechanically disturbing the sample. We therefore started systematic investigations on further mixed colloidal solids. Preliminary data on systems of larger size and charge ratio indicate that this compositional ordering may be a continuous process as the particle number density is increased. Further, in mixtures of a crystallite-forming and a glass-forming sample definite composition dependencies were also observed for plateau heights and morphological parameters. This raises the interesting question of how the different properties relate to each other and to the morphological details of the samples.

Acknowledgements Financial support was given by Sonderforschungsbereich 262 and by the Material wissenschaftliches Forschungszentrum, Mainz. This work was further supported by the Deutsche Forschungsgemeinschaft (grants Pa459/8-1, 2 and Pa459/9-1); this is gratefully acknowledged.

References

1. Lekkerkerker HNW, Dhont JKG, Verduin H, Smits C, van Duijneveldt JS (1995) Physica A 213:18
2. Poon WCK, Pirie AD, Haw MD, Pusey PN (1997) Physica A 235:110
3. Kaplan PD, Rouke JL, Yodh AG, Pine DJ (1994) Phys Rev Lett 72:582
4. Meller A, Stavans J (1992) Phys Rev Lett 68:3645
5. Okubo T, Fujita H (1996) Colloid Polym Sci 274:368
6. (a) Hunt WJ, Zukoski CF (1999) J Colloid Interface Sci 210:332; (b) Hunt WJ, Zukoski CF (1999) J Colloid Interface Sci 210:343
7. Imhof A, Dhont JKG (1995) Phys Rev B 52:6344
8. Yu H, Barentin C (1995) J Phys Condens Matter 7:L13
9. Schöpe H-J, Decker T, Palberg T (1999) J Chem Phys 109:10068
10. Evers M, Garbow N, Hessinger D, Palberg T (1998) Phys Rev E 57:6774
11. Schöpe H-J, Palberg TJ (2001) Colloid Interface Sci 233:149–161
12. Hessinger D, Evers M, Palberg T (2000) Phys Rev E 61:5493
13. Lindsay HM, Chaikin PM (1982) J Chem Phys 76:3774
14. Pusey PN, van Megen W (1989) Physica A 157:507
15. Hammond C (1998) The basics of crystallography and diffraction. Oxford University Press, Oxford

Progr Colloid Polym Sci (2001) 118: 266–275
© Springer-Verlag 2001

C. Sinn
A. Heymann
A. Stipp
T. Palberg

Solidification kinetics of hard-sphere colloidal suspensions

C. Sinn · A. Heymann · A. Stipp
T. Palberg
Institut für Physik der Johannes-
Gutenberg-Universitat, Staudingerweg 7
55099 Mainz, Germany

C. Sinn (✉)
PMS Optik AE/PT, Kölner Strazze 42
60327 Frankfurt, Germany
e-mail: christian.sinn@pms-optik.de
Tel.: +49-65-75001619
Fax: +49-69-75001617

Abstract We investigate the solidification dynamics of hard-sphere colloidal suspensions applying simultaneously small-angle and Bragg light scattering. These experiments allow a consistent picture of nucleation and crystal growth on the level of large-scale density fluctuations and of density fluctuations on the level of individual crystallites. We observe a temporally almost constant nucleation rate after an induction time that decreases with supersaturation. The classical expectation for the nucleation rate density as a function of supersaturation is in accordance with our data. We investigate the validity of the Wilson–Frenkel growth law for hard-sphere systems, which also fits our data satisfactorily. The kinetic prefactors are found to be only 2 orders of magnitude smaller (for nucleation) and 1 order of magnitude larger (for crystal growth) than expected from classical considerations. We discuss the influences of particle size polydispersity on our data.

Key words Colloidal suspensions ·
Hard spheres · Nucleation and
growth · Crystallization kinetics

Introduction

Colloidal suspensions of particles with different pair-interaction potentials have attracted considerable attention because of their possibly generic value. As the particle interaction potential can be tuned experimentally, very different aspects of condensed matter behaviour may be studied on convenient spatial and temporal scales [1]. From a theoretical point of view, a hard-sphere system [2] is beneficial as temperature is not a relevant parameter. The particle concentration (the volume fraction, ϕ) fully determines the equation of state, $\phi Z(\phi) \propto \Pi(\phi)/k_B T$, where Π is the (osmotic) pressure of the system. The equation of state is known in detail over the whole concentration range [3–8]. Experimentally, a hard-sphere suspension has been realized by grafting short polymer hairs on a polymer [9] or silica [10] core and suspending the colloidal particles in a good solvent for the polymer hairs. The van der Waals attraction is reduced by optically matching particles and solvent.

Above the freezing volume fraction $\phi_F = 0.494$, a hard-sphere suspension exhibits coexistence between a fluid and a crystalline phase. Upon increasing the particle volume fraction of the shear-molten fluid, ϕ_{fl}, the crystalline fraction of the sample, X, increase linearly up to the melting point, $\phi_M = 0.545$; $(\phi_{fl} - \phi_F)/(\phi_M - \phi_F)$ is the supersaturation of the system. With higher particle content, the crystalline phase is compressed beyond ϕ_M. The crystalline structure is observed to be face-centred cubic with a considerable amount of stacking faults present. The degree of randomness, however, is not consistent up to now [11, 12].

Whereas the (osmotic) pressure in the coexisting phases can be assumed to be equal at every instant of time $[\phi_{cr}Z(\phi_{cr}) = \phi_{fl}Z(\phi_{fl})$, where ϕ_{cr} is the (metastable) crystal density], the equilibrium of chemical potential $(\mu_{cr} = \mu_{fl})$ is only established within a finite period of time, which can be rather long (about 1 h) for the comparably slow colloidal dynamics. In this period, crystal nuclei form and subsequently grow. Several investigations have taken advantage of this behaviour in order to determine the crystallization kinetics and, in particular, the temporal evolution of crystal nucleation and growth [13–17]. The results have been interpreted in terms of the classical nucleation theory, an approach first suggested by Russel [18]. A review has been given by one of us of the results obtained with purely repulsive systems recently [19].

In the present work, we demonstrate experiments where we simultaneously perform small-angle and Bragg light scattering. We obtain data sets for the temporal evolution of the number of crystals and their size, and these are consistent on the level of the individual crystallites and on the level of long-range density fluctuations as well. We determine the nucleation rate densities at different supersaturations and compare them with classical nucleation theory. We also report crystal growth data in view of the Wilson–Frenkel growth scenario. Our data allow a numerical comparison with results from atomic and molecular systems. In addition, we assess the influence that polydispersity has on our data.

Classical crystallization theory

Nucleation kinetics

The classical picture of nucleation assumes a statistical particle density fluctuation in the metastable melt that forms an ordered cluster in the surrounding fluid [20, 21]. Whether this initial cluster is stable (and grows) or unstable (and disappears) is governed by competition between the surface energy of the cluster, given by the (macroscopic) fluid–crystal surface tension, γ, and the gain of energy upon crystallization. The radius of the critical cluster, R_c, can be calculated by minimizing the free energy. Following Kelton [22], the molar nucleation rate, j, can be written as

$$j = 8\left(\frac{6}{\pi}\right)^{1/3}\frac{DN_L\bar{V}^{1/3}}{\xi^2}\left(\frac{\gamma}{k_BT}\right)^{1/2}\exp\left(-\frac{16\pi\gamma^3\bar{V}^2}{3k_BT(\Delta\mu)^2}\right),$$
(1)

where D is the diffusion coefficient of the particles and ξ is the mean path length for the diffusion of the particle to the cluster surface. N_L is Avogadro's number and $\bar{V}$ is the average volume per molecule. $\Delta\mu$ is the chemical potential difference between the initial phase and the nucleated phase and k_BT is the thermal energy. Equation (1) has been used in a recent numerical study of nucleation and growth in methanol [23]. This allows a detailed comparison between the behaviour of molecular and colloidal matter. For hard-sphere colloidal crystals, the nucleation rate density, J, is the appropriate quantity that is experimentally accessible [18, 24]. We modify

Eq. (1) correspondingly. For the mean path length of diffusion, we adopt the mean interparticle distance in the metastable fluid, $\xi = 2r/\phi_{fl}^{1/3}$, with r the radius of the colloidal particles. The average volume per particle in the growing nucleus can also be expressed in terms of the cluster particle volume fraction, ϕ_{cl}, $\bar{V} = 4\pi r^3/3\phi_{cl}$. This gives

$$J = K_N\frac{D\phi_{fl}^{5/3}}{r^5}\exp\left(-\frac{4\pi^3\gamma^{*3}}{27\phi_{cl}^2(\Delta\mu^*)^2}\right),$$
(2)

where $K_N = 6\gamma^{*1/2}/\pi\phi_{cl}^{1/3}$, $\Delta\mu^* = \Delta\mu/k_BT$ and $\gamma^* = 4r^2\gamma/k_BT$. From the equation of state, the density of the crystals, ϕ_{cr}, as well as the chemical potential difference, $\Delta\mu^*$, can be calculated as a function of the density of the metastable fluid. Transient compression effects are not included since we set $\phi_{cl} = \phi_{cr}$. For the fluid phase, we use the equation given by Carnahan and Starling [4], whereas for the crystalline phase we use that given by Young and Alder [6]. The combination of both equations, yielding the equilibrium properties, can be solved only numerically. We therefore approximate the exact results with an error of less than 1% by the following quadratic forms [25]:

$$\phi_{cr}(\phi_{fl}) = 0.5455 + 1.308(\phi_{fl} - \phi_F) - 2.93(\phi_{fl} - \phi_F)^2$$
(3)

$$\Delta\mu^*(\phi_{fl}) = -10.354(\phi_{fl} - \phi_F) - 56.23(\phi_{fl} - \phi_F)^2$$
(4)

These equations are used together with Eq. (2) for the data evaluation. $\phi_{cr}^{1/3}$ exhibits no significant variation as a function of ϕ_{fl}. Accordingly, we regard K_N as constant in the evaluation process. Besides K_N, the only parameter that governs the nucleation rate in hard-sphere colloidal systems is expected to be the crystal–fluid surface tension, γ^*, which we regard here as a fitting parameter. Theoretical studies predict either $\gamma^* \approx 0.3$ [26, 27] or $\gamma^* \approx 0.6$ [28–30]. Experimental results [13, 31] seem to indicate that the appropriate value may be close to $\gamma^* \approx 0.55$.

Wilson–Frenkel growth

For the quantitative description of the growth kinetics of a cluster that exceeds the critical radius, we use the growth law proposed by Wilson and Frenkel [21]

$$v = v_0[1 - \exp(\Delta\mu^*)],$$
(5)

which has been successfully used already to describe the crystal growth process in charged colloidal samples [32, 33]. The user of Eq. (5) implies the following restrictions with regard to the growth process. First, the growth is assumed to be linear in time, that is, we have interface- (or reaction-) limited growth. Accordingly, the maximum growth velocity is determined by the incorporation of the particles into the crystal surface. In contrast, diffusion-limited growth predicts a growth law proportional to the square root of time. This would slow the crystal growth in the coexistence region because of the density difference between the liquid and the crystal, which is absent above ϕ_M. We shall come back to this point later. Furthermore, we assume normal growth for the crystal growth process, meaning that the growth kinetics is independent of the surface site. This necessarily implies a rough surface, leading to nonfaceted crystallites. This is in line with observations of He et al. [34], who found rough surfaces of hard-sphere crystals with the aid of light microscopy; however, they also observed growth instabilities that lead to anisotropic, nonspherical crystals. A dendritic growth process has been observed under conditions of microgravity [35].

The factor v_0 can be interpreted as a typical velocity for particles to be incorporated into the crystal surface, while the second term is the probability for them to be incorporated into the growing crystal. As colloidal particle dynamics is overdamped, we assume v_0 to be governed by the diffusion of the particles over a certain distance, $v_0 = D/\xi$. As before, we take the mean interparticle distance in the metastable fluid, $\xi = 2r/\phi_{fl}^{1/3}$.[1]

In order to describe the experimental data within the framework of the theory, one has to decide which diffusion coefficient to use. Whereas it seems to be evident that incorporation particle dynamics has to be described by self-diffusion, it is not immediately clear whether to regard the self-diffusion in the limit of short or long times. This question has been discussed in detail by van Duijneveldt [24], who suggested using a diffusion coefficient that is affected by the nearest-neighbour cage, because ξ is of the order of the dimension of the particles. In addition, he suggested using the long-time self-diffusion coefficient, because it seemed to be consistent with the experimental data. We follow his argumentation and choose the long-time self-diffusion coefficient,

$$D_s^* = \frac{D_s^L(\phi_{fl})}{D_0} = \left(1 - \frac{\phi_{fl}}{0.58}\right)^\delta , \qquad (6)$$

normalised by $D_0 = k_B T/(6\pi\eta r)$, the Stokes–Einstein diffusion coefficient of an isolated colloidal particle in a medium of viscosity η. Van Duijneveldt used $\delta = 1.76$, whereas the theoretical prediction of Tokuyama and Oppenheim [36] yields $\delta = 2$ at high volume fractions. Our choice of $\delta = 2.58$ provides the advantage that Eq. (6) has been obtained experimentally with hard-sphere colloidal suspensions very similar to ours at volume fractions close to the glass transition. Hence, our growth law reads

---

[1] In Ref. [19], a slightly different definition of ξ was used, and this leads to numerically different values for the kinetic prefactors

$$v = K_G \frac{D_s^* \phi_{fl}^{1/3}}{r} [1 - \exp(\Delta\mu^*)] \text{ with } K_G = \frac{1}{2} . \qquad (7)$$

As before, the chemical potential difference is completely determined by the equation of state in the form of Eqs. (3) and (4), which means that only one fitting parameter (K_G) remains.

Experimental

Light scattering apparatus

The light scattering apparatus was constructed specifically for our approach [25]. A scheme of the setup can be found elsewhere [19]. The vertically polarized beam of He–Ne laser (632.8 nm) is spatially filtered by a combination of a microscope objective (10x), a pinhole (diameter: 40 μm) and an asymmetric biconvex lens ($f' = 100$ mm). The distance between the pinhole and the biconvex lens is adjusted such that the transmitted beam is not collimated, but is focused onto the beam stop of the small-angle detector, located at a distance of about 150 cm. The cuvette with the colloidal suspension under study is located in a sample cell filled with a decahydronaphthalene (DHN) 1, 2, 3, 4-tetrahydronaphthalene (THN) mixture. The light beam enters the sample cell through a glass window. The Bragg scattered light leaves the sample cell through a spherical lens, which consists of a sector of a sphere (radius 50 mm, thickness 40 mm). The vortex of the lens was ground to form a plane exit window perpendicular to the optical axis of 12.5-mm diameter. The entrance window and the spherical lens are made from K10 (Schott), which has a refractive index close to the suspending fluid (see later). The centre of the cuvette is located at the front focal point of the spherical refracting surface. This optical setup guarantees that light scattered under equal angles in the illuminated volume is collected in a single point on the back focal plane of the spherical surface, where the Bragg detectors are located. On the Bragg detector post, six photodiode detector units are arranged concentrically around the spherical lens at a distance of 10 cm. Three units are arranged linearly on the small-angle detector post. Both posts are mounted onto a large aluminum frame, which is rotated about the optical axis by a stepper motor that performs a full rotation with 6272 steps. In this way, the complete two-dimensional scattering pattern is accessible. After each step (approximately 0.06°), the scattered intensity measured by the photodiodes is digitized and stored for further evaluation by a personal computer. A single measurement takes about 40 s. In order to enhance the statistical accuracy of the results, we angularly average in this contribution the intensities for every individual q. Angularly resolved measurements have been applied earlier to study heterogeneous nucleation at the cuvette walls [37]. We note that polarization effects in the Debye–Scherrer rings are superimposed by multiply scattered light. Corrections for multiple scattering were generally found to be unnecessary in the q region investigated here [25]. We calibrated the apparatus using an optical grid that was immersed into the index-matching bath. The accessible Bragg scattering vector range ($n = 1.5$) is $q = 5.5$–$18.9 \mu m^{-1}$ with a resolution of $\delta q = 0.15 \mu m^{-1}$. This comparatively large angular range allows the time evolution of different Bragg peaks to be measured. For small-angle scattering the range is $q = 0.02$–$0.81 \mu m^{-1}$ with a resolution of $\delta q = 0.012 \mu m^{-1}$.

Before starting a series of individual measurements, the colloidal suspensions were shear-molten by a slow rotation of the cuvette for several hours or days. This should destroy any crystalline order within the suspension. Stopping the rotation and inserting the cuvette into the sample cell defines $t = 0$ for the measurement series. The sample cell is thermostatted with Peltier elements to better than 0.1 K. This stability is necessary in order to maintain the refractive index matching in the course of the measurements.

Sample preparation

The particles used comprised a poly(methyl methacrylate) (PMMA) core with short hairs of poly(hydroxy stearic acid) (PHSA) grafted onto the surface. They were synthesized by Underwood (SMU28, [38]) and have been carefully characterized for their radius and the corresponding size polydispersity by static light scattering [39]. We obtained a particle core radius of $r = (435 \pm 4)$ nm and a remarkably low size polydispersity of 2.5%. The effective hard-sphere radius of these particles ($r = 445$ nm, determined from the phase boundaries) is consistent with the expectation of a hairy layer of approximately 10 nm [3].

The particles were suspended in a suitably chosen mixture of DHN (isomeric mixture) and THN (both from Merck, Germany). The mixing ratio was adjusted such that the solvent matches the refractive index of the particles. A complete match of the refractive indices of particle and solvent is virtually impossible for a core–shell system. We therefore adopted the following empirical procedure. The particles were suspended in pure DHN to yield a high volume fraction. A strongly turbid sample resulted. We added THN and observed a strong decrease in turbidity. Close to the optimal index matching point of approximately 35% THN by volume, the turbidity does not change significantly. Alternatively, we observed the dispersive behaviour of the sample. A slight surplus of THN at the matching point leads to bluish appearance of the sample when observed under white light illumination, whereas a surplus of DHN leads to a yellowish appearance. We always tried to obtain a purple transmission, indicating the matching point.

The immediate requirement for performing crystallization measurements of hard-sphere suspensions is a sample of exactly known particle volume fraction. Owing to a limited amount of colloidal material, we adopted the following procedure for fixing ϕ. We prepared a diluted suspension by adding DHN and THN to a colloidal sample until index-matching was achieved. The suspensions as well as the solvents were filtered using 5-μm polytetrafluoroethylene filters; the complete procedure was performed in a laminar flow box. The suspension was filled into a rectangular cuvette ($10 \times 5 \times 20$-mm inner volume). The particles were centrifuged to form a compact sediment of $\phi \approx 0.64$, and a sufficient amount of supernatant liquid was removed to obtain a concentrated sample ($\phi \approx 0.57$) for the first light scattering measurement. Subsequently, we diluted the sample with the supernatant liquid removed before, the amount of solvent added being controlled by weight, and investigated this diluted sample. We terminated the series at a volume fraction where nucleation becomes very slow. We then determined the effective hard-sphere volume fraction of the last sample using the sedimentation method described Paulin and Ackerson [40], where the sample volume fraction is obtained from an extrapolation of the respective volumes of the liquid and the solid phase to $t = 0$. We note that the uncertainty of this determination is mainly given by the theoretical values for ϕ_F and ϕ_M. The concentration of every sample investigated in our dilution series was calculated using $\rho_{PMMA} = 1.188$ g/cm^3, a value close to that used by Phan et al. [3].

In this investigation, we report results for two different dilution sequences. The difference in the hard-sphere volume fractions observed during these runs stems from the slight evaporation of the solvent despite careful sealing.

We finish this section by noting that THN, over periods of several weeks, slowly penetrates the particle core [3, 39]. This has already been observed for a different solvent [41]. This observation prohibited us from using calculated particle form factors for data correction purposes. We therefore used the scattering of the shear-molten state as a reference and note that in the course of a single

solidification measurement the refractive index variation within the particles remains constant.

Data evaluation and experimental results

Small-angle scattering

Small-angle light scattering experiments are plagued by parasitic light scattering from dust or optical inhomogeneities from the optical surfaces. As described before, we kept the number of transmitted surfaces low. Nevertheless, we observed strong background scattering. In order to correct for this background, we subtracted an early measurement from our data. We observed a clear-cut peak at finite q, which grew with time and became narrower. The maximum intensity shifted towards smaller q.

The observation of a peak in the small-angle scattering intensity, though in accordance with different studies [13, 15, 24], still lacks a quantitative explanation. Qualitatively, the peak results from large-scale variations of the refractive index with overall conservation of the particle number density or the mean refractive index; thus, $i(q \rightarrow 0) = 0$. Its shape is determined from details of the density fluctuations and their spatial distribution. In crystallization, it may stem from the crystal and the surrounding depletion zone, which, however, may change in the course of time [42]. In the present contribution, we focus on the temporal evolution of the peak and leave a structural explanation for further investigations.

With progressing solidification, the maximum peak intensity, $i_{max} \propto NR^6$, increases, because either the number of crystals in the scattering volume increases or the individual crystals grow. Both processes result in a shift of the peak towards smaller q values. A strong aid in the data evaluation is the observed scaling of the small-angle scattering peak [13] according to

$$i(q,t) = i_{max}(t) F[q/q_{max}(t)] \ , \qquad (8)$$

where $F(Q)$ is a universal model function for $Q = q/q_{max}(t)$. Several functional forms for $F(Q)$ have been suggested in the literature [13, 15]. For the present evaluation, we modify the standard Furukawa function [43], which harmonically adds a low-q Q^2 dependence to a large-q $Q^{-\delta}$ dependence. We add a constant, C, that is independent of Q and obtain

$$F(Q) = \frac{\delta + 2 + C}{\delta Q^{-2} + 2Q^\delta + C} \ . \qquad (9)$$

For spinodal decomposition, $d = \delta - 1$ is the fractal dimension of the surfaces, which is frequently found to be $d = 3$. As we have no a priori information about the structure of the depletion zones, we regard δ as a free parameter. The combination of Eqs. (8) and (9) yields

$$i(q)/i_1 = \left[1 + \left(\frac{q}{q_1} \right)^{-2} + \left(\frac{q}{q_2} \right)^\delta \right]^{-1} \ , \qquad (10)$$

with the fitting parameters $i_1 = i_{max}(\delta + 2 + C)$, $q_1 = q_{max}(\delta/C)^{1/2}$ and $q_2 = q_{max}(C/2)^{1/\delta}$. With great caution, $1/q_1$ may be interpreted as the characteristic extension of the crystal, and $1/q_2$ as the extension of the depletion zone. However, providing a structural model is beyond of the scope of this contribution and requires data of still improved statistical quality; therefore, we used Eq. (10) as a fitting function only and obtained from this fit i_{max} and q_{max}, from which $N(t)$ and $R(t)$ can be calculated [25]. For technical reasons, we determined $q_{1/2}$ instead of q_{max}, which is defined as the larger q value of $F(Q_{1/2}) = 1/2$ and can be evaluated very similarly to q_{max} [13]. The results are collected in Fig. 1. We shall return to a detailed discussion of the data in a subsequent section.

Bragg scattering

Owing to the close match of the refractive indices, we assume that the scattered light intensity can be calculated using the Rayleigh–Debye–Gans approximation. Accordingly, we write the scattered intensity in the Bragg range as a product of the individual particle form factor, $P(q)$, and different structural contributions

$$I(q,t) \propto P(q)\left[S_{\text{Bragg}}(q,t) + S_{\text{diff}}(q,t) + S_{\text{fluid}}(q,t)\right] \ , \tag{11}$$

where S_{fluid} is the structure factor of the fluid phase, S_{Bragg} describes the Bragg reflections of the crystals, and S_{diff} gathers contributions from grain boundaries or crystal imperfections. The respective weights of the different structures vary with time as solidification proceeds; S_{Bragg} and S_{diff} may also alter their q variation. For S_{Bragg}, this originates, for instance, from the relaxation of the initial compression of the crystals. S_{diff} on the other hand, may reflect the increasing number of crystalline imperfections in larger crystals. S_{fluid} comprises contributions of the fluid structure factor of the metastable fluid and of the structure factor of the equilibrium fluid. The latter contribution is absent for $\phi_{\text{fl}} > \phi_{\text{M}}$; the q dependence of the former could be calculated using the Percus–Yevick structure factor for ϕ_{fl} [44].

As discussed earlier, the particle form factor is not available with sufficient accuracy. Accordingly, we divide our data by an early measurement

$$i(q,t) = \frac{I(q,t)}{I(q,0)} = \frac{S_{\text{Bragg}}(q,t)}{S_{\text{fluid}}(q,0)} + B(q,t) \ , \tag{12}$$

where $B(q,t)$ is a baseline which is allowed to change only slowly and continuously as well as with time as with q. In addition, we require that the fluid structure does not lead to considerable distortions of the individual Bragg reflection. It is then possible to isolate the reflection's time dependence for further quantitative analysis.

The requirements as mentioned earlier are well established for the (2 2 0) and the (3 1 1) peaks only. As discussed in detail earlier [37], we observed that the (3 1 1) peak is influenced for our particles by scattering contributions from oriented close-packed planes that exhibit different growth kinetics. We therefore restrict ourselves here to the temporal evolution of the (2 2 0) lattice peak and note that the evolution of (3 1 1) and (1 1 1) could be determined without disturbing side effects using particles with a different size.

Because the crystalline samples under investigation consist of randomly oriented crystals, the Bragg structure factor we measure appears in the shape of Debye–Scherrer rings centred about the optical axis of the experiment. The structure factor is therefore an average over different crystallite orientations. In addition, the individual crystals are not large compared to the wavelength, meaning that the crystal form factor has to be taken into account, which leads to a finite width, Δq, of the Debye–Scherrer ring. The angular averaged intensity of the Debye–Scherrer ring is accordingly given by

$$S_{\text{Bragg}}(\Delta q,t) = \iint_{q=G+\Delta q} S_{\text{Bragg}}(|\vec{q}-\vec{G}|,t)\mathrm{d}^2 q \ , \tag{13}$$

where $\vec{G}$ is the reciprocal lattice vector. We assume that the scattering of a spherical crystal of radius R can be described by a Rayleigh–Debye–Gans form factor, centred at $\vec{q}=\vec{G}$. We evaluate Eq. (13) in the limit of $\Delta q \ll q$ and obtain

$$S_{\text{Bragg}}(\Delta q,t) \propto \left[1 - \cos(2\Delta q R) - 2\Delta q R \sin(2\Delta q R) \right.$$
$$\left. + 2\Delta q^2 R^2\right]/(2\Delta q^4 R^4) \ . \tag{14}$$

Hence, $S_{\text{Bragg}}(\Delta q,t)$ given by Eq. (14) is the shape factor of the individual crystallites, averaged in reciprocal space over every crystal orientation. In turn, the use of Eq. (14) allows us to extract the typical dimensions, R, of the crystallites from the width of the Bragg peak that we obtain by angular averaging over the Debye–Scherrer ring.

The scattered intensity is proportional to the shape factor times the crystalline volume squared times the number of crystals N times an optical constant representing the scattering strength. For the scattered intensity integrated over the Bragg peak, we thus obtain

$$i_{\text{peak}}(t) = I_0 \int_{q_1}^{q_2} N(t)R(t)^6 S_{\text{Bragg}}(\Delta q,t)\mathrm{d}\Delta q$$
$$\propto N(t)R(t)^3 \propto X(t) \ , \tag{15}$$

where the integration is restricted to those q values where the Bragg peak differs from the baseline. $X(t)$ is the crystalline fraction of the sample. The results for $N(t)$ and $R(t)$ are shown in Fig. 2, which can be compared directly with the small-angle scattering data shown in Fig. 1.

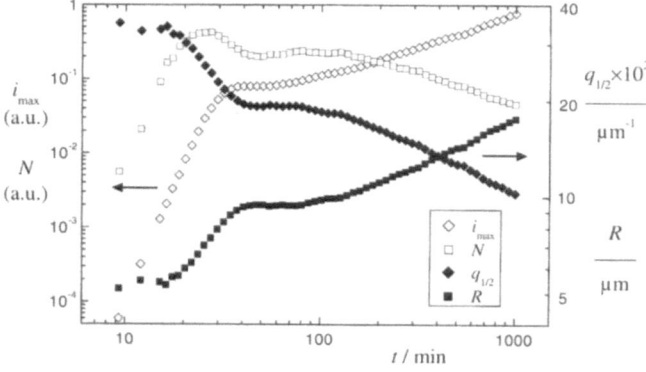

Fig. 1 Temporal behaviour of the maximum intensity, i_{max}, the characteristic scattering vector, $q_{1/2}$, the crystal number density, N, and the average crystal size, R, calculated from the small-angle scattering data. The sample volume fraction is $\phi_{\text{fl}} = 0.538$

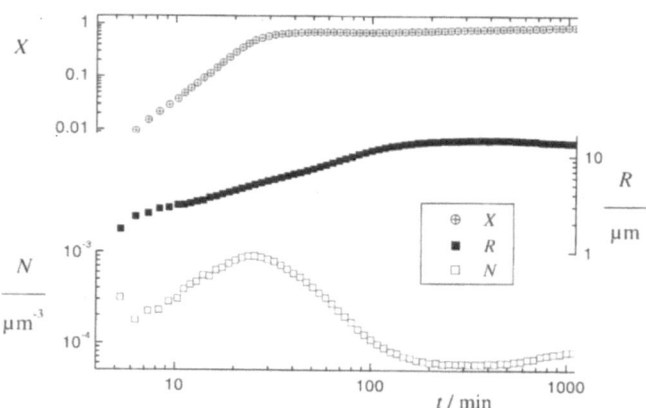

Fig. 2 Temporal behaviour of the crystalline fraction, X, R and N calculated from the Bragg scattering data. The sample volume fraction is $\phi_{\text{fl}} = 0.538$

Merging of the small-angle and Bragg scattering data

We have shown that $N(t)$ and $R(t)$ can be obtained from small-angle and Bragg scattering data independently. The data very satisfactorily exhibit the same qualitative and quantitative behaviour [25, 45]. This redundancy can be used to obtain a set of $N(t)$ and $R(t)$ of improved statistical quality upon merging these data. In addition, some experimental limitations of both individual methods can be removed from the data.

First, we observed that $R(t)$ never exceeded $R = 15 \, \mu m$ in the Bragg data, though the corresponding width of the Bragg peak is larger than the experimental resolution. No individual experimental uncertainty (such as the converging incident light beam, lens aberrations or an incorrect lens adjustment) can explain this broad reflection; however, a combination of these effects may play a role. The small-angle scattering data, on the other hand, are readily prone to saturation effects of the photodiodes, as the scattered light intensity as a function of time and of sample concentration varies over orders of magnitude. Whereas the data at large ϕ exhibit a low signal-to-noise ratio, the scattered light of the samples at small ϕ saturates the detector.

We therefore proceeded as follows. First, we compared the crystal radii, $R(t)$, as obtained from both methods and decided on the most reliable values on the basis of the standard deviation of the respective fit and the saturation limits mentioned. The results are shown in Fig. 4. The use of the model functions as described previously leads to a systematic deviation between the small-angle and the Bragg data, which can be removed by dividing the small-angle data by 1.19 [19]. The reason for this deviation for $t \to \infty$ stems from the different structural origin of the small-angle and the Bragg data: only the latter yield the radius of the crystals immediately, whereas for the former the radius is less well defined (see the discussion earlier). Next, we determined the crystalline fraction, $X(t)$, from the Bragg scattering data by normalisation of $i_{peak}(t)$ by its long-time limit, $i_{peak}(t \to \infty)$. For samples below ϕ_M, the expected crystalline fraction is taken into account.[2] Assuming that $X(t)$ is identical for the small-angle and the Bragg scattering data, we normalized the small-angle scattering data correspondingly. Again, we selected the most reliable data. From $X(t)$, we calculated $N(t)$; the results are shown in Fig. 3. From the initial slope of these data, we obtained the nucleation rate density, $J = dN/dt$, and the growth rate, $v = dR/dt$.

Comparison of theory and experiment

The results for the temporal evolution of the number of crystals are shown in Fig. 3. As a general observation, we find a roughly linear increase of $N(t)$, followed by a saturation or even a decreasing number of crystals. We interpret this behaviour as the transition from a nucleation and growth period to ripening behaviour, where the larger crystals grow at the expense of the smaller ones, and we restrict our evaluation to the former process. In addition, the nucleation seems to start with a delay for most of the samples. The data for the temporal evolution of the crystal radii, which are shown in Fig. 4, also exhibit this induction time, together with a finite

---

[2] This procedure has been cross-checked by the observation that $i_{peak}(t \to \infty)$ is constant above ϕ_M to within a systematic error 20%. If we calculate, using this value, $X(t)$ for samples below ϕ_M, we obtain the values that are predicted by the phase diagram. Interestingly, Harland and coworkers [16, 17] report $X(t \to \infty) \neq 1$ above ϕ_M, as yet unexplained

minimum crystal size. Subsequently, the data points show a more-or-less linear increase of the crystal size.

Induction time

Subsequent to the induction time, t_I, the nucleation rate is found to be almost constant. Only a double-logarithmic plot of $N(t)$ exhibits noticeable, but small, deviations from linear behaviour. Harland and Van Megen [17] observed different behaviour for similar data sets. Their number densities of crystals exhibited a temporal power-law behaviour with exponents up to 3, a behaviour they termed accelerated nucleation. A comparable behaviour is absent in our data. We therefore prefer to speak about a delayed nucleation, which gives rise to small nonlinearities in a double-logarithmic plot. From Fig. 3, we obtain t_I by a linear extrapolation of $N(t)$, that is, $N(t_I) = 0$. The corresponding data are collected in Fig. 5. The induction time is found to decrease with increasing supersaturation, saturating at approximately ϕ_M with $t_I = 4$ min. The minimal cluster radius exhibits a similar behaviour, the saturation value being approximately $2 \, \mu m$, equivalent to 1–2 lattice constants or 2–3 particle diameters.

These results agree very well with the classical picture of nucleation. The nucleation rate is assumed to be constant up to that time where ripening sets in, in accordance with our data. In addition, the initial nonlinear behaviour (the induction time) can be well understood, as the system needs some time to acquire a steady-state cluster distribution after the fast quenching into the coexistence region [22]. This period of time should decrease with increasing supersaturation, because the critical radius decreases drastically, in agreement with our data and those reported by Harland and van Megen [17]. Upon approach of the glass transition, the decreasing diffusion coefficient may lead to an again increasing induction time, as clearly visible in the data of Harland and van Megen; however, our data do not reproduce this trend [19].

Nucleation and growth rates

In proceeding further, we determined the normalized nucleation rate density, J^*, as a function of supersaturation from the data shown in Fig. 3. Where the slope changes slightly, we used the initial slope only. As shown before, classical nucleation theory predicts

$$J^*(\phi_{fl}) = \frac{Jr^5}{K_N D_0}$$
$$= D_s^*(\phi_{fl}) \phi_{fl}^{5/3} \exp\left(-\frac{4\pi^3 \gamma^{*3}}{27 \phi_{cr}^2(\phi_{fl})(\Delta\mu^*(\phi_{fl}))^2}\right),$$

(16)

Fig. 3 Temporal behaviour of the crystal number density, $N(t)$, for selected samples with the volume fraction indicated. Notice the scale variation

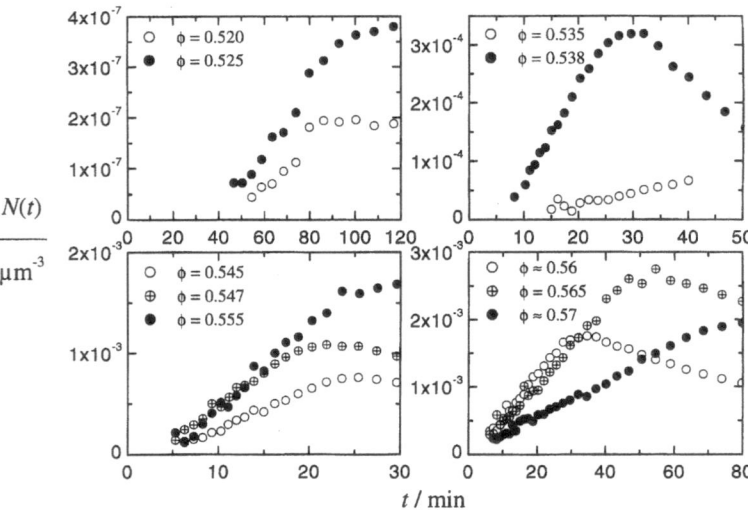

Fig. 4 Temporal behaviour of the average crystal sizes, $R(t)$, for selected samples with the volume fraction indicated. Notice the scale variation

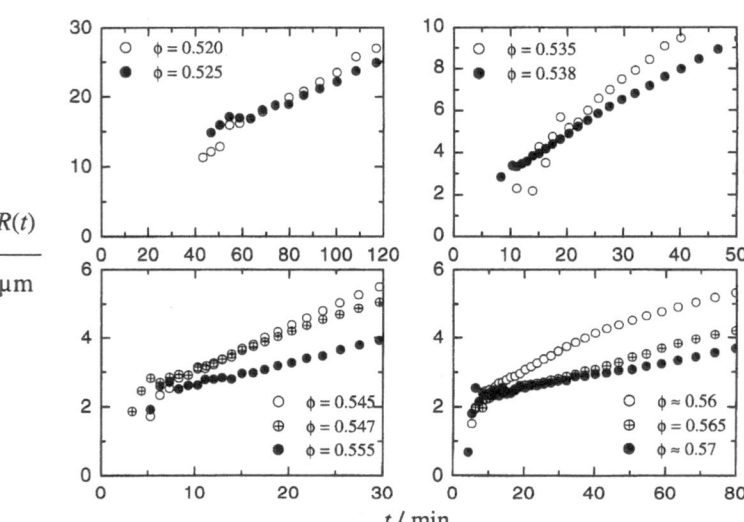

where we are left with only two adjustable parameters, that is, the surface tension γ^* and K_N. The diffusion coefficient is specified by Eq. (6) and $\phi_{cr}(\phi_{fl})$ and $\Delta\mu^*(\phi_{fl})$ by Eqs. (3) and (4). The corresponding data together with a fit of Eq. (16) are shown in Fig. 6. The expected functional dependence of $J^*(\phi_{fl})$ is in excellent agreement with our experimental data. The surface tension we determine from this fit is $\gamma^* = 0.51$, close the theoretical expectation for hard-sphere crystals reported by Marr and Gast [31] and almost identical to the value read from the data of Harland and van Megen [17]. The kinetic prefactor amounts to $K_N = 0.013$, which is remarkably close to the value used by Harland and van Megen to fit their data ($K_N = 0.01$). We keep a discussion of K_N for the end of this section. Similarly, we determined the normalized growth velocity, v^*,

$$v^*(\phi_{fl}) = \frac{vr}{K_G D_o} = D_s^*(\phi_{fl})\phi_{fl}^{1/3}[1 - \exp(\Delta\mu^*(\phi_{fl}))] \quad (17)$$

from the data shown in Fig. 4. We determined an initial linear slope from $R(t)$ and regard the radius offset as the crystal radius R_c, corresponding to the induction time t_I. The data obtained are shown in Fig. 7 together with a fit to the theory. Again, the agreement between theory and experiment is very good. The kinetic prefactor $K_G = 15$ is the only adjustable parameter.

The data at small volume fractions seem to deviate increasingly. This may have two different reasons. On the one hand, the diffusion coefficient we choose may become inadequate. This would not be noticed in the nucleation data, as these are dominated by the increase in the surface contribution at low ϕ_{fl}. We recall from the previous discussion that the choice of the functional form

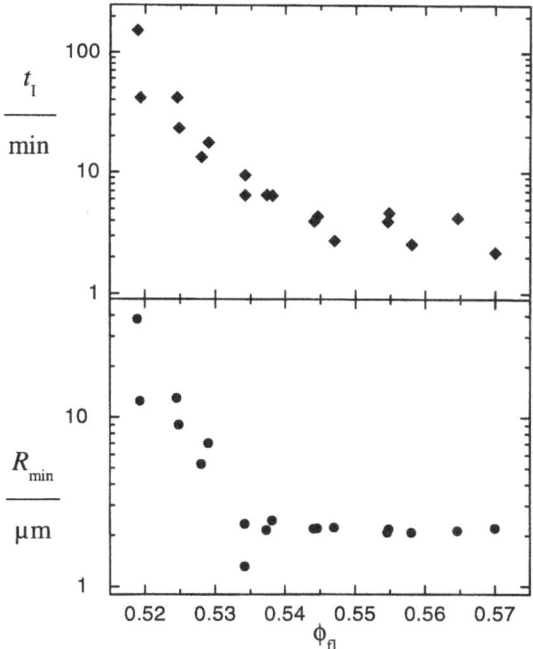

Fig. 5 Induction time, t_I, and minimal crystal size, R_{min}, as a function

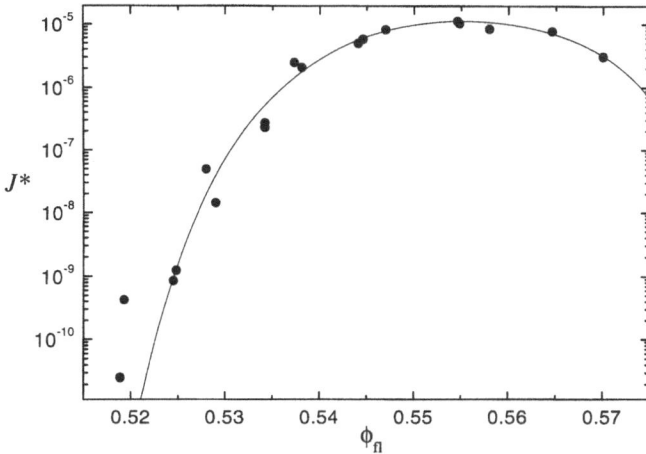

Fig. 6 Nucleation rate, J^*, as a function of the sample density, ϕ_{fl}. The *solid line* is a calculation according to the classical nucleation theory. The fitting parameters are the surface tension $\gamma^* = 0.51$ and the kinetic prefactor, $K_N = 0.013$

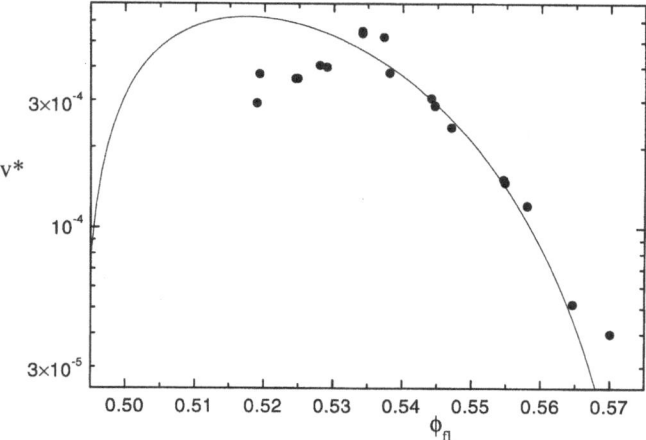

Fig. 7 Growth velocity, v^*, as a function of ϕ_{fl}. The *solid line* is a calculation according to the Wilson–Frenkel growth law. The single fitting parameter is the kinetic prefactor, $K_G = 15$

difficult to observe, because at high volume fractions the crystallization is fast and $R(t)$ is dominated by ripening. In addition, at early times the limited number of data will probably always yield a linear relationship. The growth exponents we determine lie between the extreme cases of 0.5 and 1. Diffusion-limited growth should have the most pronounced influence on the data where the crystallization is slow, that is, at low ϕ_{fl}, where we observe the data deviate.

Kinetic prefactors

Besides the results presented so far, the kinetic prefactors deserve further attention, because they appear to span orders of magnitude in experimental investigations of atomic and molecular system [22].

If we use the appropriate length scales for colloidal systems, we expect $K_N \approx 1.6$ for $\gamma^\infty = 0.51$ and $\phi_{cr} = 0.6$ (cf. Eq. 2). The experimentally determined value is approximately $K_N = 0.013$, which is 2 orders of magnitude smaller than the theoretical value. The opposite is true for K_G, where the experimentally determined value of 15 is 1 order of magnitude larger than the predicted value of 0.5.

It should now be possible to compare these results with those reported for atomic or molecular systems. Unfortunately, these data exhibit an inconsistent picture [22]. Whereas, for instance, K_N for liquid Hg is reported to be 7 orders of magnitude higher than expected, the kinetic prefactors for different n-alkanes seem to be in accordance with expectation, and the nucleation rate as a function of supersaturation is in accordance with the classical theory. The deviation for liquid Hg can be explained if the surface tension is modelled as temperature-dependent. Gránásy Iglói [46] drew a similar con-

for the self-diffusion coefficient is not unequivocal. In addition, at smaller volume fractions even collective diffusion may come into play owing to the increasing density differences between the metastable fluid and the crystal phase.

On the other hand, the growth velocities were obtained assuming $R(t) \propto t$. One might argue that a diffusion-limited growth process with $R(t) \propto t^{1/2}$ would better describe our data. This would, however, be

clusion. They compared different nonclassical nucleation theories, divided into modifications of the classical theory and density functional approaches. From a comparison with experimental data, they concluded that a modification of the surface energy assuming a diffuse interface correctly predicts the experimentally determined nucleation rates, whereas density functional approaches fail to provide agreement with experimental data.

In a recent study, Brugmans and Vos [23] investigated the nucleation and crystal growth in methanol subjected to high pressures. In their study, supersaturation was proportional to a pressure difference. They found excellent agreement of their nucleation rate data with the classical theory and of their growth rate data with the Wilson–Frenkel law. In the latter case, even K_G could be reproduced with reasonable assumptions, whereas K_N was found to be 20 orders of magnitude lower than expected, with no explanation given.

In view of these partly contradictory results for atomic and molecular systems, the agreement between classical theories and experimental data appears to be remarkably good for hard-sphere colloidal systems. Even the kinetic prefactors differ (only) by 1 or 2 orders of magnitude from expectation. In particular, the simple idea of a sharp interface with a macroscopically defined surface tension between the crystal nuclei and the surrounding fluid seems to be effective. Experimental and theoretical values for the surface tension are in good agreement with each other.

Particle polydispersity

Recently, the influence of the particle size polydispersity on hard-sphere crystallization has been discussed. In view of the results presented here, some comments seen to be appropriate.

Bolhuis and Kofke [47] predicted from Monte Carlo simulations a shift of the phase boundaries that amounts to $\phi_F = 0.502$ and $\phi_M = 0.550$ for a Gaussian-distributed 5% particle size polydispersity. The particle we used in the present study exhibited a remarkably low polydispersity of 2.5%. Accordingly, we expect that the shift of the freezing volume fraction owing to polydispersity should be rather small. As the effective hard-sphere radius, calculated from ϕ_M, is in very good agreement with expectation, we estimate that the phase boundaries we use are not prone to polydispersity effects. Again owing to the small polydispersity of our particles, we also expect no fractionated crystallization as reported by Bartlett [48]. However, the nucleation and growth rates themselves may be influenced by polydispersity. Henderson et al. [49] measured a tenfold reduction of the nucleation rate upon increasing the particle size polydispersity from 6 to 11%. They explain this effect by a (comparably slow) ejection of nonfitting particles from the crystal that is composed of the majority particle fraction; however, with small polydispersities like those of the present study, the fraction of nonfitting particles is very low and will probably influence the growth kinetics only at late stages, where the crystal attains it final composition, while the nucleation rate may not be influenced at all. From the classical theory, we can estimate the polydispersity effect on J by calculating ϕ_{cl}, assuming a cluster of a single particle with 2.5% radius deviation surrounded by four particles having nominal radii. Obviously, a smaller particle increases the nucleation rate, whereas a larger one leads to a decreased nucleation rate. In this model, the average nucleation rate therefore remains unchanged by (a symmetrical) polydispersity. In addition, the numerical effect is very small (approximately 5%) and does not explain the observed discrepancy for K_N.

A different origin for the influence of polydispersity on the crystallization has been observed by Phan et al. [50]. They found significant deviations from the crystalline equation of state with increasing polydispersity and suggested a model crystal structure, where smaller particles tend to surround larger ones, thereby reducing the crystalline packing fraction. Again, the small polydispersity of the particles in the present study prohibits us from noticing any effects related to their observation.

If we bear in mind that the choice of the correct functional form of $D(\phi_{fl})$ has a strong influence on the observed crystallization kinetics, an unequivocal determination of the effect of polydispersity on the kinetic prefactors seems to be very difficult. The important question, however, is, whether the observed deviation of the kinetic prefactors from expectation is a manifestation of a generic trend. Our data seem to favour this interpretation. As critical cluster formation as well as crystal growth are dominated by very similar particle dynamics, we are not aware of any mechanism where polydispersity causes a decrease in the nucleation rate and an increase in the growth rate at the same time.

Conclusion

We have shown that hard-sphere colloidal suspensions exhibit both, qualitatively and quantitatively, the dependence predicted by classical theories for crystal nucleation and growth as a function of supersaturation. The nucleation rate density is found to be 2 orders of magnitude slower, the crystal growth rate is found to be 1 order of magnitude faster compared to theoretical considerations. In view of the results for atomic and molecular systems, however, this deviation is small.

The small numerical deviation may be explained by the combined influence of a wrong estimate for the diffusion coefficient, the length scale involved and the particle polydispersity. Unfortunately, the results for atomic and molecular systems reported so far exhibit no clear trend to allow a unique comparison. As long as the experimental observations for atomic and molecular systems cannot be improved, the hard-sphere colloidal system can be regarded as a very effective model for crystallization processes. In view of the widespread applications of crystallization processes for industrial applications, further investigations may profit from a closer look at the dependence of the crystallization kinetics from the colloidal particle interaction potential.

Acknowledgements We are grateful to Sylvia Underwood, Bill van Megen and members of his group for the gift of the particles and for various stimulating discussions. In addition, we gratefully acknowledge valuable discussions with Bruce Ackerson. The present work was initiated and inspired by the late Klaus Schätzel. This work was supported financially by the Deutsche Forschunssgemeinschajt, (Scha389/6-1,2; Pa459/6-3) and the Materialwissenschaftliches Forchungszentrum, Mainz.

References

1. Pusey PN (1991) In: Hansen JP, Levesque D, Zinn-Justin J (eds) Liquids, freezing and glass transition. North Holland Amsterdam, pp 763–942
2. Pusey PN, van Megen W (1986) Nature 320:340–342
3. Hoover WG, Ree FH (1968) J Chem Phys 49:3609–3617
4. Carnahan NF, Starling KE (1969) J Chem Phys 51:635–636
5. Hall KR (1972) J Chem Phys 57:2252–2254
6. Young DA, Alder BJ (1979) J Chem Phys 70:473–481
7. Phan SE, Russel WB, Cheng Z, Zhu J, Chaikin PM, Dunsmuir JH, Ottewill RH (1996) Phys Rev E 54:6633–6645
8. Davidchack RL, Laird BB (1998) J Chem Phys 108:9452–9462
9. Antl L, Goodwin JW, Hill RD, Ottewill RH, Owens SM, Papworth S, Waters JA (1986) Colloids Surf 17:67–78
10. Philipse AP, Vrij A (1989) J Colloid Interface Sci 128:121–136
11. Pusey PN, van Megen W, Bartlett P, Ackerson BJ, Rarity JG, Underwood SM (1989) Phys Rev Lett 63:2573–2756
12. Woodcock LV (1997) Nature 385:141–143
13. Dhont JKG, Smits C, Lekkerkerker HNW (1992) J Colloid Interface Sci 152:386–401
14. (a) Schätzel K, Ackerson BJ (1992) Phys Rev Lett 68:337–340; (b) Schätzel K, Ackerson BJ (1993) Phys Rev E 48:3766–3777
15. He Y, Ackerson BJ, van Megen W, Underwood SM, Schätzel K (1993) Phys Rev E 54:5286–5297
16. Harland JL, Henderson SJ, Underwood SM, van Megen W (1995) Phys Rev Lett 75:3572–3575
17. Harland JL, van Megen W (1997) Phys Rev E 55:3054–3067
18. Russel WB (1990) Phase Transitions 21:127–137
19. Palberg T (1999) J Phys Condens Matter 11:323–360
20. Becker R, Döring W (1935) Ann Phys 24:719–752
21. Frenkel J (1946) Kinetic theory of liquids. Oxford University Press, Oxford
22. Kelton KF (1991) In: Ehrenreich H, Turnbull O (eds) Solid state physics, vol 45, Academic, New York, pp 75–177
23. Brugmans MJP, Vos WL (1995) J Chem Phys 103:2661–2669
24. van Duijneveldt (1994) Thesis. Utrecht
25. Heymann A (1997) Thesis. Kiel
26. Ohnesorge R, Löwen H, Wagner H (1994) Phys Rev E 50:4801–4809
27. Kyrlidis A, Brown RA (1995) Phys Rev E 51:5832–5845
28. Curtin WA (1989) Phys Rev B 39:6775–6791
29. Marr DW, Gast AP (1993) Phys Rev E 47:1212–1221
30. Marr DWM (1995) J Chem Phys 102:8283–8285
31. (a) Marr DW, Gast AP (1994) Langmuir 10:1348–1350; (b) Marr DW, East AP (1995) Phys Rev E 52:4058–4062
32. Aastuen DJW, Clark NA, Cotter LK, Ackerson BJ (1986) Phys Rev Lett 57:1733–1736
33. Würth M, Schwarz J, Culis F, Leiderer P, Palberg T (1995) Phys Rev E 52:6415–6423
34. He Y, Olivier B, Ackerson BJ (1997) Langmuir 13:1408–1412
35. Russel WB, Chaikin PM, Zhu J, Meyer WV, Rogers R (1997) Langmuir 13:3871–3881
36. Tokuyama M, Oppenheim I (1995) Physica A 216:85–119
37. Heymann A, Stipp A, Sinn C, Palberg T (1998) J Colloid Interface Sci 207:119–127
38. Underwood SM, Taylor JR, van Megen W (1994) Langmuir 10:3550–3554
39. Heymann A, Sinn C, Palberg T (2000) Phys Rev E 62:813–820
40. Paulin SE, Ackerson BJ (1990) Phys Rev Lett 64:2663–2666
41. Ottewill RH, Livsey I (1987) Polymer 28:109–113
42. Ackerson BJ, Schätzel K (1995) Phys Rev E 52:6448–6460
43. (a) Furukawa H (1984) Physica A 123:497–515; (b) Furukawa (1986) Phys Rev B 33:638–640
44. Percus JK, Yevick GY (1958) Phys Rev 110:1–13
45. Heymann A, Stipp A, Schätzel K (1995) Nouvo Cimento D 16:1149–1157
46. Gránásy L, Iglói F (1997) J Chem Phys 107:3634–3644
47. Bolhuis PG, Kofke DA (1996) Phys Rev E 54:634–643
48. Bartlett P (1998) J Chem Phys 109:10970–10975
49. Henderson SI, Mortensen TC, van Megen W (1996) Physica A 233:102–116
50. Phan SE, Russel WB, Zhu J, Chaikin PM (1998) J Chem Phys 108:9789–9795

Progr Colloid Polym Sci (2001) 118:276–279
© Springer-Verlag 2001

A. N. Rissanou
S. H. Anastasiadis
I. A. Bitsanis
J. de Joannis
C. Mujat
A. Dogariu

The information content of multiple scattering data: Monte Carlo and laboratory experiments

A. N. Rissanou · S. H. Anastasiadis
Foundation for Research and Technology –
Hellas, Institute of Electronic Structure
and Laser, P.O. Box 1527
711 10 Heraklion Crete
Greece and University of Crete
Physics Department
710 03 Heraklion Crete, Greece

I. A. Bitsanis (✉)
Foundation for Research and Technology –
Hellas, Institute of Electronic Structure
and Laser, P.O. Box 1527
711 10 Heraklion Crete, Greece

J. de Joannis
Department of Chemical Engineering
University of Florida, Gainesville
FL 32610, USA

C. Mujat · A. Dogariu
School of Optics/CREOL
University of Central Florida, Orlando,
FL 32816, USA

Abstract A new computer simulation methodology is developed which determines the structure factor, $S(q)$, of concentrated suspensions accurately and subsequently calculates $S(q)$ from the wavelength dependence of the "transport mean free path", $l^*(\lambda)$, for the same systems. Therefore our method can test directly the validity of approximations involved in the analysis of multiple scattering data. The simulation results for l^* agree closely with experimental data and explain the substantial overestimation of l^* by the "photon diffusion" formula [1, 2] observed in some experiments.

Keywords Suspension microstructure · Multiple scattering · Monte Carlo

Introduction

Knowledge of the microstructure of colloidal suspensions is crucial for understanding their physical properties; however, determination of the static structure factor, $S(q)$, by conventional light scattering is severely limited by multiple scattering events, which smear out the information content of the angular intensity distribution of scattered radiation. The traditional way to circumvent this problem, which is widely adopted in the study of polymeric and micellar systems, relies on the selection of specific solvents. This procedure (refractive index matching) recovers approximately single scattering conditions. Unfortunately, the approach is not useful for colloidal suspensions; changing the solvent alters the particle–particle potential; hence, it alters the microstructure of the suspension, the very subject of the inquiry.

Multiple scattering randomizes the direction of photon propagation after a few scattering events and over a characteristic length scale, the "transport mean free path" typically denoted as l^*. l^* is measured by various methods: diffuse transmission spectroscopy [1, 2], diffuse wave spectroscopy [3], enhanced backscattering [4–7], and photon pathlength spectroscopy [8, 9]. The suspension microstructure, i.e. the structure factor, $S(q)$, can, in principle, be determined from the wavelength dependence of l^* [1, 2].

Model and simulation method

The rigorous dependence of the transport mean free path, $l^*(\lambda)$, on the single particle form factor, $F(q,\lambda)$, and $S(q)$ would require a full solution of the associated electromagnetic problem for every suspension microstate [10–11]. Therefore, simplifications have been introduced which result in simpler relations. The most important are

1. $F(q,\lambda)$ is calculated from Mie theory in the far-field limit [11, 12].
2. $\tilde{F}(q,\lambda)$, the multiparticle form factor, is approximated by $F(q,\lambda) \times S(q)$.
3. The radial distribution function, $g(r)$, is assumed to become featureless over distances comparable to l^* [1, 2].

We have developed a computer simulation methodology that can test approximations 1 and 3. Furthermore, by comparison with experiment, it can assess the range of validity of the second approximation and supply insight not accessible by experiment.

Our simulation has three stages. The first stage consists of a Metropolis Monte Carlo simulation enhanced by the incorporation of "linked cells". The use of "linked cells" reduces substantially the computing time, which tends to increase linearly with the number of particles rather than quadratically as in a usual Monte Carlo simulation [13]. The output of this simulation is $g(r)$, $S(q)$ as well as a large collection of "snapshots" (ensemble of microstates) for the many-particle system. The method allows the simulation of fairly large systems that have reached the thermodynamic limit and allows reliable determination of $S(q)$ from the Fourier transform of $g(r)$ (Fig. 1).

The second stage is a "photon" scattering simulation, which allows the "measurement" of l^* for a variety of scattering rules, related to the functional form of $F(q,\lambda)$ and empirical accounts of interference effects. We consider collections of randomly distributed scatterers, which are configurations from stage 1. Photons interact with scatterers one by one. Photons are "inserted" in the center of the periodic box and first are scattered by the particle whose coordinates are closest to the center (core) of the system. The direction of photon propagation after the first scattering is chosen randomly from a probability distribution (phase function) based on Mie theory (far-field limit) and an approximate account of interference effects [10]. Subsequently, the program searches for the nearest scatterer (in terms of its scattering cross-section, σ) along the new direction of photon propagation and the process is repeated until the photon exits the sample (simulation box).

The photon trajectories and the number of scattering events are stored. Averaging over a large number of photons (typically 100–200 per configuration) and a collection of configurations (typically 50) yields a statistically satisfactory result for the photon mean squared displacement (MSD) versus the number of scattering events, N. l^* can now be calculated from the slope of the linear portion of the curve in Fig. 2. In analogy with the Kuhn equivalent chain [14], l^* is related to the MSD through Eq. (1)

$$l^* = \frac{\langle \Delta r^2 \rangle}{N \langle l \rangle} \ , \tag{1}$$

where N is the number of collisions, $\langle \Delta r^2 \rangle$ the mean squared displacement and $\langle l \rangle$ the mean free path between successive scattering events. The linearity of the curve after a few scattering events shows that our systems have reached the "diffusive limit" (Fig. 2).

The final stage, which is currently under development, will involve the determination of $S(q)$ from $l^*(\lambda)$ and comparison with the known and exact $S(q)$ from the first step (Metropolis Monte Carlo). This will require the "inversion" of the photon diffusion formula [1, 2]

$$\frac{1}{l^*} = \frac{\pi \rho}{(k_0 a)^4 k_0^2} \int_0^{2k_0 a} y^2 F(y, k_0 a) S(y) y \ dy, \tag{2}$$

where $y = qa$, ρ is the number density of scatterers, k_0 the wavenumber of the incident radiation and a the diameter of the scatterers. This equation is based on the diffusion approximation [1, 2] and the fact that l^* and l are linked through Eq. (3).

$$\frac{l^*}{l} = \frac{1}{1 - \langle \cos \theta \rangle} \ , \tag{3}$$

where $\langle \cos \theta \rangle$ is the average angle between successive scattering events. Equation (2) involves the serious assumption that l must exceed the distance over which structural correlations persist.

Results and conclusions

Preliminary simulation results and comparisons with experimental data are presented in Table 1, where we list values of l^* from our simulations, experimental data obtained from enhanced backscattering measurements [15] and transmission measurements [1] and the values for l^* predicted by the photon diffusion formula (Eq. 2). The structure factor used in Eq. (2) is the exact $S(q)$ determined from Metropolis Monte Carlo.

The present simulation predictions are very close to the experimental data with a typical deviation of less than 10% and a maximum discrepancy of 20%. On the other hand, the predictions of Eq. (2) in some cases diverge considerably from the experimental values.

The differences between the simulation results and the predictions of Eq. (2) are due primarily to the following two reasons: the simulation incorporates structural correlations not accounted for by Eq. (2) and it accounts for the (correct) distribution of mean free path values, while the derivation of Eq. (2) involves only the average l values.

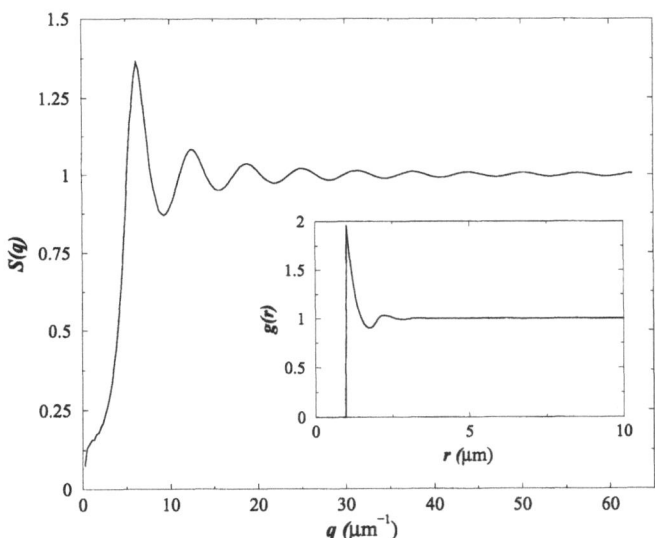

Fig. 1 Structure factor for a system of 6,912 particles with a diameter of 1.0 μm and a volume fraction $\phi = 0.25$. *Inset:* the corresponding pair correlation function for the same system

Fig. 2 Mean squared displacement of photons ($\lambda = 0.633\ \mu$m) in a dispersion of silica spheres in water ($n = n_1/n_2 = 1.095$) with a diameter $a = 1.0\ \mu$m and a volume fraction $\phi = 0.25$. The *straight line* is a linear fit in order to determine the slope and consequently a value for l^*. *Inset*: detail of the initial portion of the curve illustrating the approach to the photon diffusion limit

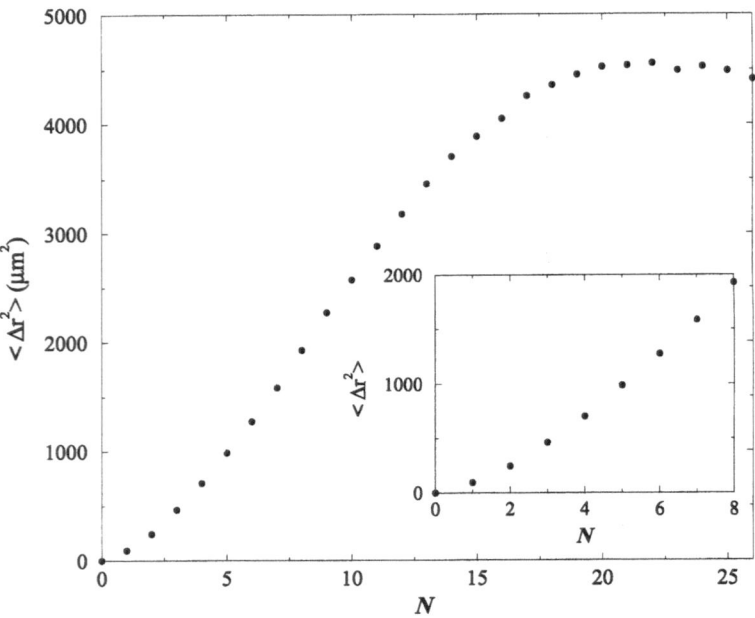

Table 1 Comparison for l^* between simulation, experiment and the photon diffusion formula

ϕ	$a\ (\mu$m$)$	$n = n_1/n_2$	$\lambda\ (\mu$m$)$	$l^*\ (\mu$m$)^a$	$l^*\ (\mu$m$)^b$	$l^*\ (\mu$m$)^c$
0.120	1.0	1.095	0.633	127.4	76.3	75.7[d]
0.250	1.0	1.095	0.633	66.7	42.7	44.3[d]
0.349	0.205	1.192	0.500	9.4	8.5	9.8[e]
0.354	0.299	1.192	0.500	8.6	8.0	9.2[e]
0.299	0.460	1.192	0.500	8.8	7.6	9.2[e]

[a] Photon diffusion formula, Eq. (2)
[b] Computer simulation
[c] Experiment
[d] Data from Ref. [15]
[e] Data from Ref. [1]

The good agreement between experiment and simulation indicates that the far-field approximation for $F(q,\lambda)$ is not crucial and that interference effects can be taken into account in a heuristic way, which is satisfactory when the refractive index contrast is not very high (less than 1.2).

In summary, we have developed a new simulation methodology that allows independent determination of $S(q)$ and l^* in concentrated suspensions. This permits the test of various assumptions involved in the interpretation of multiple light scattering measurement techniques [1–9].

Currently further and more systematic comparisons with experimental data are under way. Moreover, effort is being made to invert Eq. (2) using the $l^*(\lambda)$ from the simulation and to compare the resulting $S(q)$ with the known and exact $S(q)$ determined independently by Metropolis Monte Carlo.

Acknowledgements J.deJ., I.A.B., C.M. and A.D. acknowledge partial financial support from the US National Science Foundation through the Engineering Research Center for Particle Science and Technology at the University of Florida (grant no. EEC-94029890) during the initial stages of this work. S.H.A. and A.N.R. acknowledge that part of this research was sponsored by the Greek General Secretariat of Research and Technology ("PENED" program) and by NATO's Scientific Affairs Division (in the framework of the Science for Peace Programme). The authors thank R. Rajagopalan for many useful discussions.

References

1. Kaplan PD, Dinsmore AD, Yodh AG, Pine DJ (1994) Phys Rev E 50: 4827
2. Fraden S, Maret G (1990) Phys Rev Lett 65:512
3. Maret G, Wolf PE (1987) Z Phys B 65:97
4. Tsang L, Ishimaru A (1984) J Opt Soc Am A 1:836
5. (a) Dogariu A, Uozumi J, Asakura T (1992) Waves Random Media 2:259; (b) Dogariu A, Uozumi J, Asakura T (1996) Opt Rev 3:71
6. Wolf PE, Maret G (1985) Phys Rev Lett 55:2692
7. van Albada MP, Lagendijk A (1988) Phys Rev Lett 61:834
8. Vreeker P, van Albada MP, Sprink R, Lagendijk A (1989) Opt Commun 70:365
9. Popescu G, Dogariu A (1999) Opt Lett 24:7
10. Ishimaru A (1978) Wave propagation in random media Academic, New York
11. van de Hulst HC (1981) Light scattering by small particles, Dover, New York
12. Bohren CF, Huffman DR (1983) Adsorption and scattering of light by small particles. Wiley–Interscience, New York
13. Allen MP, Tildesley DJ (1989) Computer simulation of liquids. Oxford Science, Oxford
14. Doi M, Edwards SF (1986) The theory of polymer dynamics. Oxford Science, Oxford
15. Dogariu A, Kutsche, Likamwa P, Boreman G, Moudgil M (1997) Opt Lett 22:585

Progr Colloid Polym Sci (2001) 118: 280–284
© Springer-Verlag 2001

M. E. Fontanella
R. E. Lechner

Aggregation processes in dense reverse micelles

M. E. Fontanella (✉)
Istituto di Tecniche Spettroscopiche, CNR
via La Farina 237, 98123 Messina, Italy
e-mail: fontanel@hpits1.its.me.cnr.it
Tel.: +39-90-2939528
Fax: +39-90-2939902

R. E. Lechner
Hahn-Meitner-Institut, Berlin, Germany

Abstract The aggregative phenomena taking place in gel-forming and not gel-forming reverse micelles are investigated and compared. In particular, the different growth processes of the micelles (spherical in Aerosol OT based systems and cylindrical in lecithin-based systems) are monitored by the different diffusional properties of the inner-core water molecules, as seen by the neutronic probe in a quasielastic neutron scattering experiment. The different intermicellar aggregation processes taking place in the two kinds of systems have been evidenced through the study of the concentration dependence of the electric conductivity. The results seem to support the proposed idea that the structure of the lecithin gel consists of a percolated network of branched cylinders instead of (as previously proposed) a random entangled network of (not interconnected) polymer-like micelles.

Key words Reverse micelles · Gels · Dielectric relaxations · Electrorheology · Quasielastic neutron scattering

Introduction

The gel-forming character exhibited by a number of lecithin reverse micelle was historically ascribed to the existence of giant cylindrical aggregates that are able to entangle at high enough values of the concentration [1]. However, criticism has recently been made against such a polymer-like approach. First of all, the theoretical prediction for a micellar contour length exponentially scaling with the concentration in the dilute regime was revealed to be wrong [2]: at low values of the volume fraction the micellar size is independent of the concentration and is merely a function of the temperature and of the water content. The results from a small-angle neutron scattering experiment [3], performed on concentrated systems, do not agree with the hypothesis of a random entangled network. Some recent dielectric relaxation experiments [4] on analogous systems seem to suggest a percolated network of branched cylindrical aggregates. Such a hypothesis could solve the puzzle, but the role played by the water molecules in deter-

mining the macroscopic viscosity of the system requires further investigation. In this work tried to obtain further insight into this point by the comparative investigation of gel-forming (lecithin/cyclohexane/water) and non-gel-forming [sodium bis(2-ethylhexyl) sulfosuccinate (AOT)/cyclohexane/water] reverse micellar solutions.

Experimental procedure and discussion of the results

Materials

Soybean lecithin (Epicuron 200) was a gift from Lucas Meyer and was used as received. AOT, D_2O (100% D) and d_{12}-cyclohexane were purchased from Aldrich Chemicals. Water was deionized and bidistilled. AOT and lecithin reverse micelles were prepared at different values of the volume fraction, ϕ, and of the water content, R (R is number of water molecules per surfactant molecule).

Incoherent quasielastic neutron scattering

Incoherent quasielastic neutron scattering (QENS) measurements were performed using the NEAT instrument (BENSC), at room temperature, using an incident wavelength of 5.1 Å and a resolution $\Delta E = 217 \ \mu eV$. The QENS spectra of water confined in the micellar core were obtained by the difference between the corresponding isotopically substituted sample. The spectra obtained were deconvolved as the superposition of symmetric Lorentzian lines.

The half line widths of the resolved translational lines obtained are reported in Fig. 1 as a function of Q^2 (the R parameter refers to the water content). The continuous lines represent the results of the fitting with a jump-diffusion model

$$\Gamma_T(Q) = \frac{D_T Q^2}{1 + D_T Q^2 \tau_0} \ , \tag{1}$$

where D_T is the diffusion constant and τ_0 the residence time before the jump. Both in AOT- and in lecithin-based

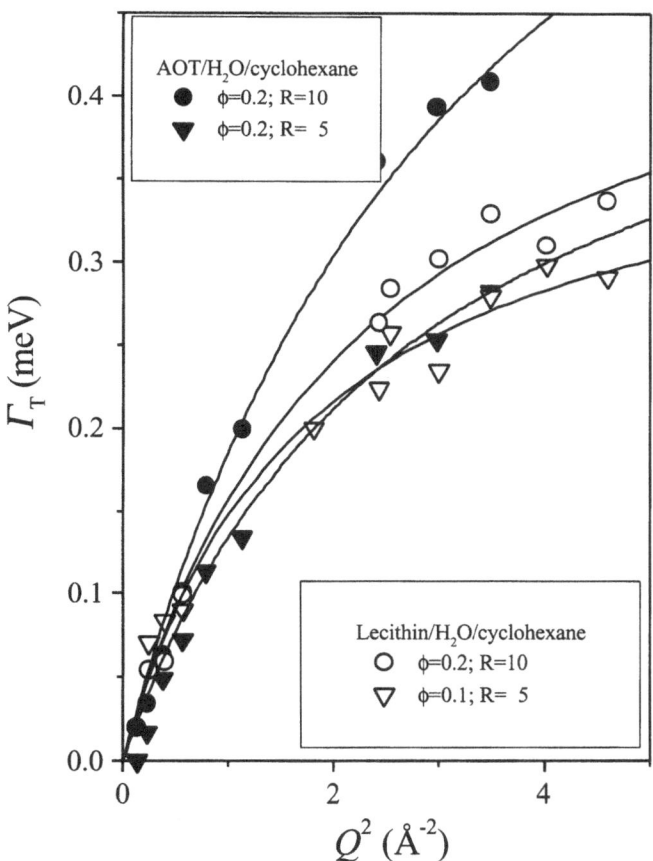

Fig. 1 Half width at half-maximum of the resolved quasielastic neutron scattering contributions from translational motions of water molecules confined in the micellar cores. The *continuous lines* are the fit results with a jump-diffusion model

systems a higher mobility of water is observed as R increases (Table 1). Only in AOT is such a result consistent with a blowing-up process of the micelles. The slight increase in the mobility observed in lecithin reverse micelles has to be interpreted as due to the formation of a second hydration shell, but all the water molecules are involved in the structure of the micellar cross-section and remain entrapped at the micellar surface. That the sol–gel transition is observed above a minimum water content suggests that addition of water promotes some local structural change (branching?).

Dielectric relaxations

Impedance spectra of the lecithin/cyclohexane/water systems were recorded by means of a Chelsea dielectric interface CDI5/L4 coupled with a Schlumberger high-frequency-response SI 1255 analyzer. The dielectric measurements were performed at room temperature, with an applied field of 6 V/cm and in the frequency range 0.01 Hz–1 MHz. The sample cell consists of two concentric stainless steel cylinders; two fused quarts windows are used as the basis. The height of the cell is 3.2 cm, while the external diameter is of 3.5 cm. The gap between the electrodes is about 0.5 cm. The dielectric data were analyzed in terms of a Cole–Cole dispersion model,

$$\varepsilon = \varepsilon' - j\varepsilon'' = \varepsilon_\infty + \frac{\varepsilon_0 - \varepsilon_\infty}{1 + (j\omega\tau)^\beta} \ , \tag{2}$$

where ω is the angular frequency, τ is the characteristic relaxation time, ε_0 and ε_∞ are the permittivities at $\omega\tau << 1$ and $\omega\tau >> 1$, respectively, and β is a parameter describing the broadness of the distribution ($0 \leq \beta \leq 1$; $\beta = 1$ means a single relaxation time). The fitting parameters determined are reported in Table 2. It

Table 1 Extracted jump-diffusion parameters for water confined in the micellar core

Surfactant (composition)	D_T (10^{-5} cm^2/s)	τ_0 (ps)
AOT ($\phi = 0.2$; $R = 5$)	2.3 ± 10%	1.6 ± 16%
AOT ($\phi = 0.2$; $R = 10$)	3.4 ± 8%	0.7 ± 13%
Lecithin ($\phi = 0.2$; $R = 5$)	3.5 ± 13%	1.6 ± 8%
Lecithin ($\phi = 0.2$; $R = 10$)	3.4 ± 11%	1.3 ± 8%

Table 2 Extracted Cole–Cole relaxation parameters for lecithin/cyclohexane/H$_2$O systems of different composition

System composition	τ (μs)	β
$\phi = 0.2$; $R = 10$	5.49	0.28
$\phi = 0.1$; $R = 10$	9.94	0.34
$\phi = 0.05$; $R = 10$	18.8	0.40
$\phi = 0.1$; $R = 6$	27.3	0.37

Fig. 2 Scheme of the experimental setup for Schieleren imaging

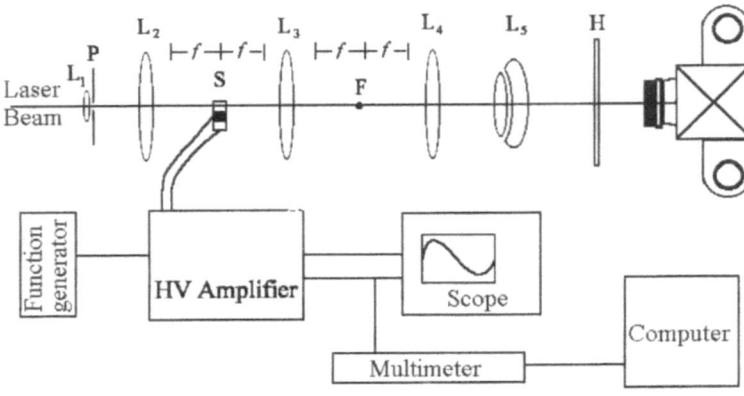

should be observed that the strongest dipoles are localized within the lecithin molecules. At low water content, these are tightly bonded among them to form the micellar interface and no relaxation process connected with the reorientation of the dipoles of the end groups can be observed. An increase in the water content can promote some reorientational relaxation. An increasing probability of the formation of branch points with ϕ could explain the observed dependence of the relaxation parameters by the system composition: at a branch point the mobility of the lecithin end groups is higher.

Impedance measurements and Schieleren imaging

The sample cell was used in connection to a Trek 664 high-voltage amplifier in order to test the effect of the application of high electric fields. The cell was inserted in the confocal point of a Schieleren optic and the changes in density induced by the electric field were recorded using a digital camera. The experimental apparatus is shown in Fig. 2.

The effects of temperature on the conductivity of the systems, at high water contents, were tested by putting the cell in a thermostatic bath. During these measurements the field strength was taken below 100 V/cm in order to maintain the intensity of the current below 200 μA.

The results are summarized in Fig. 3. The temperature dependences of the impedance look very different in the two systems. In AOT the conductivity is due to the free diffusion of ions and/or to the exchange of charge among diffusing aggregates during a collision. As the temperature increases the exchange processes become faster, the viscosity decreases and the conductivity increases. In the case of lecithin the same mechanisms occur but some other processes are probably superimposed that induce a decrease in the conductivity when the temperature increases. A possible explanation could be obtained by assuming that, at low temperature, the main channel of charge transport is represented by a connected infinite

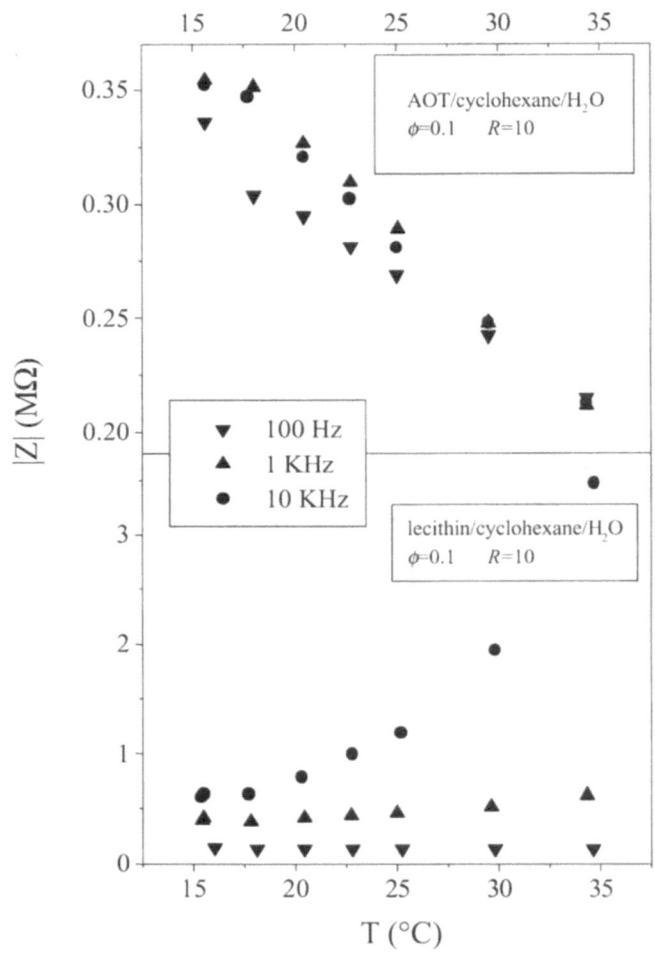

Fig. 3 Temperature evolution of the measured impedances for AOT/cyclohexane/water (*top*) and lecithin/cyclohexane/water (*bottom*) systems

cluster of branched micelles. As the temperature increases the branch points break, inducing the observed lowering of the conductivity. If this hypothesis is correcte the conductivity should scale with the number of

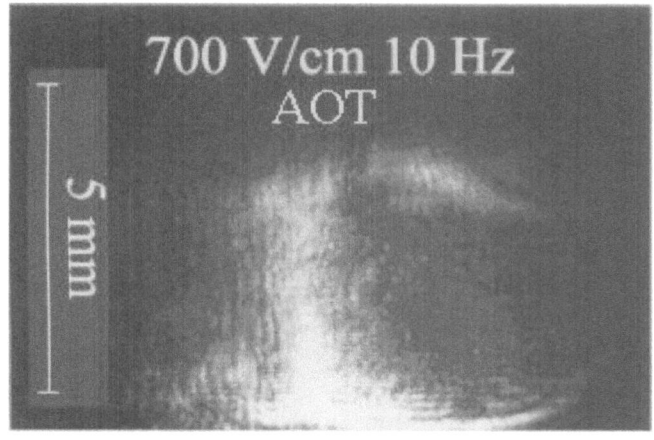

Fig. 4 Schieleren images of systems under an electric field. *Top*: AOT/cyclohexane/water ($\phi = 0.2$, $R = 10$). *Bottom*: lecithin/cyclohexane/water ($\phi = 0.2$, $R = 10$)

branching points, which scale exponentially with temperature.

Such a result was obtained at a frequency of 10 kHz, about the same value at which the loss tangent shows its maximum (see earlier).

The same conclusions can be drawn from the observation of the Schieleren images obtained when an electric field is applied to our samples. When a field at relatively low frequency is applied to the AOT-based system, some weak electrorheological structure is observed on very short times, but it is soon destroyed by the establishment of convective motions in the fluid (Fig. 4). In contrast, in the case of lecithin a well-defined ordered structure is observed on a transient of several minutes(Fig. 4). After this time the regular columnar

aggregates, evidenced for the first time, move slowly and collapse into larger aggregates and become rather disordered as time goes on. The effect could be understood by looking at the temporal behavior of the system conductivity under different electric fields. At low field strengths the conductivity of the system increases with time, suggesting that the applied field induces the formation of a connected structure bridging the electrodes, in a similar fashion to that evidenced by any conventional electrorheologic fluids. At high field strengths, the applied field induces a nonnegligible current through the sample; the temperature of the system increases destroying the interconnectivity.

The observation that the more structured sample (and the more viscous) is characterized by a higher conductivity confirms our working hypothesis of the existence of a structure built up through the formation of branch points.

Concluding remarks

In summary, the description of lecithin-based gels as a percolated network of interconnected micelles seems to be more realistic than that of entangled wormlike micelles, in the light of our results. In particular, we have shown how the differences in the properties of dense systems of spherical and cylindrical micelles come from the different interaction between the water molecules and the hydrophilic part of the surfactant molecule. The idea that the formation of branch points could be responsible for the high viscosity, observed in lecithin reverse micelles at high values of the volume fraction, is reported in the literature as common behavior for this kind of system. As an example, we report that besides the results reported for lecithin/isooctane/water systems [4], recent rheologic measurements on *n*-decane/lecithin/water micelles have pointed out the absence of any correlation between the observed increase in the shear viscosity and the growth process of micellar aggregates [6]. Also in that case the authors assumed that, at high enough R values, the formation of branch points between micelles becomes probable.

At the moment, this remains the only picture able to describe the role played by the water molecules in the establishment of the gel structure.

Acknowledgements We gratefully aknowledge the EC support through the TMR program, for the QENS experiment performed at BENSC.

References

1. Scartazzini R, Luisi PL (1988) J Phys Chem 92:829–833
2. Aliotta F, Fontanella ME, Sacchi M, Vasi C (1997) Physica A 247:247–264
3. Aliotta F, Sacchi M (1997) Colloid Polym Sci 275:910–921
4. Cirkel PA, van der Ploeg JPM, Koper GJM (1998) Phys Rev E 6:6975–6883
5. Aliotta F, Fontanella ME, Lechner RE, Pieruccini M, Ruffle B, Vasi C (1999) Phys Rev E 60:7131–7136
6. Shchipunov YA, Hoffmann H (1998) Langmuir 14:6350–6360

Progr Colloid Polym Sci (2001) 118: 285–289
© Springer-Verlag 2001

María Tirado-Miranda
Artur Schmitt
José Callejas-Fernández
Antonio Fernández-Barbero

Infuence of surface characteristics on fast-aggregating protein-coated polymer colloids

M. Tirado-Miranda (✉)
Seminar of Electricity and Magnetism
Department of Physics
University of Badajoz
Avda. de Elvas s/n
06071 Badajoz, Spain
e-mail: mtirado@ugr.es

A. Schmitt · J. Callejas-Fernández
Biocolloid and Fluid Physics Group
Department of Applied Physics
University of Granada
18071 Granada, Spain

A. Fernández-Barbero
Complex Fluid Physics Group
Department of Applied Physics
University of Almería
04120 Almería, Spain

Abstract A study of fast aggregating surface modified colloidal particles is presented. The surface characteristics of the particles were modified by adsorbing different amounts of bovine serum albumin. Since the protein charge depends on the electrochemical properties of the suspension medium, the influence of the protein net charge on both the cluster structure and the aggregation mechanisms could be studied by changing the pH of the aqueous phase. Physical parameters, such as the fractal dimension, the scaling exponent and the aggregation rate, were determined by laser light scattering. The experimental results show that the adsorbed protein molecules modify the aggregation mechanism and allow structural re-arrangement within the clusters. These effects are more pronounced when the protein molecules bear net charge. Weak flocculation could not be detected, which means that steric stabilization is totally effective. Near the isoelectric point, coagulation and bridging flocculation are the predominant aggregation mechanisms and are diffusion-controlled. Bridging flocculation, however, becomes less effective at pH 9.

Key words Colloidal aggregation mechanisms · Protein adsorption · Fractal structure · Static and dynamic light scattering

Introduction

Macromolecules, irreversibly absorbed on the particle surface, may enhance colloidal stability or induce particle flocculation through a variety of mechanisms [1, 2]. When the particles are fully coated, steric effects impede flocculation and give rise to increased colloidal stability [3, 4]. Bridging flocculation is expected for not completely covered particles since this aggregation mechanism takes place only when a covered part of the surface of one particle collides with the uncovered part of another particle [1, 5, 6]. For totally uncovered particles, the aggregation mechanism is usually referred to as coagulation. In this case, the aggregation mechanism depends on the strength of the particle interactions. For freely diffusing particles, diffusion-limited cluster aggregation (DLCA) is observed, while strongly interacting particles aggregate by reaction-limited cluster aggregation (RLCA) [7, 8]. For flocculation processes of macromolecule-coated polymer colloids, it is still not completely clear how the experimental parameters affect the different aggregation mechanisms and the structure of the resulting aggregates. Furthermore, the particles may aggregate by more than one aggregation mechanism and so it might become quite difficult to quantify the role of each of them.

In this work, we focus on the influence of the macromolecule net charge on the different aggregation mechanisms present in a given flocculation process. Different quantities of macromolecules were adsorbed onto the surface of colloidal particles. Flocculation was induced at high electrolyte concentration and monitored by two independent light scattering techniques. Static light scattering (SLS) allows the cluster fractal dimension to be obtained, while dynamic light scattering (DLS) assesses the time evolution of the mean hydrodynamic radius of the clusters. The experimental aggregation rates were calculated considering the constant kernel solution of Smoluchowski's equation. Finally, the model proposed by Schmitt et al. [9] was used to quantify the contribution of the different aggregation mechanisms to the whole flocculation process.

Materials and methods

The flocculation experiments were carried out using aqueous suspensions of polystyrene microspheres. The particle diameter was 99 ± 1 nm and the polydispersity index was 0.09 ± 0.02 as determined by photocorrelation spectroscopy. A particle surface charge density of -2.7 ± 0.1 μCcm^{-2} was measured by conductometric titration. The negative particle charge arises from dissociated sulphate groups on the surface. The colloidal stability was estimated by determining the critical coagulation concentration (ccc) from the time evolution of the scattered intensity. The value of the ccc obtained was 0.495 ± 0.007 M KCl.

Commercially available bovine serum albumin (BSA, Pentex) was chosen as the adsorbing macromolecule. Its state of aggregation was checked by native polyacrylamide gel electrophoresis. BSA dimers were found as the main component of the sample although a small fraction of high-molecular-weight aggregates could also be detected. The only treatment prior to adsorption was a cleaning step by dialysis. The protein was then used without further purification. For the protein adsorption experiment, different amounts of BSA were added to a fixed quantity of buffered colloidal suspension. The pH of the suspension was fixed close to the isoelectric point of the BSA molecules, i.e. pH 4.8, in order to facilitate adsorption. At this pH, the BSA molecules possess a compact structure. This structural organization is partially lost when the protein spreads on the sorbent surface, leading to a net increase in the system entropy [10]. This is why maximum adsorption is usually achieved near the isoelectric point of BSA [11, 12]. Furthermore, far from the isoelectric point, lateral intermolecular interactions become important and tend to impair adsorption. The corresponding adsorption isotherm showed the high affinity of BSA. At high BSA concentration, a final plateau was reached, which indicated the adsorption of a complete monolayer. This gave us the possibility to obtain particles with a well-known degree of surface coverage by simply controlling the amount of added protein. Samples with 0, 25, 50, 75 and 100% of its surface covered by BSA molecules were selected. In order to study the reversibility of adsorption, the BSA–latex complexes were centrifuged. The supernatants were filtered, using a filter of extremely low protein affinity, and measured spectrophotometrically. No protein was detected in the supernatants and, therefore, the initially adsorbed amount of BSA remains invariant.

Flocculation was induced by mixing equal amounts of buffered electrolyte solution and sample by means of a Y-shaped mixing device. The final electrolyte concentration was 0.700 M KCl. The initial particle concentration in the reaction vessel was 1.6×10^{10} cm^{-3} and the temperature was stabilized at 25 ± 1 °C. The pH was set to 4.8 and 9 by low-ionic-strength acetate and borate buffers, respectively.

The flocculation processes were monitored by performing simultaneous SLS and DLS measurements [13]. The mean scattered light intensity, I, was assessed as a function of the scattering vector, $q = (4\pi/\lambda)\sin(\theta/2)$, within an angular range from 10° to 150°. Here, λ is the medium wavelength and θ is the scattering angle. The scattered intensity function contains information about the aggregate structure in the q region corresponding to the characteristic structure length. For fully developed fractal clusters, a power law in q is predicted, $I(q) \sim q^{-d_f}$, which is valid for $\langle R_h \rangle^{-1} < q < \langle R_0 \rangle^{-1}$

[14]. Using this relationship, the fractal dimension may be determined from the slope of the double-logarithmic $I(q)$ plot.

DLS was employed for measuring the scattered intensity autocorrelation function at a scattering angle of 60°. The Siegert relationship was employed to convert it into the scattered field autocorrelation function [15]. The latter was evaluated by the cumulant method according to Koppel [16]:

$$\ln g_{field}(\tau) = -\mu_1 \tau + \mu_2 \left(\frac{\tau^2}{2}\right) + \mu_3 \left(\frac{\tau^3}{3!}\right) + \cdots \qquad (1)$$

For polydisperse systems, the first cumulant, μ_1, is related to the mean particle diffusion coefficient by $\mu_1 = \langle D \rangle q^2$. From this, the mean hydrodynamic radius, $\langle R_h \rangle$, was calculated using the Einstein–Stokes equation [15].

Theoretical background

The kinetics of aggregation processes in colloidal suspensions may be described by the time evolution of the aggregate size distribution, $N_n(t)$. For dilute colloidal systems, where only binary collisions have to be taken into account, Smoluchowski proposed the following system of rate equations [17]:

$$\frac{dN_n}{dt} = \frac{1}{2} \sum_{i+j=n} k_{ij} N_i N_j - N_n \sum_{k=1}^{\infty} k_{nk} N_k \ . \qquad (2)$$

The kernel, k_{ij}, represents the rate at which i-mers bind to j-mers. It contains all the physical information about the aggregating system and may be interpreted in terms of a sticking probability for two clusters diffusing towards one another. Most coagulation kernels used in the literature are homogeneous functions of i and j, at least for large i and j. Van Dongen and Ernst [18] introduced a classification scheme for these types of kernels, based on the relative probabilities of large clusters sticking to large clusters. Different growth kinetics and size distributions are obtained depending on which of these unions dominate. They defined the homogeneity parameter, λ, that describes the tendency of a large cluster to bind to another large cluster and governs the overall rate of aggregation. It should take the value 0 for DLCA and 1 for RLCA. In this work, λ is used to characterize the aggregation mechanism.

For long times and large clusters, the cluster size distribution exhibits dynamic scaling and approaches the limiting form [14]:

$$N_n(t) \propto s^{-2} \phi\left(\frac{n}{s}\right) \ , \qquad (3)$$

where $s = s(t)$ is related to the number-average mean cluster size. $\phi(x)$ is a universal time-independent cluster size distribution which characterizes the aggregation mechanism. This relationship indicates that the relative shape of the cluster size distribution remains constant during the whole aggregation process. Considering the dynamic scaling solution (Eq. 3) and the fractal cluster morphology, $\langle R_h \rangle$ of the aggregates is given by [19]

$$\langle R_{\rm h}(t)\rangle = \langle R_0\rangle t^{\frac{1}{d_{\rm f}(1-\lambda)}} \ , \qquad (4)$$

where $\langle R_0\rangle$ is the monomer radius and $d_{\rm f}$ is the fractal dimension. This relationship will be used to determine λ from the experimental time evolution of the mean hydrodynamic radius of the aggregates.

Using the dynamic scaling solution (Eq. 3) and considering the αth moment of the cluster size distribution, Olivier and Sorensen [20] obtained the following relation for the first cumulant of the intensity autocorrelation function:

$$\mu_1(t) = \mu_1(0)\left(1+\frac{t}{t_{\rm c}}\right)^{\frac{-1}{d_{\rm f}(1-\lambda)}} \ , \qquad (5)$$

where $t_{\rm c}=2/c_0 k_{\rm s}$ is the characteristic aggregation time. $t_{\rm c}$ is expressed as a function of the initial particle concentration, c_0, and the Smoluchowski rate constant, $k_{\rm s}$. Equation (5) allows $t_{\rm c}$ to be obtained once $d_{\rm f}$ and λ are known. After that, $k_{\rm s}$ is determined from $t_{\rm c}$ using the known initial particle concentration [21]. As we will see later, $k_{\rm s}$ helps to characterize the different aggregation mechanisms present in a sample.

Results and discussion

Cluster structure and dynamic scaling

A series of aggregation experiments was carried out for a system of particles with different degrees of surface coverage. As described earlier, all the samples were aggregated at high electrolyte concentration. $d_{\rm f}$ is shown in Fig. 1a as a function of the surface coverage, θ, for the aggregation measurements carried out at pH 4.8 and 9.

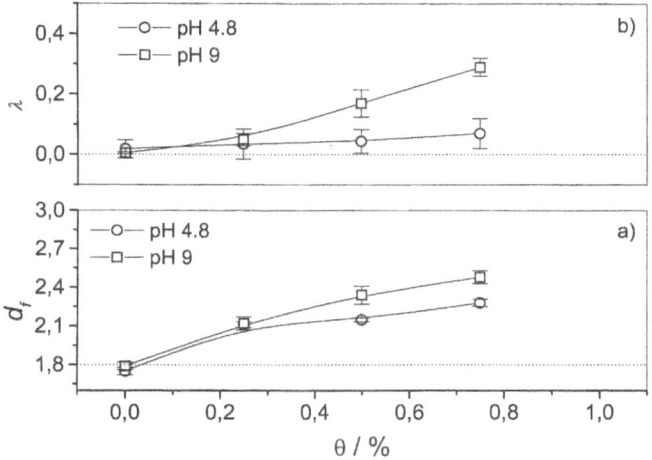

Fig. 1 a Fractal dimension and **b** homogeneity parameter as a function of the degree of surface coverage. The *continuous lines* are drawn as a guide for the eye

For uncovered particles, the experimental fractal dimension falls well within the range 1.7–1.8 expected for structures growing by pure cluster diffusion (DLCA) [22]. For the completely covered particles, the aggregation processes were so slow that the asymptotic behaviour for the mean scattered intensity was not reached; hence, the experimental fractal dimension could not be determined. This indicates that the colloidal particles are almost completely stabilized by steric effects. The experimental behaviour for intermediate degrees of surface coverage will be discussed later.

DLS was employed for monitoring the time evolution of the mean hydrodynamic radius of the clusters, $\langle R_{\rm h}\rangle(t)$, for each aggregation process. According to Eq. (4), the scaling exponent, λ, was calculated from the slope using the fractal dimension obtained by the independent SLS technique. In Fig. 1b, the homogeneity parameter is plotted versus the degree of surface coverage. For $\theta=0\%$, λ is close to 0 and thus the clusters form bonds in a pure diffusion process (DLCA). At high BSA coverage, however, no power-law dependence was found and so the measurement of λ was not possible.

As stated previously, steric stabilization predominates for particles with a high degree of surface coverage, while uncovered particles aggregate in the diffusion-limited regime. So, the question of what happens at intermediated surface coverage arises. As can be observed in Fig. 1a, the cluster fractal dimensions increase with the degree of coverage in all cases. This means that the adsorbed protein molecules affect the spatial mass distribution within the aggregates. The fractal dimensions even reach values higher than 2.1 which is normally reported in the literature for pure RLCA. In order to study this anomalous behaviour in more detail, the corresponding homogeneity parameter was also analysed. As can be seen in Fig. 1b, the behaviour of λ depends strongly on the pH of the aqueous phase. Near the isoelectric point of the BSA molecules, i.e. at pH 4.8, λ is always very close to 0 and hence the aggregation processes are almost completely diffusion controlled. The high fractal dimensions measured at pH 4.8 may be understood in terms of internal rearrangement of the clusters. In fact, a short-range steric barrier appears between the adsorbed macromolecules and prevents the particles from coming into direct contact [23]. This steric barrier allows internal rearrangement within the clusters but does not affect the aggregation mechanism because the latter is only controlled by long-range electrostatic forces. So, very compact clusters may grow even from freely diffusing particles. This structural aspect of cluster formation is discussed extensively in Ref. [24]. At pH 9, however, the homogeneity exponent grows monotonically for increasing degrees of surface coverage and reaches values up to 0.3. This implies that the aggregation processes drift away from the diffusion-limited regime and so additional repulsive interactions should be present. The origin of these may be found in the fact that at pH 9 the BSA molecules are negatively charged. Hence, additional electrostatic repulsion effects appear and the clusters formed tend to be more compact. This fact may also explain why the fractal dimensions are always higher than those observed at pH 4.8 for diffusion-controlled aggregation processes.

Aggregation mechanisms

In order to understand the results in more detail, one should take into account that the aggregation probability for a colliding pair of colloidal particles depends not only on the forces acting between them but also on whether the colliding part of the surface is covered by macromolecules or not. The probability of finding a covered surface patch is given by the fractional surface coverage, θ, and the probability of finding a bare part by $(1 - \theta)$. So, three possible aggregation mechanisms can be distinguished:

1. Coagulation, i.e. collision of two uncovered patches of the surface. This process corresponds to the case of aggregation of bare particles. The aggregation rate for this configuration is denoted k_c.
2. Weak flocculation, i.e. two covered patches of the surface collide. For this aggregation mechanism, the aggregation rate is k_{wf}.
3. Bridging flocculation. Here the collision of an uncovered part of one particle with the covered part of another particle occurs. In this configuration, a "macromolecule bridge" is formed between the particles. This mechanism is characterized by k_{bf}.

Several theoretical models have been development in order to explain the relation between the aggregation rate and the degree of surface coverage. The La Mer model [25] assumes that all particle collisions are completely diffusion controlled and stable bonds may be formed only in a bridging configuration. So, k_s should be proportional to the number of free sites on one particle and also to the number of occupied sites on the other. Consequently, k_s should be proportional to $\theta(1 - \theta)$. This relationship implies that maximum flocculation occurs at half surface coverage and that there is no flocculation at all for uncovered and totally covered particles since La Mer's model does not consider aggregation mechanisms other than bridging. More detailed models were proposed by Hogg [26], Ash and Clayfield [27], Moudgil et al. [28] and Molski [29]. All of them account for additional aggregation mechanisms and, therefore, may be considered as an extension of the La Mer model. Nevertheless, all of them show the common feature that bridging flocculation is the most efficient aggregation mechanism, independently of the experimental conditions. So, an extended model

which considers an independent collision probability for each aggregation mechanism was proposed by Schmitt et al. [9]. The total aggregation rate may be written as the sum of all contributions:

$$k_s = k_c(1 - \theta)^2 + k_{wf}\theta^2 + 2k_{bf}\theta(1 - \theta) , \qquad (6)$$

where the factor 2 considers the collision of an uncovered part of one particle with the covered part of another particle and the reverse case. This relationship allows the contribution of the different aggregation mechanisms to be quantified if k_s is measured as a function of θ.

The experimental values for k_s were obtained from the DLS data using Eq. (5). As can be seen in Fig. 2, the experimental aggregation rates decrease for increasing degrees of surface coverage and no maximum at intermediate surface coverage is observed; therefore, one intuitively suspects that no bridging flocculation takes place. For uncovered particles, k_s is very close to the value of $(6 \pm 3) \times 10^{-12}$ cm^{3}s^{-1} commonly accepted for diffusion-limited aggregation [30]. This indicates that all collisions between bare patches of the particle surface lead to aggregation and that the sticking probability in the coagulation configuration is unity.

In order to separate and quantify the contribution of the different aggregation mechanisms, Eq. (6) was used to fit the experimental data. As a boundary condition, k_c was identified with the experimental aggregation rate for $\theta = 0\%$. The best fits are included in Fig. 2 as continuous lines. The corresponding fitting parameters are given in Table 1. At pH 9, the BSA molecules are negatively charged and, hence, increase the surface charge of the colloidal particles. Under these conditions, the mechanism of coagulation is almost three times as effective as bridging flocculation. Weak flocculation is, however, not detected. So, steric stabilization should be totally effec-

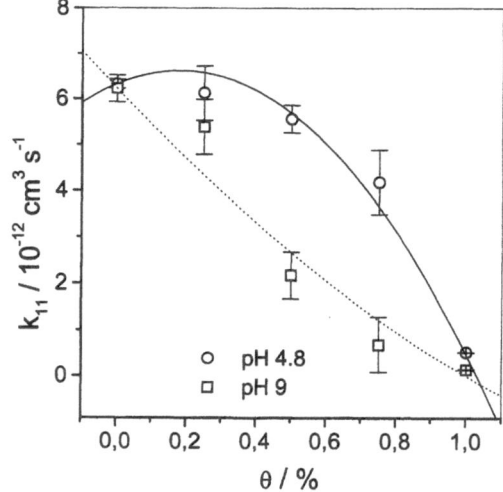

Fig. 2 Aggregation rate versus degree of surface coverage. The *lines* show the best fits according to Eq. (6)

Table 1 Aggregation rates for the different aggregation mechanisms

pH	k_c (10^{-12} cm^{3}s^{-1})	k_{wf} (10^{-12} cm^{3}s^{-1})	k_{bf} (10^{-12} cm^{3}s^{-1})
4.8	6.3 ± 0.5	0.5 ± 0.4	7.9 ± 0.8
9	6.2 ± 0.3	0.0 ± 0.4	2.2 ± 0.8

tive. At pH 4.8, the BSA molecules do not alter the net charge of the colloidal particles. Here, bridging flocculation is as effective as coagulation and even a small fraction of collisions in weak flocculation configuration leads to aggregation.

Conclusions

In this work, the effect of the net charge of the adsorbed protein molecules on the structure and the aggregation mechanism of clusters was studied. d_f, λ and k_s were determined as a function of the degree of coverage at high electrolyte concentrations and different pH. The fractal dimensions always grow with the degree of BSA coverage. This means that the spatial mass distribution within the aggregates depends on the surface characteristics of the particles. The values obtained indicate a rearrangement within the clusters owing a short-range steric barrier. The rearrangement processes are more significant when the protein molecules bear net charges. The experimental λ values show that both the amount of adsorbed protein molecules and the net charge thereof affect the aggregation regime. For uncovered particles and particles covered with uncharged protein molecules, the sticking probability is almost independent of the cluster size. Finally, k_s was determined as a function of the surface coverage and the contributions of the different aggregation mechanisms (coagulation, weak flocculation and bridging flocculation) were separated and quantified. Under the experimental conditions of this study, aggregation in the coagulation configuration is always diffusion-controlled. The bridging flocculation rate is similar to coagulation at pH 4.8 but is three times smaller at pH 9. Weak flocculation is almost completely absent.

Acknowledgement Financial support by the coordinated CICYT projects MAT2000-1550-C03-01, MAT2000-1550-C03-02 and MAT2000-1550-C03-03 is gratefully acknowledged. We thank Manuel Quesada-Pérez for kindly supplying the latex particles.

References

1. Dickinson E, Eriksson L (1991) Adv Colloid Interface Sci 34:1
2. Fernández-Barbero A, Cabrerizo-Vílchez MA, Martínez-García R, Hidalgo-Alvarez R (1993) Prog Colloid Polym Sci 93:269
3. Gregory F, Sheiham I (1974) Polym J 6:7
4. Csempesz F, Rohrsetzer S (1988) Colloids Surf 31:215
5. Pelssers EG, Cohen Stuart MA, Fleer GJ (1990) J Chem Soc Faraday Trans 86:1355
6. Ruehrwein RA, Ward A (1952) Soil Sci 73:485
7. (a) Lin HY, Lindsay HM, Weitz DA, Ball RC, Klein R Meakin P (1989) Nature 339:360; (b) (a) Lin HY, Lindsay HM, Weitz DA, Ball RC, Klein R Meakin P (1990) J Phys Condens Matter 2:3093
8. Asnaghi D, Carpineti M, Giglio M, Sozzi M (1992) Phys Rev A 45:1018
9. Schmitt A, Fernández-Barbero A, Cabrerizo-Vílchez MA, Hidalgo-Alvarez R (1998) Prog Colloid Polym Sci 110:105
10. Norde W, Lyclema J (1991) J Biomater Sci 2:183
11. Shirahama H, Takeda K, Suzawa T (1986) J Colloid Interface Sci 109:552
12. Molina-Bolivar J A, Ortega-Vinuesa J L (1999) Langmuir 15:2644
13. Tirado-Miranda M, Schmitt A, Callejas-Fernández J, Fernández-Barbero A (1997) Prog Colloid Polym Sci 104:138
14. Julien R, Botet R (1987) Aggregation and fractal aggregates. World Scientific, River Edge, NJ
15. Pecora R (ed) (1985) Dynamic light scattering. Plenum, New York
16. Koppel DEJ (1972) Chem Phys 57:4814
17. Smoluchowski MV (1917) Z Phys Chem 92:129
18. Van Dongen PGJ, Ernst MH (1985) Phys Rev Lett 54:1396
19. Ziff RM (1984) In: Family F, Landau DP (eds) Kinetics of aggregation and gelation. North-Holland, Amsterdam, pp 191-199
20. Olivier BJ, Sorensen CM (1990) J Colloid Interface Sci 134:139
21. Tirado-Miranda M, Schmitt A, Callejas-Fernández J Fernández-Barbero A (1998) Prog Colloid Polym Sci 110:110
22. Bolle G, Cametti C, Codastefano P Tartaglia P (1987) Phys Rev A 35:837
23. Vincent B, Edwards J, Emment S, Jones A (1986) Colloids Surf 18:261
24. Tirado-Miranda M, Schmitt A, Callejas-Fernández J Fernández-Barbero A (1999) Langmuir 15(10):3439
25. La Mer VK (1966) Discuss Faraday Soc 42:248
26. Hogg RJ (1984) Colloid Interface Sci 102:232
27. Ash SG, Calyfield EJ (1976) J Colloid Interface Sci 55:245
28. Moudgil BM, Shah BD, Soto HS (1987) J Colloid Interface Sci 119:466
29. Molski A (1989) Colloid Polym Sci 267:371
30. Sonntag H, Strenge K (1987) Coagulation kinetics and structure formation. Plenum, New York

Progr Colloid Polym Sci (2001) 118: 290–294
© Springer-Verlag 2001

P. Levitz
A. Delville
E. Lecolier
A. Mourchid

Liquid–solid transitions of disklike colloids: stability and jamming

P. Levitz (✉) · A. Delville · E. Lecolier
A. Mourchid
Centre de Recherche sur la
Matière Divisée (CRMD)
CNRS, 1b rue de la Ferollerie 45071
Orleans cédex 02, France

P. Levitz
LPMC, Ecole Polytechnique
91128 Palaiseau, France
e-mail: levitz@pmc.polytechnique.fr

Abstract Several suspensions of
disklike colloids undergo puzzling
liquid–solid transitions. The case
of laponite, a synthetic clay, is a
demonstrative example which has
recently generated a good deal of
experimental and theoretical study.
At relatively high ionic strength,
above 10^{-4} M, a transition from a
liquid to a soft solid appears.
Typical nematic defects are then
observed. The mechanism and the
status of this transition is currently
actively debated. In this regime, the
Debye screening length, defining the
range of electrostatic repulsion, is
always smaller than the particle size,
which controls short-range anisotropic interactions. These two contributions are then mixed in a
nontrivial way. One possible way to
uncouple these interactions is to
lower the ionic strength. In the first
part of this article, we show that a
liquid–solid transition exhibiting a
correlation peak is observed at very
low ionic strength. Such a structural
transition involves long-range electrostatic stabilization and/or jamming which is compatible with a
Wigner glass transition. However,
3D off-lattice reconstructions of
these soft solids from small-angle
scattering spectra suggest that the
particle distribution is not homogeneous in space. In the second part of
this work, the interaction potential
of different pair configurations is
computed using a basic charge distribution model. Such a stability
analysis raises interesting questions
about long-term stabilization and/or
coagulation, (very) slow structural
relaxation and nematic defects generally observed for these colloidal
suspensions at relatively high ionic
strengths.

Key words Complex fluids
Colloid glass · Simulated
annealing Laponite · Clay
suspensions

Introduction

Structural transitions of colloidal suspensions of spherical or cylindrical particles are now well documented [1,
2]. Much less work has focused on suspensions of
platelike colloids [3, 4]. Our aim is to discuss liquid–
solid transitions of charged disklike particles in relation
to particle jamming and colloidal stability. The system
under investigation is made of synthetic smectite clay
particles named laponite. Individual particles can be
considered, on average, as a negatively charged disk
1-nm thick and having an average diameter ranging from
25 to 30 nm. This material has recently generated many
experimental and theoretical studies [5–15]. The status
and the origin of a so-called phase or state diagram of
laponite in water is still under debate. Interesting enough,
above an ionic strength of 10^{-4} M, the Debye screening
length, controlling the range of electrostatic repulsion, is
always smaller than the particle diameter, which controls, in part, anisotropic excluded-volume effects [7, 12].
In the upper part of the phase diagram (Fig. 1), long-
range and anisotropic short-range interactions are then
mixed in a nontrivial way. One possible way to uncouple
these interactions is to lower the ionic strength. In such a

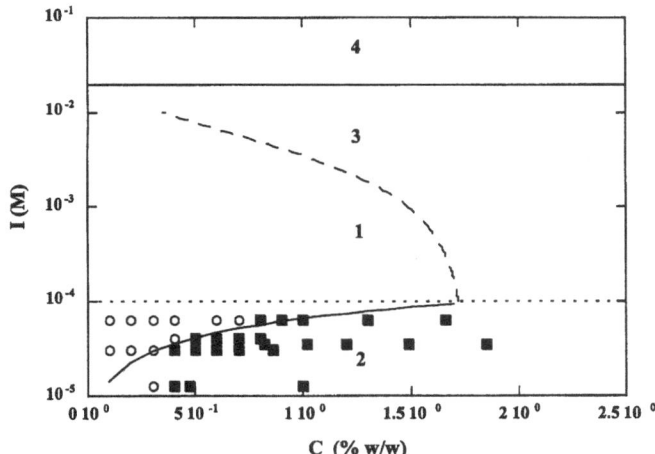

Fig. 1 Phase behaviour of laponite suspensions. As formerly published [7], region 4 is related to a macroscopic flocculation regime. The *line* separating region 1 and 3 is related to the liquid–solidlike transition observed at high ionic strength [7, 8]. The present study concerns effective ionic strengths (I) lower than 10^{-4} M. *Circles*: liquid suspensions; *squares*: solidlike suspensions. The *full line* is the prediction of the liquid–solid transition using a dressed particles model (see text)

case, long-range electrostatic repulsion should dominate and the suspension of dressed disklike particles should behave as a collection of soft repulsive spheres. The main consequence should be the appearance of a transition line at low ionic strength (below 10^{-4} M) where movements of each particle start to be strongly hindered (jamming) by mutual electrostatic repulsions. In the first part of this work, we demonstrate experimentally the existence of such a liquid–soft solid transition at very low ionic strength. In the second part, we provide a basic analysis of the colloidal stability of laponite suspensions, which raises interesting questions about long-term stabilization and/or coagulation, (very) slow structural relaxation and nematic defects generally observed for these colloidal systems at relatively high ionic strengths (above 10^{-3} M).

Experimental

Materials

We used Laponite RD, a synthetic hectorite manufactured by Laporte. The chemical analysis of the laponite was performed by the Service Central d'Analyse (Vernaison) of the CNRS by using inductively coupled plasma spectrometry. The mean chemical composition of this clay is SiO_2, 65.82%; MgO, 30.15%; Na_2O, 3.20%; LiO_2, 0.83%, corresponding to the general formula $Si_{8.00}(Mg_{5.45}Li_{0.40})H_4O_{24}Na_{0.75}$. The particles of laponite bear a negative structural charge. The measured cationic exchange capacity (CEC) was measured around 750 mM/kg [6, 7], in good agreement with the chemical analysis. Individual particles can be considered as flat disk having a thickness of 1 nm and a diameter from 25 to 30 nm. Such an effective shape permits the correct fitting of the scattering spectra of very dilute suspensions of laponite [5–7].

Sample preparation

The samples were prepared, under nitrogen, by adding the laponite powder to an aqueous solution at pH 10. To decrease the ionic strength, ion-exchange resins (Amberlyst A27 and 15, OH^- and H^+ forms, respectively) were added directly to a dilute liquid suspension of laponite. This procedure of purification is the easiest to perform and one of fastest. The possibility of dissolution of the laponite particles and/or of sample contamination is thereby markedly reduced. Once the resins had mixed with the suspension, the sample tube was agitated until the resin particles did not sediment any more. Then, the resins were separated from the colloidal suspension. At this stage, the addition of ion-exchange resins to a liquid suspension of laponite ($I = 10^{-4}$ M, pH 10) leads to a decrease in the ionic strength and promotes the appearance of a soft solid, as expected. To fix the ionic strength of our suspensions, the laponite dispersions were put in a dialysis membrane (Visking) having a molecular-weight cutoff of 12,000–14,000. The membrane was previously rinsed with deionized water. The bag containing the deionized laponite suspension was immersed in a large volume of aqueous solution at the required ionic strength (NaOH) ranging between 10^{-5} and 10^{-4} M. The approach to equilibrium of the salt chemical potential was followed through measurements of the electrical conductivity of the reservoir. The sodium hydroxide solution of the reservoir was renewed until the conductivity remained constant. Then, the dialysis was stopped. Visual inspection showed that the dialysis bags were never strongly inflated. The suspension contained in the dialysis bag was divided into two parts: the first one was used to determine, by weight loss, the final solid concentration, c(w/w), and the second part was diluted with the water of the reservoir to obtain samples with different laponite concentrations. The value of I was used as an effective ionic strength to draw the lower part of the phase diagram. Looking at the sample preparation, this value of I does not provide the salt chemical potential directly [16]. Each horizontal line in Fig. 1 represents a set of dilute suspensions from the same initial sample. In relation to the chemical dissolution problem encountered with laponite [6–14], we analysed the chemical composition of laponite particles inside a deionized suspension. These extensive analyses show that the deionization process does not alter laponite particles (same chemical composition). Dissolution of Mg^{2+} is not observed and the CEC is similar to the original one.

Rheology

Rheological measurements were performed with an imposed-stress rheometer (Cari-Med) in order to follow the liquid–soft solid transition.

Ultra-small-angle X-ray scattering

Ultra-small-angle X-ray scattering (USAXS) experiments were performed with a two-crystal multiple-reflection camera using a quasilinear collimation [17] (SCM-CEA Saclay). A modified iterative method of Lake [18] was used to desmear the experimental data. In order to minimize noise propagation, the best Pade approximant of the test function was computed at each iteration. Angular integration (Eq. 15 of Ref. [17]) of the analytical integrand is thus performed using a Gaussian quadrature.

Jamming transition at very low ionic strength

Lowering the effective ionic strength (e.g. from $6.3 \cdot 10^{-5}$ to $3.0 \cdot 10^{-5}$ M at constant solid concentration, $c = 0.4\%$) induces a transition from a viscous fluid to a soft solid having a small yield stress [15]. We checked the reversibility of this transition by dialysing the deionized dispersion against aqueous sodium hydroxide solution at pH 10. As shown in Fig. 1, a transition line from a liquid to a soft solid is observed for $I < 10^{-4}$ M; it has a positive slope. Around $I = 10^{-4}$ M, the average line separating the liquid from the soft solid nicely converges to the transition line already observed at higher ionic strengths [7]. It is possible to rationalize these results using very basic arguments. The soft repulsion owing to the overlap of the counterion diffuse layers is replaced by a renormalized hard-core repulsion with an effective radius equal to $\alpha\kappa^{-1} \cdot \kappa^{-1}$ is the Debye screening length approximated by $\kappa^{-1} = (\varepsilon kT/2e^2 I)^{1/2}$. α is a positive parameter taking into account the softness of the repulsion potential. The liquid–solid transition occurs when effective spherical particles start to strongly overlap. This jamming appears for a volume fraction close to 0.5. As shown in Fig. 1, it is possible, for $\alpha = 0.74$, to draw a line which correctly predicts the liquid–solid transition. The exact status of this transition is still unclear; however, recent work of Bonn et al. [12] on the particle dynamics of laponite suspensions prepared at $I = 10^{-4}$ M (just located at the "nose" of the transition line) gives several convincing arguments in favour of a colloidal "Wigner" glass transition.

The structure of these soft solids (degree of ordering, correlation, microsegregation) was investigated using absolute-scale USAXS. Experiments were performed for different solid suspensions at the same effective ionic strength ($I = 3 \times 10^{-5}$ M). In Fig. 2a, we have plotted the evolution of the effective structure factor, mainly the ratio $S^{\text{eff}}(q) = I_e(q)/I_p(q)$, where $I_e(q)$ is the experimental spectrum on an absolute scale and $I_p(q)$, related to the individual particle form factor, is given by Eqs.(2), (3) and (4) of Ref. [7]. Above $q = 3 \times 10^{-2}$ Å^{-1}, $S^{\text{eff}}(q)$ converge nicely to unity. This result means that absolute-scale measurements are in good agreement with the chemical analysis. From 10^{-3} to 10^{-2} Å^{-1}, we observe that the scattering intensity of the soft solid is above $I_p(q)$. A correlation peak can even be observed, compat-

ible with long-range electrostatic stabilization. Such a result is new and strongly contrasts with the evolution of the scattering spectra of laponite solids at higher ionic strength (above 10^{-4} M). In the upper part of the phase diagram, strong negative deviations from $I_p(q)$ were observed in the same q range by several authors [7, 11]. As shown elsewhere, these correlations are directly related to excluded-volume interactions between disklike particles [7]. Depending on the particle concentration, the amplitude of these correlations can certainly be associated with a more-or-less extended microsegregation.

Structure of the Wigner glass

At low ionic strength, the position of the correlation peak, q_c, shifts to high q values as the particle concentration increases. The correlation length associated with the peak, $d_c = 2\pi/q_c$, is higher than the estimated average distance, d_h, between particles when a homogeneous distribution in space is assumed. For $c = 0.82\%$, we find $d_c = 140$ nm and $d_h = 63$ nm, and for $c = 1.85\%$, $d_c = 80$ nm and $d_h = 47$ nm. For soft solids relatively far from the transition line, these preliminary results suggest a nonhomogeneous particle distribution and a

Fig. 2 Evolution of the effective structure factor, $S^{\text{eff}}(q) = I_e(q)/I_p(q)$, for different solidlike suspensions at $I = 3 \cdot 10^{-5}$ M. The particle diameter is set at 30 nm. *Squares*: $c = 0.82\%$; *circles*: $c = 1.02\%$; *triangles*: $c = 1.85\%$. The *continuous line* shows a formerly published result [7] of a solidlike suspension of laponite at higher concentration ($c = 3.85\%$) and higher ionic strength ($I = 10^{-2}$ M). **B** Ultra-small-angle X-ray scattering (*USAXS*) experimental data collected on a solidlike suspension of laponite ($c = 0.82\%$, $I = 3 \cdot 10^{-5}$ M). *Squares*: desmeared curve $I_e(q)/K_c$, where K_c is the contrast factor given by Eq. (3) of Ref. [7]. The particle diameter is set at 28 nm. The *dashed line* shows the normalized scattering intensity of a collection of independent disklike particles at the same concentration c, defined as $I_p(q)/K_c$. The *continuous line* shows the normalized scattering intensity of a 3D reconstruction of a laponite suspension using simulated annealing in the q space

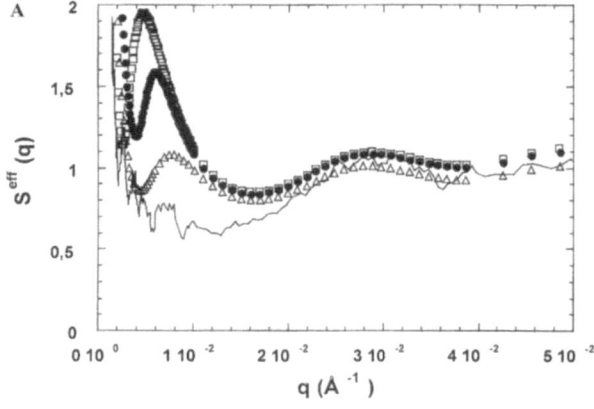

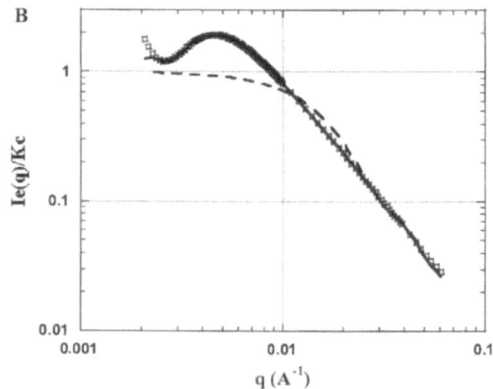

possible microsegregation. In the same line, excluded-volume interactions between disklike particles are more pronounced as the particle concentration increases. This second point can be related to the progressive attenuation of the correlation peak.

In order to assess these different (qualitative) conclusions, we generated 3D off-lattice configurations of laponite suspensions having similar USAXS spectra to the measured ones. For this purpose, a simulated annealing algorithm in the q space was successfully developed, extending recent work performed in the r space [19].

As shown in Fig. 2b, simulated annealing provides a good adjustment of the USAXS experiments. More interesting, the particle–particle pair correlation function (Fig. 3a) shows two interesting features. First of all, a first coordination shell is clearly visible at a distance ranging from 20 to 30 nm. Such a property is not observed for a random particle distribution and is directly related to a local particle microsegregation. This segregation is clearly observed in Fig. 3b. Secondly, an oscillation related to the correlation peak develops at large distances and is directly related to the correlation peak. This long-range oscillation slowly disappears as c increases.

Remarks on colloidal stability

The transition observed at very low ionic strength appears to be mainly driven by electrostatic repulsive

interactions. However, close inspection of the USAXS spectra reveals a puzzling particle microsegregation. This result is consistent with some recent observations of anomalous phase transformations in bulk suspensions of charged colloids at low or very low ionic strength [20, 21]. However, possible existence of a long-ranged attraction for system made of like charge particles is still under debate [20, 21].

Above an ionic strength close to 10^{-4} M, the Debye screening length is smaller than the particle size. As shown in Fig. 1, a liquid–solid transition is also observed at high ionic strength [7]; however, addition of salts promotes the appearance of the solid phase, in contrast to what is observed at very low ionic strength. The status of this transition is not clearly understood and several authors were recently concerned by the long-term colloidal stability and ageing of these "salted" systems [9–12]. In the following, we present a basic analysis of the colloidal stability. The interaction potential of different pair configurations is computed using a simple charge distribution model, already used for rodlike polyelectrolytes [22, 23]. At relatively high ionic strength (above 10^{-3} M), the electrostatic interaction (V_e) between two disklike particles is modelled by that between two equivalent surface charge distributions.

$$V_e(d) = \int\limits_1 \int\limits_2 u_e(r_{12}) dr_1^2 dr_2^2 ,$$

with

$$u_e(r_{12})/kT = \beta_{eff}^2 l_B \exp(-\kappa\, r_{12})/r_{12} ,$$

where d is the minimal distance between particles 1 and 2. l_B is the Bjerrum distance in water and β_{eff} is the effective charge density of the platelets due to counterion condensation. Following recent work of de Carvalho et al. [24], we take a reduction factor of the bare charge ($e/70$ Å^2) of 0.3 at $I = 5 \cdot 10^{-3}$ M. van der Waals interac-

Fig. 3 A Continuous line: pair correlation function of a 3D reconstruction of a laponite suspension using simulated annealing and having the USAXS spectrum shown in Fig. 2b ($c = 0.82\%$). *Dashed line*: pair correlation function of a random distribution of hard-core platelets at the same particle concentration. **B** Random cut through a 3D reconstruction of a laponite suspension at $c = 0.82\%$ and $I = 3 \times 10^{-5}$ M. The image is 1 μm large

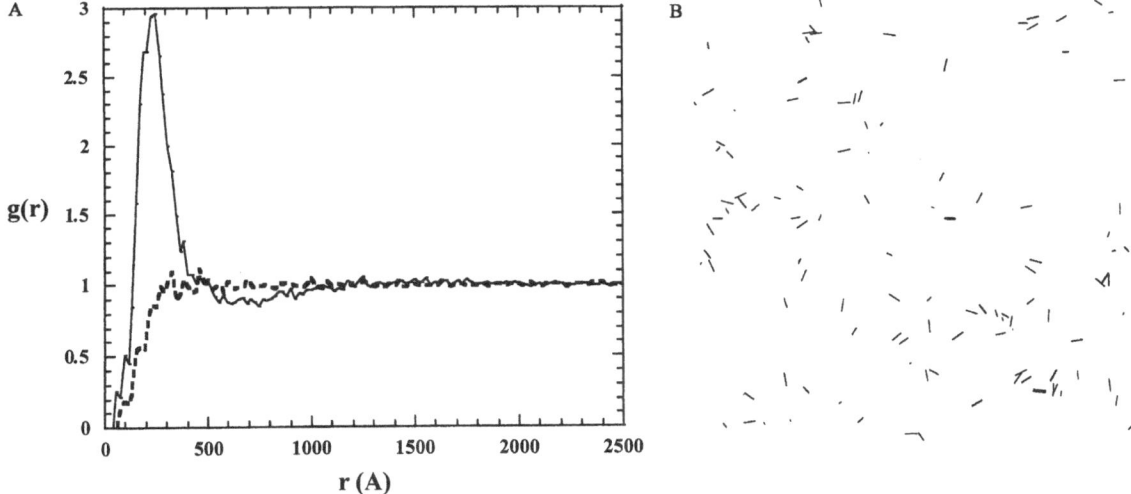

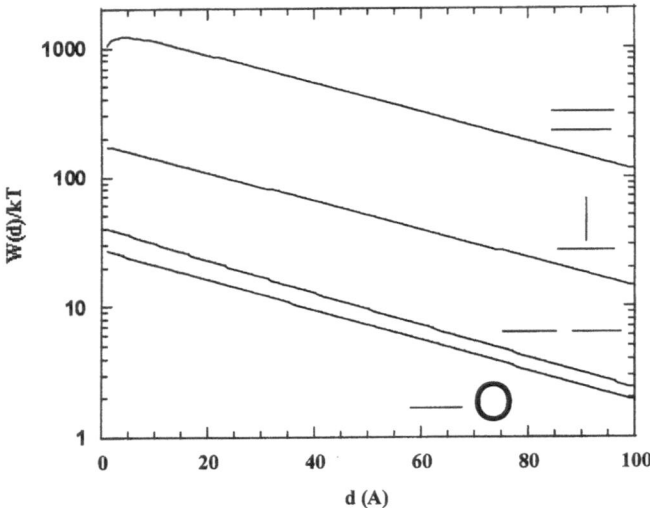

Fig. 4 Total interaction potential of two disklike particles. d is the minimal distance between the particles. The ionic strength is set at 5×10^{-3} M. The reduction factor of the bare particle charge ($e/70$ Å^2) is 0.3 [24]. From *top* to *bottom*: parallel, T-shaped, in plane (zero-twist) and $\pi/2$-twist pair configurations

tions are computed using the mica–water–mica Hamaker constant, around $5\,kT$. Four pair configurations are especially analysed: the parallel, the T-shaped, the in-plane (a zero twist) and the $\pi/2$-twist pair configurations. As shown in Fig. 4, the in-plane and the $\pi/2$-twist pair configurations are more stable than the parallel and the T-shapes ones for short minimal distances between particles. Secondly, at relatively high ionic strength and close to the particle contact, pseudo-Derjaguin–Landau–Verwey–Overbeek interactions can be of the order of a

few tens of kT. It is then possible to consider the possibility to get a slow coagulation process for some specific pair configurations. Moreover the in-plane and the twist pair configurations can be considered as good candidates in order to generate and to propagate nematic defects effectively observed in the solid phase [8, 13].

Conclusion

For charged disklike particles, we have shown the existence of a liquid–soft solid transition mainly driven by electrostatic repulsive interactions. The location of the transition line can be predicted using basic arguments. USAXS spectra do not show Bragg peaks, but a correlation peak as expected for a Wigner glass transition [12, 15]. Close inspection of these scattering experiments using simulated annealing in q space and 3D reconstructions of laponite dispersions reveals that the particle distribution is not homogeneous in space. This transition, mainly driven by long-range electrostatic repulsions, ends at an ionic strength close to 10^{-4} M. Above this value, the Debye screening length is smaller than the particle size. On the length scale of about the particle size, electrostatic interactions and other short-range interactions (possibly attractive) can compete. As the ionic strength increases, several characteristics are expected, ranging from long-term stabilization in some specific pair configurations to slow coagulation or strong flocculation.

Acknowledgements We thank S. Lyonnard for her help in performing the USAXS experiments and P.G. de Gennes for very stimulating suggestions and comments about colloidal stability.

References

1. Pusey PN (1991) In: Hansen JP, Levesque D, Zinn-Justin J (eds) Liquids, freezing and the glass transition. Elsevier, Amsterdam, p 765
2. Onsager L (1949) Ann NY Acad Sci 51:627
3. Langmuir I (1938) J Chem Phys 6:873
4. Forsyth PA, Marcelja S, Mitchell DJ, Ninham BW (1978) Adv Colloid Interface Sci 9:37
5. Ramsay JDF, Lindner PJ (1993) Chem Soc Faraday Trans 89:4207
6. Thompson DW, Butterworth JT (1992) J Colloid Interface Sci 151:236
7. Mourchid A, Delville A, Lambard J, Lécolier E, Levitz P (1995) Langmuir 11:1942
8. Gabriel JCP, Sanchez C, Davidson P, (1996) J Phys Chem 100:11139
9. (a) Kroon M, Wegdam H, Spril R (1996) Phys Rev E 54:6541; (b) Kroon M, Vos WL, Wegman RG (1998) Phys Rev E 57:1962
10. Nicolai T, Cocard S (2001) Eur Phys J E 5:221
11. Pignon F, Magnin A, Piau JM, Cabane B, Lindner P, Diat O (1997) Phys Rev Lett 56:3281
12. Bonn D, Tanaka H, Wegdam G, Kellay H, Meunier J (1998) Europhys Lett 45:52
13. Mourchid A, Lecolier E, Van Damme H, Levitz P (1998) Langmuir 14:4718
14. Mourchid A, Levitz P (1998) Phys Rev E 57:R4887
15. Levitz P, Lécolier E, Mourchid A, Delville A, Lyonnard S (2000) Europhys Lett 49:672
16. Dubois M, Zemb T, Belloni L, Delville A, Levitz P, Setton R (1992) J Chem Phys 96:2278
17. Lambard J, Lesieur P, Zemb T (1992) J Phys France 2:1191
18. Lake A (1967) Acta Crystallogr 23:191
19. Rintoul M, Torquado S (1997) J Colloid Interface Sci 186:467
20. Grier DG (2000) J Phys Condens Matter 12:A85
21. Van Roij R, Hansen JP (1998) Prog Colloid Polym Sci 110:50
22. Fixman M, Skolnick (1978) Macromolecules 11:863
23. Stoobants A, Lekkerkerker H, Odijk T (1986) Macromolecules 19:2232
24. De Carvalho R, Trizac E, Hansen JP (2000) Phys Rev E 61:1634

Progr Colloid Polym Sci (2001) 118 : 295–297
© Springer-Verlag 2001

Progr Colloid Polym Sci (2001) 118 : 298–299
© Springer-Verlag 2001

KEY WORD INDEX